Arithmetic Differential Equations

Mathematical Surveys and Monographs

Volume 118

Arithmetic Differential Equations

Alexandru Buium

American Mathematical Society

2000 *Mathematics Subject Classification.* Primary 14G20, 14G35, 30F10;
Secondary 53A55, 37F05, 14L24.

For additional information and updates on this book, visit
www.ams.org/bookpages/surv-118

Library of Congress Cataloging-in-Publication Data

Buium, Alexandru, 1955–
Arithmetic differential equations / Alexandru Buium.
p. cm. — (Mathematical surveys and monographs ; v. 118)
Includes bibliographical references and index.
ISBN 0-8218-3862-8 (alk. paper)
1. Arithmetical algebraic geometry. 2. Riemann surfaces. I. Title. II. Mathematical surveys and monographs ; no. 118.

QA242.5.B85 2005
516.3′5—dc22   2005048051

∞ The paper used in this book is acid-free and falls within the guidelines
established to ensure permanence and durability.
Visit the AMS home page at http://www.ams.org/

10 9 8 7 6 5 4 3 2 1   10 09 08 07 06 05

Contents

Preface

The main purpose of this research monograph is to develop an arithmetic analogue of the theory of ordinary differential equations. In our arithmetic theory the "time variable" t is replaced by a fixed prime integer p. Smooth real functions, $t \mapsto x(t)$, are replaced by integer numbers $a \in \mathbf{Z}$ or, more generally, by integers in various (completions of) number fields. The derivative operator on functions,

$$x(t) \mapsto \frac{dx}{dt}(t),$$

is replaced by a "Fermat quotient operator" δ which, on integer numbers, acts as $\delta : \mathbf{Z} \to \mathbf{Z}$,

$$a \mapsto \delta a := \frac{a - a^p}{p}.$$

Smooth manifolds (configuration spaces) are replaced by algebraic varieties defined over number fields. Jet spaces (higher order phase spaces) of manifolds are replaced by what can be called "arithmetic jet spaces" which we construct using δ in place of d/dt. Usual differential equations (viewed as functions on usual jet spaces) are replaced by "arithmetic differential equations" (defined as functions on our "arithmetic jet spaces"). Differential equations (Lagrangians) that are invariant under certain group actions on the configuration space are replaced by "arithmetic differential equations" that are invariant under the action of various correspondences on our varieties.

As our main application we will use the above invariant "arithmetic differential equations" to construct new quotient spaces that "do not exist" in algebraic geometry. To explain this we start with the remark that (categorical) quotients of algebraic curves by correspondences that possess infinite orbits reduce to a point in algebraic geometry. In order to address the above basic pathology we propose to "enlarge" algebraic geometry by replacing its algebraic equations with our more general arithmetic differential equations. The resulting new geometry is referred to as *$\delta-$geometry*. It then turns out that certain quotients that reduce to a point in algebraic geometry become interesting objects in $\delta-$geometry; this is because there are more invariant "arithmetic differential equations" than invariant algebraic equations. Here are 3 classes of examples for which this strategy works:

1) *Spherical case.* Quotients of the projective line $\mathbf{P}^1$ by actions of certain finitely generated groups (such as $SL_2(\mathbf{Z})$);

2) *Flat case.* Quotients of $\mathbf{P}^1$ by actions of postcritically finite maps $\mathbf{P}^1 \to \mathbf{P}^1$ with (orbifold) Euler characteristic zero;

3) *Hyperbolic case.* Quotients of modular or Shimura curves (e.g. of $\mathbf{P}^1$) by actions of Hecke correspondences.

Our results will suggest a general conjecture according to which the quotient of a curve (defined over a number field) by a correspondence is non-trivial in $\delta-$geometry for almost all primes p if and only if the correspondence has an "analytic uniformization" over the complex numbers. Then the 3 classes of examples above correspond to *spherical, flat, and hyperbolic* uniformization respectively.

Material included. The present book follows, in the initial stages of its analysis, a series of papers written by the author [**17**]-[**28**]. A substantial part of this book consists, however, of material that has never been published before; this includes our Main Theorems stated at the end of Chapter 2 and proved in the remaining Chapters of the book. The realization that the series of papers [**17**]-[**28**] consists of pieces of one and the same puzzle came relatively late in the story and the unity of the various parts of the theory is not easily grasped from reading the papers themselves; this book is an attempt at providing, among other things, a linear, unitary account of this work. Discussed are also some of the contributions to the theory due to C. Hurlburt [**71**], M. Barcau [**2**], and K. Zimmerman [**29**].

Material omitted. A problem that was left untouched in this book is that of putting together, in an adelic picture, the various $\delta-$geometric pictures, as p varies. This was addressed in our paper [**26**] where such an adelic theory was developed and then applied to providing an arithmetic differential framework for functions of the form $(p, a) \mapsto c(p, a)$, p prime, $a \in \mathbf{Z}$, where $L(a, s) = \sum_n c(n, a)n^{-s}$ are various families of $L-$functions parameterized by a. Another problem not discussed in this book is that of generalizing the theory to higher dimensions. A glimpse into what the theory might look like for higher dimensional varieties can be found in [**17**] and [**3**]. Finally, we have left aside, in this book, some of the Diophantine applications of our theory such as the new proof in [**18**] of the Manin-Mumford conjecture about torsion points on curves and the results in [**21**], [**2**] on congruences between classical modular forms.

Prerequisites. For most of the book, the only prerequisites are the basic facts of algebraic geometry (as found, for instance, in R. Hartshorne's textbook [**66**]) and algebraic number theory (as found, for instance, in Part I of S. Lang's textbook [**87**]). In later Chapters more background will be assumed and appropriate references will be given. In particular the last Chapter will assume some familiarity with the $p-$adic theory of modular and Shimura curves. From a technical point of view the book mainly addresses graduate students and researchers with an interest in algebraic geometry and / or number theory. However, the general theme of the book, its strategy, and its conclusions should appeal to a general mathematical audience.

Plan of the book. We will organize our presentation around the motivating "quotient space" theme. So quotient spaces will take center stage while "arithmetic jet spaces" and the corresponding analogies with the theory of ordinary differential equations will appear as mere tools in our proofs of $\delta-$geometric theorems. Accordingly, the Introduction starts with a general discussion of strategies to construct quotient spaces and continues with a brief outline of our $\delta-$geometric theory. We also include, in our Introduction, a discussion of links, analogies, and / or discrepancies between our theory and a number of other theories such as: differential equations on smooth manifolds [**114**], the Ritt-Kolchin differential algebra [**117**], [**84**], [**32**], [**13**], the difference algebraic work of Hrushovski and Chatzidakis [**69**], the theory of dynamical systems [**109**], Connes' non-commutative geometry [**36**], the theory of Drinfeld modules [**52**], Dwork's theory [**53**], Mochizuchi's $p-$adic Teichmüller theory [**111**], Ihara's theory of congruence relations [**73**] [**74**], and the work of Kurokawa, Soulé, Deninger, Manin, and others on the "field with one element" [**86**], [**128**], [**98**], [**46**]. In Chapter 1 we discuss some algebro-geometric preliminaries; in particular we discuss analytic uniformization of correspondences on algebraic curves. In Chapter 2 we discuss our $\delta-$geometric strategy in detail and we state our main conjectures and a sample of our main results. In Chapters 3, 4, 5 we develop the general theory of arithmetic jet spaces. The corresponding 3 Chapters deal with the global, local, and birational theory respectively. In Chapters 6, 7, 8 we are concerned with our applications of $\delta-$geometry to quotient spaces: the corresponding 3 Chapters are concerned with correspondences admitting a spherical, flat, or hyperbolic analytic uniformization respectively. Details as to the contents of the individual Chapters are given at the beginning of each Chapter. All the definitions of new concepts introduced in the book are numbered and an index of them is included after the bibliography. A list of references to the main results is given at the end of the book. Internal references of the form Theorem $x.y$, Equation $x.y$, etc. refer to Theorems, Equations, etc. belonging to Chapter x (if $x \neq 0$) or the Introduction (if $x = 0$). Theorems, Propositions, Lemmas, Corollaries, Definitions, and Examples are numbered in the same sequence; Equations are numbered in a separate sequence. Here are a few words about the dependence between the various Chapters. The impatient reader can merely skim through Chapter 1; he/she will need to read at least the (numbered) "Definitions" (some of which are not standard). Chapters 2-5 should be read in a sequence. Chapters 6-8 are largely (although not entirely) independent of one another but they depend upon Chapters 2-5.

Acknowledgments. The author wishes to thank D. Bertrand, P. Deligne, E. Hrushovski, Y. Ihara, M. Kim, Y. I. Manin, B. Mazur, S. Lang, A. Pillay, B. Poonen, F. Pop, T. Scanlon, J. Tate, D. Thakur, D. Ulmer, and J. F. Voloch for encouragement and discussions at various stages of development of these ideas. While writing this book the author was supported in part by NSF Grant #0096946.

Alexandru Buium
Albuquerque, May 2005

Introduction

We start, in Section 1, by explaining our main motivation which comes from the fact that categorical quotients of correspondences on curves tend to be trivial in algebraic geometry. In Section 2 we explain how one can fix this problem for a remarkable class of correspondences; this is done by developing a geometry, called $\delta-$*geometry*, which is obtained from usual algebraic geometry by adjoining a "Fermat quotient operator" δ. If one views δ as an analogue of a derivation with respect to a prime number then $\delta-$geometry can be viewed as obtained from usual algebraic geometry by replacing algebraic equations with "arithmetic differential equations". In Section 3 we discuss relations between our theory and some other theories.

The present Introduction is written in an informal style; a formal presentation of this material will be made in the body of the book.

0.1. Motivation and strategy

0.1.1. Correspondences and their categorical quotients. It is convenient to start in complete generality by considering an arbitrary category $\mathcal{C}$. (Morally $\mathcal{C}$ should be viewed as a category of "spaces" in some geometry.) A *correspondence* in $\mathcal{C}$ is a tuple $\mathbf{X} = (X, \tilde{X}, \sigma_1, \sigma_2)$ where X and $\tilde{X}$ are objects of $\mathcal{C}$ and $\sigma_1, \sigma_2 : \tilde{X} \to X$ are morphisms in $\mathcal{C}$. We sometimes write $\mathbf{X} = (X, \sigma)$, where $\sigma := (\tilde{X}, \sigma_1, \sigma_2)$. Following the standard terminology of geometric invariant theory **[113]** we define a *categorical quotient* for $\mathbf{X}$ to be a pair (Y, π) where Y is an object of $\mathcal{C}$ and $\pi : X \to Y$ is a morphism in $\mathcal{C}$ satisfying the following properties: 1) $\pi \circ \sigma_1 = \pi \circ \sigma_2$, 2) For any pair (Y', π') where Y' is an object of $\mathcal{C}$ and $\pi' : X \to Y'$ is a morphism such that $\pi' \circ \sigma_1 = \pi' \circ \sigma_2$ there exists a unique morphism $\gamma : Y \to Y'$ such that $\gamma \circ \pi = \pi'$. Categorical quotients are sometimes referred to as *co-equalizers*. Of course, if a categorical quotient (Y, π) exists then it is unique up to isomorphism and we shall write $Y = X/\sigma$. Correspondences form, in a natural way, a category: a morphism $\mathbf{X} \to \mathbf{X}'$ between two correspondences $\mathbf{X} = (X, \tilde{X}, \sigma_1, \sigma_2)$ and $\mathbf{X}' = (X', \tilde{X}', \sigma_1', \sigma_2')$ is, by definition, a pair of morphisms $(\pi, \tilde{\pi})$, $\pi : X \to X'$, $\tilde{\pi} : \tilde{X} \to \tilde{X}'$, such that $\pi \circ \sigma_i = \sigma_i' \circ \tilde{\pi}$, $i = 1, 2$. We will assume, for each category $\mathcal{C}$ we shall be considering, that a class of objects in $\mathcal{C}$ is given which we refer to as *trivial* objects. (Morally trivial objects should be viewed as spaces that reduce to a point.)

An important example of correspondences is provided by (discrete) *dynamical systems* (i.e. self maps). Indeed if X^* is an object in our category $\mathcal{C}$ and $s : X^* \to X^*$ is a morphism then one can attach to these data a correspondence $\mathbf{X}^* = (X^*, X^*, id_{X^*}, s)$. More generally if we assume $\mathcal{C}$ possesses fiber products then one can consider, in a natural way, the pull-back $\mathbf{X} := f^*\mathbf{X}^*$ of $\mathbf{X}^*$ via any morphism $f : X \to X^*$; cf. Equation 1.4. Categorical quotients of correspondences of the form $\mathbf{X}^*$, or more generally of the form $\mathbf{X}$, should be viewed as (categorical) *spaces of orbits* of the "dynamical system defined by s".

Another example of correspondences appearing in nature are groupoids. To review this we shall tacitly assume that all products intervening (explicitly or implicitly) in the discussion below exist in $\mathcal{C}$. A *groupoid* in $\mathcal{C}$ is a tuple

$$(X, \tilde{X}, \sigma_1, \sigma_2, \nu, \iota, \epsilon)$$

where X and $\tilde{X}$ are objects in $\mathcal{C}$ and

$$\sigma_1 : \tilde{X} \to X, \quad \sigma_2 : \tilde{X} \to X, \quad \nu : \tilde{X} \times_{\sigma_2, X, \sigma_1} \tilde{X} \to \tilde{X}, \quad \iota : \tilde{X} \to \tilde{X}, \quad \epsilon : X \to \tilde{X}$$

are morphisms such that ν satisfies the usual associativity axiom, ϵ satisfies $\sigma_1 \circ \epsilon = \sigma_2 \circ \epsilon = id_X$ plus the usual unit axiom, and ι satisfies $\sigma_1 \circ \iota = \sigma_2$, $\sigma_2 \circ \iota = \sigma_1$ plus the usual inverse axiom. The datum $(X, \tilde{X}, \sigma_1, \sigma_2)$ can then be viewed as a correspondence.

In their turn groupoids often arise from group actions. Again we tacitly assume that all products intervening (explicitly or implicitly) in the discussion below exist in $\mathcal{C}$. Let G be a group object in the category $\mathcal{C}$ and $\mu : G \times X \to X$ an action on an object X in $\mathcal{C}$. Then one can consider the groupoid

$$(X, G \times X, pr_2, \mu, \nu, \iota, \epsilon)$$

where pr_2 is the second projection, and ν, ι, ϵ are naturally induced by the group operations and the action. So in particular group actions $\mu : G \times X \to X$ define correspondences

$$(X, G \times X, pr_2, \mu).$$

Many of the examples of correspondences we shall be interested in will not come, however, from dynamical systems or groupoids.

0.1.2. Basic pathology. Typically it turns out that, in many classical geometric situations, one has interesting correspondences $\mathbf{X} = (X, \sigma) = (X, \tilde{X}, \sigma_1, \sigma_2)$ whose categorical quotient X/σ is trivial. This "basic pathology" manifests itself, for instance, in the case when $\mathcal{C}$ is the category of algebraic varieties over an algebraically closed field and the trivial objects are the points: if X and $\tilde{X}$ above are algebraic varieties and the smallest equivalence relation $\langle \sigma \rangle \subset X \times X$ containing the image of $\sigma_1 \times \sigma_2 : \tilde{X} \to X \times X$ is Zariski dense in $X \times X$ then the categorical quotient X/σ is trivial. Examples of this kind are very common. For instance assume that $dim\ X = dim\ \tilde{X} = 1$, and σ has an *infinite orbit* by which we mean that there exists an infinite sequence of points $P_1, P_2, P_3, ... \in X$ and a sequence of points $\tilde{P}_1, \tilde{P}_2, \tilde{P}_3, ... \in \tilde{X}$ such that $\sigma_1(\tilde{P}_i) = P_i$, $\sigma_2(\tilde{P}_i) = P_{i+1}$ for all $i \geq 1$. Then $\langle \sigma \rangle \subset X \times X$ is dense in $X \times X$ and hence X/σ reduces to a point in the category of varieties. Note, by the way, that in the example above it is impossible to embed $\mathbf{X}$ into a correspondence $\mathbf{X}' = (X', \tilde{X}', \sigma_1', \sigma_2')$ in the category of schemes of finite type such that $\mathbf{X}'$ admits a groupoid structure and $dim\ X' = dim\ \tilde{X}' = 1$; here, by an *embedding* we mean a morphism $(\iota, \tilde{\iota}) : \mathbf{X} \to \mathbf{X}'$ with ι and $\tilde{\iota}$ embeddings. Intuitively, $\mathbf{X}$ does not "generate" a groupoid "of finite type" and, in some sense, this is what prevents usual algebraic geometry from "controlling" the quotient.

With the above "basic pathology" in mind one is tempted to "enlarge" usual algebraic geometry so as to make X/σ non-trivial; one can then pursue the study of X/σ in this enlarged geometry. The present book is an attempt to develop such an enlarged geometry and to apply it to the study of categorical quotients.

Before proceeding to explain our approach it might be useful to put things in perspective by discussing some general aspects of quotient theory.

0.1.3. Two viewpoints in quotient theory. Assume we are given a correspondence $\mathbf{X} = (X, \sigma)$ in some category of "spaces" (e.g. smooth manifolds, varieties, schemes, etc.) There are (at least) two possible paths towards the idea of "quotient" of X by σ which we shall call here, for convenience, *invariant theory* and *groupoid theory.*

In invariant theory one seeks to construct a categorical quotient as a "genuine space" X/σ such that the (locally defined) functions on it identify with the (locally defined) "$\sigma-$invariant functions" on X. (Here a locally defined function φ on X is called $\sigma-$*invariant* if $\varphi \circ \sigma_1 = \varphi \circ \sigma_2$ on their common domain of definition). This viewpoint is actually quite successful in algebraic geometry, in case (X, σ) comes from a reductive algebraic group action $\mu : G \times X \to X$; geometric invariant theory codifies the situation in this case. On the other hand the invariant theoretic approach in algebraic geometry utterly fails for correspondences with Zariski dense equivalence relation $\langle\sigma\rangle$; these will be, by the way, the correspondences that will be at the heart of this book.

In groupoid theory one assumes that the correspondence $\mathbf{X} = (X, \tilde{X}, \sigma_1, \sigma_2)$ has a groupoid structure (ν, ι, ϵ) and one *defines* the "space X/σ" to be the groupoid $(X, \tilde{X}, \sigma_1, \sigma_2, \nu, \iota, \epsilon)$ itself regarded up to an appropriate equivalence on the class of groupoids. (This equivalence should allow roughly speaking to transfer modules from one groupoid to another in a sense that depends on the particular context we are in.) Non-commutative geometry and stack theory adopt this viewpoint. By the way, one of Connes' original motivations for developing non-commutative geometry [**36**] was to address the "basic pathology" above as it manifests itself in topology and differential geometry; this has been enormously successful and can be adapted to algebro-geometric situations [**127**], [**99**]. On the other hand passing from algebraic varieties to algebraic stacks, as one sometimes does in moduli space theory, is not sufficiently drastic to make the "basic pathology" go away !

0.1.4. Strategy of the present approach. Our approach towards the "basic pathology" above will use the viewpoint of invariant theory (as opposed to that of groupoid theory); but in order that invariant theory be non-trivial we will have to enlarge algebraic geometry by adjoining to it some new functions. Having "more functions" will increase our chances to find, as we will, interesting $\sigma-$invariant functions. Now there is a general recipe to enlarge algebraic geometry by adjoining to its functions an extra function, δ, satisfying "polynomial compatibility conditions" with respect to addition and multiplication. Cf. [**20**]; see also the last Section of Chapter 2 in our book. Under a certain genericity condition it turns out, as we shall see, that "locally" there are exactly 4 types of such δ's which can be referred to as derivation operators, difference operators, $p-$derivation operators (= "Fermat quotient operators"), and $p-$difference operators. If δ is a derivation one is led to the *differential algebra* (and corresponding geometry) of Ritt [**117**] and Kolchin [**84**]; cf. also [**32**], [**12**], [**14**], [**15**], [**16**]. If δ is a difference operator one is led to the *difference algebra* (and corresponding geometry) of Ritt-Cohn [**35**]; cf. also the work of Chatzidakis-Hrushovski [**33**] and Hrushovski [**70**]. If δ are $p-$derivation operators attached to various prime numbers p one is led to what can be called *arithmetic differential algebra* (and its corresponding geometry). The case when δ is a $p-$difference operator seems to lead to a less interesting theory. Of the above 4 types of δ, $p-$derivations are the only ones that do not vanish identically on the integers. This makes $p-$derivations especially suited for arithmetic applications; so we shall exclusively be interested, in this book, in arithmetic differential algebra. The study of arithmetic differential algebra and its associated geometry was begun in our paper [**17**] and developed in a series of papers [**18**]-[**28**], [**2**], [**3**], [**71**]. Earlier papers (cf. especially papers by Joyal [**79**] and Ihara [**75**]) contain indications that Fermat quotients should be viewed as analogues of derivations. However note that there is an important difference between our approach and the one proposed by Ihara; cf. the remarks surrounding Equation 0.31 below. Arithmetic differential algebra can be viewed as an "arithmetic" analogue of the Ritt-Kolchin differential algebra and also as a "regularized" analogue of the Cohn difference algebra; indeed arithmetic differential algebra can be viewed as obtained from difference algebra by adjoining to it certain divergent series (cf. Remark 2.6 below).

As a matter of terminology note that, in the Ritt-Kolchin differential algebra one uses the symbol δ as an abbreviation for "differential". By analogy with the tradition in the Ritt-Kolchin theory we will use the symbol δ as an abbreviation for "arithmetic differential"; in particular arithmetic differential algebra will be referred to as *$\delta-$algebra* while its "associated" geometry, which can be called *arithmetic differential geometry*, will be referred to as *$\delta-$geometry*. There will be no danger of confusion with differential algebra and difference algebra terminology because no use of the latter two types of algebra will be made in this book.

The main purpose of this book is to first develop some of the basic elements of $\delta-$geometry and then to construct (and study) interesting categorical quotients, in $\delta-$geometry, for correspondences whose categorical quotient in algebraic geometry is trivial. A conjecture will emerge to the effect that the $\delta-$geometric picture is interesting if and (essentially) only if our correspondences admit a complex analytic uniformization. Below we give a rough outline of our theory; the details of the theory will be explained in Chapter 2.

0.2. Rough outline of the theory

0.2.1. Background and notation. We start with a number field F (always assumed of finite degree over $\mathbf{Q}$). Fix a finite place $\wp$ of F which is unramified over $\mathbf{Q}$; later we will vary $\wp$. We identify $\wp$ with a maximal ideal in the ring of integers $\mathcal{O}_F$ of F. Let $p = char\ \mathcal{O}/\wp$. Let $\mathcal{O}_\wp$ be the localization of $\mathcal{O}_F$ at $\wp$, consider its completion $\hat{\mathcal{O}}_\wp$, and let $\hat{\mathcal{O}}_\wp^{ur}$ be the maximum unramified extension of $\hat{\mathcal{O}}_\wp$ (obtained by adjoining to $\hat{\mathcal{O}}_\wp$ all the roots of unity of order prime to p in an algebraic closure of its fraction field). Then $\hat{\mathcal{O}}_\wp^{ur}$ is a discrete valuation ring with maximal ideal generated by p. Consider the completion of $\hat{\mathcal{O}}_\wp^{ur}$ which we denote by

$$R_\wp := (\hat{\mathcal{O}}_\wp^{ur})\hat{}.$$

The ring $R_\wp$ has a simple well known structure: any element of it can be represented uniquely as a series $\sum_{i=0}^\infty \zeta_i p^i$ where ζ_i are roots of unity of order prime to p or 0. The ring $R_\wp$ has a unique automorphism ϕ inducing the $p-$power Frobenius map on the residue field $k_\wp := R_\wp/pR_\wp$; it is given by $\phi(\sum_{i=0}^\infty \zeta_i p^i) = \sum_{i=0}^\infty \zeta_i^p p^i$. For

$F = \mathbf{Q}$, $\wp = (p)$, we have $\hat{\mathbf{Z}}_{(p)} = \mathbf{Z}_p$, the ring of p–adic numbers, so we simply write $R_p = \hat{\mathbf{Z}}_p^{ur}$ in place of $(\hat{\mathbf{Z}}_{(p)}^{ur})\hat{}$. By the way, for a general F and $\wp$ the natural morphism $R_p \to R_\wp$ is an isomorphism so the ring $R_\wp$ depends only on p and not on F and $\wp$.

0.2.2. δ–geometry. With the notation above we let $\delta = \delta_p : R_\wp \to R_\wp$ be the "Fermat quotient operator" defined by $\delta x = (\phi(x) - x^p)/p$. We consider a "geometric category", $\mathcal{C}_\delta$, which morally underlies δ–*geometry at* $\wp$. The objects of $\mathcal{C}_\delta$ will be called δ–*ringed sets*; a δ–ringed set X_δ consists of an underlying set X_{set} equipped with a family (X_s) of subsets and a family of "structure rings" $(\mathcal{O}_s)$ of $R_\wp$–valued functions on the X_s's such that $f \in \mathcal{O}_s$ implies $\delta \circ f \in \mathcal{O}_s$. Here s runs through a monoid S and one assumes that $X_{st} = X_s \cap X_t$ for $s, t \in S$. One also assumes compatibility with restrictions i.e. $f \in \mathcal{O}_s$ implies $f_{|X_{st}} \in \mathcal{O}_{st}$. A δ–ringed set is called *trivial* if $\mathcal{O}_s = R_\wp$ for all s. There is a natural concept of morphism of δ–ringed sets; with these morphisms the δ–ringed sets form a category $\mathcal{C}_\delta$. Any correspondence in $\mathcal{C}_\delta$ has a categorical quotient. So the existence of categorical quotients is not here the issue; the issue will always be their non-triviality.

0.2.3. Passing from algebraic geometry to δ–geometry. Let $X_\wp = X$ be a smooth scheme of finite type over $R_\wp$ with irreducible geometric fibers. We will attach to the scheme X a δ–ringed set X_δ; the underlying set X_{set} of X_δ will be the set $X(R_\wp)$ of $R_\wp$–points of X and the structure rings $\mathcal{O}_s$ will consist of certain functions $P \mapsto \varphi(P)$ that locally, in coordinates $x \in R_\wp^d$, look like

$$\varphi(x) = \frac{F(x, \delta x, ..., \delta^r x)}{G(x, \delta x, ..., \delta^r x)}$$

where F, G are restricted power series with $R_\wp$–coefficients. Let us make this precise. First, a function $f : X(R_\wp) \to R_\wp$ is called a δ–*function of order* $\leq r$ if for any $P \in X(R_\wp)$ there is a Zariski open set $U \subset X$, $P \in U(R_\wp)$, and a closed embedding $U \subset \mathbf{A}^d$ such that $f_{|U(R_\wp)}$ is given in coordinates $x \in R_\wp^d$ by

$$f(x) = F(x, \delta x, ..., \delta^r x)$$

where F is a restricted power series with coefficients in $R_\wp$. Recall that a power series is called *restricted* if its coefficients converge p–adically to 0. Denote by $\mathcal{O}^r(V)$ the rings of δ–functions of order $\leq r$ on a Zariski open set $V \subset X$. Then $V \mapsto \mathcal{O}^r(V)$ defines a sheaf $\mathcal{O}^r$ of rings on X for the Zariski topology. Define a δ–line bundle to be a locally free sheaf of $\mathcal{O}^r$– modules of rank one. Consider the ring $W := \mathbf{Z}[\phi] \subset End(R_\wp)$ and let W_+ be the set of all $w = \sum a_i \phi^i \in W$ with $a_i \geq 0$. The multiplicative monoid W acts on $R^\times$ by $\lambda^w = \prod (\phi^i(\lambda))^{a_i}$, $\lambda \in R^\times$. For any δ–line bundle L one can define a δ–line bundle L^w by acting with w upon the defining cocycle of L. Let $L = K^{-1}$ be now the anticanonical bundle on X, viewed as a δ–line bundle. We consider the graded ring

$$\bigoplus_{w \in W_+} H^0(X, L^w).$$

Its homogeneous elements will be called δ–*sections*. We let S be the monoid of δ–sections of weight $\neq 0$, not divisible by p, in the above ring. For any $s \in S$ of degree w_0 we let

$$X_s := \{P \in X(R_\wp) \mid s(P) \not\equiv 0 \ \ mod \ \ p\},$$

$$\mathcal{O}_s := \{P \mapsto \varphi(P) := t(P)/s^w(P) \mid t \in H^0(X, L^{ww_0})\}.$$

The data (X_s), $(\mathcal{O}_s)$ define a δ–ringed set X_δ with underlying set $X(R_\wp)$.

Let now $\mathbf{X}_\wp = \mathbf{X} = (X, \sigma) = (X, \tilde{X}, \sigma_1, \sigma_2)$ be a correspondence in the category of schemes over $R_\wp$. Assume X and $\tilde{X}$ are smooth of finite type over $R_\wp$ and have irreducible geometric fibers. Then by the discussion above one can attach to the schemes X and $\tilde{X}$ two δ–ringed sets X_δ and $\tilde{X}_\delta$. Assume moreover that σ_1 and σ_2 are étale. Then the morphisms $\sigma_1, \sigma_2 : \tilde{X} \to X$ induce morphisms between the corresponding δ–ringed sets. So we end up with a correspondence $\mathbf{X}_\delta = (X_\delta, \sigma_\delta)$ in the category $\mathcal{C}_\delta$. One can easily describe the categorical quotient X_δ/σ_δ in $\mathcal{C}_\delta$. The underlying set of the categorical quotient X_δ/σ_δ is the set of equivalence classes of points in $X(R_\wp)$ with respect to the smallest equivalence relation containing the image of the map

$$\sigma_1 \times \sigma_2 : \tilde{X}(R_\wp) \to X(R_\wp) \times X(R_\wp).$$

Let $\tilde{L}$ be the anticanonical bundle on $\tilde{X}$. Let us say that a δ–section $s \in H^0(X, L^w)$ is *σ–invariant* if its pull-backs to $H^0(\tilde{X}, \tilde{L}^w)$ via σ_1^* and σ_2^* coincide; σ– invariant δ–sections will simply be called *δ– invariants*. Then the monoid parameterizing the structure rings of X_δ/σ_δ is the monoid of all δ–invariants $s \in H^0(X, L^w)$ which are not divisible by p. For s of degree w_0 the corresponding structure ring of X_δ/σ_δ is

$$\{[P] \mapsto \varphi(P) := t(P)/s^w(P) \mid t \in H^0(X, L^{ww_0}) \quad \text{a } \delta\text{–invariant}\}.$$

So morally X_δ/σ_σ is non-trivial if one can find at least two $R_\wp$–linearly independent δ–invariants in one of the spaces $H^0(X, L^w)$, where $0 \neq w \in W_+$. This agrees, of course, with the ideology of classical invariant theory.

0.2.4. Conjectural picture. At this point we consider a global situation. Let S be a finite set of finite places of the number field F, containing all the places ramified over $\mathbf{Q}$, assume (for simplicity) that S consists of all the places containing some integer $m \in \mathbf{Z}$, and let $\mathcal{O} := \mathcal{O}_F[1/m]$ be the ring of S–integers of F. Consider a correspondence

$$\mathbf{X} = \mathbf{X}_\mathcal{O} = (X, \sigma),$$

$\sigma = (\tilde{X}, \sigma_1, \sigma_2)$, in the category $\mathcal{C}_\mathcal{O}$ of schemes over $\mathcal{O}$. We will assume X and $\tilde{X}$ are of finite type and smooth over $\mathcal{O}$, have irreducible geometric fibers of dimension one (so they are "curves" over $\mathcal{O}$), and σ_1, σ_2 are étale. We have induced correspondences

$$\mathbf{X}_\wp = (X_\wp, \sigma_\wp), \quad \mathbf{X}_\mathbf{C} = (X_\mathbf{C}, \sigma_\mathbf{C})$$

in the category of schemes over $R_\wp$ and $\mathbf{C}$ respectively. We will always assume that the smallest equivalence relation in $X_\mathbf{C} \times X_\mathbf{C}$ containing the image of $\tilde{X}_\mathbf{C} \to X_\mathbf{C} \times X_\mathbf{C}$ is Zariski dense in $X_\mathbf{C} \times X_\mathbf{C}$ so that the categorical quotient $X_\mathbf{C}/\sigma_\mathbf{C}$ is trivial in the category of schemes over $\mathbf{C}$. As we already saw, for all $\wp \notin S$, we may consider a correspondence

$$\mathbf{X}_\delta = (X_\delta, \sigma_\delta)$$

in the category $\mathcal{C}_\delta$. Our general guiding conjecture will (roughly speaking) assert that:

The categorical quotient X_δ/σ_δ is non-trivial for almost all places $\wp$ if and (essentially) only if the correspondence $\mathbf{X}_\mathbf{C}$ admits an "analytic uniformization".

Here, by $\mathbf{X}_{\mathbf{C}}$ admitting an *analytic uniformization*, we understand that there exist morphisms of correspondences in the category of (open) Riemann surfaces (i.e. in the category of complex analytic manifolds of dimension one), $(\iota, \tilde{\iota}) : \mathbf{X}_{\mathbf{C}} \to \mathbf{Y}_{\mathbf{C}}$, $(\pi, \tilde{\pi}) : \mathbf{Y}'_{\mathbf{C}} \to \mathbf{Y}_{\mathbf{C}}$, where ι, $\tilde{\iota}$ are open immersions, π, $\tilde{\pi}$ are (possibly infinite) Galois covers unramified above X, $\tilde{X}$, and $\mathbf{Y}'_{\mathbf{C}} = (\mathbf{S}, \mathbf{S}, \tau_1, \tau_2)$, with $\mathbf{S}$ a simply connected Riemann surface and τ_1, τ_2 automorphisms of $\mathbf{S}$. We always assume here that the Galois groups of π and $\tilde{\pi}$ are *admissible* in the sense that they act properly discontinuously on $\mathbf{S}$ and are either finite (in case $\mathbf{S} = \mathbf{P}^1$) or infinite (in case $\mathbf{S} = \mathbf{C}$) or of finite covolume (if $\mathbf{S} = \mathbf{H}$, the upper half plane) respectively. According to whether $\mathbf{S}$ is $\mathbf{P}^1$, $\mathbf{C}$, or $\mathbf{H}$ there are three classes of correspondences admitting analytic uniformizations which will be called *spherical, flat, and hyperbolic* correspondences. Analytically uniformizable correspondences can be completely classified in algebraic terms; cf. Chapter 1. Up to an appropriate equivalence relation we have, very roughly speaking, the following description. (N.B. The actual situation is slightly more complicated.) The spherical uniformizable correspondences are, essentially, of the form $(\mathbf{P}^1, \mathbf{P}^1, \pi, \pi \circ \tau)$ where $\pi : \mathbf{P}^1 \to \mathbf{P}^1$ is a finite Galois cover and $\tau : \mathbf{P}^1 \to \mathbf{P}^1$ is an isomorphism; the categorical quotient in the category of sets of such a correspondence is the same as the quotient of $\mathbf{P}^1$ by the action of the group $\langle \Gamma, \tau \rangle$ generated by τ and the Galois group, Γ, of π. Of course the group $\langle \Gamma, \tau \rangle$ is generally infinite. Flat uniformizable correspondences are, essentially, of the form $(\mathbf{P}^1, \mathbf{P}^1, \sigma_1, \sigma_2)$ where σ_i are either multiplicative functions $t \mapsto t^{d_i}$ or Chebyshev polynomials, or Lattès functions [**109**] (the latter being induced by endomorphisms of elliptic curves E via isomorphisms $E/\langle \gamma \rangle \simeq \mathbf{P}^1$ where $\gamma : E \to E$ is an automorphism). The hyperbolic uniformizable correspondences (with infinite orbits) are, essentially, Hecke correspondences on modular or Shimura curves (that are classically described in terms of quaternion algebras over totally real fields). So the "if" part of our conjecture essentially says that our δ−geometric theory gives a rich picture in all the above examples; the "only if" part of the conjecture says that the above examples are, essentially, the *only* ones for which our δ−geometric theory gives a rich picture.

0.2.5. Results. Our main results show that the "if" part of the conjecture above holds (under some mild assumptions) in the spherical case, in the flat case, and in the "rational hyperbolic" case. (Here the "rational hyperbolic" case refers to the case when $\mathbf{S} = \mathbf{H}$ and the quaternion algebra describing the situation has center $\mathbf{Q}$.) For correspondences $\mathbf{X}$ admitting such a uniformization we will study the "geometry" and "cohomology" of the quotients X_δ / σ_δ. In particular we will show that the quotients X_δ / σ_δ tend to behave like "rational varieties" and the quotient maps $X_\delta \to X_\delta / \sigma_\delta$ tend to look like "pro-finite covers" whose Galois properties we shall study. The "only if" part of the conjecture is much more mysterious. We will be able to prove a local analogue of the "only if" part of our conjecture. Also we will prove a global result along the "only if" direction saying (roughly speaking) that if X_δ / σ_δ is "sufficiently non-trivial" for almost all places $\wp$ then $\mathbf{X}_{\mathbf{C}}$ is *critically finite* (in the sense of complex dynamics) and $\mathbf{X}_\wp \otimes k_\wp$ has a generically trivial *pluricanonical bundle*. The latter property will allow us to prove a version of the "only if" part of the conjecture in the "dynamical system case".

0.2.6. Proofs. Here are a few words about the strategy of our proofs. The first move is to attach to any smooth scheme of finite type $X_\wp$ over $R_\wp = \hat{\mathbf{Z}}_p^{ur}$ a projective system of formal schemes

$$(0.1) \qquad ... \to J^r(X_\wp) \to J^{r-1}(X_\wp) \to ... \to J^0(X_\wp) = \hat{X}_\wp$$

such that for each r the ring $\mathcal{O}^r(X_\wp)$ of δ–functions of order $\leq r$ on $X_\wp(R_\wp)$ identifies with the ring $\mathcal{O}(J^r(X_\wp))$ of global functions on $J^r(X_\wp)$:

$$(0.2) \qquad \mathcal{O}^r(X_\wp) \simeq \mathcal{O}(J^r(X_\wp)).$$

The formal scheme $J^r(X_\wp)$ should be viewed intuitively as an "arithmetic jet space" and will be referred to as the *p–jet space* of order r of $X_\wp$. It is an arithmetic analogue of the jet space of a manifold (relative to a submersion) in differential geometry (cf. Equation 0.6) and, in particular, it carries certain structures reminiscent of structures in classical mechanics. The elements of the rings in Equation 0.2 can then be viewed intuitively as "arithmetic differential equations" in the same way in which functions on jet spaces in differential geometry are interpreted as "differential equations". The above construction reduces the study of "δ–geometry" of $X_\wp$ to the study of (usual) algebraic geometry of $J^r(X_\wp)$. Actually the construction of p–jet spaces is quite easy to explain. Assume, for simplicity, that $X_\wp$ is affine given by

$$(0.3) \qquad X_\wp = Spec\ R_\wp[T]/(F),$$

where T is a tuple of indeterminates and F is a tuple of elements in $R_\wp[T]$. Assume furthermore, for simplicity, that the components of F have coefficients in $\mathbf{Z}_p$. Then $J^r(X_\wp)$ is, by definition, the formal spectrum

$$(0.4) \qquad J^r(X_\wp) := Spf\ R_\wp[T, T', T'', ..., T^{(r)}]\hat{}/(F, \delta F, \delta^2 F, ..., \delta^r F)$$

where $\hat{}$ means "p–adic completion", $T', T'', ...$ are new tuples of variables and $\delta F, \delta^2 F, ...$ are defined by the formulae

$$(0.5) \qquad (\delta F)(T, T') := \frac{F(T^p + pT') - F(T)^p}{p},$$

$$(\delta^2 F)(T, T', T'') := \frac{(\delta F)(T^p + pT', (T')^p + pT'') - ((\delta F)(T, T'))^p}{p},$$

etc. The polynomials $\delta F, \delta^2 F, ...$ should be viewed as arithmetic analogues of "iterated total derivatives" of F and the construction of our "arithmetic jet spaces" is then analogous to that of "differential algebraic jet spaces" to be discussed presently; cf. Equation 0.20 below. Also, like in the case of the latter, and as in the case of differential geometry, the fibers of the maps $J^r(X_\wp) \to J^{r-1}(X_\wp)$ are (p–adic completions of) affine spaces.

Having reduced our problems to problems about p–jet spaces the next step is to construct, in each of the spherical, flat, and hyperbolic cases, some remarkable *δ–invariants*. In the spherical case this will be elementary. In the flat case the construction will be essentially based on our theory of *δ–characters* [**17**]; the latter are arithmetic analogues of the Manin maps in [**97**]. In the hyperbolic case the construction will be based on our theory of *isogeny covariant δ–modular forms* [**21**], [**23**], [**24**]. We will develop the theories of δ–characters and δ–modular forms ab initio in this book; we will partly follow the above cited papers and then we will further develop these theories up to a point where we can use them for our applications.

Finally we will need methods to prove that *all* δ–invariants essentially occur via the constructions mentioned above. There will be a number of methods used for this purpose. There is, for instance, a Galois theoretic method that is sufficient to handle, in particular, the spherical case. Another method, especially useful in the flat and hyperbolic case, is close in spirit to that used by Hilbert in his original approach to invariant theory and is based on the existence of a certain system of (usual) partial differential operators acting on appropriate rings of invariants. In our case the partial differential operators in question will live on arithmetic jet spaces and will be derived by analogy with classical mechanics.

0.3. Comparison with other theories

0.3.1. Differential equations on smooth manifolds. Many of the $\delta-$ geometric concepts to be introduced and studied in this book are arithmetic analogues of concepts related to differential equations on smooth manifolds derived from classical mechanics. Let us quickly review here some of these smooth manifold concepts.

Let M be a smooth manifold (i.e. a C^{∞} real manifold) equipped with a submersion $\pi : M \to \mathbf{R}$. For each $t \in \mathbf{R}$ we think of $\pi^{-1}(t)$ as the "configuration space" at time t. (The map $M \to \mathbf{R}$ should be viewed as the object whose arithmetic analogues are the morphisms of schemes, $X \to Spec\ F$, defining varieties X over number fields F.) For any smooth map $f : \mathbf{R} \to M$ we denote by $J_0^r(f)$ the $r-$jet of f at 0. We define the *jet space of* M *(relative to* π*)* by

$$(0.6) \qquad J^r(M) := \{J_0^r(f) \mid f : \mathbf{R} \to M \text{ smooth}, \ \pi \circ f : \mathbf{R} \to \mathbf{R} \text{ a translation}\}.$$

Then $J^r(M)$ has a natural structure of smooth manifold. Any smooth section $s : \mathbf{R} \to M$ of $\pi : M \to \mathbf{R}$ lifts to a smooth map $\mathbf{pr}\ s : \mathbf{R} \to J^r(M)$ defined by

$$(\mathbf{pr}\ s)(t) := J_0^r(s \circ \tau_t),$$

where $\tau_t : \mathbf{R} \to \mathbf{R}$ is the translation $\tau_t(u) := u + t$. The smooth functions $L : J^r(M) \to \mathbf{R}$ can be thought of as (time dependent) *Lagrangians*. Alternatively we may think of such smooth functions L as *differential equations on* M. The natural projections $J^r(M) \to J^{r-1}(M)$ are fiber bundles with fiber $\mathbf{R}^n$, where $dim\ M = n + 1$.

Next, in order to simplify our discussion, we will work in coordinates. So we let $M := \mathbf{R} \times \mathbf{R}^n$ with coordinates $(t, x) = (t, x_1, ..., x_n)$ and we let $\pi : M \to \mathbf{R}$ be the first projection. Then we have a natural identification $J^r(M) = \mathbf{R} \times \mathbf{R}^{n(r+1)}$ sending the $r-$jet $J_0^r(f)$ of a map $f : \mathbf{R} \to M$, $f(t) = (t + t_0, x(t))$, into the tuple

$$\left(t_0, x(0), \frac{dx}{dt}(0), ..., \frac{d^r x}{dt^r}(0)\right).$$

If $s : \mathbf{R} \to \mathbf{R} \times \mathbf{R}^n$, $s(t) = (t, x(t))$, is a section of π then $\mathbf{pr}\ s : \mathbf{R} \to \mathbf{R} \times \mathbf{R}^{n(r+1)}$ is given by

$$(0.7) \qquad (\mathbf{pr}\ s)(t) = \left(t, x(t), \frac{dx}{dt}(t), ..., \frac{d^r x}{dt^r}(t)\right).$$

Cf. [**114**], p. 96. We denote by $(t, x, x', ..., x^{(r)})$ the coordinates on $\mathbf{R} \times \mathbf{R}^{n(r+1)}$, where $x', ..., x^{(r)}$ are $n-$tuples of variables; for each index i the variables $x_i', x_i'', ...$ are the variables *conjugate* to x_i. A (time dependent) Lagrangian is simply a smooth function

$$(0.8) \qquad L(t, x, x', ..., x^{(r)})$$

on $\mathbf{R} \times \mathbf{R}^{n(r+1)}$. Below is a list of key concepts related to the above context.

A (vertical) *vector field* on M is a derivation operator (on smooth functions of t, x) of the form

$$V := \sum_i a_i(t,x)\frac{\partial}{\partial x_i} \tag{0.9}$$

where a_i are smooth functions. More generally a (vertical) *generalized vector field* (sometimes called *Bäcklund vector field*) is a derivation operator (on smooth functions of t, x) of the form

$$V := \sum_i a_i(t,x,x',...,x^{(r)})\frac{\partial}{\partial x_i}; \tag{0.10}$$

cf. [**114**], p. 289.

The *total derivative* operator is the derivation operator (on smooth functions of $t, x, x', ..., x^{(r)}$) defined by

$$D_t := \frac{\partial}{\partial t} + \sum_{ij} x_i^{(j+1)} \cdot \frac{\partial}{\partial x_i^{(j)}}; \tag{0.11}$$

cf. [**114**], p. 109. It is characterized by the property that

$$(D_t L) \circ (\mathbf{pr}\ s) = \frac{d}{dt}(L \circ \mathbf{pr}\ s)$$

for all Lagrangians L and all sections s of π.

The *prolongation* of a (vertical) vector field as in Equation 0.9 is defined by

$$\mathbf{pr}\ V := \sum_{ij}(D_t^j a_i) \cdot \frac{\partial}{\partial x_i^{(j)}}; \tag{0.12}$$

cf. [**114**], p. 110; it is the unique (smooth) vector field on $J^r(M)$ that commutes with D_t and coincides with V on functions of t and x.

A vector field as in Equation 0.9 is an *infinitesimal symmetry* for an $\mathbf{R}$–linear space $\mathcal{L}$ of Lagrangians if

$$(\mathbf{pr}\ V)(\mathcal{L}) \subset \mathcal{L}. \tag{0.13}$$

If (L_i) is a family of Lagrangians and $\mathcal{L}$ is the ideal generated by the family $(D_t^j L_i)$ in the ring of all Lagrangians then any infinitesimal symmetry of $\mathcal{L}$ in the sense above is an infinitesimal symmetry of the "system of differential equations $L_i = 0$" in the sense of [**114**], p. 161.

The vector field V is a *variational infinitesimal symmetry* of the Lagrangian L if

$$(\mathbf{pr}\ V)(L) = 0; \tag{0.14}$$

cf. [**114**], p. 253.

The *Fréchet derivative* of a Lagrangian L is the operator on vector fields given by

$$V = \sum_i a_i \frac{\partial}{\partial x_i} \mapsto (\mathbf{pr}\ V)(L) = \langle \mathbf{pr}\ V, dL\rangle = \sum_{ij}(D_t^j a_i) \cdot \frac{\partial L}{\partial x_i^{(j)}}; \tag{0.15}$$

cf. [**114**], p. 307.

The *Euler-Lagrange form* attached to a Lagrangian L is the 1–form

$$\mathbf{EL}(L) := \sum_i \left(\sum_j (-1)^j D_t^j \left(\frac{\partial L}{\partial x_i^{(j)}} \right) \right) dx_i; \tag{0.16}$$

cf. [**114**], p. 246.

Noether's Theorem in this context is the (trivial) statement that for any Lagrangian $L(t, x, x', ..., x^{(r)})$ and any vector field V as in Equation 0.9 there exists $G = G(t, x, x', ..., x^{(2r-1)})$ such that

$$\langle V, \mathbf{EL}(L) \rangle - (\mathbf{pr}\ V)(L) = D_t G. \tag{0.17}$$

If V is a variational infinitesimal symmetry of L (i.e. Equation 0.14 holds) then G is called the *conservation law* attached to V; it is unique up to an additive constant and it has the property that if the function $t \mapsto x(t)$ satisfies the *Euler-Lagrange equations*

$$\langle \frac{\partial}{\partial x_i}, \mathbf{EL}(L) \rangle \left(t, x(t), \frac{dx}{dt}(t), ..., \frac{d^{2r}x}{dt^{2r}}(t) \right) = 0 \tag{0.18}$$

for $i = 1, ..., n$ then the function

$$t \mapsto G\left(t, x(t), \frac{dx}{dt}(t), ..., \frac{d^{2r-1}x}{dt^{2r-1}}(t) \right)$$

is constant.

Finally the *Hamiltonian vector field* attached to a function $H = H(t, x, x')$ and the vector fields $\partial/\partial x_i$ is the vector field

$$\sum_i \left(\frac{\partial H}{\partial x_i'} \frac{\partial}{\partial x_i} - \frac{\partial H}{\partial x_i} \frac{\partial}{\partial x_i'} \right). \tag{0.19}$$

All the concepts above have arithmetic analogues some of which will play a key role in the proofs of our main theorems. The arithmetic analogues of jet spaces of manifolds, of total derivatives, and of Lagrangians were already encountered in our discussion of "arithmetic jet spaces". The arithmetic analogue of prolongation of vector fields (and certain related operators) will play a key role in controlling δ–invariants in our theory. The arithmetic analogues of Fréchet derivatives will be the tangent maps in δ–geometry. In some of our basic δ–geometric examples we will prove the existence of "arithmetic (variational) infinitesimal symmetries", we will discover remarkable systems of commuting "arithmetic Hamiltonian vector fields", and we will compute "arithmetic Euler-Lagrange equations".

In order to clarify the analogy between the usual differential equations and arithmetic differential equations it will be useful to discuss an intermediate step which is provided by the Ritt-Kolchin differential algebra [**117**], [**84**].

0.3.2. Differential algebra. The presentation here follows [**13**]. In differential algebra one usually starts with a field K equipped with a derivation $\delta : K \to K$; recall that, by definition, δ is an additive map satisfying the usual Leibniz rule. We shall assume, for simplicity, that K has characteristic zero and is algebraically closed. In applications K usually contains the field $\mathbf{C}(t)$ of rational functions in one variable and δ extends the derivation d/dt. (For the "most general" case of this theory, in which K is replaced by any ring and δ is replaced by a "Hasse-Schmidt derivation" we refer to Vojta's preprint [**131**].) For any non-singular variety X over K one can then define a projective system of non-singular varieties

$$... \to J^r(X) \to J^{r-1}(X) \to ... \to J^0(X) = X$$

called the (differential algebraic) *jet spaces* of X. It is easy to review the construction of these varieties. Assume, for simplicity, that X is affine given by

$$X = Spec\ K[T]/(F)$$

where T is a tuple of indeterminates $T_1, ..., T_N$ and F is a tuple of elements in $K[T]$. Then $J^r(X)$ is, by definition, the spectrum

$$J^r(X) := Spec\ K[T, T', T'', ..., T^{(r)}]/(F, \delta F, \delta^2 F, ..., \delta^r F) \tag{0.20}$$

where $T', T'', ...$ are new tuples of variables and $\delta F, \delta^2 F, ...$ are defined by the formulae

$$(\delta F)(T, T') := F^{\delta}(T) + \sum_j \frac{\partial F}{\partial T_j}(T) \cdot T'_j, \tag{0.21}$$

$$(\delta^2 F)(T, T', T'') := (\delta F)^{\delta} + \sum_j \frac{\partial \delta F}{\partial T_j}(T, T') \cdot T'_j + \sum_j \frac{\partial \delta F}{\partial T'_j}(T, T') \cdot T''_j,$$

etc., where F^{δ}, $(\delta F)^{\delta}$,... are obtained from $F, \delta F, ...$ by applying δ to the coefficients. The polynomials $\delta F, \delta^2 F, ...$ are the *iterated total derivatives* of F. There is a clear analogy between the arithmetic jet spaces in Equation 0.4 and the differential algebraic jet spaces $J^r(X)$ in Equation 0.20. On the other hand it can be shown ([**14**], p.1443) that the $K-$points of the differential algebraic jet space $J^r(X)$ can be identified with the morphisms of $K-$schemes,

$$Spec\ K[[T]]_{e^{T\delta}}/(T^{r+1}) \to X, \tag{0.22}$$

where $K[[T]]_{e^{T\delta}}$ is the ring of power series $K[[T]]$ viewed as a $K-$algebra via the map

$$e^{T\delta} : K \to K[[T]], \quad e^{T\delta}(x) := \sum_{n=0}^{\infty} \frac{\delta^n x}{n!} T^n.$$

(The maps in Equation 0.22 should be viewed as a twisted version of *arcs* in X where the twist is given by δ. Recall that usual *arcs* of order r in X are defined as morphisms of $K-$schemes, $Spec\ K[[T]]/(T^{r+1}) \to X$, where $K[[T]]$ is viewed as a $K-$algebra via the inclusion $K \subset K[[T]]$; the set of all arcs of order r in X has a natural structure of algebraic variety over K and is referred to as the *arc space of* X of order r. Cf. [**45**]. Therefore the varieties $J^r(X)$ can be viewed as a twisted analogue of arc spaces. They identify with arc spaces in case X is defined over the field $C = \{a \in K | \delta a = 0\}$ of constants of δ. But for X not descending to C arc spaces are different from the varieties $J^r(X)$; cf. [**14**] for more on this.) Now there is a clear analogy between the maps in Equation 0.22 and the jets belonging to $J^r(M)$ in Equation 0.6. This makes differential algebraic jet spaces an analogue of jet spaces of smooth manifolds (relative to a submersion). The analogies described above can be summarized in the following diagram:

$$(1) - - - (2) - - - (3), \tag{0.23}$$

(1)=jet spaces of smooth manifolds relative to a submersion; cf. Equation 0.6
(2)=differential algebraic jet spaces; cf. Equation 0.20
(3)=arithmetic jet spaces; cf. Equation 0.4

Here are a few more remarks on differential algebraic jet spaces. As in the case of smooth manifolds, the fibers of the maps $J^r(X) \to J^{r-1}(X)$ are affine spaces. Also there is a natural map at the level of $K-$points $\nabla : X(K) \to J^r(X)(K)$ naturally induced by the map $\nabla : \mathbf{A}^1(K) = K \to J^r(\mathbf{A}^1)(K) = K^{r+1}$ defined by $\nabla(a) = (a, \delta a, ..., \delta^r a)$. One defines the ring $\mathcal{O}^r(X)$ of $\delta-$*polynomial maps* on X to be the ring of all functions $f : X(K) \to K$ that can be written as $f = \tilde{f} \circ \nabla$ where $\tilde{f} \in \mathcal{O}(J^r(X))$ is a regular function on $J^r(X)$. One gets an isomorphism $\mathcal{O}^r(X) \simeq \mathcal{O}(J^r(X))$ of which the isomorphism in Equation 0.2 is the arithmetic analogue.

Next we address the following question: are the results of this book arithmetic analogues of results involving (genuine) differential equations ? In a certain loose sense this is indeed the case as explained below.

1) The spherical case of our theory here can be loosely viewed as an arithmetic analogue of some of the classical theory of differential invariants as found, for instance, in [**135**], [**114**].

2) The flat case of our theory here can be loosely viewed as an arithmetic analogue of the theory of the Manin map [**97**] and of the differential algebraic theory we developed in [**12**], [**14**], [**15**].

3) The hyperbolic case of our theory here can be loosely viewed as an arithmetic analogue of the classical theory of differential relations among modular forms and of the differential algebraic theory we developed in [**16**]; cf. also [**8**].

Let us provide some details.

0.3.2.1. *Spherical case.* We explain here a simple situation in the classical theory of differential invariants. Start with a group G acting by algebraic automorphisms on a variety X over K (which is usually affine). Then there is an induced action of G on the jet spaces $J^r(X)$. The ring of $G-$invariant functions $\mathcal{O}^r(X)^G = \mathcal{O}(J^r(X))^G$ corresponds to a special case of *differential invariants* in classical theory. A trivial example of this would be the natural action of $G := SL_2(C)$ on the affine plane $\mathbf{A}^2 = Spec\ K[x,y]$ over K; the Wronskian $xy' - yx' \in K[x, y, x', y'] = \mathcal{O}(J^1(\mathbf{A}^1))$ is then $G-$invariant. The spherical case of our arithmetic theory involves arithmetic analogues of computations of differential invariants in the above sense.

More generally the classical differential invariant theory deals with groups G (or simply vector fields V) acting directly on jet spaces $J^r(X)$ (in such a way that the action does not necessarily come "by functoriality" from an action on X itself). The situation is described in detail in Olver's book [**114**] in the setting of jet spaces in differential geometry.

0.3.2.2. *Flat case.* As already mentioned our flat theory will involve an arithmetic analogue of the theory of Manin maps [**97**], [**12**], [**14**], [**15**]. Let us quickly review the basics of the latter in the context of differential algebra following [**12**]. Assume X is an Abelian variety over K of dimension g. Then, by functoriality, each $J^r(X)$ has a natural structure of an algebraic group and is an extension of X by a vector group (i.e. a sum of copies of the additive group $\mathbf{G}_a = Spec\ K[t]$). By the theory of extensions of Abelian varieties by vector groups there is a surjective homomorphism $J^r(X) \to H$ onto a vector group H of dimension $\geq (r-1)g$. Consequently if $r \geq 2$ we have $H \neq 0$ hence $Hom(J^r(X), \mathbf{G}_a) \neq 0$. Elements of the latter composed with ∇ give rise to homomorphisms $\psi : X(K) \to K$. These homomorphisms are called *Manin maps*. They were discovered by Manin [**97**] (via a construction different from the one above) and were used by him to provide the first proof of the Mordell conjecture over function fields. Now the strategy presented above works well in the arithmetic case. The p–jet spaces $J^r(X_\wp)$ of an Abelian scheme $X_\wp$ over $R_\wp$ continue to be group objects and, although the kernel of the projection $J^r(X_\wp) \to \hat{X}_\wp$ ceases to be a sum of copies of $\hat{\mathbf{G}}_a := Spf\ R_\wp[t]\hat{}$, one can still prove that $Hom(J^r(X_\wp), \hat{\mathbf{G}}_a) \neq 0$ provided $r \geq 2$. As before elements of the latter yield δ–functions $\psi : X(R_\wp) \to R_\wp$ which are homomorphisms; these should be viewed as arithmetic analogues of Manin's maps and play a key role in our theory in the flat case. Indeed maps of the form $\frac{\phi\circ\psi}{\psi}$ are invariant under the multiplication by integers N on our Abelian variety, $[N] : X_\wp \to X_\wp$; this invariance is the starting point for our use of ψ in our search for invariants in the flat case.

0.3.2.3. *Hyperbolic case.* We end our discussion of differential algebra by reviewing "differential algebraic invariants of isogenies" following [**16**]; this theory has an analogue in the arithmetic case that plays a key role in our treatment of the hyperbolic case. Let us consider the affine line $\mathbf{A}^1 = Spec\ K[j]$ viewed as a moduli space for elliptic curves over K. For any elliptic curve E over K we let $j(E) \in K$ be its j–invariant. On $K = \mathbf{A}^1(K)$ we have then an equivalence relation called *isogeny*: two points $a_1, a_2 \in K$ are called *isogenous* if there exists an isogeny of elliptic curves $E_1 \to E_2$ over K such that $j(E_1) = a_1$ and $j(E_2) = a_2$. The constant field $C \subset K$ is saturated with respect to isogeny hence so is $K\backslash C$. Then one can show [**16**] that there exists a rational function $U(j) \in \mathbf{Q}(j)$ such that if

$$\chi(j, j', j'', j''') := \frac{2j'j''' - 3(j'')^2}{4(j')^2} + (j')^2U(j) \in K(j)[j', (j')^{-1}, j'', j'''] \tag{0.24}$$

then the function

$$K\backslash C \to K, \quad a \mapsto \chi(a, \delta a, \delta^2 a, \delta^3 a)$$

is constant on isogeny classes. (The fraction in Equation 0.24 is, of course, the classical Schwarzian operator.) We refer to [**16**] for more on this and for a higher dimensional generalization of this. It turns out that the theory behind the function χ above has an arithmetic analogue which leads to what we call *δ–modular forms*; cf. [**21**], [**23**], [**24**]. These will be used heavily in this book.

0.3.3. Difference algebra. In the Ritt-Cohn difference algebra [**35**] one starts with a field K equipped with a ring endomorphism $\phi : K \to K$. For our discussion here we will allow K to have arbitrary characteristic. For any variety X over K one can then define a projective system of non-singular varieties

$$\ldots \to J^r_\phi(X) \to J^{r-1}_\phi(X) \to \ldots \to J^0_\phi(X) = X$$

which can be called the *difference jet spaces* of X. If X is affine given by

$$X = Spec\ K[T]/(F)$$

where T is a tuple of indeterminates and F is a tuple of elements in $K[T]$ then $J^r_\phi(X)$ is, by definition, the spectrum

$$J^r_\phi(X) := Spec\ K[T, T^\phi, T^{\phi^2}, \ldots, T^{\phi^r}]/(F, \phi F, \phi^2 F, \ldots, \phi^r F) \tag{0.25}$$

where $T^\phi, T^{\phi^2}, \ldots$ are new tuples of variables and

$$\phi : K[T, T^\phi, \ldots, T^{\phi^{r-1}}] \to K[T, T^\phi, \ldots, T^{\phi^r}]$$

is the unique ring endomorphism prolonging $\phi : K \to K$ and sending $\phi T = T^{\phi}$, $\phi T^{\phi} = T^{\phi^2}$,... e.t.c. A basic difference between difference algebra and differential algebra is that the fibers of $J^r_{\phi}(X) \to J^{r-1}_{\phi}(X)$ are *not* affine spaces in general. Actually it is plain that $J^r_{\phi}(X)$ splits as a product of varieties

$$J^r_{\phi}(X) \simeq X \times X^{\phi} \times ... \times X^{\phi^r},$$

where X^{ϕ^i} is obtained from X by twisting the coefficients of the defining equations by ϕ^i. In particular, if X is definable over the fixed field of ϕ then $J^r_{\phi}(X)$ is just the $(r+1)-$fold product of X with itself. In spite of this geometric difference between difference algebra and differential algebra a lot of analogies between the two persist. Systems of linear equations, for instance, behave similarly [**62**], p. 20. At the non-linear level the geometric model theory of the two is strikingly similar; cf. especially the work of Hrushovski and Chatzidakis explained in [**69**].

It is interesting to compare difference algebra and arithmetic differential algebra. We refer to Remark 2.6 below for an argument suggesting that arithmetic differential algebra could be viewed as obtained from difference algebra by adjoining certain divergent series. Here we close our discussion by indicating a possible bridge at the level of jet spaces. Let us assume, for simplicity, that X is an affine scheme given by the spectrum of $R_{\wp}[T]/(F)$ as in Equation 0.3. Let $K_{\wp}$ be the field of fractions of $R_{\wp}$. Then one can consider a canonical algebraisation of the $p-$jet space $J^r(X_{\wp})$ given as the scheme

$$\mathcal{J}^r(X_{\wp}) := Spec\ R_{\wp}[T, T', T'', ..., T^{(r)}]/(F, \delta F, \delta^2 F, ..., \delta^r F). \tag{0.26}$$

Then clearly the $p-$adic completion of $\mathcal{J}^r(X_{\wp})$ is $J^r(X_{\wp})$; in symbols:

$$\mathcal{J}^r(X_{\wp})\hat{} = J^r(X_{\wp}). \tag{0.27}$$

On the other hand it is trivial to see that the base change to $K_{\wp}$ of $\mathcal{J}^r(X_{\wp})$ is $J^r_{\phi}(X_{\wp} \otimes K_{\wp})$; in symbols,

$$\mathcal{J}^r(X_{\wp}) \otimes K_{\wp} = J^r_{\phi}(X_{\wp} \otimes K_{\wp}). \tag{0.28}$$

By Equations 0.27 and 0.28 we see that $\mathcal{J}^r(X_{\wp})$ might be viewed as a "bridge" between the jet spaces of difference algebra and arithmetic differential algebra. Whether this bridge can actually be crossed is an open question. We can summarize the above discussion by adjoining to the diagram in Equation 0.23 the diagram:

$$\begin{array}{ccc} (2) & --- & (3) \\ | & & | \\ | & & | \\ (4) & --- & (5) \end{array} \tag{0.29}$$

(2)=differential algebraic jet spaces; cf. Equation 0.20
(3)=arithmetic jet spaces; cf. Equation 0.4
(4)=difference jet spaces; cf. Equation 0.25
(5)=canonical algebraisation of arithmetic jet spaces; cf. Equation 0.26.

0.3.4. Dynamical systems. Let us start by reviewing some aspects of the classical iteration theory initiated by Fatou and Julia [**109**]. In this theory one starts with a rational map $s : \mathbf{P}^1 \to \mathbf{P}^1$ from the complex projective line into itself and one studies the behavior of orbits

$$z, s(z), s^2(z), ...$$

as z moves in $\mathbf{P}^1$. The z's for which the orbit is "unstable" form a closed set called the *Julia set*. The complement of the Julia set is called the *Fatou set*. Points that eventually return to themselves after a number of iterations are called *periodic*. Up to finitely many points the Julia set turns out to coincide with the closure of the set of periodic points and usually looks like a "fractal". A point $z \in \mathbf{C}$ is critical if the tangent map of s at z vanishes. The orbits of critical points hold the key to many of the dynamical properties of s. In particular the maps s for which all critical points have finite orbits have been closely investigated by Thurston [**51**]; they are called *postcritically finite* and one can attach orbifolds and Euler characteristics to them. Postcritically finite maps with Euler characteristic 0 have been classified and they are essentially those that admit an "analytic uniformization" by the complex plane $\mathbf{C}$. Let us call them here, for simplicity, *flat maps*. Flat maps can be explicitly described as being either multiplicative maps, or Chebyshev polynomials or *Lattès maps* (the latter being maps induced by endomorphisms of elliptic curves). The Julia sets of flat maps are "smooth" (i.e. not "fractals"). For multiplicative maps the Julia set is a circle. For Chebyshev maps the Julia set is a segment. For Lattès maps the Julia set is the whole of $\mathbf{P}^1$. (Note however that flat maps are not the only maps that have smooth Julia sets !)

Flat maps play a key role in our theory. Indeed let us attach to any rational map s a correspondence $\mathbf{X}^* = (\mathbf{P}^1, \mathbf{P}^1, id, s)$ in complex algebraic geometry. Assume, for simplicity, that s has rational coefficients. Fix a prime p. Then, as explained earlier, we can (roughly speaking) attach to $\mathbf{X}$ a correspondence $\mathbf{X}_\delta$ in $\delta-$geometry at p. What we will prove in this book is, roughly speaking, that the categorical quotient of $\mathbf{X}_\delta$ in $\delta-$geometry is non-trivial for almost all p if and only if s is a flat map.

More generally the concept of postcritical finiteness (and a suitable generalization of it to the case of correspondences that do not come from dynamical systems) plays a key role in the present book. This will be explained in detail in Chapters 1 and 2.

We close our discussion here by noting that complex dynamics has an interesting $p-$adic analogue. For instance, the Lubin-Tate theory of formal groups [**92**], [**93**] can be viewed as an early incarnation of a non-archimedian dynamics. For more recent work see, for instance [**93**], [**5**], [**6**], [**91**], and the bibliographies therein. Although our theory lives in the $p-$adic world and although we *will* use formal groups at periodic points in our theory, we will not need to use, in our book, the "genuine" theory of $p-$adic dynamical systems in the above cited papers. Here, by "genuine" we mean "dealing with the $p-$adic Julia set, $p-$adic wandering Fatou domains, etc." (On the other hand, as already mentioned, we *do* need to use the theory of *complex* dynamical systems through the orbifold characterization of flat maps.) Needless to say it would be very interesting to make our theory interact with the "genuine" theory of $p-$adic dynamical systems.

0.3.5. Connes' non-commutative geometry. The approach in the present book seems to be, in some sense, perpendicular to Connes' non-commutative geometric approach to quotient spaces; cf. our previous remarks about invariant versus groupoid ideology. Nevertheless, since some examples occur in both theories it is not unreasonable to expect interactions between the two approaches. In what follows we shall discuss some general principles of non-commutative geometry and we shall examine some examples. Our discussion will inevitably be extremely superficial and not entirely precise; for precise, in-depth presentations we refer to [**36**], [**99**], [**101**].

Connes' theory is formulated in the context of functional analysis and, although some parts of it have a more algebro-geometric flavor [**127**] [**99**], our presentation will follow Connes' original approach. As already mentioned the starting point in this theory is often a groupoid

$$\mathcal{G} = (X, \tilde{X}, \sigma_1, \sigma_2, \nu, \iota, \epsilon)$$

in some "geometric" category, for instance the category of smooth manifolds. Write $\gamma_1\gamma_2 = \nu(\gamma_1, \gamma_2)$. The first step is to attach to $\mathcal{G}$ a C^*-algebra, $C^*(\mathcal{G})$, called the *convolution algebra* of $\mathcal{G}$. Naively $C^*(\mathcal{G})$ might be thought of as (a norm completion of) some algebra of complex valued functions f on $\tilde{X}$ with multiplication given by "convolution":

$$(f_1 \star f_2)(\gamma) = \sum_{\gamma_1\gamma_2=\gamma} f_1(\gamma_1)f_2(\gamma_2).$$

As it is this cannot work because the sums involved are usually infinite. So one replaces, according to the context, sums by integrals, functions by "densities", etc.

The next step in Connes' theory is the realization that "non-commutative spaces" should not be defined as C^*-algebras but rather as *Morita equivalence classes* of C^*-algebras. Two C^*-algebras A and B are *Morita equivalent* if there exists a bimodule ${}_AM_B$ with certain extra data that allow one to "transfer" modules between A and B. So, roughly speaking, one defines the *non-commutative space* attached to the groupoid $\mathcal{G}$ as the C^*-algebra $C^*(\mathcal{G})$ up to Morita equivalence.

At this point one is faced with the "topological and geometrical study of non-commutative spaces"; this constitutes the heart of the theory and, for this purpose, a whole spectrum of techniques is brought into the picture such as: spectral theory, index theory, $K-$theory, etc.

In what follows we examine a number of examples in non-commutative geometry and compare them with the spherical, flat, and hyperbolic examples we are able to treat in our theory.

0.3.5.1. *Spherical case.* Consider the action of $PSL_2(\mathbf{Z})$ on $\mathbf{P}^1(\mathbf{C})$ by homographies. Note that

$$\mathbf{P}^1(\mathbf{C}) = \mathbf{H}^{\pm} \cup \mathbf{P}^1(\mathbf{R}), \quad \mathbf{H}^{\pm} := \mathbf{C}\backslash\mathbf{R} = \mathbf{H}^+ \cup \mathbf{H}^-,$$

$\mathbf{H}^+ = \mathbf{H}$ is the upper half plane in $\mathbf{C}$, $\mathbf{H}^-$ is the lower half plane, and $\mathbf{P}^1(\mathbf{R})$ is the real projective line. The action of $PSL_2(\mathbf{Z})$ on $\mathbf{H}^{\pm}$ is properly discontinuous and the quotient exists in the category of Riemann surfaces; it is the union of 2 copies of the modular curve $\mathbf{H}/PSL_2(\mathbf{Z})$ parameterizing elliptic curves over the complex numbers. However the action of $PSL_2(\mathbf{Z})$ on the "stratum" $\mathbf{P}^1(\mathbf{R})$ is not properly discontinuous and actually the categorical quotient in the category of manifolds reduces to a point. What can be done in non-commutative geometry is to consider the groupoid $\mathcal{G}$ attached to the action

$$PSL_2(\mathbf{Z}) \times \mathbf{P}^1(\mathbf{R}) \to \mathbf{P}^1(\mathbf{R})$$

and consider the C^*-algebra $C^*(\mathcal{G})$ modulo Morita equivalence. The latter is declared to *be*, by definition, the "quotient" $\mathbf{P}^1(\mathbf{R})/PSL_2(\mathbf{Z})$ and can be interpreted, in a natural way as a compactification of the modular curve $\mathbf{H}/PSL_2(\mathbf{Z})$; cf. the remarks in the next example. The non-commutative space $\mathbf{P}^1(\mathbf{R})/PSL_2(\mathbf{Z})$ has been studied by Manin and Marcolli [**100**] who proved a number of deep results about "limiting" behavior of objects from the usual modular curve $\mathbf{H}/PSL_2(\mathbf{Z})$ that tend to points on the boundary.

On the other hand the space of orbits of $\mathbf{P}^1$ under the action of $PSL_2(\mathbf{Z})$ will be one of the main examples of our theory in the spherical case; cf. the present Introduction. It would be beautiful to understand if there is a connection between the two approaches. Remark that from our perspective $\mathbf{P}^1$ modulo $PSL_2(\mathbf{Z})$ is in a "spherical situation" while in non-commutative geometry $\mathbf{P}^1(\mathbf{R})$ modulo $PSL_2(\mathbf{Z})$ is at the boundary of a "hyperbolic commutative geometric object".

0.3.5.2. *Flat case.* One of the first remarkable objects studied in non-commutative geometry were the non-commutative tori of Connes and Rieffel [**36**], [**40**]. There are analogies between these objects and the objects relevant in the flat case of our theory. Let us briefly explain this in the case of "complex dimension" one. A complex torus of dimension one is an elliptic curve E over $\mathbf{C}$. One can represent E as the quotient E_τ (in the category of Riemann surfaces) of the complex plane $\mathbf{C}$ by the action (by translation) of the subgroup $\mathbf{Z}+\tau\mathbf{Z}$, where $\tau \in \mathbf{H}$. Alternatively one can describe E_τ as the quotient (in the category of Riemann surfaces) of $\mathbf{C}^\times$ by the action of the subgroup $\langle q_\tau \rangle$ where $q_\tau = e^{2\pi i\tau}$. As τ approaches a point $\theta \in \mathbf{R}\backslash\mathbf{Q}$, q_τ will approach $q_\theta := e^{2\pi i\theta}$ which is on the unit circle $S^1 \subset \mathbf{C}$, but not a root of unity. In non-commutative geometry it is possible to define an object which can be viewed as the limit of a family of elliptic curves E_τ when $\tau \to \theta$. Here is the construction. One considers the action

$$\mu_\theta : \mathbf{Z} \times S^1 \to S^1, \quad \mu_\theta(m, z) := q_\theta^m \cdot z,$$

one attaches to this action a groupoid $\mathcal{G}_\theta$ and one defines the C^*-algebra $A_\theta := C^*(\mathcal{G}_\theta)$; the Morita equivalence class of A_θ is interpreted as a *non-commutative elliptic curve* which is the *limit of* E_τ *as* $\tau \to \theta$. One of the crucial early discoveries in non-commutative geometry was that for two irrational real numbers θ_1 and θ_2 the C^*-algebras A_{θ_1} and A_{θ_2} are Morita equivalent if and only if θ_1 and θ_2 are $PSL_2(\mathbf{Z})-$conjugate. This justifies the claim in the previous example that $\mathbf{P}^1(\mathbf{R})/PSL_2(\mathbf{Z})$ can be interpreted as a compactification of the modular curve $\mathbf{H}/PSL_2(\mathbf{Z})$. (Indeed $\mathbf{P}^1(\mathbf{Q})/PSL_2(\mathbf{Z})$ consists of one point, the classical *cusp*.)

Now let us note an analogy between the above example and what will be involved in the flat case of our theory. The quotient of S^1 by the action ρ_θ is a quotient $S^1/\langle q_\theta \rangle$ of the Lie group S^1 by a cyclic subgroup (which acts "wildly"). Accordingly one can propose to study (categorical) quotients of the form $G/\langle g_1, ..., g_n\rangle$ where G is a commutative algebraic group and $\langle g_1, ..., g_n\rangle$ is a finitely generated subgroup. Such quotients have actually been studied in differential algebra [**12**], [**13**] in relation to the Mordell conjecture over function fields and their study involves the Manin maps discussed before. As already mentioned Manin maps have arithmetic analogues that play a key role in this book.

0.3.5.3. *Hyperbolic case.* Hecke correspondences play an important role in non-commutative geometry; cf. [**39**], [**37**], [**38**], [**101**]. On the other hand Hecke correspondences will be at the heart of the hyperbolic case of the present book. We would like to compare the two theories in what follows. Let, as usual, $\mathbf{A}_f$ denote the finite adeles of $\mathbf{Q}$ i.e.

$$\mathbf{A}_f := \left(\lim_{\leftarrow} \frac{\mathbf{Z}}{n\mathbf{Z}}\right) \otimes \mathbf{Q}.$$

Then, following Shimura, one considers the space

$$Sh = GL_2(\mathbf{Q})\backslash(GL_2(\mathbf{A}_f) \times \mathbf{H}^{\pm}). \tag{0.30}$$

For any congruence subgroup $\Gamma \subset SL(2, \mathbf{Z})$ consider the modular curve

$$Sh_{\Gamma} := \mathbf{H}/\Gamma,$$

where we wrote $\mathbf{H}/\Gamma$ instead of $\Gamma\backslash\mathbf{H}$ in order to make notation agree with notation in the present book. One can consider then the inverse limit

$$Sh^0 := \lim_{\leftarrow} Sh_{\Gamma}$$

over all congruence subgroups $\Gamma \subset SL_2(\mathbf{Z})$. The space Sh^0 turns out to be a connected component of Sh. All this is classical, i.e. commutative. Now, following [**38**] one can embed the set Sh into the "much larger" set

$$Sh^{(nc)} := GL_2(\mathbf{Q})\backslash(M_2(\mathbf{A}_f) \times \mathbf{H}^{\pm}).$$

The set $Sh^{(nc)}$ does not have a nice "commutative space" structure but one can attach to it a C^*-algebra $\mathcal{A}_2$ starting from a certain (alternative !) presentation of $Sh^{(nc)}$ as a quotient by an action; cf. [**101**], p. 67. We refer to loc. cit. for an overview of the beautiful and deep theory of $\mathcal{A}_2$ and also for an overview of the Bost-Connes $\mathcal{A}_1$ that was its GL_1 prototype.

Now turning to our theory let us consider a prime number l, the matrix

$$\tau_l := \begin{bmatrix} l & 0 \\ 0 & 1 \end{bmatrix},$$

and the subgroup

$$\langle \Gamma, \tau_l \rangle \subset GL_2(\mathbf{Q})$$

(where $\Gamma \subset SL_2(\mathbf{Z})$ is any congruence subgroup). Define the set

$$Sh_{\langle \Gamma, \tau_l \rangle} := \mathbf{H}/\langle \Gamma, \tau_l \rangle.$$

This set should be viewed, roughly, as the quotient of Sh_{Γ} by a Hecke correspondence. Since the group $\langle \Gamma, \tau_l \rangle$ does not act properly discontinuously on $\mathbf{H}$ the set $Sh_{\langle \Gamma, \tau_l \rangle}$ is not an object of analytic geometry anymore. Our theory can be viewed as a way to put on the set $Sh_{\langle \Gamma, \tau_l \rangle}$ a geometric structure in $\delta-$geometry. Summarizing we have the following inclusions of sets

$$Sh^0 \subset Sh \subset Sh^{(nc)}$$

and the following projections of sets

$$Sh^0 \to Sh_{\Gamma} \to Sh_{\langle \Gamma, \tau_l \rangle}.$$

The set $Sh^{(nc)}$ is replaced, in non-commutative geometry, by the C^*-algebra $\mathcal{A}_2$ while the set $Sh_{\langle \Gamma, \tau_l \rangle}$ is replaced, in our theory, by an object of $\delta-$geometry. It is plain from the above discussion that the non-commutative theory and our theory "diverge" in this example. The set we are interested in, $Sh_{\langle \Gamma, \tau_l \rangle}$, is a "wild" quotient of the "commutative part" of $Sh^{(nc)}$.

0.3.6. Drinfeld modules. The theory of Drinfeld modules [**52**], [**62**], [**130**] exhibits analogies with our theory that deserve being understood. Let us quickly recall the definition of a Drinfeld module and make a few comments. One starts with a smooth projective geometrically connected curve X over a finite field $\mathbf{F}_{p^m}$ equipped with a point $\infty \in X$. Let $\mathbf{A}$ be the ring of regular functions on $X\backslash\{\infty\}$. Let $\iota : \mathbf{A} \to A$ be a homomorphism into a field A. Let $F : A \to A$ be the p^m-power homomorphism. Let $A[F]$ be the non-commutative ring generated by A and a variable, called F, with relations $F \cdot a = F(a) \cdot F = a^{p^m} \cdot F$, $a \in A$. (Morally $A[F]$ is a *convolution algebra* attached to the action of F on A.) Let $e : A[F] \to A$ be the ring homomorphism sending $e(\sum a_i F^i) = a_0$ and let $j : A \to A[F]$ be the inclusion. Then a *Drinfeld module* is an $\mathbf{F}_{p^m}-$algebra homomorphism $\psi : \mathbf{A} \to A[F]$ such that $e \circ \psi = \iota$, $\psi \neq j \circ \iota$. The ring $A[F]$ acts on A (with the letter F acting on A as the endomorphism F and A acting on A by multiplication). Hence $\mathbf{A}$ will act on A via ψ. The theory of Drinfeld modules can be viewed therefore as a very non-trivial example of "algebra / geometry with operators"; it entirely lives in characteristic p. The ring $\mathbf{A}$ is viewed, in this theory, as an analogue of the ring $\mathbf{Z}$ of integers. Its quotient field $\mathbf{k}$ plays the role of $\mathbf{Q}$. The completion $\mathbf{K}$ of $\mathbf{k}$ at ∞ plays the role of the real field, $\mathbf{R}$. The completion $\mathbf{C}_\infty$ of an algebraic closure of $\mathbf{K}$ plays the role of the complex field $\mathbf{C}$. The "first" examples of the theory are Drinfeld modules $\psi : \mathbf{A} \to \mathbf{C}_\infty[F]$ attached to "lattices". Drinfeld modules of ranks 1 and 2 play the role of one dimensional algebraic groups (multiplicative group and elliptic curves respectively), viewed as $\mathbf{Z}-$modules (or sometimes modules over rings of integers in imaginary quadratic fields). They can also be viewed as analogues of formal groups viewed as $\mathbf{Z}-$modules (or sometimes modules over rings of integers in $p-$adic fields). There are analogies between the exponentials in the theory of Drinfeld modules and our arithmetic Manin maps; both classes of maps have, as kernels, objects that can be viewed as "lattices". It is a suggestion of Manin (cf. private communication to the author) that maybe there exists a lifting to characteristic zero of Drinfeld module theory which is similar in spirit to the arithmetic differential algebra of the present book.

0.3.7. Dwork's theory. It is interesting to compare our *theory of arithmetic differential equations* with what is usually understood by the *arithmetic theory of differential equations* (as it appears in the work of Dwork, for instance; cf. [**55**]). It turns out that the two theories are "perpendicular" in the very precise sense that they deal with "differentiation" in two "perpendicular" directions. Indeed Dwork's theory is about *genuine* differential equations (i.e. equations involving d/dt) whereas our theory is about *analogues* of differential equations (involving, instead, the Fermat quotient operator δ). In a precise sense d/dt and δ point in two different directions. Now although the two theories are perpendicular this doesn't mean they don't interact. Indeed part of Dwork's theory can be rephrased in terms of crystalline cohomology; cf. the report by Katz [**82**]. The genuine differential equations can be read off the Gauss-Manin connection whereas our arithmetic differential equations are closely related to the action of Frobenius. On the other hand Gauss-Manin and Frobenius are, as well known, both present in the crystalline picture. We will actually heavily rely in our proofs on this crystalline interpretation when we get to discuss the hyperbolic case.

0.3.8. Mochizuchi's p–adic Teichmüller theory. An interesting problem is to find interactions between our theory here and Mochizuchi's p–adic Teichmüller theory [**111**]. Unlike the Tate and Mumford p–adic uniformizations of curves with (totally) degenerate reduction mod p both our theory and Mochizuchi's can be viewed, in a certain sense, as "uniformization" theories in the case of good reduction. As an example of what this might mean in the case of our theory we propose to see the "inverse" of our arithmetic Manin map of an elliptic curve E (cf. our discussion above) as a multivalued map from the affine line $\mathbf{A}^1$ into E. This can be interpreted, in some bold sense, as a sort of uniformization of E.

0.3.9. Ihara's congruence relations. There an interesting possibility that a profitable link can be established between our approach here and Ihara's beautiful work on congruence relations (i.e. liftings to characteristic zero of the correspondence $\Pi \cup \Pi'$ on a curve of characteristic p, where Π is the graph of the Frobenius); cf. [**73**], [**74**]. A hint as to such a link can be seen, for instance, in Lemma 8.16 of this book. On the other hand we would like to point out what we think is an important difference between our viewpoint here and the viewpoint proposed by Ihara in a related paper [**75**]. Our approach, in its simplest form, proposes to see the operator

$$\delta = \delta_p : \mathbf{Z} \to \mathbf{Z},\ a \mapsto \delta a = \frac{a - a^p}{p},$$

where p is a fixed prime, as an analogue of a derivation with respect to p. In [**75**] Ihara proposed to see the map

$$d : \mathbf{Z} \to \prod_p \mathbf{Z}/p\mathbf{Z},\ a \mapsto \left(\frac{a - a^p}{p}\ mod\ p\right) \tag{0.31}$$

as an analogue of differentiation for integers and he proposed a series of very interesting conjectures concerning the "zeroes" of the differential of an integer; these conjectures are still completely open. The main difference between Ihara's viewpoint and ours is that *we do not consider the reduction mod p of the Fermat quotients but the Fermat quotients themselves.* This allows the possibility of considering iterates δ^r of our δ which leads to the possibility of considering arithmetic analogues of higher order differential equations; and indeed most such equations relevant to our theory will have order ≥ 2. On the other hand it doesn't make sense to consider iterates of Ihara's operator d. One way out of this dilemma would be to find a map $D : \mathbf{Z} \to \mathbf{Z}$ which, composed with the canonical projection $\mathbf{Z} \to \prod_p \mathbf{Z}/p\mathbf{Z}$, yields Ihara's map d. (Then one could consider iterates D^r of D.) Voloch [**132**] proved, however, that no such map D can exist at least if one assumes the conjecture that there are only finitely many Mersenne primes.

0.3.10. Field with one element. The theory proposed in this book seems to agree with certain aspects of the "myth of the field with one element" emerging from the work of Kurokawa, Soulé, Deninger, Manin, and others; cf. [**86**], [**128**], [**98**], [**46**]. For instance the constants $a \in \hat{\mathbf{Z}}_p^{ur}$, $\delta a = 0$, of the map $\delta : \hat{\mathbf{Z}}_p^{ur} \to \hat{\mathbf{Z}}_p^{ur}$ consist of the roots of unity in $\hat{\mathbf{Z}}_p^{ur}$ together with 0; on the other hand the roots of unity and 0 are, according to the philosophy of the field with one element, precisely the $\bar{\mathbf{F}}_1$– points of the affine line where $\bar{\mathbf{F}}_1$ is the "algebraic closure of the field with one element, $\mathbf{F}_1$". Not all the predictions of the philosophy of the field with one element seem however to be in agreement with our theory. For instance there seems to be an important ideological difference between our operator

$$\delta_p : \mathbf{Z} \to \mathbf{Z}, \quad \delta_p a = \frac{a - a^p}{p},$$

and the *absolute derivation* introduced in [**86**],

$$\frac{\partial}{\partial p} : \mathbf{Z} \to \mathbf{Z}, \quad \frac{\partial}{\partial p}(p^e m) := e p^{e-1} m,$$

where $m \in \mathbf{Z}$ is prime to p. Indeed the constants of δ_p are 0 and ± 1 while the constants of $\partial / \partial p$ are 0 and all the integers coprime to p. This suggests that, according to the philosophy of $\mathbf{F}_1$, there is no theory of $\mathbf{F}_1$ that involves *one prime only*; the only theory about $\mathbf{F}_1$ would involve *all the primes at the same time*. On the contrary, in our book, there is a theory for each individual prime.

Part 1

Main concepts and results

CHAPTER 1

Preliminaries from algebraic geometry

In Section 1 of this Chapter we review some standard terminology, conventions, and notation from algebraic geometry; we also introduce some new terminology and notation that will prove useful throughout the book. In Section 2 we very briefly discuss categorical quotients in algebraic geometry, specifically in the context of algebraic varieties, algebraic stacks, affine schemes, and $p-$adic formal schemes. Our main purpose here is to offer a quick overview of some of the available approaches to quotients in algebraic geometry before we embark, in the next Chapter, onto the task of explaining the $\delta-$geometric approach proposed in this book. In Section 3 we discuss a class of correspondences on algebraic curves that will play a key role in our book; these are the correspondences that admit an analytic uniformization and will fall into 3 categories (*spherical, flat* and *hyperbolic*). The concept of analytic uniformization of a correspondence is closely connected, as we shall see, to (a generalization of) the concept of postcritical finiteness in complex dynamics.

The impatient reader might choose to simply skim through this Chapter and to later come back to the material here as needed. A good idea, nevertheless, is to read (at least) the numbered new definitions in this Chapter. With regards to proofs, our general policy, in this Chapter, was to prove only the facts for which we could find no convenient reference; these facts are all "easy". The facts (some of which are "hard") for which we could find references are not supplied with proofs.

1.1. Algebro-geometric terminology

1.1.1. Categories. Recall the following terminology already used in the Introduction.

DEFINITION 1.1. A *correspondence* in a category $\mathcal{C}$ is a tuple

$$\mathbf{X} = (X, \tilde{X}, \sigma_1, \sigma_2), \tag{1.1}$$

where $X, \tilde{X}$ are objects in $\mathcal{C}$ and $\sigma_1, \sigma_2 : \tilde{X} \to X$ are morphisms in $\mathcal{C}$. If $\sigma := (\tilde{X}, \sigma_1, \sigma_2)$ then we also say that σ is a *correspondence on* X and we write

$$\mathbf{X} = (X, \sigma). \tag{1.2}$$

We sometimes refer to X as the *base* of $\mathbf{X}$. A *categorical quotient* for $\mathbf{X}$ is a pair (Y, π) where Y is an object of $\mathcal{C}$ and $\pi : X \to Y$ is a morphism in $\mathcal{C}$ satisfying the following properties:

1) $\pi \circ \sigma_1 = \pi \circ \sigma_2$;

2) For any pair (Y', π') where Y' is an object of $\mathcal{C}$ and $\pi' : X \to Y'$ is a morphism such that $\pi' \circ \sigma_1 = \pi' \circ \sigma_2$ there exists a unique morphism $\gamma : Y \to Y'$ such that $\gamma \circ \pi = \pi'$.

If a categorical quotient (Y, π) exists then it is unique up to isomorphism and we shall write

$$Y = X/\sigma. \tag{1.3}$$

Correspondences form, in a natural way, a category: a morphism $\mathbf{X} \to \mathbf{X}'$ between two correspondences $\mathbf{X} = (X, \tilde{X}, \sigma_1, \sigma_2)$ and $\mathbf{X}' = (X', \tilde{X}', \sigma_1', \sigma_2')$ is, by definition, a pair of morphisms $(\pi, \tilde{\pi})$, $\pi : X \to X'$, $\tilde{\pi} : \tilde{X} \to \tilde{X}'$, such that $\pi \circ \sigma_i = \sigma_i' \circ \tilde{\pi}$, $i = 1, 2$. Note that if π is an isomorphism and $\tilde{\pi}$ is an epimorphism then X/σ exists if and only X'/σ' exists and, in this case, the two are isomorphic.

If fiber products exist in $\mathcal{C}$ and one is given a correspondence $\mathbf{X} = (X, \tilde{X}, \sigma_1, \sigma_2)$ and a morphism $f : Y \to X$ then one can define the *pull back correspondence*

$$f^*\mathbf{X} := (Y, (Y \times_{f,X,\sigma_1} \tilde{X}) \times_{pr_2, \tilde{X}, pr_1} (\tilde{X} \times_{\sigma_2, X, f} Y), pr_1 \circ pr_1, pr_2 \circ pr_2). \tag{1.4}$$

1.1.2. Rings. We denote by $\mathbf{Z}, \mathbf{Q}, \mathbf{R}, \mathbf{C}$ the rings of integers, rational numbers, real numbers, and complex numbers respectively; $\mathbf{Z}_+$ will denote the set of non-negative integers. We denote by $\mathbf{Z}_p$ the ring of $p-$adic integers, by $\mathbf{Q}_p$ the field of $p-$adic numbers, and by $\mathbf{F}_p = \mathbf{Z}/p\mathbf{Z}$ the prime field with p elements. We denote by

$$\mathbf{Z}_p^{ur} = \bigcup_{\zeta} \mathbf{Z}_p[\zeta]$$

the maximum unramified extension of $\mathbf{Z}_p$; here ζ runs through the set of roots of unity of order prime to p. Denote by $\hat{\mathbf{Z}}_p^{ur}$ the $p-$adic completion of $\mathbf{Z}_p^{ur}$.

Unless otherwise stated all rings will be assumed commutative with identity and all ring homomorphisms are assumed to preserve the identity. (Some non-commutative rings or commutative rings without identity will also occur in our discussion but, when this is the case, sufficient warning will be given.)

If A is a ring we denote by

$$A[\epsilon] = A \oplus \epsilon A, \quad \epsilon^2 = 0, \tag{1.5}$$

the ring of *dual numbers* on A.

If L is a field we denote by L^a an algebraic closure of L.

Number fields are subfields of $\mathbf{C}$ and are always assumed finite over $\mathbf{Q}$. A *place* in a number field F will always mean a finite place identified with a maximal ideal in the ring of integers $\mathcal{O}_F$ of F.

Let A be a ring, p a prime integer, and M a maximal ideal in A; we usually denote by

$$\hat{A}, \ A^{for} \tag{1.6}$$

the $p-$adic completion and the $M-$adic completion of A respectively. We say A is $p-$adically *complete* if $A = \hat{A}$.

DEFINITION 1.2. Let $p > 0$ be a prime number. A *local $p-$ring* is a discrete valuation ring of characteristic zero with maximal ideal generated by p. A *global $p-$ring* is a ring A such that:

1) p is an non-zero divisor in A,
2) the principal ideal (p) is prime in A,
3) A is $p-$adically separated.

(Note that A is then necessarily an integral domain.)

Any local $p-$ring is, of course, a global $p-$ring and if A is a global $p-$ring then its localization $A_{(p)}$ is a local $p-$ring. (A warning about terminology: what we call here a *local* $p-$*ring* is called, in Matsumura's book [**103**], p. 223, a $p-$*ring*. On the other hand Serre's concept of $p-$*ring* considered in his book [**120**], p. 37, does not coincide with Matsumura's! Due to these terminology discrepancies in the literature we preferred to use a different terminology.) The following basic fact about local $p-$rings will be repeatedly used in our book. Let A and B be local complete $p-$rings with residue fields k_A and k_B. Let *Hom* denote homomorphisms in the category of rings. Then the natural map

$$Hom(A,B) \rightarrow Hom(k_A, k_B)$$

is surjective; cf. [**103**], p. 224. If, in addition, k_A and k_B are perfect then the above map is an isomorphism; cf. [**120**], p. 38. In particular, if A is a complete local $p-$ring with residue field k_A then the $p-$power Frobenius endomorphism of k_A lifts to an endomorphism ϕ of A; if, in addition, k_A is perfect then ϕ is unique. (Uniqueness may fail if k_A is not perfect: e.g. one can take x a variable, $A := (\mathbf{Z}[x]_{(p)})\hat{}$, $g \in \mathbf{Z}[x]$, and $\phi = \phi_g : A \rightarrow A$, $\phi_g(x) = x^p + pg$.) Recall also that for any perfect field k there exists a complete local $p-$ring with residue field k and this ring is unique up to a unique isomorphism. In particular $\hat{\mathbf{Z}}_p^{ur}$ is the unique complete local $p-$ring with residue field $\mathbf{F}_p^a$.

1.1.3. Schemes and formal schemes. We are using the standard terminology and conventions of algebraic geometry as found, for instance, in [**66**]. Furthermore, schemes and formal schemes will always be assumed separated. Formal schemes will always be assumed locally Noetherian. A *group scheme* over a scheme X is a group object in the category of schemes over X. It is convenient to introduce the following:

DEFINITION 1.3. A *group formal scheme* over a formal scheme X is a group object in the category of formal schemes over X.

A *variety* over an algebraically closed field k will mean an irreducible reduced scheme of finite type over k; we denote, as usual, by $k(X)$ the field of rational functions on X.

If X is a scheme over a ring R and S is an $R-$algebra we denote, as usual, by

$$X(S) := Hom_R(Spec\ S, X)$$

the set of $S-$points of X; we use a similar notation for formal schemes. For a variety X over k we sometimes write X for $X(k)$. If X is a scheme and M is a locally free sheaf on it we denote by

$$\mathbf{V}(M) = Spec\ Symm(M) \tag{1.7}$$

the associated group scheme over X.

By a *formal group* or *formal group law* (of dimension n) over a ring R we will understand, as usual, an $n-$tuple tuple of formal power series in $R[[T_1, ..., T_n]]$ satisfying the usual axioms. (So formal groups over R are not the same as group formal schemes over $Spf\ \hat{R}$! Formal groups are "local objects" while group formal schemes are "global objects".)

If R is a ring we denote by

$$\mathbf{A}^N := Spec\ R[T_1, ..., T_n] \tag{1.8}$$

the affine space over R and by

$$\mathbf{G}_a := (\mathbf{A}^1, +), \quad \mathbf{G}_m := (Spec\ R[T, T^{-1}], \cdot) \tag{1.9}$$

the additive and multiplicative group schemes over R respectively. An *elliptic curve* over R is a smooth proper group scheme over R of relative dimension one with irreducible geometric fibers. (See the discussion below for a review of smoothness.)

If X is a Noetherian scheme and $P \in X$ is a closed point we denote by

$$\hat{X}, \ X^{for} \tag{1.10}$$

the $p-$adic completion of X and the completion of X at P respectively.

Recall Grothendieck's existence theorem saying in particular that if X and Y are projective schemes over a $p-$adically complete ring S then the map

$$Hom_S(X, Y) \to Hom_S(\hat{X}, \hat{Y})$$

is a bijection.

Let X be a scheme (resp. a formal scheme) and let $G \to X$ be a group scheme over X (resp. a group formal scheme over X). A *principal homogeneous space* under G is a morphism $Y \to X$ of schemes (resp. a formal schemes) together with an action $m : G \times_X Y \to Y$ such that the induced morphism

$$m \times pr_2 : G \times_X Y \to Y \times_X Y$$

is an isomorphism. A principal homogeneous space is said to be trivial if it has a section; equivalently if it is isomorphic to the principal homogeneous space $G \to X$.

By a *Galois cover* of schemes we understand a finite morphism of schemes $\pi : Y \to X$ such that $(\pi_* \mathcal{O}_Y)^\Gamma = \mathcal{O}_X$, where $\Gamma := Aut(Y/X)$ is the group of $X-$automorphisms of Y; Γ is referred to as the *Galois group* of the cover.

1.1.4. Derivations and smoothness. If B is an $A-$algebra and C is a $B-$algebra with structure homomorphism $f : B \to C$ we denote by $Der_A(B, C)$ the $B-$module of $A-$derivations $D : B \to C$ relative to f, by which we mean $A-$module homomorphisms satisfying

$$D(xy) = f(x) \cdot D(y) + f(y) \cdot D(x),$$

for $x, y \in B$. If $\mathcal{A}$ is a sheaf of rings on a topological space, $\mathcal{B}$ is a sheaf of $\mathcal{A}-$algebras, and $\mathcal{C}$ is a sheaf of $\mathcal{B}-$algebras we denote by $Der_{\mathcal{A}}(\mathcal{B}, \mathcal{C})$ the group of $\mathcal{A}-$derivations $\mathcal{B} \to \mathcal{C}$ and by $\mathcal{D}er_{\mathcal{A}}(\mathcal{B}, \mathcal{C})$ the sheaf of $\mathcal{C}-$modules of $\mathcal{A}-$derivations $\mathcal{B} \to \mathcal{C}$.

We recall here the basic definitions and properties of smoothness we are using in this book; cf. [**10**], Chapter 2. Let A be a ring and B a finitely generated $A-$algebra. One says B is *smooth* (resp. *étale*) over A if for any $A-$algebra C and any ideal $I \subset C$ with $I^2 = 0$ the canonical map

$$Hom_A(B, C) \to Hom_A(B, C/I)$$

is surjective (resp. bijective). A scheme X over A is called *smooth* (resp. *étale*) if it can be covered with open affine schemes $Spec\ B$ with B finitely generated and smooth (resp. étale) over A. A morphism $f : X \to Y$ of schemes is called *smooth* (resp. *étale*) if Y can be covered by affine open schemes $U = Spec\ A$ such that $f^{-1}(U)$ are smooth (resp. *étale*) over A. A morphism $X \to Y$ of finite type of Noetherian schemes is smooth if and only if it is flat and all geometric fibers are regular. If $Spec\ B \to Spec\ A$ is of finite type and smooth (resp. *étale*) then B is smooth (resp. *étale*) over A. If a scheme X over A is smooth then each point in X

has an affine open neighborhood $U = Spec\ B$ such that B is étale over an algebra of polynomials $A[x_1, ..., x_n]$ contained in B; $x_1, ..., x_n$ are then called *étale coordinates* on U. If $\pi : X \to Y$ is smooth then the sheaf of Kähler differentials $\Omega_{X/Y}$ is locally free and its dual, $T_{X/Y}$, identifies with the sheaf $\mathcal{D}er_{\pi^{-1}\mathcal{O}_Y}(\mathcal{O}_X, \mathcal{O}_X)$. We denote by

$$K_{X/Y} = det(\Omega_{X/Y}) \tag{1.11}$$

the *canonical bundle* of X/Y; if Y is clear from the context we write K_X instead of $K_{X/Y}$. We also denote by

$$T(X/Y) = \mathbf{V}(\Omega_{X/Y}) = Spec\ Symm(\Omega_{X/Y}) \tag{1.12}$$

the *tangent scheme* of X/Y; if Y is clear from the context we simply write $T(X)$ instead of $T(X/Y)$. Assume in the situation above that $Y = Spec\ R$ with R a local ring. Also assume $P \in X(R)$ and let $P_0 \in X$ be the image of the maximal ideal of R. Then the R-module

$$T_P X := Der_R(\mathcal{O}_{X,P_0}, R) \tag{1.13}$$

will be referred to as the *tangent space* of X at P.

1.1.5. Correspondences on schemes.

DEFINITION 1.4. Let us fix a ring R and consider the category

$$\mathcal{C} = \mathcal{C}_R := \{\text{schemes over } R\}. \tag{1.14}$$

By a *correspondence over* R we will always understand a correspondence in $\mathcal{C}_R$. An object in $\mathcal{C}_R$ is called *trivial* if it is isomorphic to *Spec* R. A correspondence $\mathbf{X} = (X, \tilde{X}, \sigma_1, \sigma_2)$ in $\mathcal{C}_R$ is said to be *non-empty* if the fibers of $\tilde{X}/R$ are non-empty. We say $\mathbf{X}$ has (relative) *dimension* d if both $\tilde{X}$ and X have pure relative dimension d over R. We say $\mathbf{X}$ has an *irreducible base* (resp. a *geometrically irreducible base*) if X/R has irreducible fibers (resp. geometrically irreducible fibers). We say that $\mathbf{X}$ is *irreducible* (resp. *geometrically irreducible*) if both $\tilde{X}/R$ and X/R have irreducible fibers (resp. geometrically irreducible fibers). We make similar definitions with the word "irreducible" replaced by the word "reduced". We say $\mathbf{X}$ is *of finite type over* R (resp. *smooth over* R) if both X and $\tilde{X}$ are of finite type (resp. smooth) over R. We say that $\mathbf{X}$ is *left étale* if σ_1 is étale. We say $\mathbf{X}$ is *étale* if both σ_1 and σ_2 are étale. A morphism

$$(\pi, \tilde{\pi}) : \mathbf{X}' = (X', \tilde{X}', \sigma_1', \sigma_2') \to \mathbf{X} = (X, \tilde{X}, \sigma_1, \sigma_2)$$

is called an *open immersion* if both π and $\tilde{\pi}$ are open immersions. We refer to $\mathbf{X}'$ as an *open subcorrespondence* of $\mathbf{X}$; by abuse we usually view $X' \subset X$ and $\tilde{X}' \subset \tilde{X}$ as open subsets and we write $\mathbf{X}' \subset \mathbf{X}$.

Note that if $\mathbf{X} = (X, \tilde{X}, \sigma_1, \sigma_2)$ is a correspondence over R and $i : Y \to X$ is an open immersion then the pull-back correspondence (cf. Equation 1.4) is given by

$$i^*\mathbf{X} = (Y, \sigma_1^{-1}(Y) \cap \sigma_2^{-1}(Y), \sigma_1, \sigma_2)$$

and $i^*\mathbf{X}$ is an open subcorrespondence of $\mathbf{X}$.

We will later need to consider the following construction. Assume $(\mathbf{X}_i)_{i \in I}$ is a family of correspondences over R with the same base, i.e. $\mathbf{X}_i = (X, \tilde{X}_i, \sigma_{1i}, \sigma_{2i})$.

Then we can consider the *disjoint union* correspondence

$$\mathbf{X} := \coprod_i \mathbf{X}_i := (X, \coprod_i \tilde{X}_i, \coprod_i \sigma_{1i}, \coprod_i \sigma_{2i}). \tag{1.15}$$

Here is another construction which will play a role later. Assume

$$\mathbf{X} = (X, \tilde{X}, \sigma_1, \sigma_2)$$

is a correspondence over R. Then, for each $n \geq 2$, we may consider the scheme

$$\tilde{X}^{\circ n} := \tilde{X} \times_X \tilde{X} \times_X \cdots \times_X \tilde{X}, \quad (n \text{ times})$$

where for each $\tilde{X} \times_X \tilde{X}$ the left $\tilde{X}$ is viewed as a scheme over X via σ_2 while the right $\tilde{X}$ is viewed as a scheme over X via σ_1. In other words

$$\tilde{X} \times_X \tilde{X} = \tilde{X} \times_{\sigma_2, X, \sigma_1} \tilde{X}.$$

E.g., for $n = 3$ we have a diagram

$$\begin{array}{ccccccc}
\tilde{X}^{\circ 3} & \rightarrow & \tilde{X}^{\circ 2} & \rightarrow & \tilde{X} & \stackrel{\sigma_2}{\rightarrow} & X \\
\downarrow & & \downarrow & & \downarrow \sigma_1 & & \\
\tilde{X}^{\circ 2} & \rightarrow & \tilde{X} & \stackrel{\sigma_2}{\rightarrow} & X & & \\
\downarrow & & \downarrow \sigma_1 & & & & \\
\tilde{X} & \stackrel{\sigma_2}{\rightarrow} & X & & & & \\
\downarrow \sigma_1 & & & & & & \\
X & & & & & &
\end{array}$$

with Cartesian squares. We have natural morphisms

$$\sigma_1^{\circ n}, \sigma_2^{\circ n} : \tilde{X}^{\circ n} \rightarrow X$$

defined, on $S-$points (S any $R-$algebra), by the formulae

$$\sigma_1^{\circ n}(\tilde{P}_1, ..., \tilde{P}_n) = \sigma_1(\tilde{P}_1),$$
$$\sigma_2^{\circ n}(\tilde{P}_1, ..., \tilde{P}_n) = \sigma_2(\tilde{P}_n).$$

E.g., in the picture above, $\sigma_1^{\circ 3}$ is the composition of the left vertical arrows while $\sigma_2^{\circ 3}$ is the composition of the top horizontal arrows. So we may consider the *$n-th$ power correspondence* in $\mathcal{C}_R$,

$$\mathbf{X}^{\circ n} := (X, \tilde{X}^{\circ n}, \sigma_1^{\circ n}, \sigma_2^{\circ n}).$$

Of course, if $\mathbf{X}$ is smooth over R and left étale (resp. étale) then $\mathbf{X}^{\circ n}$ is smooth over R and left étale (resp. étale).

We will also need to consider line bundles on correspondences. As usual a *line bundle* on a scheme X is a locally free $\mathcal{O}_X-$module of rank one. The set $Pic(X)$ of isomorphism classes of line bundles is a group under $\otimes$.

DEFINITION 1.5. If $\mathbf{X} = (X, \tilde{X}, \sigma_1, \sigma_2)$ is a correspondence over R and L is a line bundle on X then a *linearization* of L is an isomorphism $\beta : \sigma_2^* L \simeq \sigma_1^* L$. A *line bundle* on $\mathbf{X}$ is a pair $\mathbf{L} = (L, \beta)$ consisting of a line bundle L on X and a linearization β of L. Two line bundles $\mathbf{L} = (L, \beta)$ and $\mathbf{L}' = (L', \beta')$ on $\mathbf{X}$ are called *isomorphic* if there exists an isomorphism $u : L \rightarrow L'$ such that following diagram is commutative:

$$\begin{array}{ccc}
\sigma_2^* L & \stackrel{\beta}{\rightarrow} & \sigma_1^* L \\
\sigma_2^* u \downarrow & & \downarrow \sigma_1^* u \\
\sigma_2^* L' & \stackrel{\beta'}{\rightarrow} & \sigma_1^* L'
\end{array}$$

The set $Pic(\mathbf{X})$ of isomorphism classes of line bundles on $\mathbf{X}$ is an Abelian group with respect to $\otimes$. A line bundle is called *trivial* if it is isomorphic to the line bundle

$$\mathbf{O} := (\mathcal{O}_X, can)$$

where can is the isomorphism

$$\sigma_2^*\mathcal{O}_X \overset{can}{\simeq} \mathcal{O}_{\tilde{X}} \overset{can}{\simeq} \sigma_1^*\mathcal{O}_X.$$

Of course the isomorphism class of $\mathbf{O}$ is the identity element of $Pic(\mathbf{X})$. There is a natural exact sequence of groups

$$\mathcal{O}(X)^\times \to \mathcal{O}(\tilde{X})^\times \to Pic(\mathbf{X}) \to Pic(X) \tag{1.16}$$

where the left arrow is defined by

$$u \mapsto \frac{\sigma_2^* u}{\sigma_1^* u},$$

the middle arrow is defined by

$$\tilde{u} \mapsto (\mathcal{O}, \tilde{u}\cdot),$$

and the right arrow is induced by

$$(L, \beta) \mapsto L.$$

In this book a key role will be played by 2 types of line bundles on correspondences: *flat* line bundles and *pluricanonical* line bundles; cf. the Definitions below.

DEFINITION 1.6. Assume $R_0 \subset R$ is a subring, $\mathbf{X} = (X, \tilde{X}, \sigma_1, \sigma_2)$ is a correspondence over R, and λ is a locally constant $R_0^\times-$valued function on $\tilde{X}$. Define the line bundle

$$\mathbf{F}(\lambda) := (\mathcal{O}, \lambda) \tag{1.17}$$

where

$$\lambda : \sigma_2^*\mathcal{O}_X \overset{can}{\simeq} \mathcal{O}_{\tilde{X}} \overset{\lambda\cdot}{\to} \mathcal{O}_{\tilde{X}} \overset{can}{\simeq} \sigma_1^*\mathcal{O}_X.$$

We call λ the *multiplier* of $\mathbf{F}(\lambda)$. A line bundle on $\mathbf{X}$ is called R_0-*flat* if it isomorphic to a line bundle of the form $\mathbf{F}(\lambda)$. In particular $\mathbf{O} = \mathbf{F}(1)$. Clearly $\mathbf{F}(\lambda_1) \otimes \mathbf{F}(\lambda_2) \simeq \mathbf{F}(\lambda_1\lambda_2)$.

DEFINITION 1.7. If $\mathbf{X}$ is smooth over R and étale then the *canonical line bundle* is the line bundle

$$\mathbf{K} = \mathbf{K}_\mathbf{X} = (K_X, \beta) \tag{1.18}$$

where K_X is the canonical line bundle on X and β is the composition

$$\sigma_2^* K_X \overset{can}{\simeq} K_{\tilde{X}} \overset{can}{\simeq} \sigma_1^* K_X.$$

The line bundle $\mathbf{K}^{-1}$ is called the *anticanonical line bundle.* For $\nu \in \mathbf{Z}$ the line bundle $\mathbf{K}^\nu$ is called the $\nu-$*canonical line bundle* (or a *pluricanonical line bundle*).

If $\mathbf{L} = (L, \beta)$ is a line bundle on $\mathbf{X}$ then one can define the line bundle $\mathbf{L}^{\circ n}$ on $\mathbf{X}^{\circ n}$ by $\mathbf{L}^{\circ n} := (L, \beta^{\circ n})$ where $\beta^{\circ n}$ is the obvious "$n-$fold product" of various pull-backs of β. For an R_0-flat line bundle $\mathbf{L} \simeq \mathbf{F}(\lambda)$ with $\lambda \in R_0^\times$ we have $\mathbf{L}^{\circ n} \simeq \mathbf{F}(\lambda^n)$. If $\mathbf{K} = \mathbf{K}_\mathbf{X}$ is the canonical bundle on $\mathbf{X}$ then it is easy to check that

$$\mathbf{K}_\mathbf{X}^{\circ n} \simeq \mathbf{K}_{\mathbf{X}^{\circ n}}. \tag{1.19}$$

It will be important later to have a criterion for a line bundle $\mathbf{L} = (L, \beta)$ on a correspondence $\mathbf{X}$ over R to be flat. The following (trivial) Lemma gives the answer:

LEMMA 1.8. *For $\lambda \in R^\times$ the following are equivalent:*
1) $\mathbf{L} \simeq \mathbf{F}(\lambda)$,
2) There exists an invertible section $\eta \in H^0(X, L)$ such that $\beta\sigma_2^*\eta = \lambda \cdot \sigma_1^*\eta$.

PROOF. Let us check that $2 \Rightarrow 1$; the converse is proved by reversing the argument. If 2 holds then we have an isomorphism $u : \mathcal{O}_X \to L$, $1 \mapsto \eta$ inducing a commutative diagram

$$\begin{array}{ccc} \mathcal{O}_{\tilde{X}} = \sigma_2^*\mathcal{O}_X & \stackrel{\lambda}{\to} & \sigma_1^*\mathcal{O}_X = \mathcal{O}_{\tilde{X}} \\ \sigma_2^* u \downarrow & & \downarrow \sigma_1^* u \\ \sigma_2^* L & \stackrel{\beta}{\to} & \sigma_1^* L \end{array} , \qquad \begin{array}{ccc} 1 = \sigma_2^* 1 & \mapsto & \lambda = \lambda \cdot \sigma_1^* 1 \\ \downarrow & & \downarrow \\ \sigma_2^*\eta & \mapsto & \lambda \cdot \sigma_1^*\eta \end{array}$$

which shows that $\mathbf{L} \simeq \mathbf{F}(\lambda)$. □

In what follows fix k, an algebraically closed field.

DEFINITION 1.9. An irreducible reduced correspondence $\mathbf{X} = (X, \tilde{X}, \sigma_1, \sigma_2)$ of finite type over k is *dense* if the smallest equivalence relation

$$\langle\sigma\rangle \subset X(k) \times X(k) \tag{1.20}$$

on $X(k)$ containing the image of

$$\sigma_1 \times \sigma_2 : \tilde{X}(k) \to X(k) \times X(k)$$

is Zariski dense in $X(k) \times X(k)$.

Density of correspondences is, of course, a common occurrence. For instance, in the simplest case when X and $\tilde{X}$ are curves and σ has an infinite orbit, $\langle\sigma\rangle$ is dense in $X \times X$; cf. the Introduction.

DEFINITION 1.10. Two irreducible reduced correspondences $\mathbf{X}$ and $\mathbf{X}'$ of finite type over k are called *commensurable* if there exists a sequence of irreducible reduced correspondences $\mathbf{X}_1, \mathbf{X}_2, ..., \mathbf{X}_n$ of finite type over k such that $\mathbf{X} = \mathbf{X}_1$, $\mathbf{X}' = \mathbf{X}_n$, and for each i there exists either a morphism $(\pi_i, \tilde{\pi}_i) : \mathbf{X}_i \to \mathbf{X}_{i+1}$ or a morphism $(\pi_i, \tilde{\pi}_i) : \mathbf{X}_{i+1} \to \mathbf{X}_i$ with both π_i and $\tilde{\pi}_i$ dominant and quasi-finite. We say $\mathbf{X}$ and $\mathbf{X}'$ are *equivalent* if if there exists a sequence of irreducible reduced correspondences $\mathbf{X}_1, \mathbf{X}_2, ..., \mathbf{X}_n$ of finite type over k such that $\mathbf{X} = \mathbf{X}_1$, $\mathbf{X}' = \mathbf{X}_n$, and for each i there exists either a morphism $(\pi_i, \tilde{\pi}_i) : \mathbf{X}_i \to \mathbf{X}_{i+1}$ or a morphism $(\pi_i, \tilde{\pi}_i) : \mathbf{X}_{i+1} \to \mathbf{X}_i$ with π_i an isomorphism and $\tilde{\pi}_i$ dominant and quasi-finite.

DEFINITION 1.11. A morphism of irreducible reduced correspondences of finite type over k

$$(\pi, \tilde{\pi}) : (Y, \tilde{Y}, \tau_1, \tau_2) \to (X, \tilde{X}, \sigma_1, \sigma_2)$$

is a *Galois cover* if both π and $\tilde{\pi}$ are Galois covers.

If $\mathbf{L} = (L, \beta)$ is a line bundle on a correspondence $\mathbf{X} = (X, \tilde{X}, \sigma_1, \sigma_2)$ over k we may consider the isomorphism

$$\beta : H^0(\tilde{X}, \sigma_2^* L) \simeq H^0(\tilde{X}, \sigma_1^* L).$$

Then we set $\tilde{L} = \sigma_1^* L$ and we define the R–linear map

$$\Delta = \Delta^{\mathbf{L}} : H^0(X, L) \to H^0(\tilde{X}, \tilde{L}) \tag{1.21}$$

$$\Delta(f) := \sigma_1^* f - \beta\sigma_2^* f,$$

and the R–module:

$$H^0(\mathbf{X}, \mathbf{L}) = Ker\ \Delta. \tag{1.22}$$

Note that given a line bundle $\mathbf{L}$ on $\mathbf{X}$ it may very well happen that there is no non-empty open subcorrespondence $\mathbf{U} \subset \mathbf{X}$ such that the pull-back $\mathbf{L_U}$ of $\mathbf{L}$ to $\mathbf{U}$ is trivial. In relation to this we have the following trivial consequence of Lemma 1.8:

COROLLARY 1.12. *Let $\mathbf{L} = (L, \beta)$ be a line bundle on an irreducible reduced correspondence $\mathbf{X} = (X, \sigma)$ of finite type over k. The following are equivalent:*

1) There exists a non-empty open subcorrespondence $\mathbf{U} \subset \mathbf{X}$ such that $\mathbf{L_U}$ is trivial on $\mathbf{U}$.

2) There exists a non-empty open subcorrespondence $\mathbf{V} \subset \mathbf{X}$ such that

$$H^0(\mathbf{V}, \mathbf{L_V}) \neq 0.$$

3) There exists $0 \neq \eta \in L_{k(X)}$ such that $\beta\sigma_2^*\eta = \sigma_1^*\eta$ in $\tilde{L}_{k(\tilde{X})}$.

In the above statement $L_{k(X)}$ is the pull-back of L on $Spec\ k(X)$, viewed as a $k(X)$–vector space (and similarly for $\tilde{L}_{k(\tilde{X})}$).

1.2. Categorical quotients in algebraic geometry

1.2.1. Sets. We start by recording the most trivial case, that of correspondences in the category of sets

$$\mathcal{C}_{set} := \{\text{sets}\}. \tag{1.23}$$

The *trivial* objects in $\mathcal{C}_{set}$ will be, by definition, the sets of cardinality one. If $\mathbf{X} = (X, \tilde{X}, \sigma_1, \sigma_2)$ is a correspondence in the category $\mathcal{C}_{set}$ let $\langle\sigma\rangle$ be the smallest equivalence relation in $X \times X$ containing the image of $\sigma_1 \times \sigma_2 : \tilde{X} \to X \times X$. It is plain that the categorical quotient of X/σ in X exists and is given by

$$X/\sigma = X/\langle\sigma\rangle. \tag{1.24}$$

1.2.2. Algebraic varieties. Usual algebraic geometry is ill equipped to deal with categorical quotients. This is a well known phenomenon for correspondences arising from algebraic group actions (or more generally groupoids of finite type); this phenomenon is addressed by geometric invariant theory [**113**]. However things can be far worse for more general correspondences as shown by Proposition 1.13 below. Let us place ourselves in the category $\mathcal{C}_k$ of schemes over an algebraically closed field k.

PROPOSITION 1.13. *If an irreducible reduced correspondence $\mathbf{X} = (X, \sigma)$ of finite type over k is dense then X/σ exists in $\mathcal{C}_k$ and is trivial.*

PROOF. Let $\pi : X \to Y$ be a morphism in $\mathcal{C}_k$ such that $\pi \circ \sigma_1 = \pi \circ \sigma_2$. Since Y is separated over k the natural morphism $X \times_Y X \to X \times X$ is a closed immersion, in particular $X(k) \times_{Y(k)} X(k)$ is Zariski closed in $X(k) \times X(k)$. On the other hand

$$\langle\sigma\rangle \subset X(k) \times_{Y(k)} X(k) \subset X(k) \times X(k).$$

Since $\langle\sigma\rangle$ is Zariski dense in $X(k) \times X(k)$ it follows that

$$X(k) \times_{Y(k)} X(k) = X(k) \times X(k).$$

Since $X \times X$ is reduced and of finite type it follows that the closed immersion $X \times_Y X \to X \times X$ is an isomorphism. Choose any $k-$point $P \in X(k)$ and consider the morphisms $\eta : X \stackrel{can}{\to} Spec\ k \stackrel{P}{\to} X$ and $Q : Spec\ k \stackrel{P}{\to} X \stackrel{\pi}{\to}$. Then

$$(id, \eta) \in Hom_k(X, X \times X) = Hom_k(X, X \times_Y X)$$

hence

$$\pi = \pi \circ id = \pi \circ \eta = Q \circ can.$$

Moreover any two morphisms $Q_1, Q_2 : Spec\ k \to Y$ with $\pi = Q_1 \circ can = Q_2 \circ can$ must coincide. □

Unlike with algebraic group actions, the situation in Proposition 1.13 persists if one throws away "non-semistable loci" as in geometric invariant theory or if one replaces varieties by algebraic spaces or algebraic stacks; cf. Proposition 1.16 below.

In what follows we quickly review the strategy of geometric invariant theory [**113**] which is a basic instance when classical algebraic geometry can successfully deal with quotients; for details we refer to [**113**]. Our main purpose in making this review is that later on we will imitate this approach in the framework of $\delta-$geometry. Let us start with an action

$$\mu : G \times X \to X$$

of a linear algebraic group G on a projective variety X defined over an algebraically closed field k (assumed for simplicity of characteristic zero) and consider the correspondence

$$\mathbf{X} := (X, \tilde{X} := G \times X, pr_2, \mu)$$

in the category $\mathcal{C}_k$, where pr_2 is, of course, the second projection. By a $G-$*linearized line bundle* on $\mathbf{X}$ one understands a line bundle $\mathbf{L} = (L, \beta)$ on $\mathbf{X}$ where β satisfies a natural compatibility condition with respect to the group structure of G (called *cocycle condition*, for which we refer to [**113**]). Assume one is given a $G-$linearized bundle $\mathbf{L}$ with L ample and set $\tilde{L} := pr_2^* L$. Then we have

$$X = Proj \bigoplus_{m \in \mathbf{Z}_+} H^0(X, L^m), \quad \tilde{X} = Proj \bigoplus_{m \in \mathbf{Z}_+} H^0(\tilde{X}, \tilde{L}^m)$$

and G naturally acts on $H^0(X, L^m)$. Note that the space $H^0(\mathbf{X}, \mathbf{L}^m)$ in Equation 1.22 is nothing but the space of $G-$invariant sections $H^0(X, L^m)^G$ in the sense of [**113**]. One then considers the open subset X^{ss} of X, called the *semistable locus*, which is by definition the union of all principal open subsets of X defined by $G-$invariant sections; the set X^{ss} depends on L but, for convenience, we drop L from our notation. Further one defines the *semistable locus* $\tilde{X}^{ss}$ of $\tilde{X}$ as the union of all principal open sets defined by $pr_2^* f$ with f invariant as above. One gets an induced correspondence

$$\mathbf{X}^{ss} := (X^{ss}, \tilde{X}^{ss}, \sigma_1, \sigma_2) = (X^{ss}, \sigma)$$

in the category $\mathcal{C}_k$. Of course $\mathbf{X}^{ss}$ depends on $\mathbf{L}$ but, for convenience, we drop $\mathbf{L}$ from our notation. The following is a consequence of geometric invariant theory; cf [**113**]:

PROPOSITION 1.14. *If G is reductive then the categorical quotient X^{ss}/σ exists in the category $\mathcal{C}_k$, it is of finite type over k, and is given by*

$$X^{ss}/\sigma = Proj \bigoplus_{m \in \mathbf{Z}_+} H^0(\mathbf{X}, \mathbf{L}^m).$$

This quotient is trivial if and only if there is a $G-$orbit that contains an open subset of X.

REMARK 1.15. Consider the functor of $k-$points

$$\mathcal{C}_k \to \mathcal{C}_{set}, \quad X \mapsto X_{set} := X(k).$$

One gets an induced correspondence in the category of sets,

$$\mathbf{X}^{ss}_{set} := (X^{ss}_{set}, \tilde{X}^{ss}_{set}, \sigma_{1,set}, \sigma_{2,set}) = (X^{ss}_{set}, \sigma_{set}).$$

The universality property yields a natural map

$$X^{ss}_{set}/G = X^{ss}_{set}/\sigma_{set} \to (X^{ss}/\sigma)_{set} \tag{1.25}$$

from the set of $G-$orbits in X^{ss}_{set}. As is well known the map 1.25 is not a bijection in general. In particular X^{ss}/σ is generally not a *geometric quotient* in the sense of [**113**].

Note also that the theory above drastically fails if G is locally of finite type rather than of finite type over our field, for instance if G is infinite and discrete, a case of interest for us in this book.

1.2.3. Algebraic stacks. Here we assume familiarity with the definition of *algebraic stacks* in the sense of [**58**], pp. 20-22. The reader uncomfortable with algebraic stacks should simply skip the remarks below. Our aim here is to record the fact that replacing algebraic varieties by algebraic stacks does not solve the problem raised by Proposition 1.13. Let k be an algebraically closed field and consider the functor

$$\mathcal{C}_{\mathrm{ft}/k} := \{\text{schemes of finite type over } k\} \to \mathcal{C}_{stack} := \{\text{algebraic stacks over } k\}$$

$$X \mapsto X_{stack}.$$

We declare that the *trivial* objects in $\mathcal{C}_{stack}$ are, again, the points (i.e. the objects isomorphic to $(Spec\ k)_{stack}$).

PROPOSITION 1.16. *Assume $\mathbf{X} = (X, \sigma)$ is a dense irreducible reduced correspondence of finite type over k. Then X_{stack}/σ_{stack} is trivial in $\mathcal{C}_{stack}$.*

PROOF. This follows directly from Proposition 1.13 and from the definition of algebraic stacks. □

1.2.4. Affine schemes. It is worth reviewing quickly what categorical quotients mean in the category of affine schemes.

DEFINITION 1.17. For R a ring we consider the category

$$\mathcal{R} = \mathcal{R}_R := \{\text{algebras over } R\}; \tag{1.26}$$

here, as usual, algebras are assumed commutative and unitary. Let $\mathcal{R}^{op}$ be the category obtained from $\mathcal{R}$ by reversing the arrows (so $\mathcal{R}^{op}$ identifies with the category of affine $R-$schemes). An object in $\mathcal{R}^{op}$ will be called *trivial* if it is isomorphic to R.

DEFINITION 1.18. Let $\mathbf{A} = (A, \tilde{A}, \sigma_1, \sigma_2)$ be a correspondence in the category $\mathcal{R}^{op}$, so we can view σ_1, σ_2 as $R-$algebra homomorphisms $A \to \tilde{A}$. Then we can consider the $R-$module homomorphism

$$\Delta : A \to \tilde{A}, \quad \Delta(a) = \sigma_1(a) - \sigma_2(a),$$

and we define the following R–modules:

$$H^0(\mathbf{A}) := Ker\ \Delta, \quad H^1(\mathbf{A}) := Coker\ \Delta.$$

For $\tilde{a} \in \tilde{A}$ we denote its class in $H^1(\mathbf{A})$ by $[\tilde{a}]$. There are *cup product* R–linear maps

$$H^0(\mathbf{A}) \times H^0(\mathbf{A}) \to H^0(\mathbf{A}), \quad (a_1, a_2) \mapsto a_1 a_2,$$
$$H^0(\mathbf{A}) \times H^1(\mathbf{A}) \to H^1(\mathbf{A}), \quad (a_1, [\tilde{a}_2]) \mapsto [\tilde{a}_1 \tilde{a}_2],$$

where $\tilde{a} := \sigma_1(a) = \sigma_2(a)$, which make $H^0(\mathbf{A})$ into a subring of A and make $H^1(\mathbf{A})$ into an $H^0(\mathbf{A})$–module. It is trivial to check that the categorical quotient A/σ exists in $\mathcal{R}^{op}$ and is given by:

PROPOSITION 1.19. $A/\sigma = H^0(\mathbf{A})$.

So the existence of categorical quotients in $\mathcal{R}^{op}$ is a trivial matter; the main issue here is that for many interesting correspondences these categorical quotients are trivial.

We close the discussion here by recording the relation between the cohomology of a correspondence and *Ext* groups. Assume $u : A \to A$ is a ring homomorphism and consider the correspondence $\mathbf{A} = (A, A, id, u)$. Let $A[u]$ be the non-commutative ring generated by A and the letter u with relations $u \cdot a = u(a) \cdot u$, $a \in A$. (Morally $A[u]$ is a *convolution algebra* attached to the action of u on A.) Finally view A as a left $A[u]$–module. Then it is trivial to check that

$$H^i(\mathbf{A}) = Ext^i_{A[u]}(A, A), \quad i = 0, 1.$$

This can be viewed as a "shadow of a link" between categorical quotients and non-commutative geometry. If in addition u is an automorphism of A and $\mathbf{Z}$ acts on A via the powers of u then

$$H^i(\mathbf{A}) = H^i(\mathbf{Z}, A), \quad i = 0, 1,$$

where the right hand side H^i means group cohomology.

1.2.5. p–adic formal schemes. Our δ–geometry, to be introduced later, has, in particular, a p–adic nature. It is therefore useful to see that (usual) formal (p–adic) geometry does not fix the problem that was illustrated by Proposition 1.13; not only that categorical quotients in formal p–adic geometry stay, most of the times, trivial but the p–torsion in the appropriate cohomology is also trivial. To simplify our discussion we will restrict ourselves to affine formal (p–adic) schemes. The discussion below will not play any role later.

Let us fix a local p–ring R, $p > 0$, with algebraically closed residue field $k = R/pR$. If X is a smooth affine scheme over R, we set $X_0 = X \otimes k$. If X_0 is irreducible then $\mathcal{O}(\hat{X}) = \mathcal{O}(X)\hat{}$, the ring of global functions on the formal scheme $\hat{X}$, is a global p–ring.

PROPOSITION 1.20. *Let* $\mathbf{X} = (X, \tilde{X}, \sigma_1, \sigma_2)$ *be a correspondence of finite type and smooth over* R*, with* X *and* $\tilde{X}$ *affine. Consider the induced correspondence*

$$\hat{\mathbf{X}} := (\mathcal{O}(X)\hat{}, \mathcal{O}(\tilde{X})\hat{}, \sigma_1^*, \sigma_2^*)$$

in the category $\mathcal{R}^{op}$ *dual to the category of* R*–algebras and let*

$$\mathbf{X}_0 = (X_0, \tilde{X}_0, \sigma_1, \sigma_2)$$

be the induced correspondence in $\mathcal{C}_k$. Assume that $\mathbf{X}_0$ is dense. Then the following hold:

1) $H^0(\hat{\mathbf{X}}) = R$,

2) the $p-$torsion in $H^1(\hat{\mathbf{X}})$ vanishes.

In particular conclusion 1 shows that the categorical quotient of $\hat{\mathbf{X}}$ is trivial in $\mathcal{R}^{op}$ while conclusion 2 shows that there are no torsion cohomological invariants. Morally, the framework of formal geometry is unsatisfactory with regards to both H^0 and H^1.

PROOF. To check assertion 1 let $f \in H^0(\hat{\mathbf{X}})$, so $f \in \mathcal{O}(X)\hat{}$, $\sigma_1^* f = \sigma_2^* f$. Denote by $f_0 \in \mathcal{O}(X_0)$ the reduction mod p of f. So $\sigma_1^* f_0 = \sigma_2^* f_0 \in \mathcal{O}(\tilde{X}_0)$. Exactly as in the proof of Proposition 1.13 it follows that $f_0 \in k$ so we can write $f = c^{(0)} + pf^{(1)}$ with $c^{(0)} \in R$ and $f^{(1)} \in \mathcal{O}(X)\hat{}$. Since p is a non-zero divisor in $\mathcal{O}(\tilde{X})\hat{}$ it follows that $f^{(1)} \in H^0(\hat{\mathbf{X}})$. Repeating the argument we obtain a sequence $f^{(i)} = c^{(i)} + pf^{(i+1)}$, $i \geq 1$, $c^{(i)} \in R$ and $f^{(i)} \in \mathcal{O}(X)\hat{}$. Since $\mathcal{O}(X)\hat{}$ is $p-$adically separated we get

$$f = \sum_{i \geq 0} c^{(i)} p^i \in R$$

which ends the proof of assertion 1. To prove assertion 2 assume $\sigma_1^* f - \sigma_2^* f = p\tilde{g}$ for some $f \in \mathcal{O}(X)\hat{}$, $\tilde{g} \in \mathcal{O}(\tilde{X})\hat{}$. Denote, again, by $f_0 \in \mathcal{O}(X_0)$ the reduction mod p of f. Then $\sigma_1^* f_0 = \sigma_2^* f_0 \in \mathcal{O}(\tilde{X}_0)$ so, exactly as before, $f_0 \in k$ and we can write $f = c^{(0)} + pf^{(1)}$ with $c^{(0)} \in R$ and $f^{(1)} \in \mathcal{O}(X)\hat{}$. Since p is a non-zero divisor in $\mathcal{O}(\tilde{X})\hat{}$ we get

$$\sigma_1^* f^{(1)} - \sigma_2^* f^{(1)} = \tilde{g},$$

which ends the proof. □

1.3. Analytic uniformization and critical finiteness

The main conjectures and results of this book involve two types of concepts: 1) analytic uniformization of correspondences and 2) $\delta-$geometry. In this section we explain analytic uniformization. This concept is also closely related to (a generalization of) the classical concept of postcritical finiteness in complex dynamics.

1.3.1. Admissible groups. Let $\mathbf{S}$ be a simply connected Riemann surface; we may (and will) assume that $\mathbf{S}$ is either the Riemann sphere $\mathbf{P}^1$, or the complex plane $\mathbf{C}$, or the upper half plane $\mathbf{H}$.

DEFINITION 1.21. A subgroup Γ of the group $Aut(\mathbf{S})$ of complex analytic automorphisms of $\mathbf{S}$ will be called *admissible* if it acts properly discontinuously on $\mathbf{S}$ (cf. [**77**], p.29) and one of the following holds:

1) $\mathbf{S} = \mathbf{P}^1$ (hence Γ is finite);

2) $\mathbf{S} = \mathbf{C}$ and Γ is infinite;

3) $\mathbf{S} = \mathbf{H}$ and Γ has finite covolume.

Let us refer to situations 1, 2, and 3 as the *spherical, flat, and hyperbolic* situations respectively. For a complete classification of the groups Γ in the spherical situation see [**59**], pp. 223-224; this classification is classical and boils down, essentially, to the groups of the platonic solids. For a complete classification of the groups Γ in the flat situation see [**59**], pp. 225-228; these groups always contain a non-trivial normal subgroup of finite index consisting of translations of $\mathbf{C}$. Finally, the groups Γ in the hyperbolic situation are exactly the *Fuchsian lattices* i.e. the discrete subgroups of finite covolume; cf. [**77**], p.29., [**122**], p.19, [**85**], p.70.

In the flat situation we have:

PROPOSITION 1.22. *If $\Gamma \subset Aut(\mathbf{C})$ is admissible then there is an isomorphism of analytic spaces $\mathbf{C}/\Gamma \simeq G/\langle\gamma_0\rangle$ where G is either $\mathbf{G}_m$ or an elliptic curve over $\mathbf{C}$ and γ_0 is an algebraic group automorphism of G. (Hence γ_0 has order* 1 *or* 2 *if $G = \mathbf{G}_m$ and order* 1, 2, 3, 4 *or* 6 *if G is an elliptic curve.)*

PROOF. Let $T(\mathbf{C})$ denote the group of translations of $\mathbf{C}$ and $T(\Gamma) := \Gamma \cap T(\mathbf{C})$. Set $\Gamma_1 := \Gamma/T(\Gamma)$. By [**59**], pp. 225-228, Γ_1 is finite. Set $G := \mathbf{C}/T(\Gamma)$. Then $\mathbf{C}/\Gamma \simeq G/\Gamma_1$. Clearly G is either $\mathbf{G}_m$ or an elliptic curve according as the rank of $T(\Gamma)$ is 1 or 2. Also Γ_1 consists of algebraic automorphisms. There is an exact sequence

$$1 \to T(G) \to Aut(G) \to Aut_0(G) \to 1$$

where $Aut(G)$ is the group of algebraic variety automorphisms of G, $Aut_0(G)$ is the group of algebraic group automorphisms of G, and $T(G)$ is the group of all translations. We claim that

$$T(G) \cap \Gamma_1 = \{id\}. \tag{1.27}$$

Indeed, if $\gamma \in \Gamma$ induces on $\mathbf{C}/T(\Gamma)$ the same automorphism as the translation $\mathbf{C} \to \mathbf{C}$, $z \mapsto z + z_1$ then, for any $z \in \mathbf{C}$, one has

$$\gamma(z) - (z + z_1) =: \gamma_z \in Ker(\mathbf{C} \to \mathbf{C}/T(\Gamma)) \simeq T(\Gamma).$$

Since the map $\mathbf{C} \to T(\Gamma)$, $z \mapsto \gamma_z$ is continuous it must be constant, equal to $z \mapsto z_2$. So

$$\gamma(z) = z + z_1 + z_2,$$

hence $\gamma \in T(\Gamma)$ and our claim is proved. By Equation 1.27 Γ_1 is isomorphic to a subgroup of $Aut_0(G)$. But the latter is cyclic hence so is Γ_1, say $\Gamma_1 = \langle\gamma_1\rangle$. There exists $P_1 \in G(\mathbf{C})$ and $\gamma_0 \in Aut_0(G)$ such that $\gamma_1(P) = \gamma_0(P) + P_1$ for all $P \in G(\mathbf{C})$. If $\gamma_0 = id$ then, by Equation 1.27, $P_1 = 0$ so $\Gamma_1 = \{id\}$ and we are done. If $\gamma_0 \neq id$ then the map $\gamma_0 - id : G \to G$ is an isogeny. Pick $P_0 \in G(\mathbf{C})$ such that

$$(\gamma_0 - id)(P_0) = P_1.$$

Then the translation $G \to G$, $P \mapsto P + P_1$ induces an isomorphism of algebraic varieties

$$G/\langle\gamma_1\rangle \to G/\langle\gamma_0\rangle$$

and we are done. □

Note that if Γ is admissible then, in all three situations above (spherical, flat and hyperbolic), $\mathbf{S}/\Gamma$ has a structure of a smooth complex algebraic curve; this is clear in the spherical situation (where, of course, $\mathbf{P}^1/\Gamma \simeq \mathbf{P}^1$ non-canonically), it follows from Proposition 1.22 in the flat situation, and is it well known (cf., say, [**122**]) in the hyperbolic situation. This structure is unique as one can see by using, say, the classical Big Picard Theorem.

DEFINITION 1.23. Assume we are given a tuple

$$(\mathbf{S}, \Gamma, \tilde{\Gamma}, \tau_1, \tau_2) \tag{1.28}$$

where $\mathbf{S}$ is a simply connected Riemann surface, Γ and $\tilde{\Gamma}$ are admissible groups of automorphisms of $\mathbf{S}$, and τ_1, τ_2 are automorphisms of $\mathbf{S}$ such that $\tau_1\tilde{\Gamma}\tau_1^{-1}$ and

$\tau_2\tilde{\Gamma}\tau_2^{-1}$ are subgroups of finite index in Γ. Denote by $\bar{\tau}_1, \bar{\tau}_2 : \mathbf{S}/\tilde{\Gamma} \to \mathbf{S}/\Gamma$ the complex analytic maps induced by τ_1 and τ_2 respectively; by Big Picard $\bar{\tau}_1, \bar{\tau}_2$ are automatically morphisms of algebraic curves. So one can consider the following irreducible reduced correspondence of finite type and smooth over $\mathbf{C}$:

$$\mathbf{Y} := (\mathbf{S}/\Gamma, \mathbf{S}/\tilde{\Gamma}, \bar{\tau}_1, \bar{\tau}_2). \tag{1.29}$$

(This correspondence is not étale in general.) A correspondence $\mathbf{X} = (X, \tilde{X}, \sigma_1, \sigma_2)$ in the category $\mathcal{C}_{\mathbf{C}}$ is said to be *analytically uniformized* by the tuple in Equation 1.28 if there is an open immersion $\mathbf{X} \subset \mathbf{Y}$ into a correspondence $\mathbf{Y}$ of the form 1.29 such that $\mathbf{S} \to \mathbf{S}/\Gamma$ and $\mathbf{S} \to \mathbf{S}/\tilde{\Gamma}$ are unramified above X and $\tilde{X}$ respectively. (Such an $\mathbf{X}$ is, by definition, smooth over $\mathbf{C}$ and étale hence possesses a canonical bundle $\mathbf{K}$.) We say that an analytic uniformization as above is *spherical, flat, or hyperbolic*, according as $\mathbf{S}$ is $\mathbf{P}^1$, $\mathbf{C}$, or $\mathbf{H}$ respectively.

The next Proposition shows that correspondences over $\mathbf{C}$ admitting a spherical or hyperbolic analytic uniformization tend to have non-flat pluricanonical bundles $\mathbf{K}^\nu$ (for $\nu \neq 0$) while correspondences over $\mathbf{C}$ admitting a flat analytic uniformization tend to have a flat $\nu-$canonical bundle $\mathbf{K}^\nu$ for some $\nu \neq 0$. (On the other hand, as we shall see later, if $\mathbf{X}$ has a spherical or hyperbolic analytic uniformization and is defined over a number field then $\mathbf{K}$ tends to be what we shall call *formally flat at the fixed points*; cf. Lemmas 6.3 and 8.18 respectively. Moreover for almost all places $\wp$, a suitable power of $\mathbf{K}$ tends to be *generically trivial mod* $\wp$; cf. Lemmas 6.4 and 8.16 respectively.)

PROPOSITION 1.24. *Let* $\mathbf{X} = (X, \tilde{X}, \sigma_1, \sigma_2)$ *be a correspondence over* $\mathbf{C}$.

1) If $\mathbf{X}$ *is analytically uniformized by* $(\mathbf{P}^1, \Gamma, \{id\}, \pi, \pi \circ \tau)$ *where* $\langle \Gamma, \tau \rangle$ *is not solvable and* τ *has no fixed point on* $\mathbf{A}^1$ *then, for any integer* $\nu \neq 0$, *the pluricanonical bundle* $\mathbf{K}^\nu$ *is not flat.*

2) If $\mathbf{X}$ *is analytically uniformized by* $(\mathbf{C}, \Gamma, \Gamma, id, \tau)$ *where* $\Gamma = \langle T(\Gamma), z \mapsto -z \rangle$ *and* $\tau(z) = mz$, $m \in \mathbf{Z}$, $m \geq 2$, *then the* $2-$*canonical bundle* $\mathbf{K}^2$ *is flat.*

3) If $\mathbf{X}$ *is analytically uniformized by* $(\mathbf{H}, \Gamma, \tilde{\Gamma}, id, \tau)$ *where* Γ *has finite index in* $SL_2(\mathbf{Z})$ *and* $\tau(z) = mz$, $m \in \mathbf{Z}$, $m \geq 2$ *then, for any integer* $\nu \neq 0$, *the pluricanonical bundle* $\mathbf{K}^\nu$ *is not flat.*

PROOF. To prove assertion 1 assume $\mathbf{K}^\nu$ is flat for some $\nu \neq 0$ and seek a contradiction. By flatness there exists a $\nu-$tuple $1-$form η on X such that $\sigma_2^*\eta = \lambda \cdot \sigma_1^*\eta$. Set $\pi^*\eta = f(z) \cdot (dz)^\nu$ where z is an affine coordinate on $\mathbf{A}^1 \subset \mathbf{P}^1$ and $f(z) \in \mathbf{C}(z)$ is a rational fraction. Then

$$\tau^*(f(z) \cdot (dz)^\nu) = \lambda \cdot f(z) \cdot (dz)^\nu.$$

We have $\tau(z) = z + u$ for some $u \in \mathbf{C}^\times$ hence

$$f(z+u) = \lambda \cdot f(z).$$

So the divisor of f is invariant under $z \mapsto z + u$ hence f must be a constant. So $(dz)^\nu$ is $\Gamma-$invariant. By hypothesis Γ contains a transformation of the form

$$\gamma(z) = \frac{az+b}{cz+d}$$

with $c \neq 0$. We get $1 = (d\gamma/dz)^\nu = (cz+d)^{-2\nu}$, a contradiction because $\nu \neq 0$.

Assertion 2 is an easy exercise; cf. the argument in the proof of Lemma 7.4.

To prove assertion 3 assume $\mathbf{K}^{\nu}$ is flat for some $\nu \neq 0$ and seek a contradiction. By flatness there exists a non-zero $\nu-$tuple $1-$form η on X such that $\sigma_2^*\eta = \lambda \cdot \sigma_1^*\eta$. Let $f(z) \cdot (dz)^{\nu}$ be the pull-back of η on $\mathbf{H}$. Then f is a meromorphic modular function of weight 2ν for Γ satisfying

$$f(mz) = \lambda \cdot m^{-\nu} \cdot f(z).$$

By looking at Fourier expansions at ∞ we conclude that f is a constant. Hence $(dz)^{\nu}$ is $\gamma-$invariant which is again a contradiction because $\nu \neq 0$. □

In what follows we discuss the problem of describing, in "algebraic terms", all correspondences admitting spherical, flat, or hyperbolic uniformizations respectively.

1.3.2. Spherical case. We have the following description of correspondences admitting a spherical analytic uniformization. This description is, of course, in algebraic terms because $\mathbf{P}^1$ is already algebraic.

PROPOSITION 1.25. *Consider a finite subgroup* Γ *of* $Aut(\mathbf{P}^1)$ *and an automorphism* $\tau \in Aut(\mathbf{P}^1)$. *Consider the quotient map* $\pi : \mathbf{P}^1 \to \mathbf{P}^1/\Gamma$, *take a Zariski open subset* $X \subset \mathbf{P}^1/\Gamma$ *above which* π *is unramified, and set* $Y := \pi^{-1}(X)$. *Consider the correspondence over* $\mathbf{C}$,

$$\mathbf{X} = (X, Y \cap \tau^{-1}(Y), \pi, \pi \circ \tau).$$

Then $\mathbf{X}$ *admits a spherical analytic uniformization; conversely, any correspondence over* $\mathbf{C}$ *admitting a spherical analytic uniformization is equivalent to a correspondence* $\mathbf{X}$ *as above.*

PROOF. A trivial exercise. □

REMARK 1.26. It is worth noting here that equivalence relations generated by spherical correspondences are closely related to equivalence relations coming from certain infinite (discrete) group actions. To explain this consider a finite subgroup Γ of $Aut(\mathbf{P}^1)$ and an automorphism $\tau \in Aut(\mathbf{P}^1)$. Also consider the (generally infinite) subgroup $\langle \Gamma, \tau \rangle \subset Aut(\mathbf{P}^1)$ generated by Γ and τ and denote by $\pi : \mathbf{P}^1 \to \mathbf{P}^1/\Gamma$ the quotient map by Γ. On $\mathbf{P}^1$ consider the equivalence relation

$$P \sim Q \quad \text{iff} \quad P = gQ \quad \text{for some} \quad g \in \langle \Gamma, \tau \rangle.$$

On the other hand, on $\mathbf{P}^1/\Gamma$, consider the smallest equivalence relation for which

$$\pi(P) \sim \pi(\tau(P)) \quad \text{if} \quad P \in \mathbf{P}^1.$$

Then it is trivial to check that, for $P, Q \in \mathbf{P}^1$, we have

$$P \sim Q \quad \text{iff} \quad \pi(P) \sim \pi(Q).$$

This observation will morally play a key role in our study of spherical correspondences. It creates a bridge between two equivalence relations which are both "pathological" in algebraic geometry but for different reasons: $\sim$ on $\mathbf{P}^1$ is given by a group (but the group is infinite) whereas $\sim$ on $\mathbf{P}^1/\Gamma$ is given by data of finite type (but not by a group).

1.3.3. Flat case. Recall that for any algebraic group G we denote by $Aut_0(G)$ the group of algebraic group automorphisms of G. We have the following algebraic description of correspondences admitting a flat analytic uniformization:

PROPOSITION 1.27. *Consider an algebraic group G over $\mathbf{C}$ which is either $\mathbf{G}_m$ or an elliptic curve. Consider (algebraic group) automorphisms $\gamma_0, \tilde{\gamma}_0 \in Aut_0(G)$, points $P_i \in G(\mathbf{C})$, $i = 1, 2$, and non-zero (algebraic group) endomorphisms $\tau_{i0} : G \to G$, satisfying*

$$\tau_{i0}\tilde{\gamma}_0 = \gamma_0^{\nu_i}\tau_{i0}, \quad \gamma_0^{\nu_i}(P_i) = P_i, \quad i = 1, 2, \tag{1.30}$$

for suitable integers ν_i. (In particular if $\Gamma_0 := \langle\gamma_0\rangle$, $\tilde{\Gamma}_0 := \langle\tilde{\gamma}_0\rangle$, $\tau_i : G \to G$, $\tau_i(P) := \tau_{i0}(P) + P_i$, then $\tau_i\tilde{\Gamma}_0 \subset \Gamma_0\tau_i$.) Consider the maps

$$\sigma_1, \sigma_2 : G/\tilde{\Gamma}_0 \to G/\Gamma_0$$

induced by τ_1, τ_2 respectively. Pick an open subset $X \subset G/\Gamma_0$ above which $G \to G/\Gamma_0$ is unramified and consider the correspondence over $\mathbf{C}$,

$$\mathbf{X} := (X, \sigma_1^{-1}(X) \cap \sigma_2^{-1}(X), \sigma_1, \sigma_2).$$

Then $\mathbf{X}$ admits a flat analytic uniformization; conversely, any correspondence over $\mathbf{C}$ admitting a flat analytic uniformization is equivalent to a correspondence $\mathbf{X}$ as above.

PROOF. The fact that $\mathbf{X}$ admits a flat analytic uniformization is left to the reader. Assume now we are given a correspondence $\mathbf{X}' = (X', \tilde{X}', \sigma_1', \sigma_2')$ uniformized by a tuple

$$(\mathbf{C}, \Gamma, \tilde{\Gamma}, \tau_1', \tau_2').$$

We are going to use the notation in the proof of Proposition 1.22. Let $G := \mathbf{C}/T(\Gamma)$ and $\tilde{G} := \mathbf{C}/T(\tilde{\Gamma})$. Since $T(\mathbf{C})$ is normal in the group of analytic isomorphisms $Aut(\mathbf{C})$ it follows that

$$\tau_i'T(\tilde{\Gamma})(\tau_i')^{-1} \subset \Gamma \cap T(\mathbf{C}) = T(\Gamma).$$

So τ_i' induce finite morphisms of algebraic varieties

$$\tau_i' : \tilde{G} \to G.$$

By the proof of Proposition 1.22, $\mathbf{C}/\Gamma \simeq G/\Gamma_1$ where $\Gamma_1 = \langle\gamma_1\rangle = \Gamma/T(\Gamma)$ and $\mathbf{C}/\tilde{\Gamma} \simeq \tilde{G}/\tilde{\Gamma}_1$ where $\tilde{\Gamma}_1 = \langle\tilde{\gamma}_1\rangle = \tilde{\Gamma}/T(\tilde{\Gamma})$. So τ_i' induce morphisms

$$\bar{\tau}_i' : \tilde{G}/\langle\tilde{\gamma}_1\rangle \to G/\langle\gamma_1\rangle.$$

Also, by the proof of Proposition 1.22, we have an isomorphism $\bar{T}_Q : G/\langle\gamma_1\rangle \to G/\langle\gamma_0\rangle$ induced by a translation $T_Q : G \to G$, $P \mapsto P + Q$. Let $\tilde{\pi} : \tilde{G} \to \tilde{G}/\langle\tilde{\gamma}_1\rangle$ and $\pi : G \to G/\langle\gamma_0\rangle$ be the natural projections, let $v : G \to \tilde{G}$ be any isogeny, and set

$$\tau_i := T_Q \circ \tau_i' \circ v, \quad \sigma_i := \bar{T}_Q \circ \bar{\tau}_i' \circ \tilde{\pi} \circ v.$$

Then we get a commutative diagram

$$\begin{array}{ccccc} G & \overset{\tau_1}{\leftarrow} & G & \overset{\tau_2}{\rightarrow} & G \\ \pi\downarrow & & \downarrow id & & \downarrow\pi \\ G/\langle\gamma_0\rangle & \overset{\sigma_1}{\leftarrow} & G & \overset{\sigma_2}{\rightarrow} & G/\langle\gamma_0\rangle \end{array}$$

If $X = \bar{T}_Q(X')$ then clearly $\mathbf{X}'$ is equivalent to $\mathbf{X} := (X, \sigma_1^{-1}(X) \cap \sigma_2^{-1}(X), \sigma_1, \sigma_2)$ and we are done by setting $\tilde{\gamma}_0 = id$. $\square$

REMARK 1.28. If G is $\mathbf{G}_m$ or an elliptic curve without CM then the only possibilities for the data

$$\gamma_0, \tilde{\gamma}_0, \tau_{i0}, P_i$$

in the above Proposition are given by the following:

1) $\tilde{\gamma}_0 = [-1]$, $\gamma_0 = [-1]$, the multiplication by -1 in the corresponding groups, $\tau_{i0} = [N_i]$, the multiplication by an integer N_i, and P_i points with $2P_i = 0$;

2) $\tilde{\gamma}_0 = [1]$, $\gamma_0 \in \{[1], [-1]\}$, $\tau_{i0} = [N_i]$, P_i arbitrary.

Let us discuss some examples; cf. Example 1.31. These examples involve some classical rational functions which we need to recall first.

DEFINITION 1.29. *i) Multiplicative functions.* By a *multiplicative function* we will understand a rational function of the form $t^{\pm d}$, where $d \geq 2$ is an integer.

ii) Chebyshev polynomials. Recall that the *Chebyshev polynomial* of degree $d \geq 2$ is defined to be the unique polynomial $Cheb_d(t) \in \mathbf{C}[t]$ such that

$$Cheb_d(t + t^{-1}) = t^d + t^{-d}. \tag{1.31}$$

Of course $Cheb_d$ has coefficients in $\mathbf{Z}$. The first Chebyshev polynomials are

$$Cheb_2(t) = t^2 - 2, \quad Cheb_3(t) = t^3 - 3t, ...$$

iii) Lattès functions. A rational function $\sigma(t) \in \mathbf{C}(t)$, $\sigma : \mathbf{P}^1 \to \mathbf{P}^1$, will be called a *Lattès function* if σ is obtained as follows. One starts with an elliptic curve E over $\mathbf{C}$ with affine plane equation $y^2 = x^3+ax+b$. One then considers an algebraic group endomorphism $\tau_0 : E \to E$, a non-trivial algebraic group automorphism $\gamma_0 : E \to E$, and a fixed point P_0 of γ_0. Note that γ_0 has order $2, 4, 3$ or 6 and, correspondingly, one has: $\gamma_0(x, y) = (x, -y)$ or $b = 0$, $\gamma_0(x, y) = (-x, iy)$, or $a = 0$, $\gamma_0(x, y) = (\zeta_3 x, y)$, or $a = 0$, $\gamma_0(x, y) = (\zeta_3 x, -y)$. One considers the morphism $\tau : E \to E$, $\tau(P) = \tau_0(P) + P_0$ and one considers the map $\sigma : E/\Gamma \to E/\Gamma$ induced by τ, where $\Gamma := \langle \gamma_0 \rangle$ and E/Γ is identified with $\mathbf{P}^1 = Proj\ \mathbf{C}[x_0, x_1]$ via the isomorphism that sends $t := x_1/x_0$ into x, x^2, y, or y^2 according as $ord(\gamma_0)$ is $2, 4, 3$ or 6. We denote by Lat_d any Lattès function of degree d. The original examples of Lattès, cf. [**89**], are given by the above construction with $ord(\gamma_0) = 2$.

REMARK 1.30. The 3 classes of functions defined above (multiplicative, Chebyshev, and Lattès) naturally appear in a number of (a priori unrelated) contexts such as: orbifold theory [**51**], [**109**], density conjecture for hyperbolic rational maps [**96**], isospectral deformations of rational maps [**105**], permutable rational maps [**116**], measure theory related to rational maps [**136**], and Galois theory of function fields [**60**], [**65**]. Cf. also [**110**] for an overview of some of these topics. Our interest in the above examples comes from their role in the context of correspondences with flat analytic uniformization as shown in the next:

EXAMPLE 1.31. i) If, in the notations of Proposition 1.27,

$$G = \mathbf{G}_m = Spec\ \mathbf{C}[t, t^{-1}],$$

$\gamma_0 = \tilde{\gamma}_0 = [1] = id$, $P_1 = P_2 = 0$ (i.e. P_1, P_2 have coordinate 1), $\tau_{10} = id$, $\tau_{20}(t) = t^{\pm d}$, then $G/\tilde{\Gamma}_0 = G/\Gamma_0 = \mathbf{G}_m$, $\sigma_1 = id$, $\sigma_2(t) = t^{\pm d}$, a multiplicative function, and $\mathbf{X}$ is analytically uniformized by $(\mathbf{C}, \Gamma, \Gamma, id, \tau)$ where $\Gamma := \langle z \mapsto z+1 \rangle$ and $\tau(z) = \pm d \cdot z$.

ii) If, in the notations of Proposition 1.27, $G = \mathbf{G}_m$, $\gamma_0 = \tilde{\gamma}_0 = [-1]$, $\tau_{10} = id$, $P_1 = 0$, $\tau_{20}(t) = t^d$, P_1 has coordinate $\epsilon \in \{1, -1\}$ in $\mathbf{C}$, and $X = \mathbf{A}^1 \backslash \{-2, 2\}$

then we have identifications $G/\tilde{\Gamma}_0 = G/\Gamma_0 = \mathbf{A}^1$, such that $\sigma_1 = id$ and $\sigma_2(t) = \epsilon \cdot Cheb_d(t)$. There is a commutative diagram

$$\begin{array}{ccc} \mathbf{G}_m & \stackrel{\epsilon t^d}{\to} & \mathbf{G}_m \\ t+t^{-1}\downarrow & & \downarrow t+t^{-1} \\ \mathbf{A}^1 & \stackrel{\sigma_2}{\to} & \mathbf{A}^1 \end{array}$$

with $t+t^{-1}$ unramified above X. So, upon letting

$$\Gamma := \langle z \mapsto z+1, z \mapsto -z \rangle \subset Aut(\mathbf{C})$$

and upon identifying $\mathbf{A}^1$ with $\mathbf{C}/\Gamma$ via the map

$$\mathbf{C}/\Gamma \to \mathbf{A}^1 = \mathbf{C}, \quad z \ mod\ \Gamma \quad \to \quad e^{2\pi i z} + e^{-2\pi i z},$$

we have that σ_2 is induced by $\tau : \mathbf{C} \to \mathbf{C}$ where $\tau(z)$ is either $d \cdot z$ or $d \cdot z + \frac{1}{2}$ according as ϵ is 1 or -1. In particular $\mathbf{X}$ is analytically uniformized by

$$(\mathbf{C}, \Gamma, \Gamma, id, \tau).$$

iii) If, in the notations of Proposition 1.27, G is an elliptic curve, $\tau_{10} = id$, $P_1 = 0$, and $\gamma_0 = \tilde{\gamma}_0 \neq id$ then there are identifications $G/\tilde{\Gamma}_0 = G/\Gamma_0 = \mathbf{P}^1$ such that $\sigma_1 = id$ and $\sigma_2(t) : \mathbf{P}^1 \to \mathbf{P}^1$ is a Lattès function.

1.3.4. Hyperbolic case. It turns out that dense correspondences admitting a hyperbolic analytic uniformization can also be described in purely algebraic terms. To explain this we start by recalling the following standard definition. A Fuchsian lattice

$$\Gamma \subset Aut(\mathbf{H}) = PSL_2(\mathbf{R}) = SL_2(\mathbf{R})/\langle -id \rangle$$

is called *arithmetic* if there exists a semisimple algebraic group H over $\mathbf{Q}$ and a surjective homomorphism

$$u : H(R)^\circ \to PSL_2(\mathbf{R})$$

with compact kernel such that $u(H(\mathbf{Z}) \cap H(R)^\circ)$ is commensurable with Γ; cf. [**95**], p. 318. Here the upper $\circ$ stands for taking the connected component of a real Lie group.

The following is an easy consequence of a deep result of Margulis:

PROPOSITION 1.32. *Let* $\mathbf{X}$ *be a dense correspondence over* $\mathbf{C}$ *possessing a hyperbolic analytic uniformization by* $(\mathbf{H}, \Gamma, \tilde{\Gamma}, \tau_1, \tau_2)$. *Then* Γ *and* $\tilde{\Gamma}$ *are arithmetic.*

PROOF. We may assume $\tau_1 = id$ and we set $\tau_2 = \tau$. Assume Γ (and hence $\tilde{\Gamma}$) is not arithmetic. By Margulis [**102**], p. 298 (cf. also [**95**], p. 318) any non-arithmetic Fuchsian lattice has finite index in its commensurator. Let Γ^c be the commensurator of Γ; hence Γ^c contains Γ as a subgroup of finite index and $\tilde{\Gamma} \subset \Gamma^c$, $\tau \in \Gamma^c$. We claim that the correspondence in Equation 1.29 is not dense; this implies that $\mathbf{X}$ is not dense, a contradiction. To check our claim denote by E the smallest equivalence relation containing the image of the map

$$\mathbf{H}/\tilde{\Gamma} \to \mathbf{H}/\Gamma \ \times \ \mathbf{H}/\Gamma = (\mathbf{H} \times \mathbf{H})/(\Gamma \times \Gamma)$$

and note that E is contained in the equivalence relation

$$E^c := \{(z, \gamma z) \in \mathbf{H} \times \mathbf{H} | \gamma \in \Gamma^c\}/(\Gamma \times \Gamma) \subset \mathbf{H}/\Gamma \ \times \ \mathbf{H}/\Gamma.$$

Now the orbits of E^c are clearly finite hence the orbits of E are finite which proves our claim. $\square$

On the other hand all arithmetic Fuchsian lattices arise, via a classical construction, from quaternion algebras as follows; cf. [**95**], p. 253. Start with a totally real number field $E \subset \mathbf{R}$ of degree $g \geq 1$ over $\mathbf{Q}$ and let B be a quaternion algebra over E, i.e. an algebra such that $B \otimes_E E^a \simeq Mat_2(E^a)$, where E^a is the algebraic closure of E. Let $\rho_1, ..., \rho_g$ be the embeddings of E into $\mathbf{R}$, where ρ_1 is the inclusion, assume that $B \otimes_F \mathbf{R} \simeq Mat_2(\mathbf{R})$, and assume that for all $i = 2, ..., g$ we have $B \otimes_{F,\rho_i} \mathbf{R} \not\simeq Mat_2(\mathbf{R})$ (hence $B \otimes_{F,\rho_i} \mathbf{R}$ is isomorphic to the Hamilton algebra of quaternions). Let B^+ be the set of all members of B with positive determinant (when we view B as a subset of $Mat_2(\mathbf{R})$); so B^+ acts naturally on $\mathbf{H}$. Let $\mathcal{O}_E$ be the ring of integers of E. Recall that an *order* of B (over $\mathcal{O}_E$) is a subring $\mathcal{O}_B$ of B, containing 1, which is a finitely generated $\mathcal{O}_E-$submodule of B such that $\mathcal{O}_B \otimes_{\mathcal{O}_E} E = B$; we have $det(\mathcal{O}_B) \subset \mathcal{O}_E$. Fix an order $\mathcal{O}_B$ of B and consider the group

$$\mathcal{O}_B^1 = \{b \in \mathcal{O}_B | det(b) = 1\} \subset \mathcal{O}_B^\times \cap B^+.$$

It is convenient to introduce the following ad hoc

DEFINITION 1.33. A subgroup $\Gamma \subset Aut(\mathbf{H}) = PSL_2(\mathbf{R})$ is called *$E-$arithmetic* if it is commensurable with a conjugate in $PSL_2(\mathbf{R})$ of the image of a group of the form $\mathcal{O}_B^1$.

Recall:

PROPOSITION 1.34. *Any $E-$arithmetic subgroup is a Fuchsian lattice. A Fuchsian lattice in $PSL_2(\mathbf{R})$ is arithmetic if and only if it is $E-$ arithmetic for some totally real E.*

PROOF. The first assertion is proved in [**95**], p. 255. The second assertion is stated in [**95**] p.253; the proof [**95**] is given only the in case of Kleinian groups but the proof in the Fuchsian case is similar. □

REMARK 1.35. Sometimes $\mathbf{Q}-$arithmetic Fuchsian lattices are called *rational arithmetic Fuchsian lattices*; cf. [**85**], p. 73. If Γ is a $\mathbf{Q}-$arithmetic subgroup of $PSL_2(\mathbf{R})$ then the Riemann surface $\mathbf{H}/\Gamma$ is sometimes called a *modular* (resp. *Shimura*) curve according as the quaternion $\mathbf{Q}-$algebra in question is split or not. These types of curves have natural coverings which are moduli spaces for elliptic (resp. "false elliptic") curves equipped with certain structures. We refer to Chapter 8 for a review of this.

For further developments it is also useful to introduce the following terminology.

DEFINITION 1.36. A correspondence $\mathbf{X}$ over $\mathbf{C}$ will be said to have a *$E-$ arithmetic analytic uniformization* if it is uniformized by a tuple as in Equation 1.28 with $\mathbf{S} = \mathbf{H}$, such that $\Gamma, \tilde{\Gamma}$ are $E-$ arithmetic.

Putting together Proposition 1.32, Proposition 1.34 and Remark 1.35 we see that, roughly speaking:

Dense correspondences with hyperbolic analytic uniformization "come from" quaternion algebras over totally real fields E. In case $E = \mathbf{Q}$ these correspondences are commensurable with correspondences between certain moduli spaces of elliptic (resp. false elliptic) curves.

Conversely it is well known that

Quaternion algebras over totally real fields naturally lead to dense correspondences admitting hyperbolic analytic uniformizations.

Let us recall how this can be done. Indeed let $G(1) := \mathcal{O}_B^1$ and let $\Gamma(1) \subset PSL_2(\mathbf{R})$ be its image. For any integer N we let

$$G(N) := \{g \in G(1) | g - 1 \in N\mathcal{O}_B\}$$

and we let $\Gamma(N) \subset \Gamma(1)$ be the image of $G(N)$; clearly $G(N)$ has finite index in $G(1)$ so $\Gamma(N)$ and $\Gamma(1)$ are commensurable E−arithmetic Fuchsian lattices. Note that the commensurator of $G(1)$ in the group B^+ is the whole of B^+. Indeed if $Mb, Mb^{-1} \in \mathcal{O}_B$ for some positive $M \in \mathbf{Z}$ then, for any positive $P \in \mathbf{Z}$

$$b \cdot G(M^2P) \cdot b^{-1} \subset G(P), \quad b^{-1} \cdot G(M^2P) \cdot b \subset G(P).$$

In particular

$$G(M^4) \subset b \cdot G(M^2) \cdot b^{-1} \subset G. \tag{1.32}$$

So $b \cdot G(M^2) \cdot b^{-1}$ is commensurable with G; since it is also commensurable with $b \cdot G \cdot b^{-1}$ it follows that G and $b \cdot G \cdot b^{-1}$ are commensurable.

Fix an integer N and an element $b \in B^+$. We shall attach to b and N a correspondence admitting an E−arithmetic analytic uniformization as follows. Let $\Gamma(N)$ be the image of $G(N)$ in $PSL_2(\mathbf{R})$. Since b is in the commensurator of $G(1)$ it is in the commensurator of $G(N)$. Let $\tilde{G} := G(N) \cap b^{-1} \cdot G(N) \cdot b$. Then both $\tilde{G}$ and $b \cdot \tilde{G} \cdot b^{-1}$ are subgroups of finite index in $G(N)$. Let $\Gamma(N)$ and $\tilde{\Gamma}$ be the images of $G(N)$ and $\tilde{G}$ in $PSL_2(\mathbf{R})$ and let τ be image of b in $PSL_2(\mathbf{R})$. So we may consider the correspondence

$$\mathbf{X} = \mathbf{X}_{b,N} := (\mathbf{H}/\Gamma(N), \mathbf{H}/\tilde{\Gamma}, \sigma_1, \sigma_2) \tag{1.33}$$

induced by the tuple $(\Gamma(N), \tilde{\Gamma}, id, \tau)$.

PROPOSITION 1.37. *The correspondence $\mathbf{X}_{b,N}$ admits an E−arithmetic analytic uniformization. If N is large enough then $\mathbf{X}_{b,N}$ is étale. If $b^n \notin E^\times \cdot \mathcal{O}_B^1$ for $n \geq 1$ then $\mathbf{X}_{b,N}$ is dense.*

PROOF. The first claim is clear. To check the second claim note that since $G(1)$ contains at most finitely many elliptic elements up to conjugation one can choose N such that $\Gamma(N)$ has no elliptic elements; so the projection $\mathbf{H} \to \mathbf{H}/\Gamma(N)$ is étale and $\mathbf{X}$ will be étale. To check the third claim note that for all $n \neq m$, all $\lambda \in E^\times$, and all $g \in \mathcal{O}_B^1$ we have $b^m \neq \lambda g b^n$; so for all but countably many $z \in \mathbf{H}$ we have $b^m z \neq g b^n z$. Hence for such z's the images in $\mathbf{H}/\Gamma(N)$ of $b^n z$ are distinct and give rise to infinite orbits for the equivalence relation generated by our correspondence. □

1.3.5. Critically finite correspondences.

In what follows we place ourselves in the category $\mathcal{C}_{\mathbf{C}}$ of schemes over the complex field.

We first recall some standard terminology from complex dynamics; cf. [**109**]. Let $\mathbf{P}^1 = \mathbf{P}^1_{\mathbf{C}}$ be the projective line and $\sigma : \mathbf{P}^1 \to \mathbf{P}^1$ a non-constant (hence finite) morphism. One defines the *critical locus* Ω_σ and the *postcritical locus* P_σ:

$$\Omega_\sigma := \{P \in \mathbf{P}^1 \mid \sigma \text{ not étale at } P\}, \quad P_\sigma := \bigcup_{n \geq 1} \sigma^n(\Omega_\sigma). \tag{1.34}$$

DEFINITION 1.38. σ is *postcritically finite* if the postcritical locus P_σ is finite.

Postcritically finite maps are "very rare". For quadratic maps

$$\sigma(t) = t^2 + c, \quad t \in \mathbf{C} = \mathbf{P}^1 \backslash \{\infty\}, \quad c \in \mathbf{C},$$

for instance, the critical locus is

$$\Omega_{t^2+c} = \{0, \infty\}.$$

So the postcritical locus is

$$P_{t^2+c} = \{\infty, c, c^2 + c, (c^2 + c)^2 + c, ...\}.$$

In this case we have the following well known facts:

LEMMA 1.39. *If P_{t^2+c} is finite then c is an algebraic integer all of whose conjugates have absolute value ≤ 2. In particular, if F is a number field, the set*

$$\{c \in F \mid P_{t^2+c} \text{ is finite}\}$$

is finite. If $F = \mathbf{Q}$ this set is $\{-2, -1, 0\}$.

PROOF. Finiteness of P_{t^2+c} implies that two of the elements in the sequence

$$c, c^2 + c, (c^2 + c)^2 + c, ... \tag{1.35}$$

must coincide hence c is a root of a monic polynomial with $\mathbf{Z}$−coefficients. Also if P_{t^2+c} is finite then so is $P_{t^2+\tilde{c}}$ for any conjugate $\tilde{c}$ of c. If $|\tilde{c}| > 2$ then it is trivial to check that the sequence

$$\tilde{c}, \tilde{c}^2 + \tilde{c}, (\tilde{c}^2 + \tilde{c})^2 + \tilde{c}, ...$$

tends to ∞, a contradiction. So $|\tilde{c}| \leq 2$. In case $F = \mathbf{Q}$, c is in $\{-2, -1, 0, 1, 2\}$. Clearly, however, the sequence 1.35 goes to infinity for $c = 1, 2$ so c must be $-2, -1$ or 0. □

DEFINITION 1.40. Two rational fractions $\sigma_1, \sigma_2 \in \mathbf{C}(t)$, $\sigma_1, \sigma_2 : \mathbf{P}^1 \to \mathbf{P}^1$, are *conjugate* (write $\sigma_1 \sim \sigma_2$) if there exists $\sigma_3 \in Aut(\mathbf{P}^1)$ such that $\sigma_1 = \sigma_3 \circ \sigma_2 \circ \sigma_3^{-1}$.

Let us quickly explain, following [**51**], an interesting connection between flat analytic uniformization and postcritical finiteness. Given a postcritically finite $\sigma : \mathbf{P}^1 \to \mathbf{P}^1$ we let $\mu_\sigma : \mathbf{P}^1 \to \{1, 2, 3, ..., \infty\}$ be the smallest among all functions ν such that $\mu(P) = 1$ for $P \notin P_\sigma$ and such that $\mu(Q)$ is a multiple of $\mu(P) \cdot deg_P\sigma$ for each $P \in \sigma^{-1}(Q)$. One defines the *orbifold* $O_\sigma := (\mathbf{P}^1, \mu_\sigma)$ attached to σ and the *Euler characteristic*

$$\chi(O_\sigma) := 2 - \sum_{P \in P_\sigma} \left(1 - \frac{1}{\mu_\sigma(P)}\right).$$

The orbifold O_σ is called *hyperbolic* if $\chi(O_\sigma) < 0$. On the other hand one can consider the correspondence

$$(\mathbf{P}^1, \mathbf{P}^1, id, \sigma)$$

over $\mathbf{C}$; of course, this is not an étale correspondence. For any non-empty Zariski open set $X \subset \mathbf{P}^1$ we can consider the correspondence $(X, X \cap \sigma^{-1}(X), \iota, \sigma)$ where ι is the inclusion. Here is an "easy" (but remarkable) result of Thurston's orbifold theory which can be rephrased as morally saying that "finite postcritical locus plus non-hyperbolic orbifold implies flat analytic uniformization". (We refer to Example 1.29 for the definition of Chebyshev polynomials $Cheb_d$ and Lattès functions Lat_d.)

PROPOSITION 1.41. **[51]** *Assume σ has degree ≥ 2 and is postcritically finite. Then $\chi(O_\sigma) \leq 0$. Assume in addition that $\chi(O_\sigma) = 0$. Then the postcritical locus P_σ has at most 4 points and either $\sigma \sim t^{\pm d}$ or $\sigma \sim \pm Cheb_d$ or $\sigma \sim Lat_d$. In particular there exists a non-empty Zariski open set $X \subset \mathbf{P}^1$ such that the correspondence $(X, X \cap \sigma^{-1}(X), \iota, \sigma)$ admits a flat analytic uniformization.*

PROOF. This follows by combining Proposition 9.2, p. 290, Proposition 9.3, p. 290, and Corollary 2.4, p. 269, in **[51]**. □

For a result in the converse direction see Proposition 1.48 below.

We would like to generalize some of the above concepts to the case of correspondences. Consider an irreducible one dimensional correspondence of finite type and smooth over $\mathbf{C}$,

$$\mathbf{Y} := (Y, \tilde{Y}, \sigma_1, \sigma_2). \tag{1.36}$$

(Note that in applications such $\mathbf{Y}$'s will appear as (partial) compactifications of the étale correspondences $\mathbf{X}$ which we are going to study; this is why we are using here a different letter to denote our correspondences.) In the discussion below we identify, for simplicity, the varieties with their sets of $\mathbf{C}-$points. For $i = 1, 2$ we define the *critical loci*

$$\Omega_{\sigma_i} = \{\tilde{P} \in \tilde{Y} \mid \sigma_i \text{ is not étale at } \tilde{P}\}. \tag{1.37}$$

Points in these critical loci will be referred to as *critical points*. Next, by a *chain* in $\mathbf{Y}$ we understand a sequence

$$\tilde{P}_1, ..., \tilde{P}_n \in \tilde{Y}, \quad n \geq 1$$

such that

$$\sigma_2(\tilde{P}_i) = \sigma_1(\tilde{P}_{i+1}), \quad 1 \leq i \leq n-1.$$

We say that a point $\tilde{P} \in \tilde{Y}$ *precedes* (resp. *strictly precedes*) a point $\tilde{Q} \in \tilde{Y}$ if there exists a chain $\tilde{P}_1, ..., \tilde{P}_n$ with $n \geq 1$ (resp. $n \geq 2$) such that $\tilde{P}_1 = \tilde{P}$ and $\tilde{Q} = \tilde{P}_n$. For any $\tilde{P} \in \tilde{Y}$ we define the *forward orbit*

$$O^+_{\sigma_2}(\tilde{P}) = \{Q \in Y \mid Q = \sigma_2(\tilde{Q}), \tilde{P} \text{ preceeds } \tilde{Q}\} \tag{1.38}$$

and the *backward orbit*

$$O^-_{\sigma_2}(\tilde{P}) = \{Q \in Y \mid Q = \sigma_1(\tilde{Q}), \tilde{Q} \text{ preceeds } \tilde{P}\}. \tag{1.39}$$

Say that a point $\tilde{P} \in \Omega_{\sigma_2}$ is *non-degenerate* if it is not preceded and does not precede any point in Ω_{σ_1}. Say that a point $\tilde{P} \in \Omega_{\sigma_2}$ is *primitive* if it is not strictly preceded by any point in Ω_{σ_2}.

DEFINITION 1.42. 1) We say $\mathbf{Y}$ is *postcritically finite* if for any non-degenerate $\tilde{P} \in \Omega_{\sigma_2}$ the orbit $O^+_{\sigma_2}(\tilde{P})$ is finite.

2) We say $\mathbf{Y}$ is *critically finite* if for any non-degenerate $\tilde{P} \in \Omega_{\sigma_2}$ either $O^+_{\sigma_2}(\tilde{P})$ or $O^-_{\sigma_2}(\tilde{P})$ is finite.

3) We say that $\mathbf{Y}$ is *primitively postcritically finite* if for any non-degenerate primitive $\tilde{P} \in \Omega_{\sigma_2}$ the orbit $O^+_{\sigma_2}(\tilde{P})$ is finite.

So the first of these properties implies the other two.

REMARK 1.43. A postcritically finite correspondence $\mathbf{Y}$ may possess points $\tilde{P} \in \Omega_{\sigma_2}$ for which $O^+_{\sigma_2}(\tilde{P})$ is infinite; such $\tilde{P}$'s are, of course degenerate (i.e. not non-degenerate). Here is a typical example. Consider the correspondence $\mathbf{Y} = (\mathbf{P}^1, \mathbf{P}^1, \sigma_1, \sigma_2)$ over $\mathbf{C}$ where

$$\sigma_1(t) = t^2 + c, \quad \sigma_2(t) = \sigma_1(t+1) = t^2 + 2t + 1 + c, \quad t \in \mathbf{C} = \mathbf{P}^1 \backslash \{\infty\},$$

$c \in \mathbf{C}$. Clearly

$$\Omega_{\sigma_2} = \{-1, \infty\}, \quad \Omega_{\sigma_1} = \{0, \infty\}.$$

Clearly ∞ is degenerate and

$$O^+_{\sigma_2}(\infty) = \{\infty\}.$$

On the other hand -1 is degenerate because

$$\sigma_2(-1) = c = \sigma_1(0), \quad 0 \in \Omega_{\sigma_1}.$$

So $\tilde{Y}$ is postcritically finite. Furthermore it is trivial to check that

$$O^+_{\sigma_2}(0) = \{c, c+1, c+4, c+9, c+16, ...\}$$

hence $O^+_{\sigma_2}(0)$ is infinite.

Incidentally, note that if

$$X := \mathbf{P}^1 \backslash \{0, \infty\}, \quad \tilde{X} := \mathbf{P}^1 \backslash \{0, -1, \infty\},$$

then the correspondence

$$\mathbf{X} := (X, \tilde{X}, \sigma_1, \sigma_2)$$

has a spherical analytic uniformization.

In the classical case of complex dynamics the 3 properties in Definition 1.42 mean the same thing as we show now. Indeed assume $Y = \tilde{Y} = \mathbf{P}^1$, $\sigma_1 = id$, $\sigma_2 =: \sigma$ a finite map. Then we have

$$O^+_{\sigma_2}(P) = \{\sigma^n(P) \mid n \geq 1\}, \tag{1.40}$$

$$O^-_{\sigma_2}(P) = \bigcup_{n \geq 0} \sigma^{-n}(P). \tag{1.41}$$

It is trivial to check ([4], p. 66) that if the backward orbit of a point $\bigcup_{n\geq 0} \sigma^{-n}(P)$ is finite then the forward orbit $\{\sigma^n(P) \mid n \geq 1\}$ is also finite. Hence we have:

PROPOSITION 1.44. *Assume $\sigma : \mathbf{P}^1 \to \mathbf{P}^1$ is a non-constant morphism and $\mathbf{Y} := (\mathbf{P}^1, \mathbf{P}^1, id, \sigma)$. Then the following are equivalent:*

1) $\mathbf{Y}$ is postcritically finite;
2) $\mathbf{Y}$ is critically finite;
3) $\mathbf{Y}$ is primitively postcritically finite;
4) σ is postcritically finite.

More generally, we have the following:

PROPOSITION 1.45. *Assume $\mathbf{Y} = (Y, \tilde{Y}, \sigma_1, \sigma_2)$ is smooth over $\mathbf{C}$, irreducible, critically finite, with σ_2 proper and*

$$deg\ \sigma_1 < deg\ \sigma_2.$$

Then $\mathbf{Y}$ is primitively postcritically finite.

PROOF. Let $\tilde{P} \in \Omega_{\sigma_2}$ be non-degenerate and primitive and set $Z := O^{-}_{\sigma_2}(\tilde{P})$. Then, clearly,

$$\sigma_1(\sigma_2^{-1}(Z)) \subset Z. \tag{1.42}$$

If Z is infinite then $O^{+}_{\sigma_2}(\tilde{P})$ is finite and we are done. Assume Z is finite. Since $\tilde{P}$ is primitive it follows that for any $Q \in Z$ and any $\tilde{Q} \in \sigma_2^{-1}(Q)$ we have $\tilde{Q} \notin \Omega_{\sigma_2}$. Since σ_2 is proper,

$$\sharp(\sigma_2^{-1}(Z)) = (deg\ \sigma_2) \cdot (\sharp Z).$$

But then

$$\sharp(\sigma_1(\sigma_2^{-1}(Z))) \geq \frac{deg\ \sigma_2}{\deg\ \sigma_1} \cdot \sharp(Z) > \sharp(Z),$$

which contradicts Equation 1.42. □

Next we give two general recipes to construct postcritically finite correspondences; cf the next two Propositions.

PROPOSITION 1.46. *Assume $\mathbf{Y}' \to \mathbf{Y}$ is a Galois cover of one dimensional irreducible correspondences of finite type and smooth over $\mathbf{C}$ with $\mathbf{Y}'$ étale. Then $\mathbf{Y}$ is postcritically finite.*

PROOF. Denote, as usual, by Ω with index a map the set of critical points of that map. Let

$$(\pi, \tilde{\pi}) : \mathbf{Y}' := (Y', \tilde{Y}', \tau_1, \tau_2) \to \mathbf{Y} := (Y, \tilde{Y}, \sigma_1, \sigma_2)$$

be the Galois cover and set

$$Z := \tilde{\pi}(\tau_2^{-1}(\Omega_\pi)).$$

Claim 1. $\Omega_{\sigma_2} \subset Z$. Indeed if $\tilde{P} \in \Omega_{\sigma_2}$ and $\tilde{P} = \tilde{\pi}(\tilde{Q})$ then

$$\tilde{Q} \in \Omega_{\sigma_2 \circ \tilde{\pi}} = \Omega_{\pi \circ \tau_2} \tag{1.43}$$

hence, since τ_2 is étale, $\tau_2(\tilde{Q}) \in \Omega_\pi$, so $\tilde{P} \in \tilde{\pi}(\tau_2^{-1}(\Omega_\pi)) = Z$.

Claim 2. $\tau_2(\Omega_{\tilde{\pi}}) \subset \Omega_\pi$. Indeed if $\tilde{Q} \in \Omega_{\tilde{\pi}}$ then Equation 1.43 holds again hence $\tau_2(\tilde{Q}) \in \Omega_\pi$.

Claim 3. $\pi^{-1}(\pi(\Omega_\pi)) = \Omega_\pi$. This is clear because π is Galois.

Claim 4. If $\tilde{P}_1, ..., \tilde{P}_n$ is a chain in $\mathbf{Y}$ with $\tilde{P}_1 \in \Omega_{\sigma_2}$ non-degenerate then $\tilde{P}_n \in Z$. Note that Claim 4 ends the proof of the Proposition because if $P \in \Omega_{\sigma_2}$ is non-degenerate then $O^{+}_{\sigma_2}(\tilde{P}) \subset \sigma_2(Z)$. To check Claim 4 we proceed by induction on n. If $n = 1$ we are done by Claim 1. Now assume we have a chain $\tilde{P}_1, ..., \tilde{P}_n$ in $\mathbf{Y}$ with $\tilde{P}_1 \in \Omega_{\sigma_2}$ non-degenerate. By induction $\tilde{P}_{n-1} \in Z$. So

$$\tilde{P}_{n-1} = \tilde{\pi}(\tilde{Q}_{n-1}), \quad \tau_2(\tilde{Q}_{n-1}) \in \Omega_\pi.$$

On the other hand we have

$$\sigma_1(\tilde{P}_n) = \sigma_2(\tilde{P}_{n-1}) = \sigma_2(\tilde{\pi}(\tilde{Q}_{n-1})) = \pi(\tau_2(\tilde{Q}_{n-1})) \in \pi(\Omega_\pi).$$

Write $\tilde{P}_n = \tilde{\pi}(\tilde{Q}_n)$. Hence

$$\sigma_1(\tilde{P}_n) = \sigma_1(\tilde{\pi}(\tilde{Q}_n)) = \pi(\tau_1(\tilde{Q}_n)).$$

Therefore

$$\tau_1(\tilde{Q}_n) \in \pi^{-1}(\pi(\Omega_\pi)) = \Omega_\pi,$$

by Claim 3. In particular we get

$$\tilde{Q}_n \in \Omega_{\pi \circ \tau_1} = \Omega_{\sigma_1 \circ \tilde{\pi}}.$$

Since $\tilde{P}_1$ is non-degenerate

$$\tilde{P}_n = \tilde{\pi}(\tilde{Q}_n) \notin \Omega_{\sigma_1}.$$

We get $\tilde{Q}_n \in \Omega_{\tilde{\pi}}$ and hence

$$\tilde{P}_n \in \tilde{\pi}(\Omega_{\tilde{\pi}}) \subset \tilde{\pi}(\tau_2^{-1}(\Omega_{\pi})) = Z,$$

by Claim 2, so we are done. □

PROPOSITION 1.47. *Assume* $\mathbf{Y}^* = (Y^*, \tilde{Y}^*, \sigma_1, \sigma_2)$ *is a one dimensional irreducible correspondence of finite type and smooth over* $\mathbf{C}$ *and assume* $Y \subset Y^*$ *is an open set such that*

$$\sigma_1^{-1}(Y) = \sigma_2^{-1}(Y) =: \tilde{Y}$$

and such that the correspondence

$$\mathbf{Y} := (Y, \tilde{Y}, \sigma_1, \sigma_2)$$

is critically finite (postcritically finite, primitively postcritically finite respectively). Then $\mathbf{Y}^*$ *is critically finite (postcritically finite, primitively postcritically finite respectively).*

PROOF. If $\tilde{P} \in \tilde{Y}^* \backslash \tilde{Y}$ then any chain in $\mathbf{Y}^*$ containing $\tilde{P}$ is contained in $\tilde{Y}^* \backslash \tilde{Y}$ so $O^+_{\sigma_2}(\tilde{P})$ and $O^-_{\sigma_2}(\tilde{P})$ are finite. If $\tilde{P} \in \tilde{Y}$ then any chain in $\mathbf{Y}^*$ containing $\tilde{P}$ is contained in $\tilde{Y}$ and we are done again. □

The Proposition below essentially says that "analytic uniformization implies postcritical finiteness". To explain this we go back to the general picture of analytic uniformization by considering a tuple

$$(\mathbf{S}, \Gamma, \tilde{\Gamma}, \tau_1, \tau_2) \tag{1.44}$$

where $\mathbf{S}$ is a simply connected Riemann surface, Γ and $\tilde{\Gamma}$ are admissible groups of automorphisms of $\mathbf{S}$, and τ_1, τ_2 are automorphisms of $\mathbf{S}$ such that $\tau_1 \tilde{\Gamma} \tau_1^{-1}$ and $\tau_2 \tilde{\Gamma} \tau_2^{-1}$ are subgroups of finite index in Γ. In the hyperbolic case $\mathbf{S} = \mathbf{H}$ we also assume Γ and $\tilde{\Gamma}$ are arithmetic. Denote by $\bar{\tau}_1, \bar{\tau}_2 : \mathbf{S}/\tilde{\Gamma} \to \mathbf{S}/\Gamma$ the maps induced by τ_1 and τ_2 respectively.

PROPOSITION 1.48. *The correspondence*

$$\mathbf{Y} := (\mathbf{S}/\Gamma, \mathbf{S}/\tilde{\Gamma}, \bar{\tau}_1, \bar{\tau}_2).$$

is postcritically finite. Moreover the following hold:

1) If in the flat case $\mathbf{S} = \mathbf{C}$ *we assume* $\tau = id$, $\Gamma = \tilde{\Gamma}$, *and* Γ *does not consist of translations only then* $\mathbf{C}/\Gamma$ *is an open set in* $\mathbf{P}^1$ *and the map* $\sigma : \mathbf{P}^1 \to \mathbf{P}^1$ *induced by* σ_2 *is postcritically finite.*

2) If, in the hyperbolic case $\mathbf{S} = \mathbf{H}$, $\mathbf{H}/\Gamma$ *is not compact and if* $\mathbf{H}^* := \mathbf{H} \cup \mathbf{P}^1(\mathbf{Q})$ *is the upper half plane with cusps added and*

$$\mathbf{Y}^* := (\mathbf{H}^*/\Gamma, \mathbf{H}^*/\tilde{\Gamma}, \bar{\tau}_1, \bar{\tau}_2),$$

then $\mathbf{Y}^*$ *is postcritically finite.*

PROOF. We may assume $\tau_1 = id$ so $\tilde{\Gamma} \subset \Gamma$ has finite index and $\tau_2 \tilde{\Gamma} \tau_2^{-1} \subset \Gamma$ of finite index. One can find a normal subgroup $\Gamma' \subset \Gamma$ of finite index which acts freely on $\mathbf{S}$. Then

$$\tilde{\Gamma}' := \tilde{\Gamma} \cap \Gamma'$$

has finite index in both Γ' and $\tilde{\Gamma}$, in particular $\tau_2\tilde{\Gamma}'\tau_2^{-1}$ has finite index in Γ'. So there is a Galois cover $\mathbf{Y}' \to \mathbf{Y}$ with $\mathbf{Y}'$ étale given by

$$\mathbf{Y}' := (\mathbf{S}/\Gamma', \mathbf{S}/\tilde{\Gamma}', \tau_1', \tau_2')$$

so, by Proposition 1.46, $\mathbf{Y}$ is postcritically finite. The additional assertion 2 follows by applying Proposition 1.47 directly. To check the additional assertion 1 note that $\mathbf{C}/\Gamma$ is either $\mathbf{P}^1$ (and then there is nothing to prove) or $\mathbf{C}^\times = \mathbf{C}\backslash\{0\}$ or $\mathbf{C}$. Then assertion 2 follows by applying Proposition 1.47 to $Y^* = \tilde{Y}^* = \mathbf{P}^1$. □

1.3.6. Triviality of pluricanonical bundles mod p. Let k be an algebraically closed field; in our applications k will be either the complex field $\mathbf{C}$ or the algebraic closure of a finite field. Let $k(t)$ be the field of rational functions in the variable t and let $\sigma(t) \in k(t)$ be a non-constant rational function.

DEFINITION 1.49. An *invariant form of weight* $\nu \in \mathbf{Z}$ for $\sigma(t)$ is a rational function $f(t) \in k(t)$ such that

$$f(\sigma(t)) = \left(\frac{d\sigma}{dt}(t)\right)^{-\nu} \cdot f(t) \tag{1.45}$$

If we consider the ν–tuple differential form $\omega_f := f(t)\cdot(dt)^\nu$ then Equation 1.45 can be written as $\sigma^*\omega_f = \omega_f$ which actually justifies our terminology. Alternatively we will refer to ω_f as an *invariant form of weight* $\nu \in \mathbf{Z}$.

On the other hand for $\sigma(t) \in k(t)$ non-constant, $\sigma : \mathbf{P}^1 \to \mathbf{P}^1$, and for $X \subset \mathbf{P}^1$ an open set, we may consider the correspondence over k

$$\mathbf{X} := (X, \sigma^{-1}(X)\cap X, \iota, \sigma),$$

where ι is the inclusion. Assume $\mathbf{X}$ is étale so one can consider the canonical bundle $\mathbf{K}$ on $\mathbf{X}$. Then, by Lemma 1.12, the following are equivalent:

1) There exists a non-empty open subcorrespondence $\mathbf{U} \subset \mathbf{X}$ such that the ν–canonical bundle $\mathbf{K}_\mathbf{U}^\nu$ is trivial;

2) σ has a non-zero invariant form of weight ν.

The next Proposition shows that there are no non-zero invariant forms of non-zero weight in characteristic zero.

PROPOSITION 1.50. *If $k = \mathbf{C}$ is the complex field and σ has degree ≥ 2 then σ does not admit any non-zero invariant form of non-zero weight. In particular if $0 \neq \nu \in \mathbf{Z}$ the pluricanonical bundle $\mathbf{K}^\nu$ of $\mathbf{X}$ is not trivial.*

PROOF. Indeed if Equation 1.45 holds for some $f \neq 0$ and $\nu \neq 0$ then the same equation holds with σ replaced by the n–th iterate σ^n for any n. Now this equation for the iterates implies that any non-parabolic periodic point $\neq \infty$ of σ is either a zero or a pole of f; recall from [**109**], p. 99, that a point $\alpha \in \mathbf{C}$ is a *parabolic periodic point* if there exists $n \geq 1$ such that

$$\sigma^n(\alpha) = \alpha, \quad |(d\sigma^n/dt)(\alpha)| = 1.$$

On the other hand by [**109**], pp. 47 and 143, there are infinitely many non-parabolic periodic points, a contradiction. □

In contrast with the above Proposition there are interesting situations when one has "invariant forms modulo p" for almost all primes p. To explain this let $F \subset \mathbf{C}$ be a number field (always assumed of finite degree over the rationals) and let $\wp$ be a place of F (always assumed finite). Let $\mathcal{O}_\wp$ denote the localization of the ring of integers $\mathcal{O}_F$ at (the maximal ideal corresponding to) $\wp$, let $\kappa_\wp$ be the residue field of $\mathcal{O}_\wp$ and let $k_\wp = \kappa_\wp^a$ be an algebraic closure of $\kappa_\wp$. Now if $\sigma(t) \in \mathbf{C}(t)$ has coefficients in F then for all except finitely many places $\wp$ of F we may consider the rational function $\bar{\sigma}_\wp(t) \in \kappa_\wp(t) \subset k_\wp(t)$ obtained by writing $\sigma(t) = P(t)/Q(t)$, $P(t), Q(t) \in \mathcal{O}_\wp[t]$, with $Q(t)$ primitive, and then reducing the coefficients of $P(t)$ and $Q(t)$ modulo the maximal ideal of $\mathcal{O}_\wp$. It turns out that there are non-trivial examples of $\sigma(t)$'s such that $\bar{\sigma}_\wp(t)$ has non-zero invariant forms of non-zero weight for all but finitely many places $\wp$ of F. Indeed we have the following:

EXAMPLE 1.51. 1) *Multiplicative functions.* Let $\sigma(t) = t^{\pm d}$, $d \geq 2$. Then for all prime integers p not dividing d the rational function $\bar{\sigma}_p(t) \in \mathbf{F}_p^a(t)$ possesses a non-zero invariant form of weight $p-1$,

$$\omega_f = \left(\frac{dt}{t}\right)^{p-1}.$$

2) *Chebyshev polynomials.* Let $Cheb_d(t) \in \mathbf{C}[t]$ be the Chebyshev polynomial of degree d; cf. Definition 1.29. Let $\sigma(t) = \pm Cheb_d(t)$. Then for all odd prime integers p not dividing d the polynomial $\bar{\sigma}_p(t) \in \mathbf{F}_p^a[t]$ possesses a non-zero invariant form of weight $p-1$,

$$\omega_f = \frac{(dt)^{p-1}}{(t^2-4)^{(p-1)/2}}.$$

3) *Lattès functions.* Let $\sigma(t) \in \mathbf{C}(t)$ be a Lattès function attached to an elliptic curve defined over a number field F; cf. Cf Definition 1.29. Then we claim $\bar{\sigma}_\wp(t)$ has a non-zero invariant form of non-zero weight for all but finitely many places $\wp$ of F. Indeed, using the notations in Definition 1.29, if ω is a non-zero global 1−form on E defined over F and $e = \sharp(\Gamma)$ then $\omega^e = \pi^*\eta$ where $\pi : E \to E/\Gamma$ is the canonical projection and $\eta \in F(t) \cdot (dt)^e$. Now

$$\pi^*\sigma^*\eta = \tau^*\pi^*\eta = (\tau^*\omega)^e = \lambda\omega^e = \pi^*(\lambda\eta)$$

for some $\lambda \in F^\times$. Hence $\sigma^*\eta = \lambda\eta$ hence

$$\bar{\sigma}_\wp^*\eta_\wp^{q_\wp - 1} = \eta_\wp^{q_\wp - 1}$$

for all but finitely many $\wp$'s, where $q_\wp$ is the size of the residue field $\kappa_\wp$.

Conversely we have the following:

THEOREM 1.52. [**27**] *Assume $\sigma(t) \in \mathbf{C}(t)$ has coefficients in a number field F and has degree $d \geq 2$. Assume that $\bar{\sigma}_\wp(t) \in k_\wp(t)$ admits a non-zero invariant form of non-zero weight for infinitely many places $\wp$ of F. Then either $\sigma \sim t^{\pm d}$ or $\sigma \sim \pm Cheb_d$ or $\sigma \sim Lat_d$, where $Cheb_d$ and Lat_d are the Chebyshev polynomial and a Lattès function respectively.*

The idea of the proof in [**27**] is to show that if $\bar{\sigma}_\wp(t) \in k_\wp(t)$ admits a non-zero invariant form of non-zero weight for infinitely many places $\wp$ then $\bar{\sigma}_\wp(t)$ has a very special kind of ramification. Then $\sigma(t)$ itself will have the same kind of ramification. This implies in particular that $\sigma(t)$ is postcritically finite with orbifold satisfying $\chi(O_\sigma) = 0$ and one concludes by Proposition 1.41.

CHAPTER 2

Outline of δ−geometry

In Section 1 of this Chapter we introduce the basic concepts and foundational results of arithmetic differential algebra and its associated δ−geometry, by sometimes referring for proofs to later Chapters. We begin by defining *δ−rings* and *δ−ringed sets*. Our *δ−rings* are simply rings equipped with an operator that satisfies rules identical to that of a "Fermat quotient operator"; such operators will be referred to as *p−derivations*. On the other hand *δ−ringed sets* are a naive δ−analogue of ringed spaces. (We could have developed a more "sophisticated" approach based on *δ−ringed spaces* i.e. topological spaces equipped with a sheaf of δ−rings; but such an undertaking would have led us into a lot of general nonsense. We opted for the naive concept of *δ−ringed set* for the sake of economy and focus.) The next move will be to associate *δ−ringed sets* to schemes; to achieve this we first introduce *δ−functions* and *δ−line bundles* on schemes, then we construct certain natural *δ−graded rings* of *δ−sections* of δ−line bundles and, finally, we attach δ−ringed sets to these δ−graded rings in a way similar to the way one attaches algebraic varieties to graded rings in usual algebraic geometry. In Section 2 we state our main conjectures which relate δ−geometry and analytic uniformization. In Section 3 we state a sample of our main results. In Section 4 (which is not essential for the rest of the book) we prove that arithmetic differential algebra, along with differential algebra, difference algebra, and one more type of algebra (to be described there) are "locally" the only 4 possible ways to "adjoin an operator" to usual algebra (in a sense to be made precise).

2.1. Main concepts of δ−geometry

2.1.1. δ−rings. Fix a prime integer p. In view of our applications we will always assume $p \geq 5$. Later we shall vary p. Consider the polynomial with $\mathbf{Z}$−coefficients

$$C_p(X,Y) := \frac{X^p + Y^p - (X+Y)^p}{p} \in \mathbf{Z}[X,Y].$$

DEFINITION 2.1. A *p−derivation* $\delta : A \to B$ relative to a ring homomorphism $f : A \to B$ is a (set theoretic) map satisfying $\delta(1) = 0$ and

$$\delta(x+y) = \delta(x) + \delta(y) + C_p(f(x), f(y)),$$
$$\delta(xy) - f(x)^p \cdot \delta(y) + f(y)^p \cdot \delta(x) + p \cdot \delta(x) \cdot \delta(y)$$

for all $x, y \in A$.

The above definition (proposed independently by Joyal [**79**] and the author [**17**]) might seem, at first sight, rather arbitrary; in Section 4 we will show that, actually, this definition is "forced upon us" by some very natural and general axioms.

A map $\delta : A \to B$ is a p-derivation relative to a ring homomorphism $f : A \to B$ if and only if the map

$$(f, \delta) : A \to W_2(B) = B \times B, \; x \mapsto (f(x), \delta(x))$$

into the ring of Witt vectors of length 2 on B is a ring homomorphism. If $\delta : A \to B$ is a p-derivation then we always denote by $\phi : A \to B$, the map

$$\phi(x) := f(x)^p + p \cdot \delta(x);$$

clearly ϕ is a ring homomorphism satisfying $\phi(x) \equiv f(x)^p$ mod p in B for all $x \in A$. Conversely if one is given an A-algebra $f : A \to B$ with p a non-zero divisor in B and if one is given a ring homomorphism $\phi : A \to B$ which satisfies $\phi(x) \equiv f(x)^p$ mod p in B for all $x \in A$ then the formula $p \cdot \delta(x) := \phi(x) - f(x)^p$ uniquely defines a p-derivation δ relative to f. (So p-derivations δ with target in which p is a non-zero divisor are in bijection with "liftings of Frobenius" ϕ; due to a "renormalization" process to be discussed presently it is however important to stick with p-derivations rather than pass to liftings of Frobenius.) We sometimes write

$$x^\phi := \phi(x), \quad \delta x := \delta(x).$$

Let $R = R_p := \hat{\mathbf{Z}}_p^{ur}$ be the completion of the maximum unramified extension of $\mathbf{Z}_p$; this is the unique complete local p-ring with residue field $k := R/pR = \mathbf{F}_p^a$. This ring has a unique automorphism $\phi : R \to R$ lifting the p-power Frobenius automorphism of k hence R has a unique p-derivation; this p-derivation is defined by

$$\delta(x) = \frac{\phi(x) - x^p}{p}, \quad x \in R. \tag{2.1}$$

We let $K = R[1/p]$ be the quotient field of R.

From now on, unless stated otherwise, R, k, K will have the above meaning; this convention will be valid throughout the book.

DEFINITION 2.2. A *δ-ring* is a pair (A, δ) consisting of a ring A and a p-derivation $\delta : A \to A$ (relative to the identity).

Sometimes we abbreviate (A, δ) by A if δ is clear from the context. The p-adic completion of a δ-ring has a natural structure of δ-ring. By a δ-subring of a δ-ring A we understand a subring $B \subset A$ such that $\delta B \subset B$. A *δ-ring homomorphism* between two δ-rings will mean a morphisms of rings $u : A \to B$ that commutes with δ, i.e. $u \circ \delta = \delta \circ u$. We have at our disposal the δ-ring R which we refer to as our *ground* δ-ring. By a *δ-ring over R* we understand a δ-ring A together with a δ-ring homomorphism $R \to A$. A morphisms of δ-rings over R is a δ-ring homomorphism which is a morphism of R-algebras. We may consider the category

$$\mathcal{R}_\delta := \{\delta\text{-rings over } R\}; \tag{2.2}$$

this category has direct limits and inverse limits.

All δ-rings and δ-ring homomorphisms will be assumed, from now on, to be "over R"; all δ-subrings will also be "over R", i.e. the inclusion will be "over R".

EXAMPLE 2.3. The ring

$$\mathcal{F}(X) := \{f : X \to R\} \tag{2.3}$$

of all functions f from a set X into R is a $\delta-$ring via the rule $\delta f := \delta \circ f$. As usual we set $\mathcal{F}(\emptyset) = 0$.

For the next definition recall that $R[T_1, ..., T_n]\hat{}$, the $p-$adic completion of the ring $R[T_1, ..., T_n]$ of polynomials, can be identified with the ring of all power series in $R[[T_1, ..., T_n]]$ which are restricted (i.e. whose coefficients tend to 0 $p-$adically).

DEFINITION 2.4. A function $g : R^d \to R$ is called a *$\delta-$polynomial function* of order $\leq r$ (respectively a *restricted $\delta-$function* of order $\leq r$) if there exists a polynomial

$$G \in R[T_1, ..., T_{d(r+1)}]$$

(respectively a restricted power series

$$G \in R[T_1, ..., T_{d(r+1)}]\hat{}\ )$$

such that for all $(x_1, ..., x_d) \in R^d$ we have

$$g(x_1, ..., x_d) = G(x_1, ..., x_d, \delta x_1, ..., \delta x_d, ..., \delta^r x_1, ..., \delta^r x_d). \tag{2.4}$$

We will show later that:

PROPOSITION 2.5. *G is uniquely determined by g. Also a function $g : R^d \to R$ is a restricted $\delta-$function if and only if it is a $p-$adic limit, in $\mathcal{F}(R^d)$, of $\delta-$polynomial functions.*

PROOF. Cf. Corollary 3.21 below. □

A function $R^d \to R$ will be called a *$\delta-$polynomial function* (respectively a *restricted $\delta-$ function*) if it is a $\delta-$polynomial function (respectively a restricted $\delta-$ function) of order $\leq r$ for some $r \geq 0$. Then the ring of $\delta-$polynomial functions and the ring of restricted $\delta-$functions $R^d \to R$ are easily checked to be $\delta-$subrings of the $\delta-$ring $\mathcal{F}(R^d)$ of all functions $R^d \to R$.

The above concept of $\delta-$polynomial function is a direct analogue of the Ritt-Kolchin concept of $\delta-$polynomial function in the context of differential algebra [**117**], [**84**]. On the other hand $\delta-$polynomial functions, and more generally restricted $\delta-$functions, are analogues of Lagrangians in classical mechanics.

REMARK 2.6. Here are some comments on the relationship between the arithmetic differential algebraic concepts above and the *difference algebra* of Ritt - Cohn [**35**] and Chatzidakis - Hrushovski [**33**]. Assume R and $\phi : R \to R$ are as above. Following difference algebra terminology let us say that a function $h : R^d \to R$ is a *$\phi-$polynomial function* (respectively a *restricted $\phi-$function*) if there exists a polynomial

$$H \in R[T_1, ..., T_{d(r+1)}]$$

(respectively a restricted power series

$$H \in R[T_1, ..., T_{d(r+1)}]\hat{}\ )$$

such that for all $(x_1, ..., x_d) \in R^d$ we have

$$h(x_1, ..., x_d) = H(x_1, ..., x_d, \phi(x_1), ..., \phi(x_d), ..., \phi^r(x_1), ..., \phi^r(x_d)). \tag{2.5}$$

We will show later, cf. Corollary 3.21, that a function $g : R^d \to R$ is a $\delta-$polynomial function if and only if there exists $\nu \in \mathbf{Z}_+$ such that $p^\nu g$ is a $\phi-$polynomial function.

But restricted $\delta-$functions cannot, in general, be multiplied by a power of p to get a restricted $\phi-$function. Morally this is what happens. If one considers a restricted $\delta-$function $g : R^N \to R$ as in Equation 2.4 and one "formally" replaces the values of $\delta^i x_j$ in G by polynomials with $K-$coefficients in

$$x_j, \phi(x_j), ..., \phi^i(x_j)$$

then one gets a "formal" equality

$$g(x_1, ..., x_d) = \tilde{G}(x_1, ..., x_d, \phi(x_1), ..., \phi(x_d), ..., \phi^r(x_1), ..., \phi^r(x_d)) \tag{2.6}$$

where $\tilde{G} \in K[[T_1, ..., T_{d(r+1)}]]$ is a formal power series which may be (and in all interesting examples is) divergent on $R^{d(r+1)}$. (Of course, in view of divergence, Equation 2.6 is meaningless as an equality on R^d.) So, in some sense, arithmetic differential algebra is a "regularized version" of difference algebra: it is obtained from difference algebra by adjoining certain "divergent series".

DEFINITION 2.7. Let $End(R)$ be the (non-commutative) ring of additive endomorphism of R and denote by $W = \mathbf{Z}[\phi]$ the (commutative) $\mathbf{Z}-$subalgebra of $End(R)$ generated by ϕ; we refer to W as the ring of *weights*.

It turns out that ϕ is algebraically independent over $\mathbf{Z}$. For $w = w(\phi) = \sum a_i\phi^i \in W$ and $s \in \mathbf{Z}$ we set $w(s) = \sum a_i s^i$. We set

$$deg(w) := \sum a_i = w(1). \tag{2.7}$$

(So *deg* here is *not* the degree of w as a polynomial but rather as an element in a semigroup ring.) We define the *order* of $w \neq 0$ by:

$$ord(w) := \max\{i | a_i \neq 0\}; \tag{2.8}$$

if $w = 0$ we set $ord(w) = 0$. We also set, for $w \in W \backslash pW$,

$$ord(\bar{w}) = \max\{i | a_i \not\equiv 0 \mod p\}; \tag{2.9}$$

we set $ord(\bar{w}) = 0$ if $w \in pW$. We define

$$W(r) := \{w \in W | ord(w) \leq r\}, \tag{2.10}$$

$$W_+ := \{\sum a_i\phi^i \in W | a_i \geq 0\}. \tag{2.11}$$

Note that W_+ has a structure of additive monoid and also a structure of multiplicative monoid. If $w = \sum a_i\phi^i \in W_+$ and $\lambda \in A$ where A is a ring (possibly without $1-$element) on which ϕ acts as a ring endomorphism then define

$$\lambda^w = \prod_{i=0}^{s} (\phi^i(\lambda))^{a_i}; \tag{2.12}$$

the same definition makes sense for $w \in W$ if we assume (A has a $1-$element and) $\lambda \in A^\times$ is invertible.

DEFINITION 2.8. By a *local* (respectively *global*) $\delta - p-$*ring* we understand a $\delta-$ring which is a local (respectively global) $p-$ring.

Note that R is a local $\delta - p-$ring. Also we shall prove that the $\delta-$ring of restricted $\delta-$functions $R^N \to R$ is a global $\delta - p-$ring; cf. Corollary 3.21 below.

REMARK 2.9. Let A be a global $\delta - p$–ring. Then the localization $A_{(p)}$ has a natural structure of a local $\delta - p$–ring defined by the formula

$$\delta\left(\frac{a}{s}\right) := \frac{s^p \delta a - a^p \delta s}{s^{\phi+p}}.$$

DEFINITION 2.10. Let A be a local $\delta - p$–ring (as usual, over R), with residue field $k_A = A/pA$, and let $u \in A^n$ be an n–tuple of elements in A, $n \geq 1$. We say u is δ–*independent* if the reductions mod p,

$$\bar{u}_0, \bar{u}_1, \bar{u}_2, ... \in k_A^n \tag{2.13}$$

of the tuples $u, \delta u, \delta^2 u, ... \in A^n$ form an algebraically independent family over $k = R/pR$. We say that u δ–*generates* k_A if k_A is generated as a field extension of k by the elements 2.13. We say u is a δ–*basis* of A if it is δ–independent and δ–generates k_A. We also say that the empty set is a δ–basis of A if and only if $k_A = k$. If u is a δ–basis of A then $u, \delta u, \delta^2 u, ...$ are algebraically independent over R and if we set

$$R\{u\} := R[u, \delta u, \delta^2 u, ...]$$

then $R\{u\}$ is a δ–subring of A, it is a global $\delta - p$–ring and we have a natural isomorphism of local $\delta - p$–rings

$$(R\{u\}_{(p)})\hat{} \simeq \hat{A}.$$

DEFINITION 2.11. Assume $f : A \to B$ is a δ–ring homomorphism. By a δ–*derivation* $D : A \to B$ we mean an R–derivation satisfying

$$D\delta x = p\delta Dx + (Dx)^p - f(x)^{p-1} Dx \tag{2.14}$$

for all $x \in A$. Assume in addition B is flat over R. Then condition 2.14 is equivalent to the condition

$$D \circ \phi = p \cdot \phi \circ D. \tag{2.15}$$

In particular the set $Der_\delta(A, B)$ of all δ–derivations is an Abelian subgroup of $Der_R(A, B)$. Note also that the ring of dual numbers $B[\epsilon] = B \oplus \epsilon B$ has a unique structure of δ–ring such that B is a δ–subring of $B[\epsilon]$ and $\delta\epsilon = \epsilon$; a map $D : A \to B$ is a δ–derivation if and only if the map $f + \epsilon D : A \to B[\epsilon]$ is a δ–ring homomorphism.

DEFINITION 2.12. Let A be a δ–ring. By a ϕ–module over A we understand an A–module M together with an additive map $\phi : M \to M$ such that $\phi(am) = \phi(a)\phi(m)$ for $a \in A$ and $m \in M$.

REMARK 2.13. Any correspondence $\mathbf{A}$ in $\mathcal{R}_\delta^{op}$ has a categorical quotient. Indeed if $\mathbf{A} = (A, \tilde{A}, \sigma_1, \sigma_2)$ then we can view $\mathbf{A}$ as a correspondence in $\mathcal{R}^{op}$ (where $\mathcal{R}$ is the category of R–algebras) and hence we can consider the ring $H^0(\mathbf{A})$ and the $H^0(\mathbf{A})$–module $H^1(\mathbf{A})$; cf. Definition 1.18. Then it is clear that $H^0(\mathbf{A})$ is a δ–subring of A and $H^1(\mathbf{A})$ is a ϕ–module over $H^0(\mathbf{A})$. Moreover $\mathbf{A}$ has categorical quotient $A/\sigma = H^0(\mathbf{A})$ in $\mathcal{R}_\delta^{op}$.

2.1.2. δ-ringed sets. In what follows we would like to globalize the concept of δ-ring. Recall that a *ringed space over* R is a topological space equipped with a sheaf in the category $\mathcal{R}$ of rings over R. A natural way to proceed would be to consider δ-*ringed spaces* defined to be topological spaces equipped with a sheaf in the category $\mathcal{R}_\delta$ of δ-rings over R. However this approach would lead us into a lot of general nonsense. In what follows we adopt a more naive (and more economical) approach, which seems nevertheless sufficient for all practical purposes. Here is the key definition.

DEFINITION 2.14. A δ-*ringed set* X_δ consists of a set X_{set} together with the following data:

1) a family $(X_s)_{s\in S}$ of non-empty subsets of X_{set}, where S is a monoid and $X_{st} = X_s \cap X_t$ for all $s, t \in S$;

2) a family $(\mathcal{O}_s)_{s\in S}$, where $\mathcal{O}_s$ are δ-subrings of the ring, $\mathcal{F}(X_s)$, of R-valued functions on X_s such that for any $s, t \in S$ the restriction map $\mathcal{F}(X_s) \to \mathcal{F}(X_{st})$ maps $\mathcal{O}_s$ into $\mathcal{O}_{st}$.

The set X_{set} is called the *underlying set* of X_δ and its elements are called *points.* We also say X_δ is a δ-*ringed set structure* on X_{set}. The monoid S is called (for reasons to become apparent later) the monoid of *sections* of X_δ. The sets X_s are called the *principal sets.* The δ-rings $\mathcal{O}_s$ are called the *structure* δ-*rings.* If $S \neq \emptyset$ we define the δ-*ring of* δ-*rational functions* on a δ-ringed set X_δ by

$$R\langle X_\delta\rangle := \lim_{\substack{\rightarrow \\ s\in S}} \mathcal{O}_s, \tag{2.16}$$

where S is viewed as pre-ordered by divisibilty and we set

$$k\langle X_\delta\rangle := R\langle X_\delta\rangle \otimes k. \tag{2.17}$$

Furthermore, if $P \in X_{set}$, we let S_P be the submonoid of S consisting of all $s \in S$ such that $P \in X_s$. If $S_P = \emptyset$ we say P is a δ-*base point* of X_δ. The set $B(X_\delta)$ of δ-base points will be referred to as the δ-*base locus* of X_δ. If P is not a δ-base point we define the δ-*ring of germs of* X_δ *at* P,

$$\mathcal{O}_{X_\delta,P} := \lim_{\substack{\rightarrow \\ s\in S_P}} \mathcal{O}_s. \tag{2.18}$$

Here is more useful terminology:

1) X_δ is δ-*trivial* if $\mathcal{O}_s = R$ for all $s \in S$; X_δ is *totally* δ-*trivial* if $S = \emptyset$. So "totally δ-trivial" implies "δ-trivial".

2) X_δ is δ-*localized* if the maps

$$\mathcal{O}_s \otimes k \to \mathcal{F}(X_s) \otimes k$$

are injective and, whenever $P \in X_s$, $f \in \mathcal{O}_s$, $f(P) \in R^\times$, there exists $t \in S$ such that $P \in X_t$ and $f_{|X_{st}} \in \mathcal{O}_{st}^\times$. It is easy to check that if X_δ is δ-localized then the following hold: a) $\mathcal{O}_s$ are global $\delta-p$-rings, b) $R\langle X_\delta\rangle$ is a local $\delta-p$-ring hence $k\langle X_\delta\rangle$ is a field, c) the maps $\mathcal{O}_s \to \mathcal{O}_{st}$, $\mathcal{O}_s \otimes k \to \mathcal{O}_{st} \otimes k$ are injective for $s, t \in S$, and d) the rings $\mathcal{O}_{X_\delta,P}$ (for P not a δ-base point) are local with maximal ideal consisting of all $f \in \mathcal{O}_P$ such that $f(P) \in pR$.

3) X_δ is δ-*rational* if X_δ is not totally δ-trivial, X_δ is δ-localized, and the local $\delta-p$-ring $R\langle X_\delta\rangle\hat{}$ has a δ-basis.

4) X_δ is *globally* δ-*separated* if for any points $P, Q \in X_{set}$ with $P \neq Q$ there exists $s \in S$ such that $P \in X_s$, $Q \notin X_s$; X_δ is *locally* δ-*separated* if for any $s \in S$ and any points $P, Q \in X_s$ with $P \neq Q$ there exists $f \in \mathcal{O}_s$ such that $f(P) \neq f(Q)$.

5) If P is not a δ–base point the δ–*tangent space* of X_δ at P is the Abelian group

$$T_P X_\delta := Der_\delta(\mathcal{O}_{X_\delta,P}, R).$$

Next we would like to define morphisms of δ–ringed sets. To do that it is convenient to extend any monoid S to a monoid $S_{ext} := S \cup \{0\}$, by considering an extra element $0 \notin S$ such that $0s = 0$ for all $s \in S_{ext}$. Also, for any δ–ringed set X_δ we extend the families $(X_s)_{s\in S}$ and $(\mathcal{O}_s)_{s\in S}$ to families $(X_s)_{s\in S_{ext}}$ and $(\mathcal{O}_s)_{s\in S_{ext}}$ by setting $X_0 = \emptyset$ and $\mathcal{O}_0 = 0$.

DEFINITION 2.15. A *morphism* $X'_\delta \to X_\delta$ between two δ–ringed sets (with monoids of sections S, S', and structure δ–rings $\mathcal{O}_s$, $\mathcal{O}'_{s'}$) is a pair

$$\sigma_\delta = (\sigma_{set}, \sigma^\sharp)$$

where $\sigma^\sharp : S_{ext} \to S'_{ext}$ is a morphism of monoids and $\sigma_{set} : X'_{set} \to X_{set}$ is a map of sets such that $\sigma_{set}^{-1}(X_s) = X'_{\sigma^\sharp(s)}$ and such that the induced pull-back map $\sigma^* : \mathcal{F}(X_s) \to \mathcal{F}(X'_{\sigma^\sharp(s)})$ maps $\mathcal{O}_s$ into $\mathcal{O}'_{\sigma^\sharp(s)}$. A morphism σ_δ will be called δ–*dominant* if $\sigma^\sharp(S) \subset S'$.

(The complication arising from the extended monoids S_{ext} is therefore due to our desire to consider, as we should, morphisms which are not necessarily δ–dominant.) We may consider the category

$$\mathcal{C}_\delta := \{\delta\text{–ringed sets}\}. \tag{2.19}$$

The *trivial* objects of $\mathcal{C}_\delta$ are declared to be the δ–trivial δ–ringed sets. Note that any δ–dominant morphism $X'_\delta \to X_\delta$ of δ–localized δ–ringed sets induces a local $\delta - p$–ring extension

$$R\langle X_\delta\rangle \quad \subset \quad R\langle X'_\delta\rangle,$$

hence a field extension

$$k\langle X_\delta\rangle \quad \subset \quad k\langle X'_\delta\rangle.$$

Also, for any morphism $\sigma_\delta : X'_\delta \to X_\delta$ of δ–localized δ–ringed sets and any $P' \in X'_{set}$ with $P := \sigma_{set}(P')$ not a δ–base point we have a naturally induced Abelian group homomorphism, called δ–*tangent map*,

$$T_{P'}\sigma_\delta : T_{P'}X'_\delta \to T_P X_\delta. \tag{2.20}$$

Next we show that any correspondence in $\mathcal{C}_\delta$ has a categorical quotient. Indeed let

$$\mathbf{X}_\delta := (X_\delta, \tilde{X}_\delta, \sigma_{1,\delta}, \sigma_{2,\delta}) = (X_\delta, \sigma_\delta)$$

be a correspondence in $\mathcal{C}_\delta$; let S and $\tilde{S}$ the corresponding monoids of sections, and $\mathcal{O}_s$, $\tilde{\mathcal{O}}_{\tilde{s}}$ the corresponding structure δ–rings. Define

$$S' := \{s \in S | \sigma_1^\sharp(s) = \sigma_2^\sharp(s)\},\ \sigma^\sharp := \sigma_1^\sharp|_{S'} = \sigma_2^\sharp|_{S'} : S' \to \tilde{S}_{ext},\ \sigma^\sharp(0) = 0,$$

and consider the correspondences

$$\mathbf{A}_s := (\mathcal{O}_s, \tilde{\mathcal{O}}_{\sigma^\sharp(s)}, \sigma_1^*, \sigma_2^*)$$

in $\mathcal{R}_\delta^{op}$, for $s \in S'$. So we may consider the family of δ–rings:

$$(H^0(\mathbf{A}_s))_{s\in S'}.$$

On the other hand we have a correspondence

$$\mathbf{X}_{set} := (X_{set}, \tilde{X}_{set}, \sigma_{1,set}, \sigma_{2,set}) = (X_{set}, \sigma_{set})$$

in the category $\mathcal{C}_{set}$ of sets so we may consider the categorical quotient

$$Y_{set} := X_{set}/\sigma_{set} = X_{set}/\langle\sigma_{set}\rangle \tag{2.21}$$

in $\mathcal{C}_{set}$; cf. Equation 1.24. Moreover, for any $s \in S'$, we define

$$Y_s := X_s/\sigma_{set}. \tag{2.22}$$

We have a natural injective δ-ring homomorphism

$$H^0(\mathbf{A}_s) \to \mathcal{F}(Y_s)$$

and we set

$$\mathcal{O}'_s := Im(H^0(\mathbf{A}_s) \to \mathcal{F}(Y_s)), \quad s \in S'. \tag{2.23}$$

The data in Equations 2.21, 2.22 and 2.23 define a δ-ringed set structure Y_δ on Y_{set} and it is trivial to check that the categorical quotient of $\mathbf{X}_\delta$ in $\mathcal{C}_\delta$ is given by:

PROPOSITION 2.16. $X_\delta/\sigma_\delta = Y_\delta$.

Note that if, in the above Proposition, X_δ and $\tilde{X}_\delta$ are both δ-localized (or globally δ-separated, or locally δ-separated) it doesn't follow a priori that Y_δ is δ-localized (or globally δ-separated, or locally δ-separated respectively). In all our applications, however, it *will* happen that *all three* δ-sets X_δ, $\tilde{X}_\delta$, and Y_δ will be δ-localized; cf Remark 2.43.

2.1.3. δ-graded rings. In what follows we will introduce a graded version of δ-rings called δ-*graded rings* and we will attach to any δ-graded ring A^* (plus an extra structure called a *representation*) a δ-ringed set $Proj_\delta A^*$. We start by considering graded rings $A = \bigoplus_{\gamma\in\Gamma} A_\gamma$ graded by (Abelian) semigroups Γ; in applications Γ will always be either $\mathbf{Z}_+$ or W_+ (with their additive semigroup structure). We will sometimes use upper indices (A^γ instead of A_γ) if we want to distinguish between various gradings on the same ring; later, in applications, upper indices will be associated with W_+ while lower indices will be associated to $\mathbf{Z}_+$. If A is a global p-ring that has a structure of graded ring then we let

$$A_{((p))} := (S(A)^{-1}A)_0 \tag{2.24}$$

denote, as usual, the degree 0 component of the graded ring $S(A)^{-1}A$ where $S(A)$ consists of all homogeneous elements of A lying outside the prime ideal $(p) = pA$; then $A_{((p))}$ is a local p-ring. (In particular if $\Gamma = 0$ then $A_{((p))}$ coincides with the local ring $A_{(p)}$ of A at the prime ideal (p); when a ring will be given without any specification as to a grading on it we will view it as a graded ring trivially graded by $\Gamma = 0$.) Let us assume from now on that all our graded rings are graded R-algebras.

Let $\epsilon : \Gamma \to \Gamma'$ be a semigroup homomorphism and let

$$A = \bigoplus_{\gamma\in\Gamma} A_\gamma \to B = \bigoplus_{\gamma'\in\Gamma'} B_{\gamma'} \tag{2.25}$$

be a ring homomorphism which is *graded* with respect to ϵ (i.e. sends A_γ into $B_{\epsilon(\gamma)}$ for all γ).

DEFINITION 2.17. We say that the morphism 2.25 is *telescopic* if for any $\gamma \in \Gamma$ the following hold:

i) $A_\gamma \to B_{\epsilon(\gamma)}$ is injective;

ii) $A_\gamma \otimes k \to B_{\epsilon(\gamma)} \otimes k$ is injective.

If the morphism 2.25 is telescopic then one easily checks that if B is a global p-ring then A is also a global p-ring and we have an induced injective homomorphism $A_{((p))} \to B_{((p))}$. Also note that if $\Gamma = \Gamma' = 0$ then a telescopic morphism is simply an injective morphism which stays injective after tensorization with k. But for arbitrary Γ and Γ' a telescopic morphism doesn't need, of course, to be injective.

DEFINITION 2.18. A δ-*graded ring* is a pair (A^*, ϕ) where $A^* = \bigoplus_{w \in W_+} A^w$ is a W_+-graded R-algebra, which is a global p-ring, and

$$\phi : A^{>0} := \bigoplus_{w \neq 0} A^w \to A^{>0}$$

is a map satisfying the following properties:

1) ϕ is additive and multiplicative,
2) $\phi(\lambda x) = \phi(\lambda)\phi(x)$ for any $\lambda \in R$ and $x \in A^{>0}$,
3) $\phi A^w \subset A^{\phi w}$ and $\phi \otimes k : A^w \otimes k \to A^{\phi w} \otimes k$ is injective for all $w \neq 0$
4) for any $f, g \in A^w$, $w \neq 0$, we have

$$g^p \cdot f^\phi - g^\phi \cdot f^p \in pA^{w(\phi+p)}.$$

Since p is a non-zero divisor in A^* we may then define

$$\{g, f\}_\delta := \frac{g^p \cdot f^\phi - g^\phi \cdot f^p}{p} \in A^{w(\phi+p)} \tag{2.26}$$

for $f, g \in A^w$. Note that, if we view $A^{>0}$ as a ring without 1-element, then condition 1) says that ϕ is an endomorphism of $A^{>0}$. If we consider the subring (with 1-element) $R \oplus A^{>0}$ of $A^* = A^0 \oplus A^{>0}$ then conditions 1) and 2) above say that

$$\phi \oplus \phi : R \oplus A^{>0} \to R \oplus A^{>0}$$

is a ring endomorphism; condition 3) says that this endomorphism is graded with respect to $\epsilon : W_+ \to W_+$, $\epsilon(w) = \phi w$, and is telescopic.

REMARK 2.19. Note that a δ-graded ring does not have a natural structure δ-ring ! The link between δ-graded rings and δ-rings will be explained presently.

A morphism $u : (A^*, \phi) \to (B^*, \phi)$ of δ-graded rings is, by definition, an R-algebra homomorphism $u : A^* \to B^*$ such that there exists $0 \neq w_0 \in W_+$ with the property that $u(A^w) \subset B^{ww_0}$ for all $w \in W_+$ and such that $\phi \circ u = u \circ \phi$ on $A^{>0}$. We may consider the category

$$\mathcal{R}^*_\delta := \{\delta\text{-graded rings}\}. \tag{2.27}$$

Let (A^*, ϕ) be a δ-graded ring which we sometimes refer to simply as A^*. By a graded ϕ-module over A^* we mean a W_+-graded A^*-module $M^* = \bigoplus_{w \in W_+} M^w$ equipped with an additive map $\phi : M^{>0} \to M^{>0}$ such that $\phi(M^w) \subset M^{\phi w}$ and such that $\phi(am) = \phi(a)\phi(m)$ for $a \in A^{>0}$ and $m \in M^{>0}$.

Since any A^* as above is a global p-ring we may consider the multiplicative system $S(A^*)$ of all homogeneous elements of A^* lying outside the ideal $(p) = pA^*$. Recall that we denoted by $A^*_{((p))}$ the degree zero component, $(S(A^*)^{-1}A^*)^0$, of the ring $S(A^*)^{-1}A^*$; so $A^*_{((p))}$ is a local p-ring. Let $S(A^{>0})$ denote the set of all elements of $S(A^*)$ of degree $w \neq 0$; since A^* is p-adically separated, we have that $S(A^{>0}) \neq \emptyset$ if and only if $A^{>0} \neq 0$; moreover if this is the case then $A^*_{((p))}$ is also the degree zero component of the ring $S(A^{>0})^{-1}A^*$. Assume $S(A^{>0}) \neq \emptyset$; then the formula

$$\delta(f/g) = \frac{\{g, f\}_\delta}{g^{\phi+p}}, \quad g \in S(A^{>0}), \tag{2.28}$$

defines a p–derivation δ on $A^*_{((p))}$, so $A^*_{((p))}$ is a local $\delta - p$–ring. If $s \in S(A^{>0})$ we let

$$A^*_{\langle s \rangle} = ((s^{W_+})^{-1}A^*)^0 \tag{2.29}$$

be the degree zero component of the ring $(s^{W_+})^{-1}A^*$ where $s^{W_+} = \{s^w | w \in W_+\}$. Clearly $A^*_{\langle s \rangle}$ is a δ–subring of $A^*_{((p))}$ and it is a global $\delta - p$–ring; moreover, for $s, t \in S(A^{>0})$ the map $A^*_{\langle s \rangle} \otimes k \to A^*_{\langle st \rangle} \otimes k$ is injective. (More generally if M^* is a graded ϕ–module over A^* we let $M^*_{\langle s \rangle}$ be the degree zero component of the module $(s^{W_+})^{-1}M^*$.) We can view the multiplicative monoid $S := S(A^{>0})$ as pre-ordered by divisibilty. Assume $S \neq \emptyset$. We may consider then the directed family of δ–rings

$$(A^*_{\langle s \rangle})_{s \in S}. \tag{2.30}$$

whose direct limit is

$$\lim_{\to} A^*_{\langle s \rangle} = A^*_{((p))}.$$

If one is given a telescopic morphism $u : (A^*, \phi) \to (B^*, \phi)$ in $\mathcal{R}^*_\delta$ then one has an induced map $u : S(A^{>0}) \to S(B^{>0})$ and a morphism of directed families of δ–rings

$$A^*_{\langle s \rangle} \to B^*_{\langle u(s) \rangle} \tag{2.31}$$

indexed by $S(A^{>0})$; these morphisms are immediately seen to be telescopic.

Let us explain, in what follows, how to attach δ–ringed sets to δ–graded rings equipped with extra structures called *representations*; cf. the Definitions below.

DEFINITION 2.20. A *representation* of a δ–graded ring A^* on a set X_{set} is a family

$$(e_P)_{P \in X_{set}}$$

of (non-unitary) R–algebra homomorphisms (referred to as *evaluation maps*),

$$e_P : A^{>0} \to R, \quad f \mapsto f(P) := e_P(f),$$

that commute with ϕ:

$$\phi \circ e_P = e_P \circ \phi : A^{>0} \to R.$$

A representation is called *faithful* if whenever $f \in A^w$ satisfies

$$f(P) \equiv 0 \quad mod \quad p \quad \text{for all } P \in X_{set}$$

it follows that $f \in pA^w$. Two representations (e_P) and (e'_P) are *equivalent* if there exists a family (λ_P) of elements $\lambda_P \in R^\times$ such that for all $0 \neq w \in W_+$ and all $P \in X_{set}$ we have

$$e'_{P|A^w} = \lambda_P^w \cdot e_{P|A^w}.$$

Hence a representation equivalent to a faithful one is faithful.

DEFINITION 2.21. Given a faithful representation of a δ–graded ring A^* on a set X_{set} we define a δ–ringed set

$$X_\delta = Proj_\delta A^*$$

(that depends on the equivalence class of the representation only) as follows.

The underlying set of X_δ is X_{set}. The monoid of sections of X_δ is $S := S(A^{>0})$. For $s \in S$ we define the principal set

$$X_s := \{P \in X_{set} | s(P) \not\equiv 0 \mod p\}.$$

(Note that $X_s \neq \emptyset$ because the representation is faithful.) We then consider the δ–ring homomorphisms

$$A^*_{\langle s \rangle} \to \mathcal{F}(X_s)$$

$$\frac{f}{s^w} \mapsto \left(P \mapsto \frac{f(P)}{s^w(P)} \right).$$

We claim that these homomorphisms are injective and stay injective after tensorization with k. Indeed if f/s^w maps to a function in $p \cdot \mathcal{F}(X_s)$ then $(f \cdot s)(P) \equiv 0 \bmod p$ for all $P \in X_{set}$ hence, by faithfulness of the representation, $f \cdot s \in pA^*$. Since pA^* is a prime ideal we get $f \in pA^*$ and our claim is proved. Then we define the structure δ–rings

$$\mathcal{O}_s := Im(A^*_{\langle s \rangle} \to \mathcal{F}(X_s)) \simeq A^*_{\langle s \rangle}.$$

Clearly X_δ is δ–localized. If $S \neq \emptyset$ the δ–ring of δ–rational functions is given by

$$R\langle X_\delta \rangle = A^*_{((p))}.$$

Also, if $P \in X_{set}$ and $S_P \neq \emptyset$ the δ–ring of germs satisfies

$$\mathcal{O}_{X_\delta, P} = A^*_{\langle P \rangle} := \bigcup_{s \in S_P} A^*_{\langle s \rangle}.$$

2.1.4. δ–functions and δ–line bundles on schemes. We next explain how to enrich the rings of functions of algebraic geometry by "adjoining" δ. This will lead us to the concept of δ–*function* on (the points of) a scheme. Using the sheaf of δ–functions we define δ–*line bundles* and their δ–*sections*. Start by recalling that if X is a scheme over R we denote by $X(R)$ the set of R–points of X (i.e. sections of $X \to Spec\ R$) and (when no confusion should arise) we denote by X_0 the special fiber $X \otimes k$ of $X \to Spec\ R$.

DEFINITION 2.22. [**18**], [**17**]. Let X be a smooth scheme over R. A map $f : X(R) \to \mathbf{A}^1(R) = R$ is called a δ–*function* of order $\leq r$ if for any $Q \in X(R)$ one can find

1) an affine Zariski open set $U \subset X$ with $Q \in U(R)$,

2) a morphism of schemes $u : U \to \mathbf{A}^d$, and

3) a restricted δ–function $g : \mathbf{A}^d(R) = R^d \to \mathbf{A}^1(R) = R$ of order $\leq r$, such that $f(P) = g(u(P))$ for all $P \in U(R)$.

We denote by

$$\mathcal{O}^r(X) := \{f : X(R) \to R \mid f \text{ a } \delta\text{–function of order } \leq r\}; \tag{2.32}$$

then $\mathcal{O}^\infty(X) := \bigcup_r \mathcal{O}^r(X)$ is clearly a δ–subring of the ring $\mathcal{F}(X(R))$ of all functions $X(R) \to R$. Note that $\mathcal{O}^0(X) = \mathcal{O}(\hat{X})$ where $\hat{X}$ is the p–adic completion of X.

The rule that attaches to any Zariski open set $U \subset X$ the ring $\mathcal{O}^r(U)$ gives rise to a sheaf on X for the Zariski topology, denoted by $\mathcal{O}^r = \mathcal{O}^r_X$. We also set

$$\mathcal{O}^\infty = \mathcal{O}^\infty_X := \bigcup_{r=0}^\infty \mathcal{O}^r_X. \tag{2.33}$$

(This notation is at variance with that in [**18**]: the $\mathcal{O}^\infty$ there is the p-adic completion of the $\mathcal{O}^\infty$ here.)

Later we will prove:

PROPOSITION 2.23. *If X has a non-empty irreducible closed fiber $X \otimes k$ then the rings $\mathcal{O}^r(X)$ are global p-rings and the ring $\mathcal{O}^\infty(X)$ is a global $\delta - p$-ring.*

PROOF. Cf. Corollary 3.24. $\square$

If $\xi \in \mathcal{O}(X)$ is a regular function on a smooth scheme X and if $w \in W_+$ then we have an induced δ-function $\xi^w \in \mathcal{O}^r(X)$ defined by $\xi^w(P) = (\xi(P))^w$; the same definition makes sense if $\xi \in \mathcal{O}(X)^\times$ and $w \in W$ in which case $\xi^w \in \mathcal{O}^r(X)^\times$.

We next introduce line bundles into the picture. Recall that a *line bundle*, L, on a smooth scheme X over R is a locally free sheaf of $\mathcal{O}_X$-modules of rank one.

DEFINITION 2.24. By a *δ-line bundle (of order r)* on X we will understand a locally free sheaf $L \in Mod\ \mathcal{O}^r_X$ of $\mathcal{O}^r_X$-modules of rank one on X (for the Zariski topology). The sections $f \in H^0(X, L)$ of a δ-line bundle L will sometimes be referred to as *δ-sections*.

Any δ-line bundle defines (in a non-unique way) and is defined (up to isomorphism) by a Zariski open covering (U_i) of X together with a system of functions (ξ_{ij}), $\xi_{ij} \in \mathcal{O}^r(U_i \cap U_j)^\times$ satisfying the cocycle condition. A line bundle gives rise to a δ-line bundle of order 0 denoted by the same letter. If $L \in Mod\ \mathcal{O}^r_X$ is a δ-line bundle of order r on X and $w = \sum_{i=0}^s a_i\phi^i \in W$, $ord(w) = s$, then one can define a δ-line bundle

$$L^w \in Mod\ \mathcal{O}^{r+s}_X \tag{2.34}$$

of order $r+s$ on X by the formula

$$L^w := \bigotimes_{i=0}^{s} \left(L^{a_i} \otimes_{\mathcal{O}^r_X, \phi^i} \mathcal{O}^{r+s}_X\right),$$

where the notation indicates that $\mathcal{O}^{r+s}_X$ is viewed as an $\mathcal{O}^r_X$-module via ϕ^i. If L is defined by a cocycle (ξ_{ij}) then L^w is defined by the cocycle (ξ^w_{ij}) on the same covering. If $w_1, w_2 \in W\backslash\{0\}$ then

$$(L^{w_1})^{w_2} \simeq L^{w_1 w_2}$$

canonically. If $s \in \mathbf{Z}_+$ then one can define a δ-line bundle

$$L^{[s]} \in Mod\ \mathcal{O}^{r+s}_X \tag{2.35}$$

of order $r+s$ on X by the formula

$$L^{[s]} := L \otimes_{\mathcal{O}^r_X} \mathcal{O}^{r+s}_X$$

where $\mathcal{O}^{r+s}_X$ is viewed as an $\mathcal{O}^r_X$-algebra via the inclusion. If L is defined by a cocycle (ξ_{ij}) then $L^{[s]}$ is defined by the same cocycle (ξ_{ij}) viewed as a cocycle of δ-functions of order $\leq r+s$. If $u : Y \to X$ is a morphism of smooth schemes over R and L is a δ-line bundle on X then we can define in an obvious way a δ-line bundle of order r on Y,

$$f^*L \in Mod\ \mathcal{O}^r_Y; \tag{2.36}$$

if L is defined by a cocycle (ξ_{ij}) then f^*L is defined by the cocycle $(\xi_{ij} \circ u)$. If L and M are δ–line bundles on X of order r and s respectively then one can define a δ–line bundle $L \otimes M$ of order $t := max\{r, s\}$ on X by the formula

$$L \otimes M := L^{[t-r]} \otimes_{\mathcal{O}_X^t} M^{[t-s]},$$

where $\otimes$ in the right hand side is the usual tensor product of $\mathcal{O}_X^t$–modules; clearly $L \otimes M$ corresponds to taking the usual product of cocycles on a common refinement of the coverings. One checks that

$$L^{w_1} \otimes L^{w_2} \simeq L^{w_1+w_2}$$

canonically whenever $ord(w_1 + w_2) = max\{ord(w_1), ord(w_2)\}$; this is the case, for instance, if $w_1, w_2 \in W_+$. Also, if L has order r then $L^0 \simeq \mathcal{O}^r$. Finally, if $f \in H^0(X, L)$ then one has an obviously induced section $f \in H^0(X, L^{[s]})$ and one can define a section $f^w \in H^0(X, L^w)$ (by taking trivialisations on the sets U_i of an open covering, taking the functions $\xi_i \in \mathcal{O}^r(U_i)$ corresponding to f and letting f^w correspond to the functions $\xi_i^w \in \mathcal{O}^{r+ord(w)}(U_i)$).

Assume now $\mathbf{X}$ is a correspondence and let L be a δ–line bundle of order r on X. By analogy with geometric invariant theory [**113**] we make the following:

DEFINITION 2.25. A *linearization* of L is an isomorphism $\beta : \sigma_2^* L \simeq \sigma_1^* L$. A δ*–line bundle* of order r on $\mathbf{X}$ is a pair

$$\mathbf{L} = (L, \beta) \tag{2.37}$$

consisting of a δ–line bundle of order r, L, on X and a linearization β of L.

Any line bundle on $\mathbf{X}$ gives rise to a δ–line bundle of order 0, denoted by the same letter. If $\mathbf{L} = (L, \beta)$, $\mathbf{M} = (M, \nu)$ are δ–line bundles and $w \in W$, $s \in \mathbf{Z}_+$, then one defines δ–line bundles

$$\mathbf{L}^w = (L^w, \beta^w),\ \mathbf{L}^{[s]} = (L^{[s]}, \beta^{[s]}),\ \mathbf{L} \otimes \mathbf{M} = (L \otimes M, \beta \otimes \nu), \tag{2.38}$$

in the obvious way. We have an obvious notion of isomorphism of δ–line bundles similar to the one for line bundles. The isomorphism classes of δ–line bundles of order r on $\mathbf{X}$ form an Abelian group whose identity element is $\mathbf{O}^{[r]}$, where we recall that

$$\mathbf{O} := (\mathcal{O}, can) \tag{2.39}$$

and can is the canonical isomorphism

$$\sigma_2^* \mathcal{O}_X \overset{can}{\simeq} \mathcal{O}_{\tilde{X}} \overset{can}{\simeq} \sigma_1^* \mathcal{O}_X.$$

If $\mathbf{L}$ is a δ–line bundle of order r then $\mathbf{L}^0 \simeq \mathbf{O}^{[r]}$. Again, $\mathbf{L}^{w_1} \otimes \mathbf{L}^{w_2} \simeq \mathbf{L}^{w_1+w_2}$ whenever $ord(w_1 + w_2) = max\{ord(w_1), ord(w_2)\}$, e.g. if $w_1, w_2 \in W_+$. Also $(\mathbf{L}^{w_1})^{w_2} \simeq \mathbf{L}^{w_1 w_2}$ if $w_1, w_2 \neq 0$. The δ–line bundles $\mathbf{L}_1$ and $\mathbf{L}_2$ of orders r_1 and r_2 respectively will be called *polarization equivalent* if there exist $w_1, w_2 \in W_+ \backslash \phi W_+$ such that

$$ord(w_1) + r_1 = ord(w_2) + r_2,\ \ \mathbf{L}_1^{w_1} \simeq \mathbf{L}_2^{w_2}.$$

DEFINITION 2.26. A line bundle on $\mathbf{X}$ is called *flat* if it is $\mathbf{Z}_p$–flat in the sense of Definition 1.6 i.e. if it is isomorphic to a line bundle of the form $\mathbf{F}(\lambda)$ where λ is a locally constant $\mathbf{Z}_p^\times$–valued function on $\tilde{X}$. A δ–line bundle will be called *flat* if it is isomorphic to $\mathbf{F}^w$ where $\mathbf{F}$ is a flat line bundle and $w \in W$. Note that

$\mathbf{F}^w \simeq \mathbf{F}^{deg(w)[ord(w)]}$ so any flat δ-line bundle has the form $\mathbf{F}^{[r]}$ for some flat line bundle $\mathbf{F}$; we write

$$\mathbf{F}(r, \lambda) := (\mathbf{F}(\lambda))^{[r]} \tag{2.40}$$

and we call λ the *multiplier* of $\mathbf{F}(r, \lambda)$.

Recall that for $\mathbf{X}$ étale one defines the *canonical bundle* $\mathbf{K} = \mathbf{K_X}$, the anti-canonical bundle $\mathbf{K}^{-1}$, and, more generally, for $\nu \in \mathbf{Z}$, the ν-canonical bundles $\mathbf{K}^{\nu}$. A key role will be played by δ-line bundles that are polarization equivalent to $\mathbf{K}^{-1}$. More generally we have the following useful concept:

DEFINITION 2.27. A δ-line bundle $\mathbf{L}$ is called $\mathbf{Q}$- *canonical* (respectively $\mathbf{Q}$-*anticanonical*) if $\mathbf{L}^n \simeq \mathbf{K}^w$ for some integer $n \geq 1$ and some $w \in W$ with $deg(w) > 0$ (respectively $deg(w) < 0$).

Let us record the following analogue of Lemma 1.8:

LEMMA 2.28. *For $\lambda \in \mathbf{Z}_p^{\times}$, $w \in W$, $r = ord(w)$, the following are equivalent:*
1) $\mathbf{K}^w \simeq \mathbf{F}(r, \lambda)$,
2) There exists an invertible section $\eta \in H^0(X, K_X^w)$ such that $\sigma_2^* \eta = \lambda \cdot \sigma_1^* \eta$.

PROOF. Similar to that of Lemma 1.8. □

REMARK 2.29. If $\mathbf{L}$ is a flat line bundle on a correspondence $\mathbf{X}$ over R then the pull-back $\mathbf{L}_k$ on $\mathbf{X}_k := \mathbf{X} \otimes k$ has the property that its $(p-1)$-st power, $\mathbf{L}_k^{p-1}$ is trivial.

2.1.5. δ-graded rings attached to schemes. If X is a smooth scheme over R and L is a δ-line bundle on X then we may consider the ring

$$R(X, L)^* := \bigoplus_{w \in W_+} H^0(X, L^w). \tag{2.41}$$

Later we will prove:

PROPOSITION 2.30. *If X has a non-empty irreducible closed fiber $X \otimes k$ then the ring $R(X, L)^*$, equipped with the map ϕ, is a δ-graded ring.*

PROOF. Cf. Corollary 3.24. □

Consequently, under the hypotheses of the Proposition, we can consider the rings

$$R\langle X, L\rangle := (R(X, L)^*)_{((p))}, \quad k\langle X, L\rangle := (R(X, L)^*)_{((p))} \otimes k. \tag{2.42}$$

So $R\langle X, L\rangle$ is a local $\delta-p$-ring and $k\langle X, L\rangle$ is its residue field.

Now assume $\mathbf{X} = (X, \tilde{X}, \sigma_1, \sigma_2)$ is a correspondence of finite type and smooth over R. Assume $\mathbf{X}$ has an irreducible base. Let $\mathbf{L} = (L, \beta)$ be a δ-line bundle on $\mathbf{X}$. We may consider the isomorphism

$$\beta : H^0(\tilde{X}, \sigma_2^* L) \simeq H^0(\tilde{X}, \sigma_1^* L).$$

Then we set $\tilde{L} = \sigma_1^* L$ and we define the R-linear map

$$\begin{aligned} \Delta = \Delta^{\mathbf{L}} : H^0(X, L) &\to H^0(\tilde{X}, \tilde{L}) \\ \Delta(f) &:= \sigma_1^* f - \beta \sigma_2^* f, \end{aligned} \tag{2.43}$$

and the R-modules:

$$H^0(\mathbf{X}, \mathbf{L}) = Ker\ \Delta, \quad H^1(\mathbf{X}, \mathbf{L}) = Coker\ \Delta. \tag{2.44}$$

DEFINITION 2.31. The elements of $H^0(\mathbf{X},\mathbf{L})$ are called *invariant δ–sections* (or simply *δ–invariants*) of $(\mathbf{X},\mathbf{L})$. The module $H^1(\mathbf{X},\mathbf{L})$ above is called the *δ–cohomology* module of $(\mathbf{X},\mathbf{L})$.

We shall denote by

$$[\tilde{f}] \in H^1(\mathbf{X},\mathbf{L}) \tag{2.45}$$

the class of an element $\tilde{f} \in H^0(\tilde{X},\tilde{L})$. If $\mathbf{L}_1$ and $\mathbf{L}_2$ are δ–line bundles on $\mathbf{X}$ then there are natural *cup product* maps

$$H^0(\mathbf{X},\mathbf{L}_1)\times H^0(\mathbf{X},\mathbf{L}_2)\to H^0(\mathbf{X},\mathbf{L}_1\otimes\mathbf{L}_2),\ \ (f_1,f_2)\mapsto f_1\cup f_2:=f_1f_2, \tag{2.46}$$

$$H^0(\mathbf{X},\mathbf{L}_1)\times H^1(\mathbf{X},\mathbf{L}_2)\to H^1(\mathbf{X},\mathbf{L}_1\otimes\mathbf{L}_2),\ \ (f_1,[\tilde{f}_2])\mapsto f_1\cup[\tilde{f}_2]:=[\tilde{f}_1\tilde{f}_2],$$

where $\tilde{f}_1:=\sigma_1^*f_1$. Moreover one has natural maps

$$H^0(\mathbf{X},\mathbf{L}_1)\to H^0(\mathbf{X},\mathbf{L}_1^{\phi}),\ \ f\mapsto f^{\phi},$$

$$H^1(\mathbf{X},\mathbf{L}_2)\to H^1(\mathbf{X},\mathbf{L}_2^{\phi}),\ \ [f]\mapsto [f]^{\phi}:=[f^{\phi}],$$

satisfying $(f_1\cup[f_2])^{\phi}=f_1^{\phi}\cup[f_2]^{\phi}$. In particular, if $\mathbf{L}_1=\mathbf{L}$ and $\mathbf{L}_2=\mathbf{O}$ then we get a natural map

$$\cup[1]:H^0(\mathbf{X},\mathbf{L})\to H^1(\mathbf{X},\mathbf{L}). \tag{2.47}$$

Using the cup product we can define the W_+–graded ring

$$R(\mathbf{X},\mathbf{L})^*:=\bigoplus_{v\in W_+} H^0(\mathbf{X},\mathbf{L}^v); \tag{2.48}$$

this ring will be, as we shall see, the *basic object of this book.* It is, of course, a subring of the W_+–graded ring $R(X,L)^*$ and has a natural structure of object in the category, $\mathcal{R}_{\delta}^*$, of δ–graded rings. Note that the inclusion $R(\mathbf{X},\mathbf{L})^*\subset R(X,L)^*$ is telescopic. We set

$$R\langle\mathbf{X},\mathbf{L}\rangle:=(R(\mathbf{X},\mathbf{L})^*)_{((p))},\ \ k\langle\mathbf{X},\mathbf{L}\rangle:=(R(\mathbf{X},\mathbf{L})^*)_{((p))}\otimes k. \tag{2.49}$$

So $R\langle\mathbf{X},\mathbf{L}\rangle$ is a local $\delta-p$–ring and $k\langle\mathbf{X},\mathbf{L}\rangle$ is its residue field. Furthermore

$$R^1(\mathbf{X},\mathbf{L})^*:=\bigoplus_{v\in W_+} H^1(\mathbf{X},\mathbf{L}^v) \tag{2.50}$$

has a natural structure of graded ϕ–module over $R(\mathbf{X},\mathbf{L})^*$. If $\mathbf{A}^*$ is the correspondence in $\mathcal{R}^{op}$ defined by

$$\mathbf{A}^*:=(R(X,L)^*,R(\tilde{X},\tilde{L})^*,\sigma_1^*,\beta\sigma_2^*),$$

then, of course,

$$R(\mathbf{X},\mathbf{L})^*=H^0(\mathbf{A}^*),\ \ R^1(\mathbf{X},\mathbf{L})^*=H^1(\mathbf{A}^*).$$

We can define a "relative" version of H^0 by considering any R–algebra B and setting:

$$H^0(\mathbf{X},\mathbf{L};B):=Ker(\Delta^{\mathbf{L}}\otimes B:H^0(X,L)\otimes B\to H^0(\tilde{X},\tilde{L})\otimes B); \tag{2.51}$$

intuitively one should view these spaces as consisting of "invariants of $(\mathbf{X},\mathbf{L})$ over B". The case $B=k$ is particularly significant. Indeed for any R–module M let $M[p]$ denote the k–linear space of all $x\in M$ such that $px=0$. Then we have the following:

LEMMA 2.32. *There is a canonical exact sequence*

$$0 \to H^0(\mathbf{X},\mathbf{L}) \otimes k \to H^0(\mathbf{X},\mathbf{L};k) \to H^1(\mathbf{X},\mathbf{L})[p] \to 0. \tag{2.52}$$

PROOF. This follows from the general fact that, given an arbitrary R−module homomorphism $\Delta : M \to M'$ with M' torsion free, the natural map $(Ker\ \Delta)\otimes k \to M \otimes k$ is injective and we have a natural isomorphism

$$(Coker\ \Delta)[p] \simeq \frac{Ker(\Delta \otimes k : M \otimes k \to M' \otimes k)}{Im((Ker\ \Delta) \otimes k \subset M \otimes k)} \tag{2.53}$$

defined as follows. Consider an element α in the right hand side of equation 2.53 and represent α by some element $m \in M$ such that $\Delta(m) = pm'$ with $m' \in M'$. Let $\alpha' \in (Coker\ \Delta)[p]$ be the class of m'. Then the map that sends α into α' is an isomorphism from the right hand side of equation 2.53 onto the left hand side of equation 2.53. □

REMARK 2.33. The spaces $H^0(\mathbf{X},\mathbf{L};k)$ have a remarkable "continuity" property. Indeed if $\mathbf{X}_1$ and $\mathbf{X}_2$ are correspondences equipped with δ−line bundles $\mathbf{L}_1$ and $\mathbf{L}_2$ of order r and if

$$(\mathbf{X}_1,\mathbf{L}_1) \otimes R/p^{r+1}R \simeq (\mathbf{X}_2,\mathbf{L}_2) \otimes R/p^{r+1}R$$

(in the obvious sense) then

$$H^0(\mathbf{X}_1,\mathbf{L}_1;k) \simeq H^0(\mathbf{X}_2,\mathbf{L}_2;k).$$

This is interesting because, as we shall see later (in what we shall call our *converse theorems*), if one tries to deform p−adically a pair $(\mathbf{X},\mathbf{L})$ for which $H^0(\mathbf{X},\mathbf{L}) \neq 0$ one tends to obtain pairs $(\mathbf{X}',\mathbf{L}')$ for which $H^0(\mathbf{X}',\mathbf{L}') = 0$.

REMARK 2.34. To understand the significance of the map 2.47 let's see what it boils down to in a very simple case; the discussion below will not be used later. Assume

$$\mathbf{X} = (X, X, id, \sigma), \tag{2.54}$$

where σ is an automorphism of X and $\mathbf{L} = (L,\beta)$. View $\mathbf{Z}$ as acting on $H^0(X,L)$ by letting $1 \in \mathbf{Z}$ act as

$$\beta\sigma^* : H^0(X,L) \to H^0(X,\sigma^*L) \to H^0(X,L).$$

Then, clearly, we have

$$H^i(\mathbf{X},\mathbf{L}) \simeq H^i(\mathbf{Z}, H^0(X,L)),\ \ i = 0,1, \tag{2.55}$$

where the H^i in the right hand side stands for *group cohomology*. Now, more generally, if M is an R−module equipped with an automorphism s and $\mathbf{Z}$ acts on M by letting 1 act as s, then there is a natural map

$$H^0(\mathbf{Z},M) = Ker(s-1 : M \to M) \to H^1(\mathbf{Z},M) = M/((s-1)M), \tag{2.56}$$

$x \mapsto [x]$; this map is of course compatible, via the identification 2.55, with the map 2.47. Let's assume M is torsion free. Then the following are trivially checked:

1) If $s^{\circ n} = id$ there is an isomorphism

$$H^1(\mathbf{Z},M)_{tors} \simeq H^1(\mathbf{Z}/n\mathbf{Z},M). \tag{2.57}$$

2) If s is semisimple (in the sense that $M \otimes K^a$ has a K^a−basis consisting of eigenvectors of s, where K^a is an algebraic closure of K) then the map 2.56 induces an isomorphism

$$H^0(\mathbf{Z}, M) \otimes K \simeq H^1(\mathbf{Z}, M) \otimes K. \tag{2.58}$$

Here semisimplicity is essential; cf. the remark below.

3) If $Ker(s-1) \neq Ker((s-1)^2)$ then the map

$$H^0(\mathbf{Z}, M) \to H^1(\mathbf{Z}, M) \tag{2.59}$$

is not injective; indeed, if $(s-1)m \neq 0$ and $(s-1)^2 m = 0$ for some $m \in M$ then $(s-1)m$ is in the kernel of the map 2.59.

The above remarks give a hint as to what the torsion of H^1 and the map 2.47 should intuitively mean in the general case when $\mathbf{X}$ is not of the form 2.54; of course, in this general case the equations 2.57 and 2.58 should not be expected to hold and the degree of their failure should be viewed as a measure of how far one is from the situation 2.54 and from "semisimplicity".

REMARK 2.35. For flat δ−line bundles a ring more natural than (but closely related to) $R(\mathbf{X}, \mathbf{L})^*$ can be considered as follows. Let $\mathbf{X}$ be a correspondence of finite type and smooth over R, with an irreducible base and let $\mathbf{L} = (L, \beta)$ be a flat δ−line bundle on $\mathbf{X}$, of order r. We have a natural identification $H^0(X, L) \simeq \mathcal{O}^r(X)$. Define a ($\mathbf{Z}_+$−graded) global p−ring $R(\mathbf{X}, \mathbf{L})_*$ (cf. [**22**], [**29**]) by the formula

$$R(\mathbf{X}, \mathbf{L})_* := \bigoplus_{m \in \mathbf{Z}_+} H^0(\mathbf{X}, \mathbf{L}^{m[\infty]}) \tag{2.60}$$

where

$$H^0(\mathbf{X}, \mathbf{L}^{m[\infty]}) := \bigcup_{s \in \mathbf{Z}_+} H^0(\mathbf{X}, \mathbf{L}^{m[s]}).$$

We have a natural telescopic map

$$R(\mathbf{X}, \mathbf{L})_* \to \mathcal{O}^\infty(X)$$

and the target ring is a global p−ring. So $R(\mathbf{X}, \mathbf{L})_*$ is a global p−ring. Note that, for any $v \in W_+$ we have $\mathbf{L}^v \simeq \mathbf{L}^{deg(v)[ord(v)]}$ and hence we have an induced morphism

$$R(\mathbf{X}, \mathbf{L})^* \to R(\mathbf{X}, \mathbf{L})_*$$

which is graded with respect to $deg : W_+ \to \mathbf{Z}_+$. This homomorphism is telescopic so it induces an injective ring homomorphism

$$R\langle \mathbf{X}, \mathbf{L}\rangle := (R(\mathbf{X}, \mathbf{L})^*)_{((p))} \to (R(\mathbf{X}, \mathbf{L})_*)_{((p))}. \tag{2.61}$$

It is trivial to see that if $R(\mathbf{X}, \mathbf{L})_*$ does not reduce to its component of degree 0 then the map 2.61 is an isomorphism. Also it is clear that one has a telescopic map

$$R(X, L)^* \to \mathcal{O}^\infty(X),$$

inducing an isomorphism

$$(R(X, L)^*)_{((p))} \simeq \mathcal{O}^\infty(X)_{(p)},$$

hence an isomorphism between $k\langle X, L\rangle$ and the field of fractions of $\mathcal{O}^\infty(X) \otimes k$.

REMARK 2.36. Let us quickly explain the connection with [**29**] where we defined a *dynamical system* (of dimension one) to be a triple

$$\mathbf{X} = (X^*, X, \sigma)$$

where X^* is a scheme of finite type over R, $X \subset X^*$ is an open subscheme, smooth of relative dimension one over R, with geometrically irreducible fibers, and $\sigma : X^* \to X^*$ is a finite morphism. One can attach to a dynamical system as above a correspondence (still denoted by)

$$\mathbf{X} = (X, \sigma^{-1}(X) \cap X, \iota, \sigma),$$

where ι is the inclusion. Clearly $\mathbf{X}$ is an open subcorrespondence of

$$\mathbf{X}^* := (X^*, X^*, id, \sigma).$$

Moreover we considered, in [**29**], pairs $\mathbf{L} = (r, b) \in \mathbf{Z}_+ \times \mathbf{Z}_p^\times$; to any such $\mathbf{L}$ we can attach a flat δ-line bundle in the sense of the present work (still denoted by) $\mathbf{L} = \mathbf{F}(r, b^{-1})$. Then the space $H^0(\mathbf{X}, \mathbf{L})$ attached to the correspondence $\mathbf{X}$ here coincides with the space $H^0(\mathbf{X}, \mathbf{L})$ attached to the dynamical system $\mathbf{X}$ in [**29**]. In this way, the theory in [**29**] fits into the present formalism.

REMARK 2.37. Assume $(\mathbf{X}_i)_{i \in I}$ is a family of correspondences $\mathbf{X}_i = (X, \sigma_i)$ with the same base, X, and consider their disjoint union $\mathbf{X} := \coprod_i \mathbf{X}_i$. Assume moreover that we are given a family $(\mathbf{L}_i)_{i \in I}$ where $\mathbf{L}_i = (L, \beta_i)$ are δ-line bundles of order $\leq r$, with L and r not depending on i. Then one can consider the naturally induced *disjoint union* δ-line bundle of order $\leq r$ on $\mathbf{X}$,

$$\mathbf{L} := \coprod_i \mathbf{L}_i;$$

clearly

$$H^0(\mathbf{X}, \mathbf{L}) = \bigcap_i H^0(\mathbf{X}_i, \mathbf{L}_i).$$

2.1.6. δ-ringed sets attached to schemes. Let $\mathcal{C} = \mathcal{C}_R$ be the category of schemes over R. Our next goal is to attach to any pair (X, L), where X is a smooth scheme over R with irreducible non-empty closed fiber and L is a δ-line bundle on X, a δ-ringed set $X_\delta := X_\delta(L)$. This will allow us to attach to any pair $(\mathbf{X}, \mathbf{L})$, where $\mathbf{X}$ is an irreducible étale correspondence in $\mathcal{C}$ of finite type and smooth over R and $\mathbf{L}$ is a δ-line bundle on $\mathbf{X}$, a correspondence $\mathbf{X}_\delta$ in $\mathcal{C}_\delta$. Then we will show how to compute the categorical quotient of the latter.

Start with a pair (X, L) where X is a smooth scheme over R with non-empty irreducible closed fiber $X \otimes k$ and L is a δ-line bundle on X of order r; then we define a δ-ringed set $X_\delta := X_\delta(L)$ as follows. One can consider the δ-graded ring $R(X, L)^*$ in Equation 2.41. We note that $R(X, L)^*$ has a representation on the set

$$X_{set} := X(R) \tag{2.62}$$

in the sense of Definition 2.20 given as follows. For each $P \in X(R)$ choose an affine open subset $U \subset X$ with $P \in U(R)$ such that L has a basis η on U; so $L = \mathcal{O}^r \cdot \eta$ on U. Hence for $0 \neq w \in W_+$ of order s, $L^w = \mathcal{O}^{r+s} \cdot \eta^w$ on U. For any $f \in H^0(X, L^w)$ consider the function $f/\eta^w \in \mathcal{O}^{r+s}(U)$. Define the evaluation map

$$e_P : H^0(X, L^w) \to R, \quad f \mapsto f(P) := \frac{f}{\eta^w}(P) \in R.$$

Clearly the equivalence class of this representation does not depend on the choices made (for U and η).

Later we will show:

PROPOSITION 2.38. *The above representation is faithful.*

PROOF. Cf. Proposition 3.19 below. $\square$

So, by Definition 2.21, we may consider the δ–ringed set

$$X_\delta = X_\delta(L) := Proj_\delta R(X,L)^*. \tag{2.63}$$

Recall that the underlying set of X_δ is $X_{set} := X(R)$. The monoid of sections of X_δ is

$$S :=: S(R(X,L)^{>0}). \tag{2.64}$$

The principal subsets are:

$$X_s := \{P \in X(R) \mid s(P) \not\equiv 0 \;\; mod \;\; p\}, \tag{2.65}$$

The structure δ–rings are given by

$$\mathcal{O}_s := Im(R(X,L)^*_{\langle s\rangle} \to \mathcal{F}(X_s)) \simeq R(X,L)^*_{\langle s\rangle}. \tag{2.66}$$

So for $s \in H^0(X, L^{w_0})$, $\mathcal{O}_s$ consists of all functions $X_s \to R$ of the form

$$P \mapsto \frac{f(P)}{s^w(P)},$$

where $f \in H^0(X, L^{ww_0})$. Note that X_δ is δ–localized and, if $S \neq \emptyset$,

$$R\langle X_\delta\rangle = R\langle X,L\rangle, \quad k\langle X_\delta\rangle = k\langle X,L\rangle.$$

Also, if $P \in X(R)$ and $S_P \neq \emptyset$ then the δ–ring of germs, $\mathcal{O}_{X_\delta,P}$, satisfies

$$\mathcal{O}_{X_\delta,P} \simeq R(X,L)^*_{\langle P\rangle} := \bigcup_{s\in S_P} R(X,L)^*_{\langle s\rangle}. \tag{2.67}$$

Note also that X_δ is totally δ–trivial if and only if $S = \emptyset$, hence if and only if $R(X,L)^{>0} = 0$. Finally note that X_δ is often locally and globally δ–separated; this is the case, for instance, if $L^w = M^v$ for some ample line bundle M on X and some $w, v \in W_+\backslash\{0\}$.

REMARK 2.39. Assume L is a δ–line bundle on X and assume

$$s_0, ..., s_n \in H^0(X,L)$$

are δ–sections not belonging to $pH^0(X,L)$. Let

$$X(R)' := \bigcup_{i=0}^{n} X_{s_i} \subset X(R).$$

We have a natural map of sets

$$\pi : X(R)' \to \mathbf{P}^n(R), \quad P \mapsto (s_0(P) : ... : s_n(P)). \tag{2.68}$$

We want to explain how the map π above is "part of" a morphism of δ–ringed sets. One can consider the δ–ringed set X'_δ with underlying set $X'_{set} := X(R)'$, monoid of sections $S := S(R(X,L)^{>0})$, principal sets $X'_s := X'_{set} \cap X_s$, and structure δ–rings

$$\mathcal{O}_s := Im(R(X,L)^*_{\langle s\rangle} \to \mathcal{F}(X'_s))$$

(which continue to be isomorphic with $R(X,L)^*_{\langle s\rangle}$ by Proposition 3.19). On the other hand, as we shall see later (cf. Remark 6.9), we have a natural homomorphism

$$\pi^* : R(\mathbf{P}^n, \mathcal{O}(1))^* \to R(X,L)^*.$$

This homomorphism induces a homomorphism of monoids

$$\pi^\sharp : S(R(\mathbf{P}^n, \mathcal{O}(1))^{>0})_{ext} \to S(R(X,L)^{>0})_{ext}$$

as follows. First $\pi^\sharp(0) = 0$. Then, for $f \in S(R(\mathbf{P}^n, \mathcal{O}(1))^{>0})$ we set $\pi^\sharp(f) = \pi^*(f)$ if $\pi^*(f) \in S(R(X,L)^{>0})$ and $\pi^\sharp(f) = 0$ otherwise. Then the pair $(\pi, \pi^\sharp)$ defines a morphism of δ-ringed sets

$$\pi_\delta : X'_\delta \to \mathbf{P}^n_\delta = Proj_\delta R(\mathbf{P}^n, \mathcal{O}(1))^*. \tag{2.69}$$

DEFINITION 2.40. We will say that (X,L) is δ-*trivial* if X_δ is δ-trivial, equivalently if

$$\textit{either} \quad R(X,L)^{>0} = 0 \quad \textit{or} \quad R\langle X,L\rangle = R.$$

(Also δ-*non-trivial* will always mean *not* δ-*trivial.*) We will say that (X,L) is δ-*rational* if X_δ is δ-rational, equivalently if

$$R(X,L)^{>0} \neq 0 \quad \textit{and} \quad R\langle X,L\rangle\hat{} \quad \textit{has a } \delta-\textit{basis}.$$

REMARK 2.41. If $\mathbf{P}^1 := Proj\ R[x_0,x_1]$ is the projective line over R then one can show that $(\mathbf{P}^1, \mathcal{O}(1))$ is δ-rational; this is far from trivial and will be proved in Corollary 6.8.

Now we come back to correspondences. Let $\mathbf{X} = (X, \tilde{X}, \sigma_1, \sigma_2) = (X, \sigma)$ be an irreducible étale correspondence of finite type and smooth over R and let $\mathbf{L}$ be a δ-line bundle on $\mathbf{X}$. Then the maps

$$\sigma_1^* : R(X,L)^* \to R(\tilde{X},\tilde{L})^*,\ \beta\sigma_2^* : R(X,L)^* \to R(\tilde{X},\tilde{L})^* \tag{2.70}$$

turn out to be telescopic; cf. Corollary 3.25. So we have an induced correspondence in the category $\mathcal{C}_\delta$ of δ-ringed sets,

$$\mathbf{X}_\delta := (X_\delta, \tilde{X}_\delta, \sigma_{1,\delta}, \sigma_{2,\delta}) = (X_\delta, \sigma_\delta) \tag{2.71}$$

with $\sigma_{\delta 1}$ and $\sigma_{\delta 2}$ δ-dominant. We show, in what follows, how to construct the categorical quotient X_δ/σ_δ of $\mathbf{X}_\delta$. We consider first the correspondence

$$\mathbf{X}_{set} := (X_{set}, \tilde{X}_{set}, \sigma_{1,set}, \sigma_{2,set}) = (X_{set}, \sigma_{set}) \tag{2.72}$$

in the category of sets and its categorical quotient

$$Y_{set} := X_{set}/\sigma_{set} = X_{set}/\langle\sigma_{set}\rangle, \tag{2.73}$$

cf. Equation 1.24. There is a faithful representation of the δ-graded ring $R(\mathbf{X}, \mathbf{L})^*$ on the set Y_{set} which is "natural up to equivalence". So we may consider the δ-ringed set

$$Y_\delta = Proj_\delta R(\mathbf{X}, \mathbf{L})^*. \tag{2.74}$$

Recall that, according to our Definition 2.21 the underlying set of Y_δ is Y_{set}, the monoid of sections of Y_δ is

$$\mathbf{S} := S(R(\mathbf{X}, \mathbf{L})^{>0}), \tag{2.75}$$

the principal sets of Y_δ are given by

$$Y_s := X_s/\sigma_{set} = X_s/\langle\sigma_{set}\rangle, \tag{2.76}$$

and the structure δ-rings are given by

$$\mathcal{O}'_s := Im(R(\mathbf{X},\mathbf{L})^*_{\langle s\rangle} \to \mathcal{F}(Y_s)). \tag{2.77}$$

So for $s \in H^0(\mathbf{X},\mathbf{L}^{w_0})$, $\mathcal{O}'_s$ consists of all functions $X_s/\langle\sigma_{set}\rangle \to R$ of the form

$$[P] \mapsto \frac{f}{s^w}(P), \quad P \in X_s$$

where $f \in H^0(\mathbf{X},\mathbf{L}^{ww_0})$. Using Proposition 2.16 it is trivial to check that, in the category $\mathcal{C}_\delta$ of δ-ringed sets, the categorical quotient X_δ/σ_δ of $\mathbf{X}_\delta$ exists and is given by:

PROPOSITION 2.42. $X_\delta/\sigma_\delta = Y_\delta = Proj_\delta R(\mathbf{X},\mathbf{L})^*$.

REMARK 2.43. Note that X_δ/σ_δ is δ-localized and, if $\mathbf{S} \neq \emptyset$,

$$R\langle X_\delta/\sigma_\delta\rangle = R\langle\mathbf{X},\mathbf{L}\rangle,$$
$$k\langle X_\delta/\sigma_\delta\rangle = k\langle\mathbf{X},\mathbf{L}\rangle.$$

Note also that X_δ/σ_δ is totally δ-trivial if and only if $\mathbf{S} = \emptyset$ hence if and only if $R(\mathbf{X},\mathbf{L})^{>0} = 0$. Note finally that, *most of the times* in our applications, X_δ/σ_δ is neither locally nor globally δ-separated. Failure of local δ-separation should be viewed as an expression of the fact that our categorical quotients fail to be "geometric quotients"; failure of local δ-separation is less dramatic and is related to "semistability" issues. Indeed note that, unlike in the classical geometric invariant theory, there was no need to consider semistable loci; morally this is possible because the restriction to semistable loci is, in some sense, built-in our δ-ringed set formalism. Cf. Definition 2.46 below.

REMARK 2.44. As in Remark 2.39 if one is given δ-sections $s_0, ..., s_n \in H^0(\mathbf{X},\mathbf{L})$ not belonging to $pH^0(\mathbf{X},\mathbf{L})$ then the associated morphism of δ-ringed sets in Equation 2.69 factors through a morphism of δ-ringed sets

$$Y'_\delta \to \mathbf{P}^n_\delta,$$

where Y'_δ is the set X'_{set}/σ_{set} with a structure of δ-ringed set naturally induced from that of Y_δ.

As in the case of pairs (X, L) it is useful to introduce the following:

DEFINITION 2.45. We will say that $(\mathbf{X},\mathbf{L})$ is δ-*trivial* if X_δ/σ_δ is trivial, equivalently if

$$\textit{either} \quad R(\mathbf{X},\mathbf{L})^{>0} = 0 \quad \textit{or} \quad R\langle\mathbf{X},\mathbf{L}\rangle = R.$$

(Also δ-*non trivial* will always mean *not* δ-*trivial*.) We will say that $(\mathbf{X},\mathbf{L})$ is δ-*rational* if X_δ/σ_δ is δ-rational, equivalently if

$$R(\mathbf{X},\mathbf{L})^{>0} \neq 0 \quad \textit{and} \quad R\langle\mathbf{X},\mathbf{L}\rangle\hat{} \quad \textit{has a } \delta - \textit{basis}.$$

DEFINITION 2.46. A point $P \in X(R)$ is called a δ-*base point* (resp. a *strict* δ-*base point*) of $(\mathbf{X},\mathbf{L})$ if for any $w \in W_+$ with $w \neq 0$ and for any $f \in H^0(\mathbf{X},\mathbf{L}^w)$ we have $f(P) \equiv 0 \bmod p$ (resp. $f(P) = 0$). We denote by $B(\mathbf{X},\mathbf{L})$ (resp. $BS(\mathbf{X},\mathbf{L})$) the sets of δ-base points (resp. strict δ-base points) of $(\mathbf{X},\mathbf{L})$; we refer to the sets $B(\mathbf{X},\mathbf{L})$ and $BS(\mathbf{X},\mathbf{L})$ as δ-*base loci*. We have

$$BS(\mathbf{X},\mathbf{L}) \subset B(\mathbf{X},\mathbf{L}),$$
$$B(\mathbf{X},\mathbf{L}) = X_{set} \backslash \bigcup_{s\in\mathbf{S}} X_s$$

where $\mathbf{S} = S(R(\mathbf{X}, \mathbf{L})^{>0})$; furthermore $B(\mathbf{X}, \mathbf{L})$ is the pull-back, via the projection $X_{set} \to X_{set}/\sigma_{set}$ of the δ−base locus $B(X_\delta/\sigma_\delta)$. The δ−base loci are analogues, in our theory, of the non-semistable loci of geometric invariant theory [**113**].

REMARK 2.47. Assume $\mathbf{L}_1 = (L_1, \beta_1)$ and $\mathbf{L}_2 = (L_2, \beta_2)$ are polarization equivalent δ−line bundles on $\mathbf{X}$ of orders r_1 and r_2 respectively. Then, clearly, $(\mathbf{X}, \mathbf{L}_1)$ is δ−trivial if and only if $(\mathbf{X}, \mathbf{L}_2)$ is δ−trivial. Assume in what follows that $(\mathbf{X}, \mathbf{L}_1)$ and $(\mathbf{X}, \mathbf{L}_2)$ are δ−non-trivial. We claim that

$$R\langle \mathbf{X}, \mathbf{L}_1\rangle \simeq R\langle \mathbf{X}, \mathbf{L}_2\rangle, \quad R\langle X, L_1\rangle \simeq R\langle X, L_2\rangle; \tag{2.78}$$

in particular

$$k\langle \mathbf{X}, \mathbf{L}_1\rangle \simeq k\langle \mathbf{X}, \mathbf{L}_2\rangle, \quad k\langle X, L_1\rangle \simeq k\langle X, L_2\rangle, \tag{2.79}$$

so $(\mathbf{X}, \mathbf{L}_1)$ is δ−rational if and only if so $(\mathbf{X}, \mathbf{L}_2)$ is δ−rational. To verify the isomorphisms in Equation 2.78 write $\mathbf{L}_1^{w_1} \simeq \mathbf{L}_2^{w_2}$ for $w_1, w_2 \in W_+ \backslash \phi W_+$. Let

$$f = \frac{a}{s^w} \in R(\mathbf{X}, \mathbf{L}_1)^*_{\langle s\rangle},$$

$$s \in H^0(\mathbf{X}, \mathbf{L}_1^{w_0}), \ a \in H^0(\mathbf{X}, \mathbf{L}_1^{w_0 w}),$$

where $w_0, w \in W_+ \backslash \{0\}$. Since $w_1 - 1 \in W_+$ we can write

$$f = \frac{a s^{w(w_1 - 1)}}{(s^{w_1})^w},$$

$$s^{w_1} \in H^0(\mathbf{X}, \mathbf{L}_1^{w_0 w_1}) = H^0(\mathbf{X}, \mathbf{L}_2^{w_0 w_2}),$$

$$a s^{w(w_1-1)} \in H^0(\mathbf{X}, \mathbf{L}_1^{w_0 w + w_0 w(w_1 - 1)}) = H^0(\mathbf{X}, \mathbf{L}_2^{w_0 w_2 w}).$$

So $f \in R(\mathbf{X}, \mathbf{L}_2)^*_{\langle s^{w_1}\rangle}$ and our claim about $\mathbf{X}$ is proved. The claim about X can be proved similarly.

REMARK 2.48. Assume $R(\mathbf{X}, \mathbf{L})^{>0} \neq 0$. The morphism

$$X_\delta \to X_\delta/\sigma_\delta \tag{2.80}$$

is δ−dominant hence induces a field extension

$$k\langle X_\delta/\sigma_\delta\rangle \quad \subset \quad k\langle X_\delta\rangle$$

which identifies with the field extension

$$k\langle \mathbf{X}, \mathbf{L}\rangle \quad \subset \quad k\langle X, L\rangle. \tag{2.81}$$

Assume in addition that

1) $L = M^v$, $v \in W_+ \backslash \phi W_+$, with M a line bundle on X,

2) either X is affine or X is projective and M is ample on X.

(These assumptions are true in all applications we have in mind.) Then we claim that there is a natural embedding of the field $k(X)$ of rational functions on $X \otimes k$ into the field $k\langle X_\delta\rangle = k\langle X, L\rangle$. Indeed any element of $k(X)$ can be written in the form f/g mod p with $f, g \in H^0(X, M^n)$ for some $n \geq 1$. Since $v - 1 \in W_+$ we can then write $f/g = fg^{v-1}/g^v$ and note that $fg^{v-1}, g^v \in H^0(X, L^n)$ which proves our claim. So in the situation above we can consider the field extension

$$k(X) \cdot k\langle X_\delta/\sigma_\delta\rangle \subset k\langle X_\delta\rangle$$

which identifies with the extension

$$k(X) \cdot k\langle \mathbf{X}, \mathbf{L}\rangle \quad \subset \quad k\langle X, L\rangle, \tag{2.82}$$

where "·" means "compositum".

DEFINITION 2.49. The field extension 2.81 is called the *generic δ–fiber* of the morphism 2.80. The field extension 2.82 is called the *relative generic δ–fiber* of the morphism 2.80.

We will later investigate Galois theoretic properties of the generic and relative generic δ–fibers.

REMARK 2.50. Recall from Equation 1.13 that we defined the *tangent space of X at $P \in X(R)$* as $T_P X := Der_R(\mathcal{O}_{X,P_0}, R)$ where $P_0 \in X(k)$ is the reduction mod p of P identified with the corresponding closed point of X. Then, if P is not a δ–base point of X_δ, there is a natural Abelian group homomorphism

$$T_P X \to T_P X_\delta \tag{2.83}$$

to the δ–tangent space of X_δ at P; cf. Equations 3.43 below and 2.67. If in addition conditions 1 and 2 in Remark 2.48 are satisfied then one can easily check that the map in Equation 2.83 is an isomorphism. In particular $T_P X_\delta$ has a natural structure of R–module. However the δ–tangent maps between such $T_P X_\delta$'s will generally *not* be R–linear but merely $\mathbf{Z}_p$–linear.

REMARK 2.51. Definitions 2.40, 2.45, and 2.49 allow us to replace the language of δ–ringed sets X_δ and categorical quotients X_δ/σ_δ by the language of pairs (X, L), $(\mathbf{X}, \mathbf{L})$, and their associated rings

$$R(X, L)^*, \ R\langle X, L\rangle, \ k\langle X, L\rangle, \ R(\mathbf{X}, \mathbf{L})^*, \ R\langle \mathbf{X}, \mathbf{L}\rangle, \ k\langle \mathbf{X}, \mathbf{L}\rangle.$$

Indeed these pairs and rings are the only objects that are technically relevant; but, of course, their significance only comes through the formalism of δ–ringed sets and their categorical quotients, and this is why we briefly developed this formalism in the first place. Beginning with the next Chapter the language of δ–ringed sets and their categorical quotients will not be used anymore.

REMARK 2.52. Throughout this subsection our correspondences $\mathbf{X}$ were assumed to be irreducible. The whole discussion generalizes, in an entirely obvious way, to the case when $\mathbf{X}$ is a disjoint union of irreducible correspondences and $\mathbf{L}$ are, correspondingly, disjoint unions of δ–line bundles.

2.1.7. Correspondences over number fields. In order to state our main conjectures in the next Section we need to introduce some terminology and notation related to correspondences over number fields.

Let $F \subset \mathbf{C}$ be a number field and S a finite set of finite places of F containing the ramified places of F over $\mathbf{Q}$. For simplicity we always assume S is *saturated* in the sense that it consists of all places containing some rational integer m. Consider the ring $\mathcal{O}$ of S–integers in F,

$$\mathcal{O} = \mathcal{O}_F[1/m], \tag{2.84}$$

where $\mathcal{O}_F$ is the ring of integers of F and m is divisible by all the ramified rational primes. Assume that

$$\mathbf{X}_\mathcal{O} = \mathbf{X} = (X, \tilde{X}, \sigma_1, \sigma_2) \tag{2.85}$$

is a one dimensional geometrically irreducible étale correspondence of finite type and smooth over $\mathcal{O}$. For any finite place

$$2, 3 \notin \wp \notin S$$

let $\mathcal{O}_\wp$ be the localization of $\mathcal{O}_F$ at $\wp$, consider the completion $\hat{\mathcal{O}}_\wp$, let $\hat{\mathcal{O}}_\wp^{ur}$ be the maximum unramified extension of $\hat{\mathcal{O}}_\wp$, and further complete to get:

$$R_\wp := (\hat{\mathcal{O}}_\wp^{ur})\hat{}.$$

Let $k_\wp := R_\wp/pR_\wp$, $p\mathbf{Z} = \wp \cap \mathbf{Z}$. Then $R_\wp \simeq \hat{\mathbf{Z}}_p^{ur}$ has a unique structure of local $\delta - p$–ring with p–derivation $\delta = \delta_\wp$. Denote by

$$\mathbf{X}_F = (X_F, \sigma_F), \quad \mathbf{X}_\mathbf{C} = (X_\mathbf{C}, \sigma_\mathbf{C}), \quad \mathbf{X}_\wp = (X_\wp, \sigma_\wp) \tag{2.86}$$

the geometrically irreducible correspondences over F, $\mathbf{C}$, and $R_\wp$ obtained by base change respectively. For each $\wp$ we view $\mathbf{X}_\wp$ equipped with the anticanonical bundle $\mathbf{K}_\wp^{-1}$ and we consider the correspondence

$$\mathbf{X}_\delta = (X_\delta, \sigma_\delta)$$

attached to $(\mathbf{X}_\wp, \mathbf{K}_\wp^{-1})$; $\mathbf{X}_\delta$ implicitly depends, of course, on $\wp$ but for simplicity we do not make this dependence explicit in our notation.

By the way, note that for each $\wp$ one can consider the δ–base loci $BS(\mathbf{X}_\wp, \mathbf{K}_\wp^{-1})$ and $B(\mathbf{X}_\wp, \mathbf{K}_\wp^{-1})$; cf. Definition 2.46.

DEFINITION 2.53. Define the following sets still referred to as *δ–base loci*:

$$\overline{BS}(\mathbf{X}, \mathbf{K}^{-1}) := \bigcap_{2,3 \notin \wp \notin S} (X(\mathcal{O}_\wp) \cap BS(\mathbf{X}_\wp, \mathbf{K}_\wp^{-1})) \subset X(F),$$

$$\underline{BS}(\mathbf{X}, \mathbf{K}^{-1}) := \bigcup_{\sharp T < \infty} \bigcap_{\wp \notin S \cup T} (X(\mathcal{O}_\wp) \cap BS(\mathbf{X}_\wp, \mathbf{K}_\wp^{-1})) \subset X(F),$$

$$\underline{B}(\mathbf{X}, \mathbf{K}^{-1}) := \bigcup_{\sharp T < \infty} \bigcap_{\wp \notin S \cup T} (X(\mathcal{O}_\wp) \cap B(\mathbf{X}_\wp, \mathbf{K}_\wp^{-1})) \subset X(F).$$

In the above formulae, T is any set of places of F and the intersections

$$X(\mathcal{O}_\wp) \cap BS(\mathbf{X}_\wp, \mathbf{K}_\wp^{-1}), \quad X(\mathcal{O}_\wp) \cap B(\mathbf{X}_\wp, \mathbf{K}_\wp^{-1})$$

are taken inside $X_\wp(R_\wp)$ and are viewed as subsets of $X(F)$ via the inclusion $X(\mathcal{O}_\wp) \subset X(F)$. Also note that

$$\underline{BS}(\mathbf{X}, \mathbf{K}^{-1}) \subset \overline{BS}(\mathbf{X}, \mathbf{K}^{-1}) \cap \underline{B}(\mathbf{X}, \mathbf{K}^{-1}).$$

Clearly, $\underline{BS}(\mathbf{X}, \mathbf{K}^{-1})$ and $\underline{B}(\mathbf{X}, \mathbf{K}^{-1})$ depend only on $\mathbf{X}_F$ (and not on the model $\mathbf{X} = \mathbf{X}_\mathcal{O}$). If we want to stress this we denote them by $\underline{BS}(\mathbf{X}_F, \mathbf{K}_F^{-1})$ and $\underline{B}(\mathbf{X}_F, \mathbf{K}_F^{-1})$ respectively; in particular, if F' is a finite extension of F we may consider the δ–base loci $\underline{BS}(\mathbf{X}_{F'}, \mathbf{K}_{F'}^{-1})$ and $\underline{B}(\mathbf{X}_{F'}, \mathbf{K}_{F'}^{-1})$. Also we may consider the δ–base loci

$$\underline{BS}(\mathbf{X}_{F^a}, \mathbf{K}_{F^a}^{-1}) := \bigcup_{F'} \underline{BS}(\mathbf{X}_{F'}, \mathbf{K}_{F'}^{-1})$$

and

$$\underline{B}(\mathbf{X}_{F^a}, \mathbf{K}_{F^a}^{-1}) := \bigcup_{F'} \underline{B}(\mathbf{X}_{F'}, \mathbf{K}_{F'}^{-1})$$

where F' runs through the set of all finite extensions of F contained in an algebraic closure F^a.

On the other hand, if we want to stress the dependence of $\overline{BS}(\mathbf{X}, \mathbf{K}^{-1})$ on $\mathcal{O}$ we denoted it by $\overline{BS}(\mathbf{X}_\mathcal{O}, \mathbf{K}_\mathcal{O}^{-1})$.

Morally the smaller a δ–base locus is the more δ–invariants we have.

2.2. Main conjectures

We are in a position to state our main conjectures. Assume in what follows that $\mathbf{X} = \mathbf{X}_{\mathcal{O}}$ is the correspondence in Equation 2.85 and we assume, for simplicity that X is affine.

CONJECTURE 2.54. *Assume $\mathbf{X}_{\mathbf{C}}$ is dense and admits an analytic uniformization. Then the following hold:*

1) For almost all $\wp$, X_δ/σ_δ is $\delta-$non-trivial and $\delta-$rational;

2) For almost all $\wp$ the field extension $k_\wp(X_\wp)\cdot k_\wp\langle X_\delta/\sigma_\delta\rangle \subset k_\wp\langle X_\delta\rangle$ is algebraic;

3) In the flat or hyperbolic case the $\delta-$base loci $\overline{BS}(\mathbf{X},\mathbf{K}^{-1})$ and $\underline{B}(\mathbf{X},\mathbf{K}^{-1})$ are finite; in particular the $\delta-$base locus $\underline{BS}(\mathbf{X},\mathbf{K}^{-1})$ is finite.

Here *almost all* means "all except finitely many". There are really 3 conjectures in the above Conjecture, corresponding to the conclusions 1, 2, 3 respectively. In view of Definitions 2.40, 2.45, and 2.49 we can rephrase conditions 1 and 2 in the above Conjecture as saying that for almost all $\wp$:

1) $(\mathbf{X}_\wp,\mathbf{K}_\wp^{-1})$ is $\delta-$non-trivial and $\delta-$rational;

2) The field extension $k_\wp(X_\wp)\cdot k_\wp\langle\mathbf{X}_\wp,\mathbf{K}_\wp^{-1}\rangle \subset k_\wp\langle X_\wp,K_\wp^{-1}\rangle$ is algebraic.

With regards to conclusion 2 one should be able, in various special cases, to analyze the Galois theoretic properties of the field extension there.

Note that if $\mathbf{X}_{\mathbf{C}}$ is dense and admits a hyperbolic analytic uniformization then, by Proposition 1.32, the uniformization is an $E-$arithmetic (hyperbolic) uniformization. Our results will provide rather strong evidence for Conjecture 2.54 for spherical, flat, and hyperbolic $\mathbf{Q}-$arithmetic uniformizations; we have no example when this conjecture holds for hyperbolic $E-$arithmetic uniformizations with $E \neq \mathbf{Q}$. The reason why one might hope that Conjecture 2.54 is true in case $E \neq \mathbf{Q}$ as well is that, although Shimura curves arising from quaternion algebras over totally real fields $E \neq \mathbf{Q}$ do not have a "direct" modular interpretation, they are nevertheless "closely connected" to some moduli problems for Abelian varieties with certain extra structures; cf. [**31**], p. 185. One can hope then to be able to use the results in [**3**] about Siegel differential modular forms.

In the converse direction we propose the following:

CONJECTURE 2.55. *Assume $\mathbf{X}_{\mathbf{C}}$ is dense and assume that* **one** *of the following holds:*

1) For almost all places $\wp$, X_δ/σ_δ is $\delta-$non-trivial;

2) For almost all $\wp$ the field extension $k_\wp(X_\wp)\cdot k_\wp\langle X_\delta/\sigma_\delta\rangle \subset k_\wp\langle X_\delta\rangle$ is algebraic;

3) The $\delta-$base locus $\underline{BS}(\mathbf{X},\mathbf{K}^{-1})$ is finite.

Then $\mathbf{X}_{\mathbf{C}}$ is commensurable with a correspondence that admits an analytic uniformization.

We really have 3 conjectures in this Conjecture, corresponding to the hypotheses 1, 2, 3 respectively. Note also that condition 1 is (easily seen to be) weaker than condition 2. On the other hand it might be reasonable to expect that, in many cases, the conclusion of the Conjecture can be strengthened by replacing the word "commensurable" with the word "equivalent".

Results along Conjecture 2.54 will be referred to as *direct theorems*. Results along the line of Conjecture 2.55 will be referred to as *converse theorems*. More generally we will use the name of *converse theorem* to refer to any theorem claiming the vanishing (or smallness) of a space of the form $H^0(\mathbf{X}_\wp,\mathbf{L}_\wp)$.

REMARK 2.56. Note that in our Conjecture 2.54 we fix our correspondence $\mathbf{X}_{\mathcal{O}} = \mathbf{X} = (X, \tilde{X}, \sigma_1, \sigma_2)$ and we vary the prime p; let us refer to this setting as being "horizontal". There is an alternative, "vertical" setting which we are going to consider in the body of the book in which we fix p and X and we vary the curve $\tilde{X}$. We will not discuss the vertical problem here. Note however that the "horizontal" problems will be attacked by looking at the corresponding "vertical" problems first.

REMARK 2.57. Note that our Conjectures predict (and our results will put forward) a link, in the context of correspondences on curves, between:

1) existence of analytic uniformizations and

2) existence of "enough" δ–invariants for almost all primes.

This link, however, will never appear, in our book, as a direct "logical" link between the two sets of concepts. The link that we will see will manifest itself at the "phenomenological" level in the sense that the examples of correspondences for which 1) is known to hold tend to be the same as the ones for which 2) is known to hold. This "phenomenological" (rather than "logical") link is not worse than what we usually see in number theory / Diophantine geometry when it comes to links between complex analytic and arithmetic properties of geometric objects.

2.3. Main results: a sample

In what follows we state our 3 main "direct" results on the "horizontal" Conjecture 2.54, referring to the spherical, flat, and hyperbolic cases respectively. Then we state 2 general results along the lines of the "converse", "horizontal" Conjecture 2.55. We end with a few comments on various types of "vertical" problems we are going to study in this book. For more specific converse theorems in the spherical, flat, and hyperbolic case we refer to the last sections of our last 3 Chapters respectively. These 3 Chapters will also extensively address the "vertical" counterpart. For local analogues of Conjectures 2.54 and 2.55 and local results we refer to Section 1 in Chapter 4.

2.3.1. Direct results: spherical case. Let $\mathcal{O} \subset \mathbf{C}$ be the ring of S–integers in a number field F, S a saturated finite set of finite places of F containing all the ramified places and all places above 2 and 3. Let $\Gamma \subset SL_2(\mathcal{O})$ be a finite subgroup and let $\tau \in SL_2(\mathcal{O})$. View $SL_2(\mathcal{O})$ as acting on the projective line $\mathbf{P}^1 = Proj\ \mathcal{O}[x_0, x_1]$ over $\mathcal{O}$ in the canonical way. Let $Y \subset \mathbf{P}^1$ be an affine open Γ–invariant subscheme such that $X := Y/\Gamma$ is smooth over $\mathcal{O}$ and the canonical projection $\pi : Y \to X$ is étale. Set $\tilde{X} := Y \cap \tau^{-1}(Y)$ and consider the correspondence

$$\mathbf{X}_{\mathcal{O}} := \mathbf{X} = (X, \tilde{X}, \pi, \pi \circ \tau)$$

in $\mathcal{C}_{\mathcal{O}}$.

Of course, the induced correspondence $\mathbf{X}_{\mathbf{C}}$ admits a spherical analytic uniformization. Conversely, by Proposition 1.25, any correspondence over $\mathbf{C}$ that admits a spherical analytic uniformization "defined over a number field" is equivalent to one of the form $\mathbf{X}_{\mathbf{C}}$ above.

DEFINITION 2.58. A subgroup G of $SL_2(\mathcal{O})$ is *full* if for almost all unramified places $\wp \subset \mathcal{O}$ of degree 1 (i.e. with $\mathcal{O}/\wp = \mathbf{Z}/p\mathbf{Z}$, $p\mathbf{Z} = \wp \cap \mathbf{Z}$) and for all $n \geq 0$ the homomorphism

$$G \to PSL_2(\mathbf{Z}/p^{n+1}\mathbf{Z}) := SL_2(\mathbf{Z}/p^{n+1}\mathbf{Z}))/\langle -id \rangle$$

is surjective

For instance, for any $\mathcal{O}$, $SL_2(\mathbf{Z}) \subset SL_2(\mathcal{O})$ is full.

THEOREM 2.59. *For infinitely many places* $\wp$, $(\mathbf{X}_\wp, \mathbf{K}_\wp^{-1})$ *is* $\delta-$*non-trivial. If in addition the group* $\langle \Gamma, \tau \rangle$ *is full then, for infinitely many* $\wp$:

1) $(\mathbf{X}_\wp, \mathbf{K}_\wp^{-1})$ *is* $\delta-$*rational;*

2) $k_\wp\langle X_\wp, K_\wp^{-1}\rangle$ *can be embedded into an algebraic Galois extension of the field* $k_\wp\langle \mathbf{X}_\wp, \mathbf{K}_\wp^{-1}\rangle$ *with Galois group* $PSL_2(\mathbf{Z}_p)$ *or* $PGL_2(\mathbf{Z}_p)$.

The issue of when the $\delta-$base loci are finite is left open. We actually expect that in general the $\delta-$base locus $\underline{BS}(\mathbf{X}, \mathbf{K}^{-1})$ is infinite. On the other hand there exist examples when $\overline{BS}(\mathbf{X}, \mathbf{K}^{-1})$ is finite. For an analysis of the case when $\langle \Gamma, \tau \rangle$ is not full we refer to the body of the book; cf. Theorems 6.30 and 6.31.

2.3.2. Direct results: flat case. Again let $\mathcal{O} \subset \mathbf{C}$ be the ring of $S-$integers in a number field F, S a saturated finite set of finite places containing all the ramified places of F and all places above 2 and 3. Let G be either the multiplicative group $\mathbf{G}_m$ or an elliptic curve over $\mathcal{O}$, let $N \in \mathbf{Z}$ be invertible in $\mathcal{O}$ with $N \notin \{1, -1\}$, let $[N] : G \to G$ be the multiplication by N endomorphism, and let $\epsilon \in \{1, -1\}$. Then $[N]$ induces a morphism of schemes $\sigma : G/\langle [\epsilon] \rangle \to G/\langle [\epsilon] \rangle$. Note that $G/\langle [\epsilon] \rangle$ is smooth over $\mathcal{O}$. Let $X \subset G/\langle [\epsilon] \rangle$ be an affine Zariski open set such that the natural projection $\pi : G \to G/\langle [\epsilon] \rangle$ is étale above X, let $\tilde{X} = X \cap \sigma^{-1}(X)$ and consider the étale irreducible correspondence in $\mathcal{C}_\mathcal{O}$:

$$\mathbf{X}_\mathcal{O} = \mathbf{X} = (X, \tilde{X}, \iota, \sigma),$$

where ι is the inclusion.

Clearly, $\mathbf{X}_\mathbf{C}$ admits a flat analytic uniformization. Also, by Proposition 1.27 any correspondence over $\mathbf{C}$ that admits a flat analytic uniformization and is defined over a number field "looks very much like" our $\mathbf{X}_\mathbf{C}$ above. (The special case treated here corresponds to taking, in Proposition 1.27,

$$\gamma_0 = \tilde{\gamma}_0 = [1] \text{ or } \gamma_0 = \tilde{\gamma}_0 = [-1],$$

$$\tau_{10} = [1], \quad \tau_{20} = [N], \quad P_1 = P_2 = 0.$$

The general case can be treated by similar methods but, in order to avoid technicalities, we will stick to the above special case.)

THEOREM 2.60. *The following hold:*

1) For all $\wp \notin S$, $(\mathbf{X}_\wp, \mathbf{K}_\wp^{-1})$ *is* $\delta-$*non-trivial and* $\delta-$*rational;*

2) For all $\wp \notin S$ *the field extension* $k_\wp(X_\wp) \cdot k_\wp\langle \mathbf{X}_\wp, \mathbf{K}_\wp^{-1}\rangle \subset k_\wp\langle X_\wp, K_\wp^{-1}\rangle$ *is algebraic. If* $\epsilon = 1$ *and* $G = \mathbf{G}_m$ *the above extension is Galois, with Galois group* $\mathbf{Z}_p^\times$.

3) The $\delta-$*base locus* $\overline{BS}(\mathbf{X}, \mathbf{K}^{-1})$ *is finite. If* $\epsilon = 1$, $G = \mathbf{G}_m$, $F = \mathbf{Q}$, *the* $\delta-$*base locus* $\underline{B}(\mathbf{X}, \mathbf{K}^{-1})$ *is contained in the set*

$$\mathcal{W} := \{a \in \mathbf{Q}^\times \mid a^p \equiv a \text{ mod } p^2 \text{ in } \mathbf{Z}_{(p)} \text{ for almost all } p\}.$$

For the (somewhat technical results on the) Galois properties of the field extension in assertion 2 above in case G is an elliptic curve we refer to Chapter 7. Note also that the set $\mathcal{W}$, which could be called the *Wieferich set*, is actually conjectured to consist of 1 and -1. Silverman [**126**] proved that this conjecture is implied by the *abc* conjecture. Cf. also [**133**] for an elliptic analogue of $\mathcal{W}$.

REMARK 2.61. Theorem 2.60 implies an a priori stronger version of itself which, in assertion 3, states that:

Let F' be any finite extension of F, let $\mathcal{O}'$ be the integral closure of $\mathcal{O}$ in F' and let $\mathbf{X}_{\mathcal{O}'} := \mathbf{X} \otimes \mathcal{O}'$. Then the δ-base locus $\overline{BS}(\mathbf{X}_{\mathcal{O}'}, \mathbf{K}_{\mathcal{O}'}^{-1})$ is finite. In particular the δ-base locus $\underline{BS}(\mathbf{X}_{F'}, \mathbf{K}_{F'})$ is finite.

Indeed this follows by applying the Theorem to $\mathbf{X}_{\mathcal{O}'}$.

On the other hand it will follow from our proof of Theorem 2.60 that

The δ-base locus $\underline{BS}(\mathbf{X}_{F^a}, \mathbf{K}_{F^a}^{-1})$ is infinite.

2.3.3. Direct results: hyperbolic case. Rational arithmetic Fuchsian lattices come in two flavors: they are either attached to the split quaternion $\mathbf{Q}$-algebra $M_2(\mathbf{Q})$ or to a non-split quaternion $\mathbf{Q}$-algebra. The first type is related to modular curves (parameterizing elliptic curves) whereas the second type is related to Shimura curves (parameterizing false elliptic curves). We only discuss below the split case; the non-split case will be discussed in Chapter 8.

For the standard definitions and properties of the modular curves considered below we refer to [**83**] pp. 87, 117, 125, 129, and [**47**], pp. 69-72. Fix an integer $N \geq 4$ and a prime l not dividing N. Let

$$Y_1(N)/\mathbf{Z}[1/N] \tag{2.87}$$

be the scheme whose S-points (S any scheme over $\mathbf{Z}[1/N]$) identify with isomorphism classes of pairs (A, α) where A/S is an elliptic curve and $\alpha : (\mathbf{Z}/N\mathbf{Z})_S \to A$ is a closed immersion of group schemes. Recall that $Y_1(N)$ is smooth affine of relative dimension one over $\mathbf{Z}[1/N]$, with geometrically irreducible fibers. If (A, α), (A', α') are two pairs as above then by an *isogeny* of degree l,

$$u : (A, \alpha) \to (A', \alpha'),$$

we mean an isogeny $u : A \to A'$ of degree l such that $u \circ \alpha = \alpha'$. Let

$$Y_1(N, l)/\mathbf{Z}[1/Nl] \tag{2.88}$$

be the scheme whose S-points identify with isomorphism classes of isogenies of degree l as above. Recall that $Y_1(N, l)$ is smooth of relative dimension one over $\mathbf{Z}[1/Nl]$, with geometrically irreducible fibers. Let $\mathcal{O} := \mathbf{Z}[1/Nl]$ and view $Y_1(N)$ as a scheme over $\mathcal{O}$ via base change. We may consider the following correspondence in $\mathcal{C}_{\mathcal{O}}$:

$$\mathbf{X}_{\mathcal{O}} := \mathbf{X} = (X := Y_1(N), \tilde{X} := Y_1(N, l), \sigma_1, \sigma_2) \tag{2.89}$$

where σ_1, σ_2 are defined on S-points by

$$\sigma_1((A, \alpha) \to (A', \alpha')) = (A, [l] \circ \alpha), \quad \sigma_2((A, \alpha) \to (A', \alpha')) = (A', \alpha');$$

clearly σ_1 and σ_2 are étale above $\mathcal{O}$.

Note that $\mathbf{X}_{\mathbf{C}}$ admits a hyperbolic $\mathbf{Q}$-arithmetic uniformization. Conversely, any correspondence over $\mathbf{C}$ admitting a hyperbolic $\mathbf{Q}$-arithmetic uniformization is not very much different from our $\mathbf{X}_{\mathbf{C}}$ above, at least from the viewpoint of our theory. To avoid technical complications we stick to our special case above.

THEOREM 2.62. *The following hold:*

1) For almost all p, $(\mathbf{X}_p, \mathbf{K}_p^{-1})$ *is* δ*−non-trivial; moreover, for infinitely many* p, $(\mathbf{X}_p, \mathbf{K}_p^{-1})$ *is* δ*−rational;*

2) For infinitely many p *the field extension* $k_p(X_p)\cdot k_p\langle \mathbf{X}_p, \mathbf{K}_p^{-1}\rangle \subset k_p\langle X_p, K_p^{-1}\rangle$ *is algebraic, prosolvable.*

3) The δ*−base locus* $\overline{BS}(\mathbf{X}, \mathbf{K}^{-1})$ *is finite; the* δ*−base locus* $\underline{B}(\mathbf{X}, \mathbf{K}^{-1})$ *is empty.*

Here a field extension $L \subset F$ is called *prosolvable* if F can be embedded into a (possibly infinite, algebraic) Galois extension of L whose Galois group is a projective limit of finite solvable groups.

For an analogue, for Shimura curves, of assertion 1 in Theorem 2.62 we refer to Theorem 8.103 in the last Chapter.

REMARK 2.63. We will actually prove a stronger version of assertion 3 above as follows.

Let F' *be any number field, let* $\mathcal{O}'$ *be the integral closure of* $\mathcal{O}$ *in* F' *and let* $\mathbf{X}_{\mathcal{O}'} := \mathbf{X} \otimes \mathcal{O}'$. *Then the* δ*−base locus* $\overline{BS}(\mathbf{X}_{\mathcal{O}'}, \mathbf{K}_{\mathcal{O}'}^{-1})$ *is finite and the* δ*−base locus* $\underline{BS}(\mathbf{X}_{F'}, \mathbf{K}_{F'}^{-1})$ *is empty. In particular the* δ*−base locus* $\underline{BS}(\mathbf{X}_{F^a}, \mathbf{K}_{F^a}^{-1})$ *is empty. If in addition* F' *has at least one real embedding then the* δ*−base locus* $\underline{B}(\mathbf{X}_{F'}, \mathbf{K}_{F'}^{-1})$ *is empty.*

2.3.4. Converse results: base loci. Assume $\mathcal{O}$ is the ring of $S-$ integers in a number field $F \subset \mathbf{C}$ and S is a saturated finite set of finite places of F containing all ramified places and all places above 2 and 3. Consider a one dimensional correspondence of finite type and smooth over $\mathcal{O}$,

$$\mathbf{Y}_{\mathcal{O}} := (Y, \tilde{Y}, \sigma_1, \sigma_2), \tag{2.90}$$

with geometrically irreducible base. Let

$$\mathbf{X}_{\mathcal{O}} := (X, \tilde{X}, \sigma_1, \sigma_2) \tag{2.91}$$

be an open étale subcorrespondence such that the geometric fibers of $\tilde{X}$ are dense in the corresponding geometric fibers of $\tilde{Y}$. We may consider the δ−base loci $\underline{BS}(\mathbf{X}_{F'}, \mathbf{K}_{F'}^{-1})$ for all finite extensions F' of F and the $\delta-$ base locus $\underline{BS}(\mathbf{X}_{F^a}, \mathbf{K}_{F^a}^{-1})$; cf. Definition 2.53.

THEOREM 2.64. *Assume* **one of the following** *holds:*

1) The δ*−base locus* $\underline{BS}(\mathbf{X}_{F^a}, \mathbf{K}_{F^a}^{-1})$ *is finite;*

2) $\sigma_1 : \tilde{Y} \to Y$ *is an isomorphism and the* δ*−base loci* $\underline{BS}(\mathbf{X}_{F'}, \mathbf{K}_{F'}^{-1})$ *are finite for all finite extensions* F' *of* F.

Then $\mathbf{Y}_{\mathbf{C}}$ *is critically finite.*

Critical finiteness is a rather strong conclusion; indeed recall from Proposition 1.41, that if a correspondence attached to a map $\sigma : \mathbf{P}^1 \to \mathbf{P}^1$ is critically finite and has a non-hyperbolic associated orbifold then we have a flat analytic uniformization. In *this* sense Theorem 2.64 can be viewed as a "converse result".

2.3.5. Converse results: triviality of pluricanonical bundles mod p**.** Assume again that $\mathcal{O}$ is the ring of $S-$ integers in a number field $F \subset \mathbf{C}$ and S is a saturated finite set of finite places of F containing all ramified places and all places above 2 and 3.

THEOREM 2.65. *1) Assume* $\mathbf{X}_{\mathcal{O}}$ *is a one dimensional geometrically irreducible étale correspondence of finite type and smooth over* $\mathcal{O}$ *and assume the following condition holds for some place* $\wp$ *of residual characteristic* p*:*

$$\text{(2.92)} \quad \text{There exists } v_{\wp} \in W \text{ with } v_{\wp}(0) \not\equiv 0 \text{ mod } p \text{ such that } H^0(\mathbf{X}_{\wp}, \mathbf{K}_{\wp}^{v_{\wp}}) \neq 0$$

Then there exists an open non-empty subcorrespondence $\mathbf{U}_{\wp} \subset \mathbf{X}_{\wp}$ *and an integer* $\nu \neq 0$ *such that the* $\nu-$*canonical bundle* $\mathbf{K}^{\nu}_{\mathbf{U}_{\wp} \otimes k_{\wp}}$ *of* $\mathbf{U}_{\wp} \otimes k_{\wp}$ *is trivial.*

2) Consider the correspondence $\mathbf{Y} := (\mathbf{P}^1, \mathbf{P}^1, id, \sigma)$ *over* $\mathcal{O}$, *where* $\sigma \in \mathbf{C}(t)$ *is a rational function of degree* $d \geq 2$ *and let* $\mathbf{X} \subset \mathbf{Y}$ *be an étale open subcorrespondence over* $\mathcal{O}$. *Assume condition 2.92 holds for infinitely many places* $\wp$. *Then either* $\sigma \sim t^{\pm d}$ *or* $\sigma \sim \pm Cheb_d$ *or* $\sigma \sim Lat_d$. *In particular* $\mathbf{X}_{\mathbf{C}}$ *has a non-empty open subcorrespondence that admits a flat analytic uniformization.*

Recall that we denoted by $Cheb_d$ the Chebyshev polynomial of degree d, Lat_d is any Lattès function of degree d, and $\sim$ denotes conjugation via an automorphism of the complex projective line $\mathbf{P}^1_{\mathbf{C}}$. With regards to assertion 1 the generic triviality of the $\nu-$canonical bundle mod p can also be viewed as reminiscent of "analytic uniformization". Indeed, as we will see in the last 3 Chapters generic triviality of the $\nu-$canonical bundle mod p can be proved "directly" in each of the spherical, flat, and hyperbolic cases; cf. Lemmas 6.4, 7.4, and 8.16. With regards to assertion 2 let us note that, as we shall see, if $\sigma = t^{\pm d}$ or $\sigma = \pm Cheb_d$ or $\sigma \sim Lat_d$ and if $\mathbf{X} = (X, \tilde{X}, \sigma_1, \sigma_2)$ is a suitable open étale subcorrespondence of $(\mathbf{A}^1, \mathbf{A}^1, id, \sigma)$ then condition 2.92 holds for almost all $\wp$.

2.3.6. Comments on other results. As already remarked, the horizontal results will be proved by looking at the vertical picture first. What we shall essentially do is consider correspondences $\mathbf{X}$ over $R = \hat{\mathbf{Z}}_p^{ur}$ which naturally arise from correspondences over number fields admitting a spherical, flat, or hyperbolic uniformization; we shall equip these $\mathbf{X}$'s with natural $\delta-$line bundles $\mathbf{L}$ (usually polarization equivalent to $\mathbf{K}^{-1}$ but in some sense more natural than $\mathbf{K}^{-1}$) and study the following types of problems.

1) *$\delta-$invariants.* We will address the non-vanishing and finite generation of the $R-$module $H^0(\mathbf{X}, \mathbf{L})$. We will study the structure of the ring $R(\mathbf{X}, \mathbf{L})^*$; this ring is often seen to be closely related to a ring of "homogeneous $\delta-$polynomials" $R\{x\}_*$ or $R\{x\}^*$ or a variant of these. (Cf. Chapter 5 for the definition of the latter rings.) We will then investigate $\delta-$rationality of $(\mathbf{X}, \mathbf{L})$.

2) *relative generic $\delta-$ fiber.* We will investigate the algebraicity and Galois theoretic properties of the extension

$$k(X) \cdot k\langle \mathbf{X}, \mathbf{L} \rangle \subset k\langle X, L \rangle.$$

The study of problems 1 and 2 above will be directly applied to the horizontal case. In addition to problems 1 and 2 we will also look at the following problems.

3) *$\delta-$cohomology.* We will study the space $H^1(\mathbf{X}, \mathbf{L})$, especially its $p-$torsion; typically we will show that the $p-$torsion of $H^1(\mathbf{X}, \mathbf{L})$ "tends to be" built up from certain classes that "come from" spaces of the form $H^0(\mathbf{X}, \mathbf{L}^m)$. This allows to view $H^1(\mathbf{X}, \mathbf{L})$ as being, to a certain extent, a "cohomology group of the quotient" rather than "of $(\mathbf{X}, \mathbf{L})$".

4) *Infinitesimal theory.* We will compute the $\delta-$ tangent maps of various remarkable maps to projective spaces. Also we will introduce the concept of *infinitesimal symmetry* of a $\delta-$section by analogy with the corresponding concept in mechanics (cf. Definition 3.48 below) and we shall investigate infinitesimal symmetries in the spherical, flat, and hyperbolic case. The presence of infinitesimal symmetries for our $\delta-$invariants should be viewed as a surprise; indeed our $\delta-$invariants are originally introduced as invariants for correspondences which can be thought of as *discrete* symmetries. Finally, we will introduce and study, in our examples, the concepts of *Hamiltonian flow* and *Euler-Lagrange equations.*

5) *Converse theorems for deformations.* Typically we will show that certain $p-$adic "deformations" $(\mathbf{X}', \mathbf{L}')$ of $(\mathbf{X}, \mathbf{L})$ satisfy $H^0(\mathbf{X}', \mathbf{L}') = 0$. This will be done by showing that the presence of $\delta-$invariants has an impact on the *multipliers* of the *cycles* of our correspondence. For $\mathbf{Q}-$anticanonical $\delta-$line bundles we will show that the existence of certain $\delta-$invariants implies what we will call "quadratic $\phi-$dependence" of multipliers; this is a "rationality" property which is rather rare. For $\mathbf{Q}-$canonical (and flat) $\delta-$line bundles, the existence of $\delta-$invariants will (very roughly speaking) imply a "symmetry" property of the *multiplier set.* One can view this symmetry property as an indication of the possible presence of a group behind the picture and, hence, of a possible analytic uniformization.

6) *Local analogues of the Conjectures.* Conjectures 2.54 and 2.55 above have a "local"(="formal") analogue in which étale correspondences are replaced by power series in $T \cdot R[[T]]^{\times}$ and "analytic uniformization" is replaced by "uniformization by a formal group". We will prove a number of results along these local conjectures.

The reader is referred to the "List of results" at the end of the book for some "loose statements" of some of the main results together with references to the precise statements in the text.

2.4. Appendix: Axiomatic characterization of $\delta-$geometries

This Appendix is not essential for the sequel and may be skipped. The aim of this Appendix is to show, following [**20**], that there are, essentially, only 4 ways to enlarge algebraic geometry by adjoining to it an operator that satisfies certain natural axioms with respect to addition and multiplication; the 4 geometries correspond to the Ritt-Kolchin differential algebra, the Ritt-Cohn difference algebra, the arithmetic differential algebra of this book, and one more type of algebra that we shall describe presently. So the moral of this Appendix is that the $\delta-$geometry developed in this book is essentially "forced upon us" once we seek an arithmetic analogue of the Ritt-Kolchin theory.

Here is our plan. We shall first introduce the concept of $\delta-$*rule.* Then we shall give four types of examples of such rules which correspond to the 4 types of algebra referred to above. Finally we shall define what we mean by *generic* $\delta-$rules and by *equivalent* $\delta-$rules and we shall prove that, "locally", any generic $\delta-$rule is equivalent to a $\delta-$rule of one of the 4 types.

DEFINITION 2.66. A $\delta-$*rule* is a pair (Σ, Π) of polynomials in 4 variables with integer coefficients,

$$\Sigma, \Pi \in \mathbf{Z}[X_0, Y_0, X_1, Y_1].$$

Let A be a ring and $f : A \to B$ an A–algebra. A δ–*operator* (relative to f, belonging to the δ–rule (Σ, Π)) is a map of sets $\delta : A \to B$ satisfying

$$\delta(0) = \delta(1) = 0, \tag{2.93}$$

$$\delta(x+y) = \Sigma(f(x), f(y), \delta(x), \delta(y)), \tag{2.94}$$

$$\delta(xy) = \Pi(f(x), f(y), \delta(x), \delta(y)), \tag{2.95}$$

for all $x, y \in A$. We will usually write δx instead of $\delta(x)$. Let p be any prime integer and $\nu \geq 1$ any integer. Consider the following δ–rules:

Type I:

$$\Sigma = X_1 + Y_1,$$
$$\Pi = X_0 Y_1 + Y_0 X_1.$$

Type II:

$$\Sigma = X_1 + Y_1,$$
$$\Pi = X_0 Y_1 + Y_0 X_1 + X_1 Y_1.$$

Type III:

$$\Sigma = X_1 + Y_1 + p^{-1}(X_0^{p^\nu} + Y_0^{p^\nu} - (X_0 + Y_0)^{p^\nu}),$$
$$\Pi = X_0^{p^\nu} Y_1 + Y_0^{p^\nu} X_1 + p X_1 Y_1,$$

Type IV:

$$\Sigma = X_1 + Y_1,$$
$$\Pi = X_0 Y_1 + Y_0 X_1 + p^\nu X_1 Y_1.$$

A δ–operator belonging to a rule of Type I or II is called a *derivation operator* or a *difference operator* respectively. A δ–operator belonging to a rule of Type III with $\nu = 1$ was called, in this book, a p–*derivation operator*; δ–operators belonging to a rule of Type IV can be called p–*difference operators*.

If $\delta : A \to B$ is a δ–operator belonging to a δ–rule of type II, III, or IV then we always denote by $\phi : A \to B$ the map defined by

$$\phi(x) = x + \delta x,$$
$$\phi(x) = x^{p^\nu} + p\delta x,$$
$$\phi(x) = x + p^\nu \delta x,$$

respectively. Then ϕ is trivially seen to be a ring homomorphism.

DEFINITION 2.67. Fix any δ–rule of Type I, II, III, or IV and let $\delta : R \to R$ be a δ–operator relative to the identity, belonging to this δ–rule, where R is an integral domain of characteristic zero. Let T be a family of indeterminates (T_j). We define the *ring of δ–polynomials* $R\{T\}$ as follows. Let $T', T'', ..., T^{(r)}, ...$ be families of indeterminates indexed by the same set as T and set

$$R\{T\} := R[T, T', T'', ..., T^{(r)}, ...], \tag{2.96}$$

the ring of polynomials in the variables $T, T', T'', ..., T^{(r)}, ...$ There is a unique operator $\delta : R\{T\} \to R\{T\}$, extending $\delta : R \to R$, such that $\delta T_j^{(i)} = T_j^{(i+1)}$. Indeed, for Type I, we must take

$$\delta F = F^\delta + \sum_{i=0}^{\infty} \sum_j T_j^{(i+1)} \frac{\partial F}{\partial T_j^{(i)}},$$

where F^δ is obtained from F by applying δ to its coefficients. For types II, III, IV we consider the unique ring homomorphism $\phi : R\{T\} \to R\{T\}$, extending $\phi : R \to R$, and satisfying:

$$\phi(T_j^{(i)}) = T_j^{(i)} + T_j^{(i+1)},$$

$$\phi(T_j^{(i)}) = (T_j^{(i)})^{p^\nu} + pT_j^{(i+1)},$$

$$\phi(T_j^{(i)}) = T_j^{(i)} + p^\nu T_j^{(i+1)},$$

respectively and define $\delta : R\{T\} \to R\{T\}$ by the formulae:

$$\delta F = \phi(F) - F,$$

$$\delta F = \frac{\phi(F) - F^{p^\nu}}{p},$$

$$\delta F = \frac{\phi(F) - F}{p^\nu},$$

respectively.

DEFINITION 2.68. Let A be a ring. A set theoretic map $\delta : A \to A$ is *generic* if, whenever $G \in A[X_0, X_1]$ is a polynomial such that $G(x, \delta x) = 0$ for all $x \in A$, we must have $G = 0$. A δ–rule (Σ, Π) will be called *generic* if there exists a ring A of characteristic zero and there exists a δ– operator $\delta : A \to A$ (relative to the identity, belonging to (Σ, Π)) which is generic.

REMARK 2.69. Any δ–rule of Type I, II, III, IV is generic. Indeed, start with any δ–operator $\delta : R \to R$ on an integral domain of characteristic zero; e.g. one can take $R = \mathbf{Z}$ and $\delta = 0$ for Types I, II, IV respectively $\delta x = (x - x^p)/p$ for Type III. Then one can take T to be one indeterminate and take A in Definition 2.68 to be the ring of δ–polynomials $R\{T\}$ with its canonical δ–operator.

DEFINITION 2.70. Two δ–rules (Σ_1, Π_1) and (Σ_2, Π_2) are called *equivalent* over a ring R if there exists an invertible element $\lambda \in R^\times$ and a polynomial $F \in R[X_0]$ with $F(0) = F(1) = 0$ such that, for any R–algebra A and any δ–operator δ_1 : $A \to A$ belonging to (Σ_1, Π_1), the map $\delta_2 : A \to A$ defined by $\delta_2 x = \lambda\delta_1 x + F(x)$ is a δ–operator belonging to (Σ_2, Π_2).

The have the following classification result:

THEOREM 2.71. **[20]** *Let R be a local p–ring with residue field K.*

1) Any generic δ–rule is equivalent, over R, to one of the δ–rules of Type I, II, III, or IV.

2) Any generic δ–rule is equivalent over K to one of the δ–rules of Type I or II.

For the proof we need a series of preliminary remarks.

REMARK 2.72. 1) Let A be a ring and let $\delta : A \to A$ be a generic set theoretic map. Assume $G \in A[X_{10}, ..., X_{n0}, X_{11}, ...X_{n1}]$ is a polynomial such that

$$G(x_1, ..., x_n, \delta x_1, ..., \delta x_n) = 0$$

for all $x_1, ..., x_n \in A$. Then $G = 0$. Indeed this follows immediately by induction on n. As a consequence any generic δ–operator belongs to exactly one δ–rule.

2) Let (Σ, Π) be a generic δ–rule. Then, for *any* ring A, the set $A \times A$ becomes a ring with respect to the composition laws

$$(x_0, x_1) \oplus (y_0, y_1) = (x_0 + y_0, \Sigma(x_0, y_0, x_1, y_1)) \tag{2.97}$$

$$(x_0, x_1) \otimes (y_0, y_1) = (x_0 y_0, \Pi(x_0, y_0, x_1, y_1)); \tag{2.98}$$

the zero element is $(0, 0)$ and the unit element is $(1, 0)$. Indeed, let's check for instance that the addition on $A \times A$ is associative. We must check that the following two polynomials in 6 variables

$$\Sigma(X_0 + Y_0, Z_0, \Sigma(X_0, Y_0, X_1, Y_1), Z_1) \quad \textit{and} \quad \Sigma(X_0, Y_0 + Z_0, X_1, \Sigma(Y_0, Z_0, Y_1, Z_1))$$

are equal. By part 1 of our Remark it is sufficient to check that the above polynomials become equal when $X_0, Y_0, Z_0, X_1, Y_1, Z_1$ are replaced by $x, y, z, \delta x, \delta y, \delta z$, for any $x, y, z \in B$, where B is a ring of characteristic zero equipped with an operator $\delta : B \to B$ which is generic. But, under this substitution, both polynomials become $\delta(x + y + z)$. The rest of the ring axioms, including the statement about the zero and the unit being $(0, 0)$ and $(1, 0)$, can be proved similarly. By the way the inverse for addition is given by

$$-(x_0, x_1) = (-x_0, \Pi(-1, x_0, \delta(-1), x_1)).$$

3) Two generic δ–rules (Σ_1, Π_1) and (Σ_2, Π_2) are equivalent over a ring R if and only if there exists an invertible element $\lambda \in R^\times$ and a polynomial $F \in R[X_0]$ with $F(0) = F(1) = 0$ such that for any R–algebra A the map

$$(A \times A, \oplus_1, \otimes_1) \to (A \times A, \oplus_2, \otimes_2),$$

$$(x_0, x_1) \mapsto (x_0, \lambda x_1 + F(x_0))$$

is a ring isomorphism; here $\oplus_i, \otimes_i$ are the corresponding composition laws. To check this one imitates the argument in part 2 of this Remark.

4) A map $\delta : A \to A$ is a δ–operator belonging to a generic δ–rule (Σ, Π) if and only if the map

$$id \times \delta : A \to (A \times A, \oplus, \otimes)$$

is a ring homomorphism. Also note that if (Σ, Π) is of one of the Types I, II, III, IV then the corresponding ring structure on $A \times A$ is a familiar one. For instance, for Type I,

$$(A \times A, \oplus, \otimes) \simeq A[\epsilon],$$

the ring of dual numbers. For Type II,

$$(A \times A, \oplus, \otimes) \simeq (A, +, \cdot) \times (A, +, \cdot),$$

the twofold product of the ring A. For Type III (and $\nu = 1$),

$$(A \times A, \oplus, \otimes) \simeq W_2(A),$$

the ring of Witt vectors of length 2 on A. A similar picture holds for Type III (general ν) and Type IV.

5) Let A be any ring and let $\mathbf{A}^1_A = \mathit{Spec}\ A[X_0]$, $\mathbf{A}^2_A = \mathit{Spec}\ A[X_0, X_1]$ be the affine line and the affine plane over A. Denote by $\Theta(A)$ the set of all ring A–scheme structures on $\mathbf{A}^2_A$ such that the first projection $pr_1 : \mathbf{A}^2_A \to \mathbf{A}^1_A$ is a ring A–scheme homomorphism and such that the zero and the unit elements in $\mathbf{A}^2_A$ correspond to $(0, 0), (1, 0)$. We can identify $\Theta(A)$ with the set of all pairs (Σ, Π) where $\Sigma, \Pi \in A[X_0, X_1, Y_0, Y_1]$ such that for any A–algebra B the formulae in Equations 2.97 and 2.98 define a ring structure on $B \times B$ with zero and unit

equal to $(0,0)$ and $(1,0)$ respectively. On the other hand consider the group $\Gamma(A)$ of all A-scheme automorphisms σ of $\mathbf{A}_A^2$ that fix $(0,0)$ and $(1,0)$, and for which $pr_1 \circ \sigma = pr_1$. Assume in addition that A is an integral domain of characteristic zero. Then looking at Jacobian matrices, one immediately sees that any such σ is given by

$$(x_0, x_1) \mapsto (x_0, \lambda x_1 + f(x_0))$$

for some $\lambda \in A^\times$ and some $f \in A[X]$ with $f(0) = f(1) = 0$. So we can identify $\Gamma(A)$ with $A^\times \times (X^2 - X)A[X]$, the latter equipped with the semidirect product group structure

$$(\lambda_1, f_1) \cdot (\lambda_2, f_2) = (\lambda_1\lambda_2, f_1 + \lambda_1 f_2)$$

Now $\Gamma(A)$ acts on $\Theta(A)$ by transport of structure. Explicitly, if $(\lambda, f) \in \Gamma(A)$ and $(\Sigma, \Pi) \in \Theta(A)$ then we have

$$(\lambda, f) \cdot (\Sigma, \Pi) = (\Sigma', \Pi')$$

where

$$\Sigma' = \lambda \cdot \Sigma(X_0, Y_0, \lambda^{-1}[X_1 - f(X_0)], \lambda^{-1}[Y_1 - f(Y_0)]) + f(X_0 + Y_0)$$
$$\Pi' = \lambda \cdot \Pi(X_0, Y_0, \lambda^{-1}[X_1 - f(X_0)], \lambda^{-1}[Y_1 - f(Y_0)]) + f(X_0 Y_0).$$

As was previously noted any generic $\delta-$ rule gives an element of $\Theta(\mathbf{Z})$. Also, two generic $\delta-$ rules are equivalent over a ring A if and only if, as elements of $\Theta(A)$, they are $\Gamma(A)-$ conjugate.

We are ready to prove Theorem 2.71.

PROOF. We first prove assertion 2 of the Theorem. Note that any element of $\Theta(K)$ defines, in particular, a commutative algebraic group structure on $\mathbf{A}_K^2$ with zero element $(0,0)$ such that the first projection to $\mathbf{A}_K^1$ is a morphism of algebraic groups. Since the kernel of the first projection is isomorphic, as a variety, to $\mathbf{A}_K^1$, and since, by the theory of algebraic groups, any algebraic group structure on $\mathbf{A}_K^1$ is isomorphic to the usual additive group structure, it follows that our algebraic group $\mathbf{A}_K^2$ is an extension of the additive group by itself. By [**119**] p.171, this extension is trivial. So any pair $(\Sigma, \Pi) \in \Theta(K)$ is conjugate, under the action of $\Gamma(K)$, to a pair (Σ', Π'), where $\Sigma' = X_1 + Y_1$. By distributivity and commutativity in the ring structure axioms Π' must be a bilinear symmetric form i.e.

$$\Pi' = \alpha X_0 Y_0 + \beta(X_0 Y_1 + X_1 Y_0) + \gamma X_1 Y_1$$

Since $\Pi'(1, Y_0, 0, Y_1) = Y_1$ we get $\alpha = 0$ and $\beta = 1$. If $\gamma = 0$ then $(\Sigma', \Pi') = (X_1 + Y_1, X_0 Y_1 + Y_0 X_1)$ and we are done. If $\gamma \neq 0$ note that

$$(\Sigma', \Pi') = (\gamma^{-1}, 0)(X_1 + Y_1, X_0 Y_1 + Y_0 X_1 + X_1 Y_1)$$

and we are done again.

Next we prove assertion 1 of the Theorem. There are two cases to examine for a pair $(\Sigma, \Pi) \in \Theta(R)$:

Case 1. (Σ, Π) is $\Gamma(K)-$conjugate to $(X_1 + Y_1, X_0 Y_1 + Y_0 X_1)$.

Case 2. (Σ, Π) is $\Gamma(K)-$ conjugate to $(X_1 + Y_1, X_0 Y_1 + Y_0 X_1 + X_1 Y_1)$.

Assume we are in Case 1. In this case we claim that (Σ, Π) is actually $\Gamma(R)-$conjugate to

$$(X_1 + Y_1, X_0 Y_1 + Y_0 X_1)$$

(which ends the proof in Case 1.) Indeed write

$$(\Sigma, \Pi) = (\lambda, f)(X_1 + Y_1, X_0 Y_1 + Y_0 X_1)$$

where $\lambda \in K^\times$ and $f \in (X^2 - X)K[X]$. By the form of the action

$$\Sigma = X_1 + Y_1 + f(X_0 + Y_0) - f(X_0) - f(Y_0),$$
$$\Pi = X_0Y_1 + Y_0X_1 + f(X_0Y_0) - X_0f(Y_0) - Y_0f(X_0).$$

The above formulae show that we may assume $\lambda = 1$. Also, since Π has coefficients in R, it follows immediately that f has coefficients in R. So $(\lambda, f) \in \Gamma(R)$ and our claim is proved.

Assume now we are in Case 2. Write

$$(\Sigma, \Pi) = (\lambda, f)(X_1 + Y_1, X_0Y_1 + Y_0X_1 + X_1Y_1)$$

where $\lambda \in K^\times$ and $f \in (X^2 - X)K[X]$. Hence

$$\Sigma = X_1 + Y_1 + f(X_0 + Y_0) - f(X_0) - f(Y_0) \tag{2.99}$$
$$\Pi = X_0Y_1 + Y_0X_1 - X_0f(Y_0) - Y_0f(X_0)$$
$$+\lambda^{-1}[X_1Y_1 - X_1f(Y_0) - Y_1f(X_0) + f(X_0)f(Y_0)] + f(X_0Y_0).$$

Setting $f^*(X) = f(X) - \lambda X$ we obtain

$$\Pi = \lambda^{-1}[X_1Y_1 - X_1f^*(Y_0) - Y_1f^*(X_0) + f^*(X_0)f^*(Y_0)] + f^*(X_0Y_0). \tag{2.100}$$

Since Π has coefficients in R, looking at the coefficients of X_1Y_1 and $X_1Y_0^i$ we get that $\lambda^{-1} \in R$ and $\lambda^{-1}f^* \in XR[X]$. If $\lambda \in R$ we have $(\lambda, f) \in \Gamma(R)$ hence (Σ, Π) is $\Gamma(R)$-conjugate to a δ-rule of Type I or II and we are done. Hence we may assume from now on that $\lambda \notin R$, i.e. $\pi := \lambda^{-1} \in pR$. Let us write $f = f_1X + f_2X^2 + ... + f_mX^m$. Assume for a moment that $f_i \in R$ for all $i \geq 2$. Since $f(1) = 0$ it follows that $f_1 \in R$ as well. In this case note that

$$(1, -f)(\Sigma, \Pi) = (X_1 + Y_1, X_0Y_1 + Y_0X_1 + \pi X_1Y_1). \tag{2.101}$$

Since $(1, -f) \in \Gamma(R)$ and since the right hand side of Equation 2.101 is $\Gamma(R)$-conjugate to a δ-rule of Type IV we are done in this case. So from now on we may assume that the following condition holds:

(*) There exists an index $i_0 \geq 2$ such that $f_{i_0} \notin R$.

Recall that by a *cocycle* $G \in R[X, Y]$ one understands a polynomial G that satisfies

$$G(Y, Z) - G(X + Y, Z) + G(X, Y + Z) - G(X, Y) = 0.$$

By the *coboundary* associated to a polynomial $\phi \in R[X]$ one understands the polynomial in two variables

$$(\partial\phi)(X, Y) := \phi(X + Y) - \phi(X) - \phi(Y).$$

Since Σ has coefficients in R, it follows that

$$G(X_0, Y_0) := f(X_0 + Y_0) - f(X_0) - f(Y_0)$$

is a cocycle in $R[X_0, Y_0]$. By the result on the structure of cocycles in [**90**], there exist a polynomial $\phi \in R[X]$ and there exist $a_0, a_1, a_2, ... \in R$ such that

$$G(X_0, Y_0) = \phi(X_0 + Y_0) - \phi(X_0) - \phi(Y_0) + \sum_{j \geq 1} a_jp^{-1}[(X_0 + Y_0)^{p^j} - X_0^{p^j} - Y_0^{p^j}].$$

It follows that the polynomial $f - \phi - \sum_{j\geq 1} a_jp^{-1}X^{p^j}$ is additive, hence it is a monomial of degree one. Hence we have $pf_i \in R$ for all $i \geq 2$. Let $\pi := \lambda^{-1}$. Since, by Equation 2.100, the coefficient $\pi f_{i_0}^2 + f_{i_0}$ of $X_0^{i_0}Y_0^{i_0}$ in Π is in R we must have $\pi f_{i_0} \in R^\times$ so we get $pR = \pi R$. Set $F = -\pi f^*$. As we have seen $F \in R[X]$ and note that the reduction of F modulo p, $\bar{F} \in (R/pR)[X]$, has degree

≥ 2 because the coefficient of X^{i_0} in F is invertible. Multiplying Equations 2.99 and 2.100 by π and reducing modulo p we find that $\bar{F}(X_0 + Y_0) = \bar{F}(X_0) + \bar{F}(Y_0)$ and $\bar{F}(X_0 Y_0) = \bar{F}(X_0)\bar{F}(Y_0)$. The latter (multiplicativity) relation implies that $\bar{F} = X^m$ for some m. Using the additivity relation for $\bar{F}$ we get that $m = q \neq 1$ is an integer power of p. Consequently we may write $F = X^q + \pi g$, $g \in R[X]$. Then a straightforward computation shows that (Σ, Π) equals

$$(1, -g)(X_1 + Y_1 + \pi^{-1}[X_0^q + Y_0^q - (X_0 + Y_0)^q], X_0^q Y_1 + Y_0^q X_1 + \pi X_1 Y_1).$$

The latter is clearly $\Gamma(R)-$ conjugate to a $\delta-$rule of Type III. This ends the proof of the Theorem. □

Part 2

General theory

CHAPTER 3

Global theory

The ring $\mathcal{O}^r(X)$ of δ–functions of order $\leq r$ on a smooth scheme X over $R := \hat{\mathbf{Z}}_p^{ur}$ naturally identifies with the ring of global functions, $\mathcal{O}(J^r(X))$, on a certain formal scheme, $J^r(X)$, called the r–th p–jet space of X; cf. [**17**], [**18**], [**21**]. The geometry of these p–jet spaces is responsible for most results involving δ–functions. In this Chapter we provide an exposition of p–jet theory partly following (and complementing) the above mentioned papers.

The Chapter is organized as follows. In Section 1 we review the construction of p–jet spaces. In Section 2 we examine the behavior of p–jet spaces with respect to étale maps. Based on this analysis we will be able, in Section 3, to prove the isomorphism $\mathcal{O}^r(X) \simeq \mathcal{O}(J^r(X))$ and some related results. We will then go back to analyzing étale covers in Section 4 and prove that "taking invariants under fixed point free finite group actions" commutes with "taking p–jet spaces". In Section 5 we examine the interaction between δ–base loci and "postcritical loci" and we prove the global "converse" Theorem 2.64. Section 6 is devoted to the infinitesimal theory: we shall introduce and discuss δ–tangent spaces and maps, δ–differentials of δ–functions, infinitesimal symmetries, etc. This infinitesimal theory will play a key role in our applications.

3.1. p–jet spaces of schemes

We start by considering a basic δ–ring of the theory.

DEFINITION 3.1. Let T be the set of variables $T_1, ..., T_N$. We define a δ–ring $R\{T\}$, called the *ring of δ–polynomials* in T, as follows. Let $T', T'', ..., T^{(r)}, ...$ be families of indeterminates indexed by the same set as T and set

$$R\{T\} := R[T, T', T'', ..., T^{(r)}, ...], \tag{3.1}$$

the ring of polynomials in the variables $T, T', T'', ..., T^{(r)}, ...$ One defines operators

$$\phi, \delta : R\{T\} \to R\{T\}$$

by the formulae

$$\phi(f) = f^{(\phi)}(T^p + pT', (T')^p + pT'', ..., T^{(r)p} + pT^{(r+1)}) \tag{3.2}$$

$$\delta f = \frac{\phi(f) - f^p}{p}$$

where $(T^{(i)})^p$ denotes the N–tuple $(T_1^{(i)})^p, ..., (T_N^{(i)})^p$ and $f^{(\phi)}(T, ..., T^{(r)})$ is obtained from $f(T, ..., T^{(r)})$ by acting with ϕ on the coefficients. As usual we also use the notation

$$\phi(f) = f^\phi = f(T, ..., T^{(r)})^\phi.$$

Then ϕ is a ring homomorphism lifting Frobenius and δ is a $p-$derivation. Clearly $R\{T\}$ is a global $\delta - p-$ring. Hence $R\{T\}_{(p)}$ and its completion $(R\{T\}_{(p)})\hat{}$ are local $\delta - p-$rings.

DEFINITION 3.2. A *prolongation sequence* S^* is a sequence

$$S^0 \stackrel{\varphi_0}{\rightarrow} S^1 \stackrel{\varphi_1}{\rightarrow} S^2 \stackrel{\varphi_2}{\rightarrow} \dots$$

of ring homomorphisms together with a sequence of maps

$$S^0 \stackrel{\delta_0}{\rightarrow} S^1 \stackrel{\delta_1}{\rightarrow} S^2 \stackrel{\delta_2}{\rightarrow} \dots$$

such that each δ_n is a $p-$derivation relative to φ_n and each δ_n prolongs δ_{n-1}, i.e. $\delta_n \circ \varphi_{n-1} = \varphi_n \circ \delta_{n-1}$.

By abuse we denote all φ_n by φ and all δ_n by δ. The φ's are referred to as *structure homomorphisms* and we shall usually drop them from our notation altogether; the δ's are referred to as the *structure $p-$derivations*. By a morphism of prolongation sequences $S^* \to U^*$ we understand a sequence of ring homomorphisms $u_n : S^n \to U^n$ such that $u_n \circ \varphi = \varphi \circ u_{n-1}$ and $u_n \circ \delta = \delta \circ u_{n-1}$ for all $n \geq 1$. A prolongation sequence *over* R will mean a prolongation sequence all of whose rings and algebra maps are $R-$algebra maps and all of whose δ_n are compatible with the δ on R. Prolongation sequences form, in an obvious way, a category. A basic example of a prolongation sequence of $p-$adically complete rings can be obtained as follows. Let T be the set of variables $T_1, \dots, T_N$ and let $T', T'', \dots, T^{(r)}, \dots$ be new sets of N variables as above. The $p-$derivation $\delta : R\{T\} \to R\{T\}$ on the ring of $\delta-$polynomials induces $p-$derivations

$$\delta : R[T, T', \dots, T^{(r)}]\hat{} \to R[T, T', \dots, T^{(r+1)}]\hat{}$$

relative to the inclusions. Then one can consider the prolongation sequence

$$(R[T, T', \dots, T^{(r)}]\hat{})$$

with structure homomorphisms given by the inclusions and structure $p-$derivations δ induced by δ in Equation 3.2.

Now if X is affine, equal, say, to *Spec* $R[T]/I$ for some ideal I then one defines the *$r-th$ $p-jet$ space* of X as being the formal scheme

$$J^r(X) := Spf\ R[T, T', \dots, T^{(r)}]\hat{}/(I, \delta I, \dots, \delta^r I). \tag{3.3}$$

In particular $J^0(X) = \hat{X}$. The sequence of rings of global regular functions $(\mathcal{O}(J^n(X)))$ has a natural structure of prolongation sequence $\mathcal{O}(J^*(X))$ induced by the operators δ above. This prolongation sequence has the following universality property.

PROPOSITION 3.3. *Assume X is affine. Then for any prolongation sequence S^* over R, consisting of $p-$adically complete rings, and any homomorphism $u : \mathcal{O}(X) \to S^0$ over R there is a unique morphism of prolongation sequences over R,*

$$u_* = (u_n) : \mathcal{O}(J^*(X)) \to S^*,$$

with $u_0 = u$.

PROOF. The case $I = 0$ of the Proposition clearly implies the general case. Assume $I = 0$ and define

$$u_n : R[T, \dots, T^{(n)}]\hat{} \to S^n$$

by the rule

$$u_n(T^{(i)}) = \delta^i(u(T))$$

for $i \leq n$. To check that $u_{m+1} \circ \delta = \delta \circ u_m$ for $m \geq 0$ one simply notes that the set

$$\{F \in R[T, ..., T^{(m)}]\hat{}\,|u_{m+1}(\delta(F)) = \delta(u_m(F))\}$$

is a $p-$adically closed subring of $R[T, ..., T^{(m)}]\hat{}$ containing R and $T, T', ..., T^{(m)}$, therefore it must coincide with $R[T, ..., T^{(m)}]\hat{}$. □

As a consequence we get that the construction $X \mapsto J^r(X)$, for affine X, is compatible with localization in the following sense:

COROLLARY 3.4. *If $X = Spec\ A$ and $U = Spec\ A_f$, $f \in A$ then $\mathcal{O}(J^r(U)) = (\mathcal{O}(J^r(X))_f)\hat{}$; equivalently $J^r(U) = J^r(X) \times_{\hat{X}} \hat{U}$.*

PROOF. First note that the sequence of rings $(\mathcal{O}(J^*(X))_f)\hat{}$ has a natural structure of prolongation sequence given by

$$\delta\left(\frac{g}{f^n}\right) = \frac{f^{np}\delta g - g^p\delta(f^n)}{f^{2np}} \cdot \left[1 - p\frac{\delta f}{f^p} + p^2\left(\frac{\delta f}{f^p}\right)^2 - p^3\left(\frac{\delta f}{f^p}\right)^3 + ...\right].$$

The prolongation sequence $(\mathcal{O}(J^*(X))_f)\hat{}$ is then easily seen to satisfy the universality property of $\mathcal{O}(J^*(U))$. □

Consequently, if X is not necessarily affine then one can define a formal scheme $J^r(X)$ by gluing the various $J^r(U_i)$ for $X = \bigcup U_i$ an affine Zariski covering.

DEFINITION 3.5. The formal scheme $J^r(X)$ is called the *$r-$th $p-$jet space* of X.

We clearly get a projective system of formal schemes

$$... \to J^r(X) \to J^{r-1}(X) \to ... \to J^1(X) \to J^0(X) = \hat{X}. \tag{3.4}$$

This projective system (together with the underlying $p-$derivations that exist locally) satisfies an obvious universality property which globalizes the one in Proposition 3.3. By this universality property one obtains functors $X \mapsto J^r(X)$ from the category of schemes of finite type over R to the category of formal schemes over R. Also, by functoriality, if G is a group scheme of finite type over R then $J^r(G)$ have naturally induced structures of group formal schemes over R and the projections in Equation 3.4 are group homomorphisms in the category of formal schemes.

The formal schemes $J^r(X)$ are analogues of the jet spaces in classical mechanics; cf. Equation 0.6. The functions in $\mathcal{O}(J^r(X))$ play the role of differential functions (or Lagrangians); cf. Equation 0.8. The $p-$derivations

$$\delta : \mathcal{O}(J^r(X)) \to \mathcal{O}(J^{r+1}(X))$$

play the role of total derivatives of Lagrangians; cf. Equation 0.11.

For any scheme (or formal scheme) X over R we will usually denote by X_0 the $k-$scheme $X \otimes k$.

DEFINITION 3.6. A scheme X over R has a *lift of Frobenius* if there exists a commutative diagram of schemes

$$\begin{array}{ccc} X & \stackrel{\tilde{\phi}}{\rightarrow} & X \\ \downarrow & & \downarrow \\ Spec\ R & \stackrel{\phi}{\rightarrow} & Spec\ R \end{array}$$

such that $\tilde{\phi} \otimes k : X_0 \to X_0$ is the p-power Frobenius. We say $\hat{X}$ has a *lift of Frobenius* if there is a similar diagram of formal schemes as above with X replaced by $\hat{X}$.

Note that if X is smooth and affine then, as a direct consequence of the definition of smoothness, $\hat{X}$ has a lift of Frobenius. By the universality property of the sequence of p-jet spaces we get

COROLLARY 3.7. *The projection $J^1(X) \to \hat{X}$ has a section if and only if $\hat{X}$ has a lift of Frobenius. In particular, if X is smooth over R, $J^1(X) \to \hat{X}$ admits sections locally in the Zariski topology.*

DEFINITION 3.8. In what follows, for any k-scheme Y we denote by $F_Y : Y \to Y$ the absolute Frobenius endomorphism (hence F_Y is the identity on the topological space of Y and is the p-power map on regular functions) and by

$$FT(Y/k) = Spec\ Symm(F_Y^* \Omega_{Y/k}) \tag{3.5}$$

the *Frobenius tangent scheme* (*Frobenius tangent bundle*, if Y is smooth).

PROPOSITION 3.9. *Let X be a smooth scheme of finite type over R. Then the morphism*

$$J^1(X)_0 \to X_0$$

is a (Zariski locally trivial) principal homogeneous space for the Frobenius tangent bundle $FT(X_0/k) \to X_0$. In particular, if X is a group scheme over R then $Ker(J^1(X)_0 \to X_0)$ is canonically isomorphic as an algebraic group to the Lie algebra of X_0.

PROOF. We construct the structure of principal homogeneous space locally but in a functorial way; this will allow us to glue the local pictures. So assume $X = Spec\ A$ is affine. Then, for any k-algebra B, to give a morphism $Spec\ B \to J^1(X) \otimes k$ is easily seen to be the same as to give a morphism $u : Spec\ B \to X$ together with a p-derivation $\delta : A \to B$ relative to $u^* : A \to B$. On the other hand to give a morphism $Spec\ B \to FT(X_0/k)$ is easily seen to be the same as to give a morphism $u : Spec\ B \to X$ together with a derivation $D : A \to B$ relative to $F_B \circ u^* : A \to B$ where F_B is the p-power Frobenius of B. Now we have the following simple fact. Let $f : A \to B$ be a ring homomorphism where B has characteristic p. Then the set of p-derivations relative to f is either empty or a principal homogeneous space for the group of all derivations relative to $F_B \circ f$. So the functor of points of $J^1(X)_0$ is a principal homogeneous space for the functor of points of $FT(X_0/k)$. We conclude that $J^1(X) \otimes k$ is a principal homogeneous space for $FT(X_0/k)$. Note that $J^1(X)_0 \to X_0$ has local sections in the Zariski topology by Corollary 3.7. □

DEFINITION 3.10. The cohomology class in $H^1(X_0, F^*T_{X_0/k})$ of the principal homogeneous space in Proposition 3.9 is called the *(internal) Kodaira-Spencer class* of X and is denoted by $KS(X)$.

As a rule the word *internal* will be dropped in what follows without notice. The usual Kodaira-Spencer class encountered in deformation theory (which can be thought of to be *external*) will also play an implicit role later in this work but no confusion should arise. The class in Definition 3.10 can be introduced, of course, directly, without the use of $p-$jet spaces and plays a role in the work of Deligne-Illusie [**43**] and Raynaud. In any case, it is computed explicitly as follows. One covers X by affine open subsets U_i, one lifts the $p-$derivation δ on R to $p-$derivations δ_i on $\mathcal{O}(U_i)\hat{}$, one considers the differences

$$(\delta_i - \delta_j) \otimes k$$

which turn out to induce usual derivations relative to the Frobenius map of

$$\mathcal{O}(U_i \cap U_j) \otimes k,$$

satisfying the cocycle condition, and one takes $KS(X)$ to be the cohomology class defined by this cocycle.

REMARK 3.11. It can be shown [**18**] that if X is a smooth scheme over R, with irreducible geometric fibers, the proalgebraic $k-$variety

$$\lim_{\leftarrow} J^r(X)_0$$

identifies with the so-called *Greenberg transform* of X. We will not need this so we will not pursue this further. For the Greenberg transform (originally introduced by S. Lang in his thesis) we refer to [**63**]. On the other hand note that our basic $p-$derivation δ on

$$\lim_{\rightarrow} \mathcal{O}(J^r(X))$$

does not preserve the ideal (p) hence it does not induce an operator on

$$\lim_{\rightarrow} \mathcal{O}(J^r(X)_0);$$

consequently the Greenberg transform "forgets" about δ hence about the differential aspect of our theory.

For the sake of completeness we close this Section by recording a result from [**18**] that will not be used in the sequel; we shall omit its proof.

THEOREM 3.12. [**18**] *Let X be a smooth projective curve over R, of genus ≥ 2. Then, for all $n \geq 1$, the formal schemes $J^n(X)$ are affine.*

As explained in [**18**] Theorem 3.12 fails if X has genus 0 or 1. In [**18**] Theorem 3.12 is shown to imply the Manin-Mumford conjecture saying that the intersection of a complex curve of genus ≥ 2 with the torsion subgroup of its Jacobian is always finite. This conjecture was first proved by Raynaud [**115**]. The method in [**18**] yields in addition an explicit bound for the cardinality of the intersection.

3.2. Behavior with respect to étale maps

PROPOSITION 3.13. [**17**] *Let X be a smooth affine scheme over R and let $u : R[x] \rightarrow \mathcal{O}(X)$ be an étale morphism where x is a finite family of indeterminates. Let $x', x'', ..., x^{(r)}$ be families of new indeterminates indexed by the same set as x. Then the natural morphism*

$$\mathcal{O}(X)[x', x'', ..., x^{(r)}]\hat{} \rightarrow \mathcal{O}(J^r(X)),$$

sending $x^{(i)}$ into $\delta^i(u(x))$ is an isomorphism.

In order to prove the above Proposition we need:

LEMMA 3.14. [17] *Let A be a Noetherian $R-$algebra, $u : A \to B$ a finitely generated étale $A-$algebra, and $v : B \to C$ a ring homomorphism into a $p-$adically complete ring C. Then any $p-$derivation relative to $v \circ u$ lifts uniquely to a $p-$derivation relative to v.*

PROOF. We may assume $p^nC = 0$ for some n. Set $f = v \circ u$, let δ be a $p-$derivation relative to f, and consider the commutative diagram of rings

$$\begin{array}{ccc} A & \stackrel{u}{\to} & B \\ (f,\delta)\downarrow & & \downarrow v \\ W_2(C) & \stackrel{pr_1}{\to} & C \end{array}$$

where $W_2(C)$ is the ring of Witt vectors of length 2 on C and pr_1 is the first projection. Since u is étale and $(Ker\ pr_1)^{n+1} = 0$, there exists a unique ring homomorphism $\sigma : B \to W_2(C)$ such that $pr_1 \circ \sigma = v$ and $\sigma \circ u = (f, \delta)$ which means exactly that δ extends uniquely to a $p-$derivation relative to v. □

PROOF. (of Proposition 3.13.) Set $X = Spec\ B = Spec\ R[T]/I$ and

$$B^n := R[T, T', ..., T^{(n)}]/(I, I', ..., I^{(n)});$$

so $(B^n)\hat{} = \mathcal{O}(J^n(X))$. Consider the prolongation sequence

$$R[x] \stackrel{\delta}{\to} R[x, x'] \stackrel{\delta}{\to} R[x, x', x''] \to ...$$

with δ sending x into x', x' into x'' a.s.o. By the existence statement in Lemma 3.14 these $p-$derivations lift to $p-$derivations

$$B \stackrel{\delta}{\to} B[x']\hat{},\quad B[x'] \stackrel{\delta}{\to} B[x', x'']\hat{}, ...$$

The latter $p-$derivations induce $p-$derivations

$$\hat{B} \stackrel{\delta}{\to} B[x']\hat{} \stackrel{\delta}{\to} B[x', x'']\hat{} \stackrel{\delta}{\to} ... \tag{3.6}$$

By the uniqueness statement in Lemma 3.14 each $p-$derivation in the sequence 3.6 prolongs the preceding one; so the sequence 3.6 is a prolongation sequence. To conclude it is enough to check that the prolongation sequence 3.6 satisfies the universality property of the sequence

$$B\hat{} \stackrel{\delta}{\to} (B^1)\hat{} \stackrel{\delta}{\to} (B^2)\hat{} \stackrel{\delta}{\to} ...$$

appearing in Proposition 3.3. But this is a trivial exercise using the uniqueness in Lemma 3.14.

□

As a consequence of Proposition 3.13 we get the following:

COROLLARY 3.15. *If X is a smooth scheme over R with non-empty irreducible closed fiber $X \otimes k$ then $\mathcal{O}(J^r(X))$ are global $p-$rings.*

PROOF. Let us show that p is a prime element in $\mathcal{O}(J^r(X))$. Assume $fg = ph$ with $f, g, h \in \mathcal{O}(J^r(X))$. Fix an affine open subset $U \subset X$. Then $fg = ph$ in $\mathcal{O}(J^r(U))$. By Proposition 3.13,

$$\mathcal{O}(J^r(U)) \otimes k \simeq \mathcal{O}(U \otimes k)[x', x'', ..., x^{(r)}]$$

while the latter is clearly an integral domain because $U \otimes k$ is integral. So either $f \otimes 1$ or $g \otimes 1$ is 0 in $\mathcal{O}(J^r(U)) \otimes k$; assume, say, that $f \otimes 1 = 0$. Take an affine open covering $X = \bigcup V_i$. Then, for any i, we will have $f \otimes 1 = 0$ in $\mathcal{O}(J^r(U \cap V_i)) \otimes k$. Using, again, Proposition 3.13, one readily sees that the maps

$$\mathcal{O}(J^r(V_i)) \otimes k \to \mathcal{O}(J^r(U \cap V_i)) \otimes k$$

are injective. In particular $f \otimes 1 = 0$ in each $\mathcal{O}(J^r(V_i)) \otimes k$; so $f = p\varphi_i$, $\varphi_i \in \mathcal{O}(J^r(V_i))$, for each i. Since, by Proposition 3.13, p is a non-zero divisor in

$$\mathcal{O}(J^r(V_i \cap V_j))$$

for all i, j, it follows that the φ_i's glue together to give a section $\varphi \in \mathcal{O}(J^r(X))$ with $f = p\varphi$. This ends the proof that p is a prime element in $\mathcal{O}(J^r(X))$. Similarly one checks that p is a non-zero divisor in this ring. Finally this ring is clearly p–adically separated, so $\mathcal{O}(J^r(X))$ is a global p–ring. □

Here is yet another consequence of Proposition 3.13:

COROLLARY 3.16. *Let $u : X \to Y$ be an étale morphism of smooth schemes over R. Then*

$$J^r(X) \simeq J^r(Y) \times_{\hat{Y}} \hat{X}.$$

Assume in addition that X and Y have non-empty irreducible fibers $X \otimes k$, $Y \otimes k$ and set

$$B^r = \mathcal{O}(J^r(Y)),\ C^r := \mathcal{O}(J^r(X)),\ \bar{B}^r := B^r \otimes k,\ \bar{C}^r := C^r \otimes k.$$

Then the following hold:

1) The maps $\bar{B}^{r-1} \to \bar{B}^r$ are injective. In particular the maps $B^{r-1} \to B^r$ are injective with torsion free cokernel. Moreover $\bar{B}^r, B^r$ are integral domains and we have an equality between their groups of units $(\bar{B}^r)^\times = \bar{B}^\times$.

2) The maps $\bar{B}^r \to \bar{C}^r$ are injective. In particular the maps $B^r \to C^r$ are injective with torsion free cokernel.

3) $\bar{C}^{r-1} \cap \bar{B}^r = \bar{B}^{r-1}$.

4) If $f, g \in B^r$, $fg \in p^n B^r$, $f \notin pB^r$ then $g \in p^n B^r$.

5) $B^r \cap p^n C^r = p^n B^r$.

Assume in addition that $X \to Y$ is an open immersion; then:

6) If $c \in (C^r)^\times$, $1 \leq v(p) \leq p^{ord(\bar{v})}$, and $c^v \in B^{r+ord(v)}$ then $c \in B^r$. Similarly, if $c \in C^r$ and $c^\phi \in B^{r+1}$ then $c \in B^r$.

PROOF. We may assume X and Y are affine and Y has étale coordinates x (which will then also be étale coordinates for X). By Proposition 3.13

$$B^r = B^{r-1}[x^{(r)}]\hat{},\ C^r = C^{r-1}[x^{(r)}]\hat{}.$$

In particular $J^r(X) \simeq J^r(Y) \times_{\hat{Y}} \hat{X}$. Assume now that X and Y have non-empty irreducible closed fibers; so $\bar{B}^0 \to \bar{C}^0$ is an inclusion of integral domains. It follows that $\bar{B}^r$ and $\bar{C}^r$ are integral domains and $(\bar{B}^r)^\times = (\bar{B}^{r-1})^\times$. Then assertions 1, 2, and 3 follow. Assertions 4 and 5 follow by induction. Let's prove assertion 6. We may assume X is a principal open set in Y so $\bar{C}^0$ is a ring of fractions of $\bar{B}^0$. Hence $\bar{C}^r$ is a ring of fractions of $\bar{B}^r$. We will only check the case c is invertible in C^r; the case c non-invertible and $v = \phi$ is similar. Let $v = \sum a_i \phi^i$, $s := ord(v)$, $t := ord(\bar{v})$. It is enough to prove that for any integer $n \geq 1$ we can write

$$c = b_n + p^n c_n, \quad \text{with } b_n \in B^r,\ c_n \in C^{r+1}. \tag{3.7}$$

(For if this is so then b_n converges to c in C^r; but b_n is a Cauchy sequence in B^r due to assertion 5 and B^r is complete so b_n converges to an element in B^r and we are done.) We shall check condition 3.7 by induction on n. Denote by $x \mapsto \bar{x}$ the canonical surjection $C^i \to \bar{C}^i$. To check the case $n = 1$ note that since $c^v \in B^{r+s}$ we have $c^{v(p)} \in \bar{B}^{r+s}$. Since $Spec\ \bar{B}^{r+s}$ is smooth over k and connected, $\bar{B}^{r+s}$ is an integrally closed domain. Also $\bar{B}^{r+s}$ and $\bar{C}^{r+s}$ have the same quotient field. Since $\bar{c}$ is integral over $\bar{B}^{r+s}$ we must have $\bar{c} \in \bar{B}^{r+s}$. By assertion 3 we get $\bar{c} \in \bar{B}^r$. So we can write $c = b_0 + pc_1$, $b_0 \in B^r$, $c_1 \in C^r$; this settles the case $n = 1$ in our induction. Assume now that condition 3.7 holds for some $n \geq 1$. In particular we must have $b_n \in (C^r)^\times$ so we may write

$$\begin{aligned} c^v &= (b_n + p^n c_n)^{a_0}(b_n^\phi + p^n c_n^\phi)^{a_1} \ldots (b_n^{\phi^s} + p^n c_n^{\phi^s})^{a_s} \\ &= b_n^v + p^n (\sum_{i=0}^{s} a_i b_n^{v-\phi^i} c_n^{\phi^i}) + p^{2n}\gamma, \end{aligned}$$

with $\gamma \in C^{r+s}$. Hence $c^v - b_n^v \in B^{r+s} \cap p^n C^{r+s} = p^n B^{r+s}$; cf. assertion 5. Write $b_n^v - c^v = p^n \beta_n$, with $\beta_n \in B^{r+s}$. We have $0 \neq \bar{c} \in \bar{C}^r \subset \bar{C}^{r+1}$, $\bar{b}_n \neq 0$, and

$$\bar{\beta}_n \bar{b}_n^{p^t - v(p)} + \sum_{i=0}^{t} \bar{a}_i \bar{b}_n^{p^t - p^i} \bar{c}_n^{p^i} = 0.$$

Now the latter equation shows that $\bar{c}_n$ is integral over $\bar{B}^{r+s}$. Hence, as before, $\bar{c}_n \in \bar{B}^{r+s}$, and hence $\bar{c}_n \in \bar{B}^r$. So there exists $b \in B^r$ and $c_{n+1} \in C^r$ such that $c_n = b + pc_{n+1}$. Hence $c = b_{n+1} + p^{n+1}c_{n+1}$, where $b_{n+1} = b_n + p^n b$ and our induction step follows. □

3.3. Link between p–jet spaces and δ–functions

If X is a scheme of finite type over R then there is a natural map between sets of R–points

$$\nabla^r : X(R) \to J^r(X)(R) \tag{3.8}$$

and an R–algebra map

$$\mathcal{O}(J^r(X)) \to \mathcal{O}^r(X), \quad f \mapsto \tilde{f}, \tag{3.9}$$

defined as follows. Assume first X is affine. If $P \in X(R)$ then P induces a homomorphism (still denoted by) $P : \mathcal{O}(X) \to R$. By the universality property in Proposition 3.3 one gets a map $\nabla^r(P) : \mathcal{O}(J^r(X)) \to R$ which identifies with an R–point of $J^r(X)$. This defines the map 3.8. Now if $f \in \mathcal{O}(J^r(X))$ one sets

$$\tilde{f}(P) := \nabla^r(P)(f);$$

this defines the map 3.9. In down to earth terms assume $u : X = Spec\ R[T]/I \subset \mathbf{A}^N = Spec\ R[T]$ is the natural embedding and f is the class of an element

$$F \in R[T, T', \ldots, T^{(r)}]\hat{}.$$

Then we have

$$\begin{aligned} \nabla^r(P) &= (u(P), \delta(u(P)), \ldots, \delta^r(u(P))), \\ \tilde{f}(P) &= F(u(P), \delta(u(P)), \ldots, \delta^r(u(P))). \end{aligned}$$

This shows in particular that $\tilde{f}$ is a δ–function of order $\leq r$. The above definition globalizes in the obvious way to the case when X is not necessarily affine.

The map ∇^n should be viewed as an analogue of the $n-$jet (or $n-$prolongation) map on functions in classical mechanics; cf. Equation 0.7.

Note that if $X = \mathbf{A}^1$ then the map $\nabla^n : R \to R^{n+1}$ is simply given by

$$\nabla^n(a) = (a, \delta a, ..., \delta^n a).$$

(Note that the components of $\nabla^n(a)$ mod p **do not coincide** with the Witt coordinates of a if $n \geq 2$.)

LEMMA 3.17. *If $f \in \mathcal{O}(J^r(X))$ then*

$$\delta(\tilde{f}(P)) = (\delta f)\tilde{}(P), \quad P \in X(R).$$

PROOF. We may assume $X = \mathbf{A}^N$ and $f = F \in R[T, T', ..., T^{(r)}]\hat{}$. We need to show that for any $a \in R^N$ we have

$$\delta(F(a, \delta a, ..., \delta^r a)) = (\delta F)(a, \delta a, ..., \delta^{r+1} a). \tag{3.10}$$

Now the set of all $F \in R[T, T', ..., T^{(r)}]\hat{}$ satisfying Equation 3.10 is clearly a $p-$adically closed subring of $R[T, T', ..., T^{(r)}]\hat{}$ containing $R, T, ..., T^{(r)}$, hence it coincides with the whole of $R[T, T', ..., T^{(r)}]\hat{}$. $\square$

We have the following basic propositions:

PROPOSITION 3.18. [**18**] *The map ∇^n induces a bijection*

$$\nabla^n_0 : X(R/p^{n+1}R) \to J^n(X)(k).$$

PROPOSITION 3.19. [**18**] *If X is smooth over R the natural maps*

$$\mathcal{O}(J^r(X)) \to \mathcal{O}^r(X)$$

are isomorphisms; hence, by Lemma 3.17, they induce an isomorphism of prolongation sequences $\mathcal{O}(J^*(X)) \simeq \mathcal{O}^*(X)$. Moreover, if $\tilde{f} \in \mathcal{O}^r(X)$ and $\tilde{f}(P) \in pR$ for all $P \in X(R)$ then $\tilde{f} \in p\mathcal{O}^r(X)$; equivalently the natural map

$$\mathcal{O}^r(X) \otimes k \to \mathcal{F}(X(R)) \otimes k$$

is injective.

In view of Proposition 3.19 we will usually identify the rings $\mathcal{O}(J^r(X))$ and $\mathcal{O}^r(X)$ and we will identify f with its image $\tilde{f}$. In order to prove these Propositions we need the following

LEMMA 3.20. *The map $\nabla^n : R \to R^{n+1}$ induces a bijection $R/p^{n+1}R \to k^{n+1}$.*

PROOF. First, for any positive integer j and any $a, b, c \in R$, one has

$$b \equiv a + p^j c \mod p^{j+1} \implies \delta b \equiv \delta a + p^{j-1} c^p \mod p^j.$$

By iteration this yields, for any $n \geq 1$ and any $i \leq n$,

$$\delta^i(a + p^n c) \equiv \delta^i a + p^{n-i} c^{p^i} \mod p^{n+1-i}. \tag{3.11}$$

Equation 3.11 shows that ∇^n_0 induces a map $R/p^{n+1}R \to k^{n+1}$. To check that the map $R/p^{n+1}R \to k^{n+1}$ is injective we proceed by induction on n. Assume $a, b \in R$ are such that $\delta^i a \equiv \delta^i b \mod p$ for $0 \leq i \leq n$. By induction $b = a + p^n c$ for some $c \in R$. Hence

$$\delta^n b = \delta^n(a + p^n c) \equiv \delta^n a + c^{p^n} \mod p$$

so $c^{p^n} \equiv 0 \mod p$, hence $c \equiv 0 \mod p$ and we are done. To check that the map $R/p^{n+1}R \to k^{n+1}$ is surjective we proceed again by induction. Assume

$(\lambda_0, ..., \lambda_n) \in R^{n+1}$. By induction there exists $a \in R$ such that $\delta^i a \equiv \lambda_i \bmod p$ for $0 \leq i \leq n-1$. Since k is perfect there exists $c \in R$ such that

$$c^{p^n} \equiv \lambda_n - \delta^n a \mod p.$$

Let $b = a + p^{n+1}c$. Then Equation 3.11 easily implies that $\delta^i b \equiv \lambda_i \bmod p$ for $0 \leq i \leq n$. □

PROOF. (of Proposition 3.18.) We may assume X is affine, say

$$X = Spec\ R[T]/I \subset \mathbf{A}^N = Spec\ R[T].$$

By Lemma 3.20 the map

$$\nabla_0^n : \mathbf{A}^N(R/p^{n+1}R) = (R/p^{n+1}R)^N \to J^n(\mathbf{A}^N)(k) = k^{N(n+1)}$$

is a bijection. It follows that we have an induced injective map

$$\nabla_0^n : X(R/p^{n+1}R) \to J^n(X)(k).$$

To show that the latter is also surjective pick a point $Q \in J^n(X)(k)$. Then $Q = \nabla_0^n(P)$ for some $P \in \mathbf{A}^N(R/p^{n+1}R)$. We need to show that if $F \in I$ then $F(P) = 0 \in R/p^{n+1}R$ so that P will factor through a $R/p^{n+1}R-$ point of X. By Lemma 3.20 again, it is enough to show that $\delta^i(F(P)) \equiv 0 \bmod p$ for all $0 \leq i \leq n$. But $\delta^i(F(P)) = (\delta^i F)(\nabla^r(P))$ and the reduction mod p of the latter equals $(\delta^i F)(Q) = 0$ which ends the argument. □

PROOF. (of Proposition 3.19.) The map $\mathcal{O}(J^r(X)) \to \mathcal{O}^r(X)$ is obtained by applying the global sections functor to the map of sheaves (on X, for the Zariski topology):

$$i_*\pi_{r*}\mathcal{O}_{J^r(X)} \to \mathcal{O}_X^r,$$

where $\pi_r : J^r(X) \to J^0(X) = \hat{X}$ is the canonical projection and $i : X_0 := X \otimes k \to X$ is the natural closed immersion (where $\hat{X}$ is identified with X_0 as a topological space). So, in order to prove the first assertion of the Proposition, it is enough to show that the above map of sheaves is an isomorphism. This map is surjective at each stalk from the very definition of $\delta-$functions. (One needs to use here the fact that any closed point of X_0 comes from a $k-$point of X_0 which lifts, by smoothness, to an $R-$point of X.) To check injectivity at the level of stalks it is enough to show that, for X affine, the map $\mathcal{O}(J^r(X)) \to \mathcal{O}^r(X)$ is injective. To check the latter it is enough to show that the map $\mathcal{O}(J^r(X)) \otimes k \to \mathcal{O}^r(X) \otimes k$ is injective. We will show more, namely that the composition

$$\mathcal{O}(J^r(X)) \otimes k \to \mathcal{O}^r(X) \otimes k \to \mathcal{F}(X(R)) \otimes k$$

is injective. This will also prove the second assertion of the Proposition. We may assume $X = Spec\ R[T]/I$. Let $f \in \mathcal{O}(J^r(X))$ be represented by $\Phi \in R[T, T', ..., T^{(r)}]\hat{}$ and assume $\tilde{f}(P) \in pR$ for all $P \in X(R)$. Let $u : X(R) \to R^N$ be the embedding defined by T. So we know that $\Phi(u(P), \delta(u(P)), ..., \delta^r(u(P))) \in pR$ for all $P \in X(R)$. Reducing modulo p we get $\Phi_0(\nabla_0^r(u(P))) = 0$ where $\Phi_0 \in k[T, T', ..., T^{(r)}]$ is the reduction of Φ mod p. Now by Proposition 3.18 $\nabla_0^r(u(P))$ runs through all of $J^r(X)(k)$. By Proposition 3.15 $J^r(X) \otimes k$ is reduced and, by Equation 3.3 its ideal in $k[T, T', ..., T^{(r)}]$ is generated by $I, \delta I, ..., \delta^r I$. By Hilbert's Nullstellensatz,

$$\Phi_0 \in (I, \delta I, ..., \delta^r I)k[T, T', ..., T^{(r)}]$$

which concludes our proof. □

COROLLARY 3.21. *We have isomorphisms*

$$\mathcal{O}^r(\mathbf{A}^N) \simeq R[T, T', ..., T^{(r)}]\hat{},$$

$$\{\delta-\textit{polynomial functions } R^N \to R \textit{ of order} \leq r\} \simeq R[T, T', ..., T^{(r)}],$$

$$\{\delta-\textit{polynomial functions } R^N \to R\} \simeq R\{T\}.$$

Consequently, for any function $g : R^N \to R$, the following are equivalent:

a) g is a $\delta-$function of order $\leq r$,

b) g is a restricted $\delta-$function of order $\leq r$,

c) g is a $p-$adic limit of $\delta-$polynomial functions of order $\leq r$.

Finally, for any function $g : R^N \to R$, the following are equivalent:

i) g is a $\delta-$polynomial function,

ii) there exists $\nu \in \mathbf{Z}_+$ such that $p^\nu g$ is a $\phi-$polynomial function.

In statement c) above the limit is taken, of course, in the $p-$adic topology of the ring $\mathcal{F}(R^N)$.

PROOF. The first 3 isomorphisms follow from Proposition 3.19 plus Equation 3.3. So a) is equivalent to b). Clearly b) implies c). To prove c) implies a) one notes that, by the injectivity of the map

$$\mathcal{O}^r(\mathbf{A}^N) \otimes k \to \mathcal{F}(R^N) \otimes k,$$

we have that

$$\text{if } f \in \mathcal{F}(R^N) \text{ and } p \cdot f \in \mathcal{O}^r(\mathbf{A}^N) \text{ then } f \in \mathcal{O}^r(\mathbf{A}^N). \tag{3.12}$$

Therefore any sequence of elements in $\mathcal{O}^r(\mathbf{A}^N)$ which is $p-$adically convergent in $\mathcal{F}(R^N)$ is also $p-$adically convergent in $\mathcal{O}^r(\mathbf{A}^N)$. Clearly i) implies ii). The fact that ii) implies i) follows from the condition 3.12 above. □

Proposition 3.19 plus Corollary 3.4 imply:

COROLLARY 3.22. *If T is one variable and $\mathbf{G}_m := Spec\ R[T, T^{-1}]$ is the multiplicative group over R then*

$$\mathcal{O}^r(\mathbf{G}_m) \simeq R[T, T^{-1}, T', ..., T^{(r)}]\hat{}.$$

REMARK 3.23. An interesting role will be played later by the following construction of roots of $W-$powers. Let $w = \sum_{j=0}^r c_j \phi^j \in W$ and let d be a positive divisor of $w(p) = \sum_{j=0}^r c_j p^j$, d not divisible by p. (For instance one can take d to be a common divisor of $p-1$ and $deg(w)$.) We can define an element

$$T^{w/d} \in \mathcal{O}^r(\mathbf{G}_m)$$

by the formula

$$T^{w/d} := T^{w(p)/d} \cdot \prod_j \left[1 + pT^{-p^j} \left(\frac{T^{\phi^j} - T^{p^j}}{p}\right)\right]^{c_j/d}; \tag{3.13}$$

where, for $\gamma \in \mathbf{Z}_p$,

$$(1 + px)^\gamma := \sum_{i \geq 0} \gamma(\gamma - 1)...(\gamma - i)\frac{p^i}{i!}x^i \in \mathbf{Z}_p[x]\hat{}.$$

One can characterize $T^{w/d}$ as the unique element in $R[T, T^{-1}, T', ..., T^{(r)}]\hat{}$ satisfying

$$(T^{w/d})^d = T^w \quad \text{and} \quad T^{w/d} \equiv T^{w(p)/d} \quad mod\ \ p.$$

Using this characterization we obtain the following properties:

1) $(T_1T_2)^{w/d} = T_1^{w/d}T_2^{w/d}$ in $R[T_1, T_2, T_1^{-1}, T_2^{-1}, T_1', T_2', ..., T_1^{(r)}, T_2^{(r)}]\hat{}$,

2) For any positive $e \in \mathbf{Z}$ we have $T^{ew/ed} = T^{w/d}$,

3) For any $v \in W$ we have $(T^v)^{w/d} = (T^{w/d})^v = T^{vw/d}$,

4) If $d = 2$ then for any $a \in \mathbf{Z}_p^{\times}$ we have the following expression:

$$a^{w/2} = a^{deg(w)/2} \cdot \left(\frac{a}{p}\right)^{c_1+c_3+c_5+\dots},$$

where $\left(-\right)$ is the Legendre symbol and

$$a^{w/2} := T^{w/2}|_{T=a, T'=\delta a, \dots, T^{(r)}=\delta^r a}.$$

Proposition 3.19 together with Corollaries 3.15 and 3.16 imply the following Corollaries.

COROLLARY 3.24. *If X is a smooth scheme over R with non-empty irreducible closed fiber $X \otimes k$ and L is a $\delta-$line bundle on X then the rings $\mathcal{O}^r(X)$ and $R(X, L)^*$ are global $p-$rings.*

PROOF. For $\mathcal{O}^r(X)$ one uses Proposition 3.19 and Corollary 3.15. For $R(X, L)^*$ one uses Proposition 3.19 plus an argument entirely similar to the one in the proof of Corollary 3.15; one only needs to take, in the proof of Corollary 3.15, $f \in H^0(X, L^w)$, $g \in H^0(X, L^v)$, $h \in H^0(X, L^{w+v})$, and to assume that L restricted to U and V_i is trivial. □

COROLLARY 3.25. *Let $u : X \to Y$ be an étale morphism of smooth schemes over R and assume X and Y have non-empty irreducible closed fibers $X \otimes k$, $Y \otimes k$. Let L be a $\delta-$line bundle L on Y. Then the morphisms*

$$H^0(Y, L) \to H^0(X, u^*L), \quad H^0(Y, L) \otimes k \to H^0(X, u^*L) \otimes k$$

are injective; in particular the homomorphism

$$R(Y, L)^* \to R(X, u^*L)^*$$

is telescopic.

PROOF. One uses Proposition 3.19 and Corollary 3.16. □

REMARK 3.26. Our discussion above trivially implies, in particular, that for a smooth X the rings $\mathcal{O}^r(X)$ depend on the $p-$adic completion of X only. Also if $\mathbf{X} = (\tilde{X}, X, \sigma_1, \sigma_2)$ is a correspondence of finite type and smooth over R and $\mathbf{L} = (L, \beta)$ is a $\delta-$line bundle on it then the spaces $H^0(\mathbf{X}, \mathbf{L})$ and $H^1(\mathbf{X}, \mathbf{L})$ depend on the $p-$adic completions of $\tilde{X}, X, L$ only.

3.4. Galois covers

PROPOSITION 3.27. [**29**] *Let $Y \to X$ be an étale Galois cover of smooth schemes over R. Assume X and Y have non-empty irreducible closed fibers. Let Γ be the Galois group. Then, for any r, the ring $\mathcal{O}^r(Y)^{\Gamma}$ of $\Gamma-$invariant functions in $\mathcal{O}^r(Y)$ identifies with $\mathcal{O}^r(X)$.*

EXAMPLE 3.28. Here is an example which illustrates the flavor of this Proposition. Assume $X = \mathbf{G}_m := Spec\ R[x, x^{-1}]$ is the multiplicative group scheme over R, $Y = Spec\ R[y, y^{-1}]$ be another copy of $\mathbf{G}_m$ and let $\pi : Y \to X$ be the étale map defined by $x \mapsto y^2$. Then π is Galois with group $\Gamma = \langle\gamma\rangle$, where γ is defined by $y \mapsto -y$. By Corollary 3.22, $\mathcal{O}^1(X) \simeq R[x, x^{-1}, x']\hat{}$, $\mathcal{O}^1(Y) \simeq R[y, y^{-1}, y']\hat{}$; we shall view $\mathcal{O}^1(X)$ as a subring of $\mathcal{O}^1(Y)$ via the identification $x = y^2$. Note that γ sends y' into

$$\delta(-y) = (\phi(-y) - (-y)^p)/p = -\delta y = -y'$$

so, for instance, the function yy' is $\Gamma-$invariant; one can ask what element in $R[x, x^{-1}, x']\hat{}$ equals yy'. One can solve this problem by a mere computation as follows. From $x = y^2$ we get

$$x' = \delta(y^2) = 2y^p y' + p(y')^2.$$

The above is a quadratic equation in y'. Solving for y' and multiplying the result by y we get

$$yy' = x^{(p+1)/2} \sum_{n \geq 0} \binom{1/2}{n} p^{n-1} \left(\frac{x'}{x^p}\right)^n \in R[x, x^{-1}, x']\hat{}. \tag{3.14}$$

This example shows that, even for simple maps as $x \mapsto y^2$, the actual formulae behind the statement of the Proposition can be rather complicated and, in any case, involve series rather than polynomials in general. The equation 3.14 should be viewed, of course, as an analogue of the equation $yy' = (1/2)x'$ (where $x = y^2$ and x' is *usual* derivative of x).

REMARK 3.29. The hypothesis that Y/X be étale cannot be dropped from Proposition 3.27. Indeed if $Y = \mathbf{A}^1 = Spec\ R[y]$, $X = \mathbf{A}^1 = Spec\ R[x]$, $x = y^2$, then $\mathcal{O}^1(Y) = R[y, y']\hat{}$, $\mathcal{O}^1(X) = R[x, x']\hat{}$ and

$$yy' \in \mathcal{O}^1(Y)^\Gamma.$$

On the other hand,

$$yy' \notin \mathcal{O}^1(X);$$

cf. Equation 3.14.

PROOF. (of Proposition 3.27). The question is local on X. So we may assume X is affine and possesses a system of étale coordinates $z = \{z_1, ..., z_n\}$, $R[z] \subset \mathcal{O}(X)$. Then, by Propositions 3.13 and 3.19 we have a natural isomorphisms

$$\mathcal{O}^r(X) \simeq \mathcal{O}(X)[z', ..., z^{(r)}]\hat{}, \quad \mathcal{O}^r(Y) \simeq \mathcal{O}(Y)[z', ..., z^{(r)}]\hat{}. \tag{3.15}$$

Note that $Y_0 := Y \otimes k \to X_0 := X \otimes k$ is still Galois with group Γ. Indeed, by flatness, the latter map has the same degree, $|\Gamma|$, as the map $Y \to X$; on the other hand we have a factorization

$$Y_0 \to Y_0/\Gamma \to X_0.$$

Now we claim that the induced map $\Gamma \to Aut(Y_0)$ is injective. Indeed if $\gamma \in \Gamma$ induces the identity on Y_0 then γ fixes all $k-$points of Y. Take one of these $k-$points; then γ induces an isomorphism on the completion of the local ring of Y at our fixed $k-$point which is the identity mod p. But it is a trivial exercise to see that any automorphism of finite order of a power series ring with $R-$coefficients that is the identity mod p must be the identity. This proves our claim. Since the map $\Gamma \to Aut(Y_0)$ is injective, $Y_0 \to Y_0/\Gamma$ has also degree $|\Gamma|$; so the map

$Y_0/\Gamma \to X_0$ must be an isomorphism, because X_0 is normal. In particular we have $\mathcal{O}(Y_0)^\Gamma = \mathcal{O}(X_0)$. This easily implies that the map $\mathcal{O}(X)\hat{} \to (\mathcal{O}(Y)\hat{})^\Gamma$ is an isomorphism. Since the action of Γ on $z', ..., z^{(r)}$ is trivial we clearly have

$$(\mathcal{O}(Y)[z', ..., z^{(r)}]\hat{})^\Gamma = \mathcal{O}(X)[z', ..., z^{(r)}]\hat{},$$

which ends our proof. □

COROLLARY 3.30. *Assume Y^* is a smooth scheme over R, $\tau : Y^* \to Y^*$ is an endomorphism, and $\Gamma \subset Aut(Y^*)$ is a finite group all of whose elements commute with τ. Set $X^* := Y^*/\Gamma$, let $\pi : Y^* \to X^*$ be the canonical projection, and let $\sigma : X^* \to X^*$ be the endomorphism induced by τ. Let $X \subset X^*$, $Y := \pi^{-1}(X)$, and assume X is smooth, $\pi : Y \to X$ is étale, and both X and Y have irreducible closed fibers. Consider the correspondences*

$$\mathbf{X} = (X, X \cap \sigma^{-1}(X), \iota, \sigma), \ \ \mathbf{Y} = (Y, Y \cap \tau^{-1}(Y), \iota, \tau),$$

where ι are the inclusions. Let $\mathbf{L} = \mathbf{F}(r, \lambda)$ be a flat $\delta-$line bundle (viewed as a $\delta-$line bundle on both $\mathbf{Y}$ and $\mathbf{X}$). Then Γ acts on $H^0(\mathbf{X}, \mathbf{L})$ and the natural map

$$H^0(\mathbf{X}, \mathbf{L}) \to H^0(\mathbf{Y}, \mathbf{L})^\Gamma$$

is an isomorphism. Assume moreover that the order of Γ is not divisible by p. Then the natural map

$$H^1(\mathbf{X}, \mathbf{L})[p] \to (H^1(\mathbf{Y}, \mathbf{L})[p])^\Gamma$$

is an isomorphism.

PROOF. The first part of the Corollary follows directly from Proposition 3.27. To check the second part of the Corollary recall from the proof of Proposition 3.27 that

$$\mathcal{O}^r(Y)^\Gamma = \mathcal{O}^r(X), \ \ \mathcal{O}^r(\tau^{-1}(Y))^\Gamma = \mathcal{O}^r(\sigma^{-1}(X)),$$

$$(\mathcal{O}^r(Y) \otimes k)^\Gamma = \mathcal{O}^r(X) \otimes k, \ \ (\mathcal{O}^r(\tau^{-1}(Y)) \otimes k)^\Gamma = \mathcal{O}^r(\sigma^{-1}(X)) \otimes k.$$

Then our second assertion follows from the following general fact. Let $\Delta : M \to M'$ be an $R-$module homomorphism where M, M' are torsion free and let Γ be a finite group acting (by $R-$module automorphisms) on M, M' such that Δ is $\Gamma-$equivariant. Assume that

$$(M \otimes k)^\Gamma = M^\Gamma \otimes k, \ (M' \otimes k)^\Gamma = (M')^\Gamma \otimes k.$$

Let $\Delta_\Gamma : M^\Gamma \to (M')^\Gamma$ be the restriction of Δ. Then the natural map

$$(Coker\ \Delta_\Gamma)[p] \to ((Coker\ \Delta)[p])^\Gamma$$

is an isomorphism. This is an easy exercise in group cohomology; indeed one needs to use formula 2.53 in the proof of Lemma 2.32 together with the fact that

$$H^1(\Gamma, Ker\ \Delta) = H^1(\Gamma, (Ker\ \Delta) \otimes k) = 0.$$

The latter vanishing conditions hold because the two H^1's are $R-$modules killed by the order of Γ; cf. [**120**], p. 130. □

3.5. δ–base loci versus postcritical loci

In this Section we prove, in particular, Theorem 2.64. We need some preliminaries.

Start with an integral domain A with field of fractions $Q(A)$. Let

$$\mathbf{Y}_A := (Y, \tilde{Y}, \sigma_1, \sigma_2)$$

be a one dimensional correspondence of finite type and smooth over A with geometrically irreducible base. Let

$$\mathbf{X}_A := (X, \tilde{X}, \sigma_1, \sigma_2)$$

be an open étale subcorrespondence of $\mathbf{Y}_A$ such that the geometric fibers of $\tilde{X}$ are dense in the corresponding geometric fibers of $\tilde{Y}$. We shall write $\mathbf{Y}$ and $\mathbf{X}$ in place of $\mathbf{Y}_A$ and $\mathbf{X}_A$ whenever A is clear from the context. Exactly as in the complex case, for $i = 1, 2$ we define the *critical loci*

$$\Omega_{\sigma_i}(\mathbf{Y}) = \{\tilde{P} \in \tilde{Y}(A) \mid \sigma_i \text{ is not étale at } \tilde{P} \otimes Q(A) \in \tilde{Y}(Q(A))\}. \tag{3.16}$$

Next, by a *chain in* $\mathbf{Y}$ *over* A we understand a sequence

$$\tilde{P}_1, ..., \tilde{P}_n \in \tilde{Y}(A), \quad n \geq 1$$

such that

$$\sigma_2(\tilde{P}_i) = \sigma_1(\tilde{P}_{i+1}), \quad 1 \leq i \leq n-1.$$

We have a corresponding concept of a point in $\tilde{Y}(A)$ preceding another point in $\tilde{Y}(A)$. For any $\tilde{P} \in \tilde{Y}(A)$ we define the *forward orbit*

$$O^+_{\sigma_2}(\tilde{P}, \mathbf{Y}) = \{Q \in Y(A) \mid Q = \sigma_2(\tilde{Q}), \ \tilde{P} \text{ preceeds } \tilde{Q}\}$$

and the *backward orbit*

$$O^-_{\sigma_2}(\tilde{P}, \mathbf{Y}) = \{Q \in Y(A) \mid Q = \sigma_1(\tilde{Q}), \ \tilde{Q} \text{ preceeds } \tilde{P}\}.$$

Reference to $\mathbf{Y}$ in the above notation is justified by the fact that we are using the symbol σ_2 in relation to different correspondences such as $\mathbf{X}$ and $\mathbf{Y}$.

LEMMA 3.31. *Assume $A = R = \hat{\mathbf{Z}}_p^{ur}$ and assume $\mathbf{Y}$ is left étale (i.e. $\sigma_1 : \tilde{Y} \to Y$ is étale). Then, for any point $\tilde{P} \in \Omega_{\sigma_2}(\mathbf{Y})$ such that $\sigma_1(\tilde{P}) \in X(R)$ we have*

$$O^+_{\sigma_2}(P, \mathbf{Y}) \cap X(R) \subset BS(\mathbf{X}, \mathbf{K}^{-1}).$$

PROOF. Set $P := \sigma_1(\tilde{P}) \in X(R)$. Let $\tilde{P}_1, ..., \tilde{P}_n \in \tilde{Y}(R)$ be a chain with $\tilde{P} = \tilde{P}_1$ and $Q = \sigma_2(\tilde{P}_n) \in X(R)$; we want to show that $Q \in BS(\mathbf{X}, \mathbf{K}^{-1})$ i.e. we want to show that $\eta(Q) = 0$ for all $\eta \in H^0(\mathbf{X}, \mathbf{K}^{-w})$, $0 \neq w \in W_+$. Fix in what follows such an η; hence

$$\sigma_2^* \eta = \sigma_1^* \eta. \tag{3.17}$$

Consider the scheme

$$\tilde{Y}^{\circ n} := \tilde{Y} \times_Y \cdots \times_Y \tilde{Y}.$$

This scheme is smooth because σ_1 is étale. It is equipped with two natural maps

$$\sigma_1^{\circ n}, \sigma_2^{\circ n} : \tilde{Y}^{\circ n} \to Y$$

with $\sigma_1^{\circ n}$ étale. One can consider the R–point

$$\tilde{P}^{(n)} := (\tilde{P}_1, ..., \tilde{P}_n) \in \tilde{Y}^{\circ n}(R),$$
$$\sigma_1^{\circ n}(\tilde{P}^{(n)}) = P, \quad \sigma_2^{\circ n}(\tilde{P}^{(n)}) = Q.$$

Since the fiber $\tilde{Y}^{\circ n} \otimes k$ is smooth $\tilde{P}^{(n)} \otimes k$ belongs to exactly one component of $\tilde{Y}^{\circ n} \otimes k$. So if Y' is obtained from $\tilde{Y}^{\circ n}$ by removing all the components of $\tilde{Y}^{\circ n} \otimes k$ which do not contain $\tilde{P}^{(n)} \otimes k$ we have that $Y' \otimes k$ is irreducible. On the other hand note that

$$\tilde{X}^{\circ n} := \tilde{X} \times_X \cdots \times_X \tilde{X}$$

is open in $\tilde{Y}^{\circ n}$. Since $\tilde{X} \otimes k$ was assumed dense in $\tilde{Y} \otimes k$ we have that

$$(\tilde{Y} \otimes k) \backslash (\tilde{X} \otimes k)$$

consists of finitely many points so

$$(\tilde{Y}^{\circ n} \otimes k) \backslash (\tilde{X}^{\circ n} \otimes k)$$

consists of finitely many points hence $\tilde{X}^{\circ n} \otimes k$ is dense in $\tilde{Y}^{\circ n} \otimes k$. (Note that both can be reducible !) In particular

$$(Y' \cap \tilde{X}^{\circ n}) \otimes k \neq \emptyset.$$

Now Equation 3.17 implies that

$$(\sigma_2^{\circ n})^* \eta = (\sigma_1^{\circ n})^* \eta$$

on $\tilde{X}^{\circ n}$ hence on $Y' \cap \tilde{X}^{\circ n}$ hence on

$$Y'_U := Y' \cap [(\sigma_1^{\circ n})^{-1}(U) \cap (\sigma_2^{\circ n})^{-1}(U)]$$

for any open set $U \subset X$; the latter holds by Corollary 3.25 in view of the fact that $Y'_U \otimes k$ and $(Y' \cap \tilde{X}^{\circ n}) \otimes k$ are non-empty and irreducible. Choose U affine such that

$$P, Q \in U$$

and such that there is an invertible 1−form ω on U. Note that $\tilde{P}^{(n)} \in Y'_U(R)$. Write

$$\eta = \varphi \cdot \omega^{-w}, \quad \varphi \in \mathcal{O}^r(U).$$

We get

$$\varphi \circ \sigma_2^{\circ n} \cdot [(\sigma_1^{\circ n})^* \omega]^w = \varphi \circ \sigma_1^{\circ n} \cdot [(\sigma_2^{\circ n})^* \omega]^w \quad on \quad Y'_U(R).$$

We now evaluate at $\tilde{P}^{(n)}$. Clearly $(\sigma_2^{\circ n})^* \omega$ vanishes at $\tilde{P}^{(n)}$ while $(\sigma_1^{\circ n})^* \omega$ is invertible at $\tilde{P}^{(n)}$. So we get

$$0 = \varphi(\sigma_2^{\circ n}(\tilde{P}^{(n)})) = \varphi(Q)$$

and we are done. □

LEMMA 3.32. *Assume $A = \mathcal{O} := \mathcal{O}_F[1/m]$ where F is a number field and $m \in \mathbf{Z}$. Assume* **one of the following** *holds:*

1) $\mathbf{Y}$ is left étale (i.e. $\sigma_1 : \tilde{Y} \to Y$ is étale) and the $\delta-$base locus $\underline{BS}(\mathbf{X}_{F^a}, \mathbf{K}_{F^a})$ is finite;

2) $\sigma_1 : \tilde{Y} \to Y$ is an isomorphism and the $\delta-$base loci $\underline{BS}(\mathbf{X}_{F'}, \mathbf{K}_{F'})$ are finite for all finite extensions F' of F.

Then for any point $\tilde{P} \in \Omega_{\sigma_2}(\mathbf{Y}_{\mathbf{C}})$ such that $\sigma_1(\tilde{P}) \in X(\mathbf{C})$ the forward orbit

$$O^+_{\sigma_2}(\tilde{P}, \mathbf{Y}_{\mathbf{C}})$$

is finite.

PROOF. Consider a chain over $\mathbf{C}$, $\tilde{P}_1, ..., \tilde{P}_n$, with $\tilde{P}_1 = \tilde{P}$ and set $Q := \sigma_2(\tilde{P}_n)$. This chain is defined over $\mathcal{O}' = \mathcal{O}_{F'}[1/m']$ where F' is a finite extension of F and $m' \in \mathbf{Z}$ is divisible by m. In case 2 one can arrange that the field F' depend only on $\tilde{P}$ and not on Q; indeed one can take the field F' to be any finite extension of F such that $\tilde{P} \in \tilde{Y}(F')$. Set $Z = Y \backslash X$.

We claim that either $Q \in Z(\mathbf{C})$ or $Q \in \underline{BS}(\mathbf{X}_{F'}, \mathbf{K}_{F'}^{-1})$. This will, of course, end our proof.

To check our Claim assume $Q \notin Z(\mathbf{C})$. Then $Q \notin Z(\mathcal{O}')$. Then $Q \in X(\mathcal{O}'_{\wp})$ for almost all places $\wp$ of $\mathcal{O}'$. Now $\tilde{P} \in \Omega_{\sigma_2}(\mathbf{Y}_{\wp})$, $\sigma_1(\tilde{P}) \in X(R_{\wp})$, and

$$Q \in O^+_{\sigma_2}(\tilde{P}, \mathbf{Y}_{\wp}) \cap X(R_{\wp})$$

for almost all $\wp$. By Lemma 3.31 we get $Q \in BS(\mathbf{X}_{\wp}, \mathbf{K}_{\wp}^{-1})$ for almost all $\wp$, so $Q \in \underline{BS}(\mathbf{X}_{F'}, \mathbf{K}_{F'}^{-1})$, and we are done. □

We are ready to prove Theorem 2.64.

PROOF. *Step 1: Case* $\mathbf{Y}$ *is left étale.* Assume $\tilde{P} \in \Omega_{\sigma_2}(\mathbf{Y}_{\mathbf{C}})$ and assume the backward orbit $O^-_{\sigma_2}(\tilde{P}, \mathbf{Y}_{\mathbf{C}})$ is infinite. We will prove that the forward orbit $O^+_{\sigma_2}(\tilde{P}, \mathbf{Y}_{\mathbf{C}})$ is finite; this will, of course, end the proof in this case.

Now, by the assumption that $O^-_{\sigma_2}(\tilde{P}, \mathbf{Y}_{\mathbf{C}})$ is infinite one can find a point

$$Q \in O^-_{\sigma_2}(\tilde{P}, \mathbf{Y}_{\mathbf{C}}) \cap X(\mathbf{C}).$$

So there is a chain $\tilde{P}_1, ..., \tilde{P}_n$ over $\mathbf{C}$ such that $\sigma_1(\tilde{P}_1) = Q$ and $\tilde{P}_n = \tilde{P}$. This chain is defined over some $\mathcal{O}' := \mathcal{O}_{F'}[1/m']$ where F' is a finite extension of F and $m' \in \mathbf{Z}$ is divisible by m. Consider the point

$$\tilde{P}^{(n)} = (\tilde{P}_1, ..., \tilde{P}_n) \in \tilde{Y}^{\circ n}(\mathcal{O}').$$

We consider the correspondence

$$\mathbf{Y}^{\circ n}_{\mathcal{O}'} = (Y_{\mathcal{O}'}, \tilde{Y}^{\circ n}_{\mathcal{O}'}, \sigma_1^{\circ n}, \sigma_2^{\circ n})$$

over $\mathcal{O}'$; recall that $\tilde{Y}^{\circ n}_{\mathcal{O}'}$ is smooth because σ_1 is étale on $\tilde{Y}$. One can consider the open étale subcorrespondence $\mathbf{X}^{\circ n}_{\mathcal{O}'}$ of $\mathbf{Y}^{\circ n}_{\mathcal{O}'}$,

$$\mathbf{X}^{\circ n}_{\mathcal{O}'} := (X_{\mathcal{O}'}, \tilde{X}^{\circ n}_{\mathcal{O}'}, \sigma_1^{\circ n}, \sigma_2^{\circ n}).$$

As in the proof of Lemma 3.31 the geometric fibers of $\tilde{X}^{\circ n}_{\mathcal{O}'}$ are dense in the corresponding geometric fibers of $\tilde{Y}^{\circ n}_{\mathcal{O}'}$. Since, clearly, we have injections

$$H^0(\mathbf{X}_{\wp}, \mathbf{K}_{\wp}^{-1}) \subset H^0(\mathbf{X}^{\circ n}_{\wp}, \mathbf{K}_{\wp}^{-1})$$

for almost all $\wp$ it follows that

$$\underline{BS}(\mathbf{X}^{\circ n}_{F''}, \mathbf{K}_{F''}^{-1}) \subset \underline{BS}(\mathbf{X}_{F''}, \mathbf{K}_{F''}^{-1})$$

for all finite extensions F'' of F'. In particular $\underline{BS}(\mathbf{X}^{\circ n}_{F^a}, \mathbf{K}_{F^a}^{-1})$ is finite (resp. $\underline{BS}(\mathbf{X}^{\circ n}_{F''}, \mathbf{K}_{F''}^{-1})$ are finite) if we are in case 1 (resp. 2). Finally it is easy to check that

$$\tilde{P}^{(n)} \in \Omega_{\sigma_2^{\circ n}}(\mathbf{Y}^{\circ n}_{\mathbf{C}})$$

and, clearly,

$$\sigma_1^{\circ n}(\tilde{P}^{(n)}) = \sigma_1(\tilde{P}_1) = Q \in X(\mathbf{C}).$$

Applying Lemma 3.32 to $\mathbf{X}^{\circ n}_{\mathcal{O}'}$, $\mathbf{Y}^{\circ n}_{\mathcal{O}'}$, and the point $\tilde{P}^{(n)}$ we get that the forward orbit

$$O^{+}_{\sigma_2^{\circ n}}(\tilde{P}^{(n)}, \mathbf{Y}^{\circ n}_{\mathbf{C}})$$

is finite. One then checks easily that

$$O^{+}_{\sigma_2}(\tilde{P}, \mathbf{Y_C}) \subset \bigcup_{i=1}^{n-1} \sigma_2^{\circ i}((\sigma_1^{\circ i})^{-1}(O^{+}_{\sigma_2^{\circ n}}(\tilde{P}^{(n)}, \mathbf{Y}^{\circ n}_{\mathbf{C}}))$$

hence the forward orbit

$$O^{+}_{\sigma_2}(\tilde{P}, \mathbf{Y_C})$$

is finite.

Step 2: General case. Consider the maximal set $\tilde{Y}^0 \subset \tilde{Y}$ where $\sigma_1 : \tilde{Y} \to Y$ is étale. Then $\tilde{Y}^0$ is open and we may consider the left étale correspondence

$$\mathbf{Y}^0 := (Y, \tilde{Y}^0, \sigma_1, \sigma_2);$$

$\mathbf{X}$ is an étale open subcorrespondence in $\mathbf{Y}^0$ and the geometric fibers of $\tilde{X}$ are dense in the corresponding geometric fibers of $\tilde{Y}^0$. By Step 1 $\mathbf{Y}^0_{\mathbf{C}}$ is critically finite. Now let $\tilde{P} \in \Omega_{\sigma_2}(\mathbf{Y_C})$ be non-degenerate. In particular $\tilde{P} \in \tilde{Y}^0$ and

$$O^{+}_{\sigma_2}(\tilde{P}, \mathbf{Y_C}) = O^{+}_{\sigma_2}(\tilde{P}, \mathbf{Y}^0_{\mathbf{C}}), \quad O^{-}_{\sigma_2}(\tilde{P}, \mathbf{Y_C}) = O^{-}_{\sigma_2}(\tilde{P}, \mathbf{Y}^0_{\mathbf{C}}).$$

So either $O^{+}_{\sigma_2}(\tilde{P}, \mathbf{Y_C})$ or $O^{-}_{\sigma_2}(\tilde{P}, \mathbf{Y_C})$ is finite and we are done. □

3.6. $\delta-$tangent maps and $\delta-$differentials of $\delta-$functions

In this Section we introduce the concepts of $\delta-$*tangent map* and $\delta-$*differential* of a $\delta-$ function and we construct a series of associated operators by analogy with the formalism of classical mechanics. Some of these concepts will play a key role in the proof of our main results.

3.6.1. $\delta-$tangent maps of $\delta-$functions. Recall that for a smooth scheme X over R we denoted by

$$T(X) := Spec\ Symm(\Omega_{X/R})$$

the tangent scheme of X. Also we set, as usual, $\mathcal{O}^{\infty}_X = \lim_{\to} \mathcal{O}^r_X$. We denote by $\pi : T(X) \to X$ the canonical projection. We will construct a natural compatible sequence of derivations $T : \mathcal{O}^r_X \to \pi_* \mathcal{O}^r_{T(X)}$ inducing an $R-$derivation

$$T : \mathcal{O}^{\infty}_X \to \pi_* \mathcal{O}^{\infty}_{T(X)}, \quad f \mapsto Tf \tag{3.18}$$

respecting the filtration by orders i.e. $T(\mathcal{O}^r_X) \subset \pi_* \mathcal{O}^r_{T(X)}$ for all r. We call Tf the $\delta-$*tangent map* of f. One should view Tf as an analogue of the Fréchet derivative in classical mechanics; cf. Equation 0.15. The construction will be local and natural so will provide a global concept. So we may assume X is affine. Recall that for any ring S we denote by $S[\epsilon]$ the ring $S \oplus \epsilon S$ of dual numbers over S. Next let us note that any prolongation sequence

$$S^* : S^0 \xrightarrow{\delta} S^1 \xrightarrow{\delta} S^2 \xrightarrow{\delta} \dots$$

can be uniquely extended to a prolongation sequence

$$S^*[\epsilon] : S^0[\epsilon] \xrightarrow{\delta} S^1[\epsilon] \xrightarrow{\delta} S^2[\epsilon] \xrightarrow{\delta} \dots$$

such that $\delta\epsilon = \epsilon$ (and, hence $\phi\epsilon = p\epsilon$). In particular we have a prolongation sequence

$$\mathcal{O}^*(T(X))[\epsilon] : \mathcal{O}^0(T(X))[\epsilon] \xrightarrow{\delta} \mathcal{O}^1(T(X))[\epsilon] \xrightarrow{\delta} \mathcal{O}^2(T(X))[\epsilon] \xrightarrow{\delta} ...$$

On the other hand we have a natural inclusion $\mathcal{O}(X) \subset \mathcal{O}(T(X))$ and a natural derivation

$$d : \mathcal{O}(X) \to \mathcal{O}(T(X)) = Symm(\Omega_{X/R}),$$

induced by the universal Kähler derivation

$$d : \mathcal{O}(X) \to \Omega_{\mathcal{O}(X)/R},$$

so we have an R-algebra map

$$\mathcal{O}(X) \to \mathcal{O}(T(X))[\epsilon], \quad f \mapsto f + \epsilon \cdot df.$$

By the universality property of $\mathcal{O}^*(X)$ there are naturally induced ring homomorphisms

$$\mathcal{O}^r(X) \to \mathcal{O}^r(T(X))[\epsilon] \tag{3.19}$$

whose composition with the first projection

$$pr_1 : \mathcal{O}^r(T(X))[\epsilon] \to \mathcal{O}^r(T(X)), \quad a + \epsilon b \mapsto a$$

is the identity. Composing the morphism 3.19 with the second projection

$$pr_2 : \mathcal{O}^r(T(X))[\epsilon] \to \mathcal{O}^r(T(X)), \quad a + \epsilon b \mapsto b$$

we get R-derivations

$$T : \mathcal{O}^r(X) \to \mathcal{O}^r(T(X)) \tag{3.20}$$

which agree with each other as r varies hence induce an R-derivation

$$T : \mathcal{O}^\infty(X) \to \mathcal{O}^\infty(T(X)).$$

This construction globalizes to a map as in equation 3.18. Note that T restricted to $\mathcal{O}_X$ equals d.

On the other hand, for any affine X let us define

$$\mathcal{O}^r(T(X))^+ := \sum_{j=0}^{r} \sum_{f \in \mathcal{O}(X)} \mathcal{O}^r(X) \cdot (df)^{\phi^j} \subset \mathcal{O}^r(T(X)). \tag{3.21}$$

Note that if $T_1, ..., T_d \in \mathcal{O}(X)$ is a system of étale coordinates then $\mathcal{O}^r(T(X))^+$ is a free $\mathcal{O}^r(X)$-module with basis

$$\{(dT_i)^{\phi^j} | 1 \leq i \leq d, \ 0 \leq j \leq r\}.$$

If X is not necessarily affine then the assignment

$$U \mapsto \mathcal{O}^r(T(U))^+$$

gives rise to a locally free sheaf of $\mathcal{O}^r_X$-modules, $(\pi_* \mathcal{O}^r_{T(X)})^+$. We also set

$$(\pi_* \mathcal{O}^\infty_{T(X)})^+ := \lim_{\to} (\pi_* \mathcal{O}^r_{T(X)})^+.$$

PROPOSITION 3.33. *If X is affine the image of the map $T : \mathcal{O}^r(X) \to \mathcal{O}^r(T(X))$ is contained in $\mathcal{O}^r(T(X))^+$. So if X is not necessarily affine we have induced maps*

$$T : \mathcal{O}^r_X \to (\pi_* \mathcal{O}^r_{T(X)})^+, \quad T : \mathcal{O}^\infty_X \to (\pi_* \mathcal{O}^\infty_{T(X)})^+. \tag{3.22}$$

PROOF. Since T is an $R-$derivation and $\mathcal{O}^r(X)$ is topologically generated by R and elements of the form $\delta^s f$ with $s \leq r$, $f \in \mathcal{O}(X)$, it is enough to check that, for any such s and f

$$T(\delta^s f) \in \mathcal{O}^s(T(X))^+.$$

On the other hand

$$T(\delta^s f) = pr_2(\delta^s(f + \epsilon df)),$$

so we conclude by assertion 1 in Lemma 3.34 below. □

LEMMA 3.34. *Let $x, x', x'', \ldots$ and $z, z', z'', \ldots$ be variables.*

1) There exist unique polynomials

$$\Lambda_{sj}(x, x', \ldots, x^{(s-1)}) \in R[x, x', \ldots, x^{(s-1)}], \quad s \geq 0, \quad j \geq 0,$$

such that

$$\delta^s(x + \epsilon z) = x^{(s)} + \epsilon \cdot \sum_{j=0}^{\infty} \Lambda_{sj}(x, x', \ldots, x^{(s-1)}) \cdot z^{\phi^j} \tag{3.23}$$

in the ring $R[\epsilon][x, x', \ldots, x^{(s)}, z, z', \ldots, z^{(s)}]$. Moreover $\Lambda_{ss} = 1$ for all $s \geq 0$ and $\Lambda_{sj} = 0$ for $j > s$.

2) For any $j \geq 0$ and $k \geq 0$ we have

$$\sum_{s=0}^{\infty} \frac{\partial x^{\phi^k}}{\partial x^{(s)}} \cdot \Lambda_{sj}(x, x', \ldots, x^{(s-1)}) = p^k \cdot \delta_{kj},$$

where δ_{kj} is the Kronecker delta.

PROOF. Let us prove assertion 1. Uniqueness is clear. To prove existence we proceed by induction on s. The case $s = 0$ is clear. Assume we constructed Λ_{tj} for all $t \leq s$ and all $j \geq 0$. Then

$$\delta^{s+1}(x + \epsilon z) = \delta(x^{(s)} + \epsilon \cdot \sum_{j=0}^{\infty} \Lambda_{sj} \cdot z^{\phi^j})$$

$$= \frac{1}{p}\{\phi(x^{(s)} + \epsilon \cdot \sum_{j=0}^{\infty} \Lambda_{sj} \cdot z^{\phi^j}) - (x^{(s)} + \epsilon \cdot \sum_{j=0}^{\infty} \Lambda_{sj} \cdot z^{\phi^j})^p\}$$

$$= \frac{1}{p}\{(x^{(s)})^p + px^{(s+1)} + p\epsilon \cdot \sum_{j=0}^{\infty} \Lambda_{sj}^{\phi} z^{\phi^{j+1}} - (x^{(s)})^p - p\epsilon (x^{(s)})^{p-1} \sum_{j=0}^{\infty} \Lambda_{sj} z^{\phi^j}\}$$

$$= x^{(s+1)} + \epsilon\{-(x^{(s)})^{p-1}\Lambda_{s0} \cdot z + \sum_{j=1}^{\infty}(\Lambda_{s,j-1}^{\phi} - (x^{(s)})^{p-1}\Lambda_{sj})z^{\phi^j}\}$$

which concludes the induction.

Let us prove assertion 2. Let D be the $R[\epsilon]-$derivation on

$$R[\epsilon][x, x', x'', \ldots, z, z', z'', \ldots]$$

defined by

$$D = \epsilon \cdot \sum_{j=0}^{\infty} \sum_{s=0}^{\infty} \Lambda_{sj}(x, x', \ldots, x^{(s-1)}) \cdot z^{\phi^j} \cdot \frac{\partial}{\partial x^{(s)}} = \sum_{s=0}^{\infty} [\delta^s(x + \epsilon z) - x^{(s)}] \frac{\partial}{\partial x^{(s)}}.$$

We claim that

$$D \circ \phi = \phi \circ D. \tag{3.24}$$

Indeed, since $D \circ \phi - \phi \circ D$ is a derivation relative to ϕ it is enough to show that $D(\phi x^{(k)}) = \phi(Dx^{(k)})$ for all $k \geq 0$. We have the following computation:

$$\begin{aligned} D(\phi x^{(k)}) &= D((x^{(k)})^p + px^{(k+1)}) = p(x^{(k)})^{p-1}D(x^{(k)}) + pD(x^{(k+1)}) \\ &= p(x^{(k)})^{p-1}D(x^{(k)}) + p[\delta^{k+1}(x+\epsilon z) - x^{(k+1)}] \\ &= p(x^{(k)})^{p-1}D(x^{(k)}) + p[\delta(x^{(k)} + D(x^{(k)})) - x^{(k+1)}] \end{aligned}$$

$$= p(x^{(k)})^{p-1}D(x^{(k)}) + p\{\frac{1}{p}[(x^{(k)})^p + px^{(k+1)} + \phi(D(x^{(k)})) - (x^{(k)} + D(x^{(k)}))^p] - x^{(k+1)}\}$$

$$= \phi(D(x^{(k)})),$$

which proves our Claim 3.24. By Equation 3.24 we get

$$D(x^{\phi^k}) = (Dx)^{\phi^k} = (\epsilon z)^{\phi^k} = p^k \epsilon z^{\phi^k}$$

for all k. On the other hand

$$D(x^{\phi^k}) = \epsilon \sum_{j=0}^{\infty} \left(\sum_{s=0}^{\infty} \Lambda_{sj}(x, x', ..., x^{(s-1)}) \frac{\partial x^{\phi^k}}{\partial x^{(s)}} \right) z^{\phi^j}$$

which concludes the proof. □

REMARK 3.35. Here are the first few values of the Λ_{sj}'s:

$$\Lambda_{00} = 1,$$

$$\Lambda_{10} = -x^{p-1}, \quad \Lambda_{11} = 1,$$

$$\Lambda_{20} = -(x')^{p-1}, \quad \Lambda_{21} = -(x^{\phi})^{p-1} - (x')^{p-1}, \quad \Lambda_{22} = 1.$$

With the notation in Proposition 3.33 and Lemma 3.34 we have

COROLLARY 3.36. *The map $T : \mathcal{O}^{\infty}(X) \to \mathcal{O}^{\infty}(T(X))^+$ is the unique R–derivation satisfying the conditions*

$$T(\delta^s f) = \sum_{j=0}^{\infty} \Lambda_{sj}(f, \delta f, ..., \delta^{s-1} f) \cdot (df)^{\phi^j}$$

for all $f \in \mathcal{O}(X)$ and all $s \geq 0$.

Assume in the next Corollary that X is affine and $T_1, ..., T_d \in \mathcal{O}(X)$ are étale coordinates on X. Identify, as usual,

$$\mathcal{O}^r(X) = \mathcal{O}(X)[T', ..., T^{(r)}]\hat{}$$

where $T = \{T_1, ..., T_d\}$. The derivations

$$\partial / \partial T_i : R[T] \to R[T]$$

uniquely lift to derivations

$$\partial / \partial T_i : \mathcal{O}^{\infty}(X) \to \mathcal{O}^{\infty}(X)$$

that vanish on $T_i', ..., T_i^{(r)}$ for all i. On the other hand we have at our disposal derivations $\frac{\partial}{\partial T_i^{(s)}}$ on $\mathcal{O}^{\infty}(X)$ for all $s \geq 1$.

COROLLARY 3.37. *For any $f \in \mathcal{O}^{\infty}(X)$ we have*

$$Tf = \sum_{i=1}^{d} \sum_{j=0}^{\infty} \sum_{s=0}^{\infty} \frac{\partial f}{\partial T_i^{(s)}} \cdot \Lambda_{sj}(T_i, T_i', ..., T_i^{(s-1)}) \cdot (dT_i)^{\phi^j}.$$

3.6.2. Contraction and adjunction operators. The space $Der_R(\mathcal{O}_X, \mathcal{O}_X)$ of R–derivations $\partial : \mathcal{O}_X \to \mathcal{O}_X$ is, of course, an analogue of the space of vector fields on a manifold. More generally, we will consider below R–derivations $\partial : \mathcal{O}_X \to \mathcal{O}_X^\infty$; such objects should be viewed as analogues of the *generalized vector fields* on manifolds in the sense of [**114**], p. 289, sometimes called *Bäcklund vector fields*. Cf. also Equation 0.10. The R–derivations $\mathcal{O}_X \to \mathcal{O}_X^1$ should be then viewed as analogues of *contact vector fields*. In the applications we have in view derivations $\partial : \mathcal{O}_X \to \mathcal{O}_X$ would, in principle, suffice. But the paradigm of R–derivations $\partial : \mathcal{O}_X \to \mathcal{O}_X^\infty$ is the most natural one so we will include it for the sake of completeness. Note that there is a natural map

$$Der_R(\mathcal{O}_X, \mathcal{O}_X) \to Der_R(\mathcal{O}_X, \mathcal{O}_X^\infty);$$

for any $\partial \in Der_R(\mathcal{O}_X, \mathcal{O}_X)$ we will continue to denote by ∂ the image of ∂ in

$$Der_R(\mathcal{O}_X, \mathcal{O}_X^\infty).$$

PROPOSITION 3.38. *For any R–derivation $\partial : \mathcal{O}_X \to \mathcal{O}_X^\infty$ and any $j \geq 0$ there exists a unique $\mathcal{O}_X^\infty$–linear map*

$$\langle \partial, \ \rangle_j : (\pi_* \mathcal{O}_{T(X)}^\infty)^+ \to \mathcal{O}_X^\infty, \tag{3.25}$$

such that for any local section f of $\mathcal{O}_X$ we have

$$\langle \partial, (df)^{\phi^s} \rangle_j = 0 \quad \text{if} \quad s \neq j,$$

$$\langle \partial, (df)^{\phi^j} \rangle_j = (\partial f)^{\phi^j}.$$

PROOF. Uniqueness is clear. To prove existence we may assume X is affine and has étale coordinates $T_1, ..., T_d$. Then define

$$\langle \partial, \sum_{i=1}^d \sum_{m=0}^\infty g_{im} (dT_i)^{\phi^m} \rangle \rangle_j := \sum_{i=1}^d g_{ij} (\partial T_i)^{\phi^j}. \tag{3.26}$$

If $df = \sum_{i=1}^d f_i dT_i$, $f_i \in \mathcal{O}(X)$, then $\partial f = \sum_{i=1}^d f_i \partial T_i$ so

$$\langle \partial, (df)^{\phi^s} \rangle_j = \langle \partial, \sum_{i=1}^d f_i^{\phi^s} (dT_i)^{\phi^s} \rangle_j = \sum_{i=1}^d \delta_{sj} f_i^{\phi^s} (\partial T_i)^{\phi^s} = \delta_{sj} (\partial f)^{\phi^s},$$

where δ_{sj} is the Kronecker symbol. □

The operators $\langle \partial, \ \rangle_j$ can be referred to as *contraction* operators.

PROPOSITION 3.39. *For each $r \geq 0$ there is a unique $\mathbf{Z}$–linear map*

$$ad_r : (\pi_* \mathcal{O}_{T(X)}^r)^+ \to (\pi_* \mathcal{O}_{T(X)}^{2r})^+$$

such that for any local sections f in $\mathcal{O}_X$ and g in $\mathcal{O}_X^r$ we have

$$ad_r (g (df)^{\phi^s}) = p^{r-s} g^{\phi^{r-s}} (df)^{\phi^r}$$

for all $0 \leq s \leq r$.

PROOF. Uniqueness is clear. To check existence we may assume X is affine with étale coordinates $T_1, ..., T_d$ and set

$$ad_r (\sum_{i=1}^d \sum_{s=0}^r g_{is} (dT)^{\phi^s}) := \sum_{i=1}^d (\sum_{s=0}^r p^{r-s} g_{is}^{\phi^{r-s}}) (dT_i)^{\phi^r}.$$

□

The operators can be referred to as *adjunction* operators; they are reminiscent of the operator, in the theory of linear differential equations, that takes a linear differential operator into its "adjoint".

3.6.3. Conjugate and prolonged operators. For any R–derivation $\partial : \mathcal{O}_X \to \mathcal{O}_X^\infty$ and any $j \geq 0$ we may consider the composition

$$\partial_j : \mathcal{O}_X^\infty \stackrel{T}{\to} (\pi_* \mathcal{O}_{T(X)}^\infty)^+ \stackrel{\langle \partial, \ \rangle_j}{\longrightarrow} \mathcal{O}_X^\infty, \tag{3.27}$$
$$f \mapsto \partial_j f := \langle \partial, Tf \rangle_j,$$

and also the sums

$$\partial_* := \sum_{j=0}^\infty \partial_j : \mathcal{O}_X^\infty \to \mathcal{O}_X^\infty,$$

$$\partial_{**} = \sum_{j=0}^\infty p^{-j} \partial_j : \mathcal{O}_X^\infty \otimes K \to \mathcal{O}_X^\infty \otimes K.$$

DEFINITION 3.40. The operators ∂_j, $j \geq 0$, are called the *conjugate operators* of ∂; the operators ∂_* and ∂_{**} are called the *(integral resp. rational) prolongations* of ∂.

The operators ∂_j morally play the role of conjugate variables in classical mechanics; the operators ∂_* and ∂_{**} morally play the role of the prolongation of a generalized vector field in classical mechanics; cf. Equation 0.12.

If one starts with an R–derivation $\partial : \mathcal{O}_X \to \mathcal{O}_X$ then one can consider the induced R–derivation $\partial : \mathcal{O}_X \to \mathcal{O}_X^\infty$ and, hence, the conjugate operators ∂_j and the prolongation operators ∂_* and ∂_{**}; these operators will then preserve the filtration by orders.

If X is affine with étale coordinates $T_1, ..., T_d$ then for any $f \in \mathcal{O}^\infty(X)$ we have

$$\partial_j f = \sum_{i=1}^d \sum_{s=0}^\infty \frac{\partial f}{\partial T_i^{(s)}} \cdot \Lambda_{sj}(T_i, T_i', ..., T_i^{(s-1)}) \cdot (\partial T_i)^{\phi^j}. \tag{3.28}$$

In particular we see that $\partial_j : \mathcal{O}_X^\infty \to \mathcal{O}_X^\infty$ are R–derivations vanishing on $\mathcal{O}_X^{j-1}$, $\partial_* : \mathcal{O}_X^\infty \to \mathcal{O}_X^\infty$ is an R–derivation lifting ∂, $\partial_{**} : \mathcal{O}_X^\infty \otimes K \to \mathcal{O}_X^\infty \otimes K$ is a K–derivation lifting ∂, and, for α any local section of $\mathcal{O}_X^\infty$ we have

$$(\alpha \cdot \partial)_j = \alpha^{\phi^j} \cdot \partial_j. \tag{3.29}$$

Also, by assertion 2 in Lemma 3.34 plus Equation 3.28 we have

$$\left(\frac{\partial}{\partial T_n}\right)_j (T_m^{\phi^k}) = p^k \cdot \delta_{nm} \delta_{jk} \tag{3.30}$$

for all $n, m \in \{1, ..., d\}$ and all $j, k \geq 0$. (Here, again, δ_{nm}, δ_{jk} are the Kronecker symbols.) Furthermore, by Corollary 3.36, we get, for any $f \in \mathcal{O}(X)$, any $j \geq 0$, and any $s \geq 0$:

$$\partial_j \delta^s f = \Lambda_{sj}(f, \delta f, ..., \delta^{s-1} f) \cdot (\partial f)^{\phi^j}. \tag{3.31}$$

In particular

$$\partial_0 f = \partial f$$

for any local section f in $\mathcal{O}_X$.

The following is also clear from Equation 3.28:

PROPOSITION 3.41. *Let $u : \tilde{X} \to X$ be an étale map of smooth schemes over R and denote by $u^* : \mathcal{O}^\infty(X) \to \mathcal{O}^\infty(\tilde{X})$ the map $u^*f := f \circ u$. Let $\partial : \mathcal{O}_X \to \mathcal{O}_X$ be an $R-$derivation and $u^*\partial : \mathcal{O}_{\tilde{X}} \to \mathcal{O}_{\tilde{X}}$ the unique derivation lifting ∂. Then, for any $f \in \mathcal{O}^\infty(X)$ we have*

$$u^*(\partial_j f) = (u^*\partial)_j(u^*f), \quad u^*(\partial_* f) = (u^*\partial)_*(u^*f), \quad u^*(\partial_{**} f) = (u^*\partial)_{**}(u^*f).$$

PROPOSITION 3.42. *For any $R-$derivation $\partial : \mathcal{O}_X \to \mathcal{O}_X^\infty$ and any $j \geq 0$ we have*

$$\partial_j \circ \phi = p \cdot \phi \circ \partial_{j-1}$$

on $\mathcal{O}_X^\infty$, where $\partial_{-1} := 0$. In particular ∂_* can be characterized as the unique $R-$derivation $\mathcal{O}_X^\infty \to \mathcal{O}_X^\infty$ whose restriction to $\mathcal{O}_X$ is ∂ and such that

$$\partial_* \circ \phi = p \cdot \phi \circ \partial_* \quad on \quad \mathcal{O}_X^\infty;$$

consequently ∂_* is a $\delta-$derivation. Also ∂_{**} can be characterized as the unique $K-$derivation $\mathcal{O}_X^\infty \otimes K \to \mathcal{O}_X^\infty \otimes K$ whose restriction to $\mathcal{O}_X$ is ∂ and such that

$$\partial_{**} \circ \phi = \phi \circ \partial_{**} \quad on \quad \mathcal{O}_X^\infty.$$

PROOF. We may assume X is affine with étale coordinates $T = \{T_1, ..., T_d\}$. Consider the derivation

$$D_j := \partial_j \circ \phi - p \cdot \phi \circ \partial_{j-1} : \mathcal{O}_X^\infty \to \mathcal{O}_X^\infty.$$

We may write

$$\partial = \sum_{m=1}^{d} \alpha_m \cdot \frac{\partial}{\partial T_m},$$

$\alpha_m \in \mathcal{O}^\infty(X)$. So we have

$$\partial_j = \sum_{m=1}^{d} \alpha_m^{\phi^j} \cdot \left(\frac{\partial}{\partial T_m}\right)_j,$$

hence

$$D_j = \sum_{m=1}^{d} \left[\alpha_m^{\phi^j} \cdot \left(\frac{\partial}{\partial T_m}\right)_j \circ \phi - p \cdot \phi \circ \left(\alpha_m^{\phi^{j-1}} \cdot \left(\frac{\partial}{\partial T_m}\right)_{j-1}\right)\right] = \sum_{m=1}^{d} \alpha_m^{\phi^j} D_{jm}$$

where

$$D_{jm} := \left(\frac{\partial}{\partial T_m}\right)_j \circ \phi - p \cdot \phi \circ \left(\frac{\partial}{\partial T_m}\right)_{j-1}.$$

It is enough to check that D_{jm} vanishes on $\mathcal{O}^\infty(X)$. Note that D_{jm} induces a derivation on $R[T, T', T'', ...]$ so it is enough to show that D_{jm} vanishes on

$$K[T, T', T'', ...] = K[T, T^\phi, T^{\phi^2}, ...].$$

But using Equation 3.30 we have

$$\begin{aligned} D_{jm}(T_n^{\phi^k}) &= \left(\frac{\partial}{\partial T_m}\right)_j (T_n^{\phi^{k+1}}) - p \cdot \phi\left(\left(\frac{\partial}{\partial T_m}\right)_{j-1} (T_n^{\phi^k})\right) \\ &= p^{k+1}\delta_{nm}\delta_{j,k+1} - p(p^k\delta_{nm}\delta_{j-1,k}) = 0, \end{aligned}$$

which ends our proof. □

PROPOSITION 3.43. *Let X/R be a smooth scheme and let $\partial : \mathcal{O}_X \to \mathcal{O}_X$ be an R–derivation. Then, for each $r \geq 1$, the map $\partial_r : \mathcal{O}^r_X \to \mathcal{O}^r_X$ can be characterized as the unique derivation with the following properties:*

1) ∂_r vanishes on $\mathcal{O}^{r-1}_X$,

2) $\partial_r \circ \delta^r = \phi^r \circ \partial$ on $\mathcal{O}_X$.

Furthermore we have

3) $\partial_r \circ \phi^r = p^r \cdot \phi^r \circ \partial$ on $\mathcal{O}_X$.

REMARK 3.44. Note that ∂_r and ∂_{r-1} are related by the formula

$$\partial_r \circ \delta = \phi \circ \partial_{r-1}$$

on $\mathcal{O}^{r-1}_X$. Indeed this follows from Proposition 3.42 and condition 1 above.

PROOF. Uniqueness is clear. Property 1 is clear too. Properties 2 and 3 are equivalent under the assumption that property 1 holds. But property 3 follows from Proposition 3.42. □

PROPOSITION 3.45. *Let X/R be a smooth scheme and let $\partial : \mathcal{O}_X \to \mathcal{O}_X$ be an R–derivation. Then, for each $j \geq 0$, the map $\partial_j : \mathcal{O}^\infty_X \to \mathcal{O}^\infty_X$ can be characterized as the unique R–derivation with the following properties:*

1) $\partial_j \circ \phi^s = 0$ on $\mathcal{O}_X$ if $j \neq s$,

2) $\partial_j \circ \phi^j = p^j \cdot \phi^j \circ \partial$ on $\mathcal{O}_X$.

PROOF. Uniqueness is clear. Condition 2 follows from Proposition 3.42. Condition 1 for $s < j$ is also clear. Let us check condition 1 for $s > j$. We may assume X is affine with étale coordinates $T_1, ..., T_d$. By Proposition 3.42 we have

$$\partial_j \circ \phi^s = p^j \cdot \phi^j \circ \partial_0 \circ \phi^{s-j}.$$

So it is enough to check that, for $m \geq 1$, the map

$$\partial_0 \circ \phi^m : \mathcal{O}(X) \to \mathcal{O}^\infty(X)$$

vanishes. Since this map is a derivation relative to $\phi^m : \mathcal{O}(X) \to \mathcal{O}^\infty(X)$ it is enough to check that this map vanishes on T_i for all i. Write

$$\partial = \sum_{k=1}^d \alpha_k \cdot \frac{\partial}{\partial T_k},$$

with $\alpha_k \in \mathcal{O}^\infty(X)$. We then have

$$\partial_0 = \sum_{k=1}^d \alpha_k \cdot \left(\frac{\partial}{\partial T_k}\right)_0,$$

on $\mathcal{O}^\infty(X)$ hence it is enough to show that

$$\left(\frac{\partial}{\partial T_k}\right)_0 (T_i^{\phi^m}) = 0;$$

this follows directly from Equation 3.30. □

The next Proposition addresses commutation issues between the various ∂_j's. Let us denote by [,] the commutator of any 2 R–derivations.

PROPOSITION 3.46. *For any $R-$derivations $\partial, \tilde{\partial} : \mathcal{O}_X \to \mathcal{O}_X$ we have*
1) $[\partial_i, \tilde{\partial}_j] = 0$ *for all* $i \neq j$,
2) $[\partial_j, \tilde{\partial}_j] = p^j \cdot [\partial, \tilde{\partial}]_j$ *for all* j,
3) $[\partial_{**}, \tilde{\partial}_{**}] = [\partial, \tilde{\partial}]_{**}$.

In particular the map

$$Der_R(\mathcal{O}_X, \mathcal{O}_X) \to Der_R(\mathcal{O}_X^\infty, \mathcal{O}_X^\infty), \ \partial \mapsto \partial_0$$

is a Lie $R-$algebra map. Also the maps

$$Der_R(\mathcal{O}_X, \mathcal{O}_X) \otimes K \to Der_R(\mathcal{O}_X^\infty, \mathcal{O}_X^\infty) \otimes K, \ \partial \mapsto p^{-j}\partial_j,$$

$$Der_R(\mathcal{O}_X, \mathcal{O}_X) \otimes K \to Der_R(\mathcal{O}_X^\infty \otimes K, \mathcal{O}_X^\infty \otimes K), \ \partial \mapsto \partial_{**}$$

are Lie $\mathbf{Q}_p-$algebra maps.

PROOF. We may assume X is affine with étale coordinates $T_1, ..., T_d$. To check assertion 1 write

$$\partial = \sum_{n=1}^{d} \alpha_n \frac{\partial}{\partial T_n}, \quad \tilde{\partial} = \sum_{m=1}^{d} \tilde{\alpha}_m \frac{\partial}{\partial T_m}, \quad \alpha_n, \tilde{\alpha}_m \in \mathcal{O}(X).$$

Then we have

$$\partial_i \tilde{\partial}_j = \sum_{n=1}^{d}\sum_{m=1}^{d} \alpha_n^{\phi^i} \cdot \left[\left(\frac{\partial}{\partial T_n}\right)_i (\tilde{\alpha}_m^{\phi^j}) \right] \cdot \left(\frac{\partial}{\partial T_m}\right)_j + \sum_{n=1}^{d}\sum_{m=1}^{d} \alpha_n^{\phi^i} \tilde{\alpha}_m^{\phi^j} \cdot \left(\frac{\partial}{\partial T_n}\right)_i \circ \left(\frac{\partial}{\partial T_m}\right)_j.$$

By Proposition 3.45 we have

$$\left(\frac{\partial}{\partial T_n}\right)_i (\tilde{\alpha}_m^{\phi^j}) = 0$$

so it is enough to check that

$$\left[\left(\frac{\partial}{\partial T_n}\right)_i, \left(\frac{\partial}{\partial T_m}\right)_j \right] = 0. \tag{3.32}$$

It is enough to check the latter holds in the ring $R[T, T', T'', ...]$ hence in the ring $K[T, T', T'', ...] = K[T, T^\phi, T^{\phi^2}, ...]$. But, in the latter ring, Equation 3.32 holds because of Equation 3.30.

To check assertion 2 of the Proposition it is enough to show that the maps

$$([\partial_j, \tilde{\partial}_j] - p^j \cdot [\partial, \tilde{\partial}]_j) \circ \phi^s$$

vanish on $\mathcal{O}(X)$ for all $s \geq 0$; this follows trivially from the commutation relations in Proposition 3.45.

Assertion 3 of the Proposition follows from assertions 1 and 2. □

PROPOSITION 3.47. *If $\partial, \tilde{\partial} : \mathcal{O}_X \to \mathcal{O}_X^\infty$ are $R-$derivations then*

$$[\partial_{**}, \tilde{\partial}_{**}] = ([\partial_{**}, \tilde{\partial}_{**}]_{|\mathcal{O}_X})_{**}.$$

PROOF. By Proposition 3.42 it is enough to check that $[\partial_{**}, \tilde{\partial}_{**}]$ commutes with ϕ; but this is clear because, by the same Proposition, both ∂_{**} and $\tilde{\partial}_{**}$ commute with ϕ. □

Proposition 3.47 permits one to introduce a Lie $\mathbf{Q}_p$–algebra structure on

$$Der_R(\mathcal{O}_X, \mathcal{O}_X^{\infty}) \otimes K = Der_R(\mathcal{O}_X, \mathcal{O}_X^{\infty} \otimes K)$$

(analogue to the Lie algebra structure on the space of Bäcklund vector fields): the bracket of two derivations $\partial, \tilde{\partial} \in Der_R(\mathcal{O}_X, \mathcal{O}_X^{\infty}) \otimes K$ is taken to be

$$[\partial, \tilde{\partial}] := p^{-2n} \cdot [(p^n\partial)_{**}, (p^n\tilde{\partial})_{**}]_{|\mathcal{O}_X},$$

where n is such that $p^n\partial, p^n\tilde{\partial} \in Der_R(\mathcal{O}_X, \mathcal{O}_X^{\infty})$. In view of Proposition 3.46 the map

$$Der_R(\mathcal{O}_X, \mathcal{O}_X) \otimes K \to Der_R(\mathcal{O}_X, \mathcal{O}_X^{\infty}) \otimes K$$

is a Lie $\mathbf{Q}_p$–algebra map.

By analogy with classical mechanics (cf. Equations 0.13 and 0.14) we make the following:

DEFINITION 3.48. If $\mathcal{L}$ is a K–linear subspace of $\mathcal{O}^{\infty}(X) \otimes K$ and if $\partial \in Der_R(\mathcal{O}_X, \mathcal{O}_X)$ (respectively $\partial \in Der_R(\mathcal{O}_X, \mathcal{O}_X^{\infty}) \otimes K$) is such that

$$\partial_{**}\mathcal{L} \subset \mathcal{L} \tag{3.33}$$

we say that ∂ is an *infinitesimal symmetry* (respectively a *generalized infinitesimal symmetry*) of $\mathcal{L}$. If $f \in \mathcal{O}^{\infty}(X) \otimes K$ and $\partial_{**}f = 0$ we say that ∂ is a (generalized) *variational infinitesimal symmetry of f*.

The infinitesimal symmetries of a fixed $\mathcal{L}$ form a Lie $\mathbf{Z}_p$–subalgebra of

$$Der_R(\mathcal{O}_X, \mathcal{O}_X).$$

PROPOSITION 3.49. *Let X/R have relative dimension d and assume the derivations*

$$\partial^1, ..., \partial^d : \mathcal{O}_X \to \mathcal{O}_X$$

form a basis of the sheaf $T_{X/R}$ of R–derivations of $\mathcal{O}_X$. Then the conjugate operators

$$\partial_0^1, ..., \partial_0^d, ..., \partial_r^1, ..., \partial_r^d : \mathcal{O}_X^r \to \mathcal{O}_X^r \tag{3.34}$$

form a basis of the $\mathcal{O}_X^r$–module $Der_R(\mathcal{O}_X^r, \mathcal{O}_X^r)$ of R–derivations of $\mathcal{O}_X^r$.

PROOF. We may assume X is affine and has étale coordinates $T_1, ..., T_d \in \mathcal{O}(X)$. Then $\Omega_{X/R} = \sum_{i=1}^d \mathcal{O}_X \cdot dT_i$ hence

$$det(\partial^i T_j) \in \mathcal{O}(X)^{\times}.$$

By Proposition 3.13

$$\frac{\partial}{\partial T_1}, ..., \frac{\partial}{\partial T_d}, ..., \frac{\partial}{\partial T_1^{(r)}}, ..., \frac{\partial}{\partial T_d^{(r)}} \tag{3.35}$$

is a basis of the $\mathcal{O}^r(X)$–module of R–derivations of $\mathcal{O}^r(X)$. Equation 3.28 shows that the derivations in Equation 3.34 restricted to $\mathcal{O}^r(X)$ can be expressed as linear combinations, with $\mathcal{O}^r(X)$–coefficients, of the derivations in Equation 3.35; the matrix of the coefficients in these linear combinations is a matrix with $r+1$ diagonal $d \times d$ blocks having determinants

$$det(\partial^i T_j), (det(\partial^i T_j))^{\phi}, ..., (det(\partial^i T_j))^{\phi^r}$$

and zero entries below these blocks so this matrix is invertible and we are done. □

3.6.4. Euler-Lagrange operators and "Noether Theorem". Recall from Proposition 3.39 that we defined, for each $r \geq 0$, an adjunction operator ad_r from $(\pi_* \mathcal{O}^r_{T(X)})^+$ into $(\pi_* \mathcal{O}^{2r}_{T(X)})^+$. We define the *Euler-Lagrange operators* as the composition

$$EL_r : \mathcal{O}^r_X \xrightarrow{T} (\pi_* \mathcal{O}^r_{T(X)})^+ \xrightarrow{ad_r} (\pi_* \mathcal{O}^{2r}_{T(X)})^+. \tag{3.36}$$

These operators should be viewed as arithmetic analogues of the classical Euler-Lagrange operator in classical mechanics; cf. Equation 0.16. If $f \in \mathcal{O}^r(X)$ and $\partial : \mathcal{O}_X \to \mathcal{O}_X$ is an R-derivation then the function

$$\langle \partial, EL_r(f) \rangle_r \in \mathcal{O}^{2r}(X)$$

can be referred to as the *Euler-Lagrange equation* attached to f and ∂.

PROPOSITION 3.50. *Assume X is smooth over R and irreducible. Let $f \in \mathcal{O}^r(X)$ and let $\partial : \mathcal{O}_X \to \mathcal{O}_X$ be an R-derivation. Then the Euler-Lagrange equation attached to f and ∂ is given by the formula*

$$\langle \partial, EL_r f \rangle_r = \sum_{j=0}^{r} p^{r-j} (\partial_j f)^{\phi^{r-j}}. \tag{3.37}$$

Moreover, if

$$g := \sum_{j=0}^{r} p^{r-j} \sum_{s=0}^{r-j-1} (\partial_j f)^{\phi^s} \in \mathcal{O}^{2r-1}(X)$$

then the following formula holds:

$$\langle \partial, EL_r f \rangle_r - p^r \partial_{**} f = g^{\phi} - g. \tag{3.38}$$

Equation 3.38 should be viewed as an arithmetic analogue of "Noether's Theorem". Cf. Equation 0.17.

PROOF. We may assume that X is affine with étale coordinates $T_1, ..., T_d$. Let $\partial^i := \partial / \partial T_i : \mathcal{O}_X \to \mathcal{O}_X$. Write $Tf = \sum_{i,j} f_{ij} (dT_i)^{\phi^j}$. Hence

$$\partial^s_j f = \langle \partial^s, Tf \rangle_j = \sum_i f_{ij} \left(\frac{\partial T_i}{\partial T_s} \right)^{\phi^j} = f_{sj}.$$

Also, clearly, $\partial = \sum_i (\partial T_i) \partial^i$. Then

$$\langle \partial, EL_r f \rangle_r = \langle \partial, ad_r(Tf) \rangle_r = \langle \partial, ad_r(\sum_{ij} (\partial^i_j f)(dT_i)^{\phi^j} \rangle_r$$

$$= \langle \partial, \sum_i (\sum_j p^{r-j} (\partial^i_j f)^{\phi^{r-j}})(dT_i)^{\phi^r} \rangle_r = \sum_{ij} p^{r-j} (\partial^i_j f)^{\phi^{r-j}} (\partial T_i)^{\phi^r}$$

$$= \sum_j p^{r-j} (\partial_j f)^{\phi^{r-j}}.$$

On the other hand, by definition,

$$p^r \partial_{**} f = \sum_j p^{r-j} \partial_j f$$

and Equation 3.38 follows. □

3.6.5. δ–tangent maps on tangent spaces. Recall from Equation 1.13 that for $P \in X(R)$ we defined the *tangent space* $T_P X := Der_R(\mathcal{O}_{X,P_0}, R)$; since X is smooth $T_P X$ is a free $R-$ module of rank equal to the relative dimension, d, of X/R. Also let $U \subset X$ be any Zariski open set with $P \in U(R)$ and let $m_P \subset \mathcal{O}^\infty(U)$ be the preimage of pR via the evaluation map

$$\mathcal{O}^\infty(U) \to R, \quad f \mapsto f(P).$$

Then the $\mathbf{Z}_p-$linear map

$$Der_R(\mathcal{O}_X, \mathcal{O}_X) \to Der_\delta(\mathcal{O}_X^\infty, \mathcal{O}_X^\infty), \quad \partial \mapsto \partial_* = \sum \partial_j$$

induces a $\mathbf{Z}_p-$linear map

$$T_P X \to Der_\delta(\mathcal{O}^\infty(U), R), \quad \partial_P \mapsto \partial_{P*}, \tag{3.39}$$

hence, by composing with the natural map

$$Der_\delta(\mathcal{O}^\infty(U), R) \to Der_\delta(\mathcal{O}^\infty(U)_{m_P}, R),$$

we get a $\mathbf{Z}_p-$linear map

$$T_P X \to Der_\delta(\mathcal{O}^\infty(U)_{m_P}, R), \quad \partial_P \mapsto \partial_{P*}. \tag{3.40}$$

Explicitly, if U is affine with étale coordinates $T_1, ..., T_d$, $T_i(P) = \alpha_i$, the map in Equation 3.39 sends $\partial : \mathcal{O}(U) \to R$ into the derivation

$$f \mapsto \sum_{i=1}^{d} \sum_{j=0}^{\infty} \sum_{s=0}^{\infty} \frac{\partial f}{\partial T_i^{(s)}}(P) \cdot \Lambda_{sj}(\alpha_i, ..., \delta^{s-1}\alpha_i) \cdot (\partial T_i)^{\phi^j},$$

for $f \in \mathcal{O}^\infty(U)$. So for any $f \in \mathcal{O}^\infty(U)_{m_P}$ we get a $\mathbf{Z}_p-$linear *$\delta-$tangent map,*

$$T_P f : T_P X \to R, \quad (T_P f)(\partial) := \partial_{P*} f. \tag{3.41}$$

Under the identification

$$T_P X \simeq R^d, \quad \partial \mapsto (\partial T_1, ..., \partial T_d),$$

the map in Equation 3.41 has the form

$$R^d \to R, \quad (\beta_1, ..., \beta_d) \mapsto \sum \lambda_{ij} \beta_i^{\phi^j}, \tag{3.42}$$

for some $\lambda_{ij} \in R$.

Here is a consequence of the above discussion. Let L be a $\delta-$line bundle on X and select U such that L is trivial on U. Then there is a natural $\delta-$ring homomorphism

$$R(X, L)^*_{\langle P \rangle} \to \mathcal{O}^\infty(U)_{m_P}$$

so we have an induced map

$$T_P X \to Der_\delta(R(X, L)^*_{\langle P \rangle}, R), \quad \partial_P \mapsto \partial_{P*}. \tag{3.43}$$

This map is easily seen to be an isomorphism if X is affine or if X is projective and $L = M^v$, M an ample line bundle on X, and $v \in W_+ \backslash \phi W_+$.

On the other hand let $s_0, ..., s_n \in H^0(X, L)$ be $\delta-$sections, consider the subset of $X(R)$ defined by

$$X(R)' := \{P \in X(R) | s_i(P) \not\equiv 0 \ mod \ p \ for \ some \ i\},$$

consider the map

$$f : X(R)' \to \mathbf{P}^n(R), \quad f(P) = (s_0(P) : ... : s_n(P)),$$

and let $P \in X(R)'$. Then the maps 3.41 induce a $\mathbf{Z}_p-$linear *$\delta-$tangent map*,

$$T_P f : T_P X \to T_{f(P)}\mathbf{P}^n \tag{3.44}$$

which, in coordinates, looks like the map in Equation 3.42. The map in Equation 3.44 is compatible, via the map in Equation 3.43, with the $\delta-$tangent map introduced in Equation 2.20.

3.6.6. Poisson structure. Recall that a *Poisson structure* on an $R-$algebra A (where R is any commutative ring) is an $R-$bilinear map

$$\{\ ,\ \} : A \times A \to A$$

satisfying the following properties:
1) $\{f_1, f_2\} = -\{f_2, f_1\}$,
2) $\{f_1, \{f_2, f_3\}\} + \{f_2, \{f_3, f_1\}\} + \{f_3, \{f_1, f_2\}\} = 0$,
3) $\{f_1 f_2, f_3\} = f_1\{f_2, f_3\} + \{f_1, f_3\}f_2$,

for all $f_i \in A$. If $D_1, ..., D_n$ are pairwise commuting $R-$derivations on A, (a_{ij}) is a skew-symmetric $n \times n$ matrix with coefficients in R and A is commutative, then the formula

$$\{f, g\} := \sum_{i,j=1}^{n} a_{ij} \cdot D_i f \cdot D_j g$$

defines a Poisson structure on A. Consequently, if X is a smooth scheme over R and $0 \leq i < j$ then for any $R-$derivation $\partial : \mathcal{O}_X \to \mathcal{O}_X$ the formula

$$\{f, g\}_\partial^{ij} := \partial_i f \cdot \partial_j g - \partial_j f \cdot \partial_i g$$

defines a Poisson structure on $\mathcal{O}_X^\infty$ (i.e. on $\mathcal{O}^\infty(U)$ for all Zariski open $U \subset X$). The Poisson structure $\{\ ,\ \}_\partial^{01}$ can be viewed as an arithmetic analogue of the usual Poisson structure in classical mechanics.

DEFINITION 3.51. If $f \in \mathcal{O}^\infty(X)$ then the derivation

$$V_f^{ij} := \{f,\ \}_\partial^{ij} = \partial_i f \cdot \partial_j - \partial_j f \cdot \partial_i \in Der_R(\mathcal{O}_X^\infty, \mathcal{O}_X^\infty)$$

will be called the $(i,j)-$*Hamiltonian flow* defined by f and ∂.

The above concept is an analogue of usual Hamiltonian flows in mechanics; cf. Equation 0.19. As the reader can see what we propose, in our formalism, is to view "Lagrangian" and "Hamiltonian" concepts as living on the same space $J^r(X)$. This is morally possible because our rule of formation of conjugate operators supplies a sort of "intrinsic simplectic structure".

3.6.7. $\delta-$differentials of $\delta-$functions. So far we looked at the $p-$jet spaces of the tangent bundle and we developed the concept of $\delta-$tangent map. There is a picture which is, in some sense, "dual" to this one in which one looks at the tangent bundle of $p-$jet spaces and one develops the concept of *$\delta-$differentials*. We now explain this dual picture.

If Y is a formal scheme which is locally the $p-$adic completion of a smooth scheme over R then we set, by abuse of notation,

$$\Omega = \Omega_{Y/R} := \lim_{\leftarrow} \Omega_{Y_n/R_n},$$

where $R_n := R/p^{n+1}R$, $Y_n := Y \otimes R_n$. Then $\Omega_{Y/R}$ is locally free, dual to $T_{Y/R}$, the sheaf of R–derivations of $\mathcal{O}_Y$. We set as usual $\Omega^n_{Y/R} := \wedge^n \Omega_{Y/R}$, the sheaf of n–forms. We have a natural de Rham complex

$$\mathcal{O}_Y \xrightarrow{d} \Omega_{Y/R} \xrightarrow{d} \Omega^2_{Y/R} \xrightarrow{d} \ldots$$

In particular this applies to $Y = J^r(X)$ where X/R is a smooth scheme over R. For each $r \geq 0$ we have an induced de Rham complex

$$\mathcal{O}^r(X) = H^0(J^r(X), \mathcal{O}) \xrightarrow{d} H^0(J^r(X), \Omega) \xrightarrow{d} H^0(J^r(X), \Omega^2) \xrightarrow{d} \ldots \tag{3.45}$$

For $f \in \mathcal{O}^r(X)$ we refer to df as the δ–*differential* of f. As usual a form is called *closed* if it is in the kernel of d.

It follows directly from definitions that the δ–differential $df \in H^0(J^r(X), \Omega)$ and the δ–tangent map $Tf \in \mathcal{O}^r(T(X))^+$ are related as follows.

PROPOSITION 3.52. *For any $f \in \mathcal{O}^r(X)$, any R–derivation $\partial : \mathcal{O}_X \to \mathcal{O}_X$, and any $j \in \{0, ..., r\}$ we have*

$$\langle \partial, Tf \rangle_j = \langle \partial_j, df \rangle.$$

The pairing $\langle\ ,\ \rangle$ in the right hand side of the above Equation is induced by the natural pairing between Ω and its dual.

Note that if $\pi_r : J^r(X) \to J^0(X) = \hat{X}$ is the natural projection and $i : X_0 = X \otimes k \to X$ is the natural closed immersion then we have a natural isomorphism of sheaves on X:

$$\mathcal{D}er_R(\mathcal{O}^r_X, \mathcal{O}^r_X) \simeq i_* \pi_{r*} T_{J^r(X)/R}.$$

PROPOSITION 3.53. *There is a natural isomorphism of sheaves*

$$i_* \pi_{r*} T_{J^r(X)/R} \simeq \bigoplus_{j=0}^{r} T^{\phi^j [r-j]}_{X/R}.$$

PROOF. This follows directly from Proposition 3.49 and Equation 3.29. $\square$

Next note that the homomorphisms $\phi : \mathcal{O}^i \to \mathcal{O}^{i+1}$, $i < r$, induce group homomorphisms

$$\phi^* : H^0(J^i(X), \Omega^j) \to H^0(J^{i+1}(X), \Omega^j) \tag{3.46}$$

which commute with d.

LEMMA 3.54. *For any $j \geq 1$ we have*

$$\phi^* H^0(J^i(X), \Omega^j) \subset p \cdot H^0(J^{i+1}(X), \Omega^j).$$

In particular, for any $i < r$ and any $\omega \in H^0(X, \Omega)$, we can define

$$\omega_i := p^{-i}(\phi^*)^i \omega \in H^0(J^i(X), \Omega) \subset H^0(J^r(X), \Omega),$$

where the last inclusion is given by pull-back via the canonical projection $J^r(X) \to J^i(X)$. We call $\omega_0, ..., \omega_r$ the *conjugates* of ω. Note that, since ϕ^* commutes with d, it follows that if ω is closed (e.g. if X/R has relative dimension one) then $\omega_0, ..., \omega_r$ are also closed.

PROOF. If $g, f \in \mathcal{O}^i(X)$ we have

$$\phi^*(gdf) = g^\phi d(f^\phi) = g^\phi d(f^p + p\delta f) = p \cdot g^\phi (f^{p-1} df + d(\delta f)),$$

which proves our claim for $j = 1$. The claim for arbitrary j follows from the one for $j = 1$. $\square$

PROPOSITION 3.55. *If $\omega^1, ..., \omega^d \in H^0(X, \Omega_{X/R})$ is a basis of $\Omega_{X/R}$ dual to a basis $\partial^1, ..., \partial^d \in H^0(X, T_{X/R})$ of $T_{X/R}$ then the conjugates*

$$\omega_0^1, ..., \omega_0^d, ..., \omega_r^1, ..., \omega_r^d \in H^0(J^r(X), \Omega)$$

form a basis of $\Omega_{J^r(X)/R}$ which is dual to the basis

$$\partial_0^1, ..., \partial_0^d, ..., \partial_r^1, ..., \partial_r^d \in H^0(J^r(X), T)$$

of $T_{J^r(X)/R}$ consisting of the conjugates of $\partial^1, ..., \partial^d$ (cf. Proposition 3.49). In particular, for any $f \in \mathcal{O}^r(X)$, we have the following formula in $H^0(J^r(X), \Omega)$:

$$df = \sum_{i=1}^{d} \sum_{j=0}^{r} (\partial_j^i f) \cdot \omega_j^i.$$

PROOF. We may assume we have étale coordinates $T_1, ..., T_d \in \mathcal{O}(X)$. Let ∂ be the column vector with entries $\partial^1, ..., \partial^d$ and let ω be the row vector with entries $\omega^1, ..., \omega^d$. Write

$$\partial = A \frac{\partial}{\partial T}$$

with $A \in GL_d(\mathcal{O}(X))$ and $\frac{\partial}{\partial T}$ the column vector with entries $\frac{\partial}{\partial T_j}$. Then

$$\partial_i = A^{\phi^i} \left(\frac{\partial}{\partial T} \right)_i.$$

On the other hand $\omega = dT \cdot A^{-1}$, where $dT = (dT_1, ..., dT_d)$, hence

$$p^{-i}(\phi^*)^i \omega = p^{-i} \cdot d(T^{\phi^i}) \cdot A^{-\phi^i}.$$

We have

$$\begin{pmatrix} \partial_0 \\ \cdot \\ \cdot \\ \partial_r \end{pmatrix} (\omega_0, ..., \omega_r) = \begin{pmatrix} A \cdot (\frac{\partial}{\partial T}) \\ \cdot \\ \cdot \\ A^{\phi^r} \cdot (\frac{\partial}{\partial T})_r \end{pmatrix} (dT \cdot A^{-1}, ..., p^{-r} \cdot d(T^{\phi^r}) \cdot A^{-\phi^r}).$$

Call B the $(r+1)d \times (r+1)d$ matrix above. Then $B = (B_{ij})$ where B_{ij} are the $d \times d$ matrices

$$B_{ij} = A^{\phi^i} \cdot \left(\frac{\partial}{\partial T} \right)_i \cdot p^{-j} \cdot d(T^{\phi^j}) \cdot A^{-\phi^j}.$$

By Equation 3.30 we have

$$B_{ij} = p^{-j} \cdot A^{\phi^i} \cdot \left[\left(\frac{\partial}{\partial T} \right)_i (T^{\phi^j}) \right] \cdot A^{-\phi^j} = p^{-j} \cdot A^{\phi^i} \cdot (p^j \delta_{ij} I_d) \cdot A^{-\phi^j} = \delta_{ij} I_d,$$

where I_d is the identity $d \times d$ matrix. So $B = I_{(r+1)d}$ and we are done. □

3.6.8. Action of ∂_j on δ−sections. Our next objective is to examine the action of the operators ∂_j on sections of δ−line bundles. So let X be a smooth scheme over R and L a line bundle on X. Consider the "physical line bundle" over X with its zero section removed:

$$\mathbf{V}^*(L) = Spec \bigoplus_{m \in \mathbf{Z}} L^{\otimes m}. \tag{3.47}$$

For any $r \geq 0$ set

$$S^r = S_X^r := \mathcal{O}^r(X), \quad M^r = M_X^r := \mathcal{O}^r(\mathbf{V}^*(L)), \tag{3.48}$$

$$M^r_X(w) := H^0(X, L^{w[r-ord(w)]}), \quad \bar{M}^r_X := M^r_X \otimes_R k, \quad \bar{M}^r_X(w) := M^r_X(w) \otimes_R k.$$

Then the direct sum $\bigoplus_{w \in W(r)} M^r_X(w)$ has a natural ring structure and so does its reduction mod p, $\bigoplus_{w \in W(r)} \bar{M}^r_X(w)$.

PROPOSITION 3.56. *For any integer $r \geq 0$ the following hold:*
i) There are natural inclusions

$$\bigoplus_{w \in W(r)} M^r_X(w) \subset M^r_X, \quad \bar{M}^r_X(w) \subset \bar{M}^r_X.$$

ii) Assume one is given an R–derivation

$$\partial : \bigoplus_{m \in \mathbf{Z}} L^{\otimes m} \to \bigoplus_{m \in \mathbf{Z}} L^{\otimes m}$$

with the property that there exists an integer $c \geq 0$ such that, for all $m \geq 0$, $\partial(L^{\otimes m}) \subset L^{\otimes(m+c)}$. Let $\partial : \mathcal{O}_{\mathbf{V}^*(L)} \to \mathcal{O}_{\mathbf{V}^*(L)}$ be the induced derivation, $\partial_j : \mathcal{O}^r_{\mathbf{V}^*(L)} \to \mathcal{O}^r_{\mathbf{V}^*(L)}$ be the conjugate operators, and $\partial_j : M^r_X \to M^r_X$ the induced operators. Then, for all $w \in W_+$,

$$\partial_j(M^r_X(w)) \subset M^r_X(w + c\phi^j).$$

iii) If an element $f \in M^r_X(w)$ is invertible in M^r_X then its inverse lies in $M^r_X(-w)$.
iv) If $X_1 \subset X$ is an affine open subscheme then

$$M^r_X \cap M^r_{X_1}(w) = M^r_X(w).$$

REMARK 3.57. As will be clear from the proof:
1) M^r_X is not the p–adic closure of $\bigoplus_{w \in W(r)} M^r_X(w)$.
2) The induced map $\bigoplus_{w \in W(r)} \bar{M}^r_X(w) \to \bar{M}^r_X$ is not injective.

We shall refer to $\partial_j : M^r_X \to M^r_X$ as the conjugates of ∂. We also have naturally induced operators

$$(3.49) \quad \partial_* : M^r_X \to M^r_X, \quad \partial_{**} : M^r_X \otimes K \to M^r_X \otimes K, \quad \langle \partial, EL_r(-) \rangle_r : M^r_X \to M^{2r}_X.$$

PROOF. (of Proposition 3.56). The Proposition is, clearly, a local statement on X. So we may, and will, assume that X is affine, L is free and, hence, upon choosing a basis x of L, we have induced identifications $M^r_X(w) = S^r \cdot x^w$,

$$(3.50) \qquad \bigoplus_{m \in \mathbf{Z}} L^{\otimes m} = \mathcal{O}(X)[x, x^{-1}] =: M,$$

$$(3.51) \qquad \bigoplus_{w \in W(r)} M^r_X(w) = M_r := S^r[x, x^{-1}, \phi x, \phi x^{-1}, ..., \phi^r x, \phi^r x^{-1}].$$

We have, at our disposal, the R–derivation $\partial : M \to M$. The ring $M^r = M^r_X$ is given by

$$(3.52) \qquad M^r = \mathcal{O}(J^r(Spec\ M)) = S^r[x, x^{-1}, x', ..., x^{(r)}]\hat{}.$$

The ring M_r in Equation 3.51 is embedded into the ring M^r in Equation 3.52 via

$$\phi x \mapsto x^p + px', \ \phi^2 x \mapsto (x^p + px')^p + p((x')^p + px''), \ ...$$

Note that if $\bar{S}^r := S^r/pS^r$ then the image of $\bar{M}^r_X(w) = \bar{S}^r \cdot \bar{x}^{\otimes w}$ in $\bar{M}^r_X$ equals $\bar{S}^r \cdot \bar{x}^{w(p)}$ so the map $\bar{M}^r_X(w) \to \bar{M}^r_X$ is injective. This proves assertion i) in our Proposition.

To prove assertion ii) we may assume X has étale coordinates $T = \{T_1, ..., T_d\}$. For $g \in S^r$ and $w = \sum a_k \phi^k$ we have

$$\partial_j(g \cdot x^w) = \sum_{i=1}^{d}\sum_{s=0}^{\infty} \frac{\partial(g \cdot x^w)}{\partial T_i^{(s)}} \cdot \Lambda_{sj}(T_i, T_i', ..., T_i^{(s)}) \cdot (\partial T_i)^{\phi^j}$$

$$+ \sum_{s=0}^{\infty} \frac{\partial(g \cdot x^w)}{\partial x^{(s)}} \Lambda_{sj}(x, x', ..., x^{(s)}) \cdot (\partial x)^{\phi^j}$$

$$= \sum_{i=1}^{d}\sum_{s=0}^{\infty} x^w \cdot \frac{\partial g}{\partial T_i^{(s)}} \cdot \Lambda_{sj}(T_i, T_i', ..., T_i^{(s)}) \cdot (\partial T_i)^{\phi^j} + g \cdot (a_j p^j x^{w-\phi^j}) \cdot (\partial x)^{\phi^j};$$

cf. assertion 2 in Lemma 3.34 for the last equality. The last expression clearly belongs to

$$S^r \cdot x^{w+c\phi^j}.$$

To prove assertion iii) we, again, note that this is a local matter so we are reduced to show that if an element $s \in S^r$ is such that $s \cdot x^w$ is invertible in the ring $S^r[x, x^{-1}, x', ..., x^{(r)}]\hat{}$ then s is invertible in S^r (and hence $s^{-1} \cdot x^{-w}$ is the inverse of $s \cdot x^w$). Write

$$s \cdot x^w \cdot (\sum s_{i_0 i_1 ... i_r} x^{i_0}(x')^{i_1}...(x^{(r)})^{i_r}) = 1,$$

with all $s_{i_0 i_1 ... i_r}$ in S^r. Denoting by upper bars classes in $\bar{S}^r := S^r/pS^r$ we obtain

$$\bar{s} \cdot (\sum \bar{s}_{i_0, i_1, ..., i_r} x^{i_0 + w(p)}(x')^{i_1}...(x^{(r)})^{i_r}) = 1,$$

hence we get

$$\bar{s} \cdot \bar{s}_{-w(p),0,...,0} = 1 \ \ in \ \ \bar{S}^r.$$

So $\bar{s}$ is invertible in $\bar{S}^r$; since S^r is $p-$adically complete it follows that s itself is invertible in S^r.

Assertion iv) is again local so we need to show that if $S_1 = \mathcal{O}(X_1)\hat{}$ and $s_1 \in S_1^r$ is such that $s_1 \cdot x^w$ belongs to $S^r[x, x^{-1}, x', ..., x^{(r)}]\hat{}$ then $s_1 \in S^r$. Write $w = \sum a_i \phi^i$,

$$s_1 \cdot x^w = s_1 \cdot x^{a_0}(x^p + px')^{a_1}...((x^{(r)})^p + px^{(r+1)})^{a_r},$$

$$s_1 \cdot x^w = \sum s_{i_0 i_1 ... i_r} x^{i_0}(x')^{i_1}...(x^{(r)})^{i_r}$$

with all $s_{i_0 i_1 ... i_r}$ in S^r. Picking out the coefficient of

$$x^{a_0}(x')^{a_1}...(x^{(r+1)})^{a_r}$$

we get that $p^{deg(w)} \cdot s_1 = s_{a_0 a_1 ... a_r} \in S^r$. Since $p^n S_1^r \cap S^r = p^n S^r$ we get that $s_1 \in S^r$. □

REMARK 3.58. The above Proposition applies in particular to the *Euler derivation*

$$\mathcal{D} : \bigoplus_{m \in \mathbf{Z}} L^{\otimes m} \to \bigoplus_{m \in \mathbf{Z}} L^{\otimes m}, \tag{3.53}$$

$$\mathcal{D}(\sum \alpha_m) = \sum m\alpha_m, \quad \alpha_m \in L^{\otimes m}.$$

If x is a basis of L as in the proof of the above Proposition then, of course,

$$\mathcal{D} = x\frac{\partial}{\partial x}.$$

The sections f of $L^{\otimes m}$ are characterized by the equation $\mathcal{D}f = m\cdot f$. More generally, if $w = \sum a_i \phi^i \in W$ and $\mathcal{D}_0, ..., \mathcal{D}_r : M_X^r \to M_X^r$ are the conjugate operators of $\mathcal{D}$ then we have

$$\mathcal{D}_j f = p^j a_j \cdot f, \tag{3.54}$$

for all $f \in L^w$.

In our main applications to be considered later X will have relative dimension one over R and we are going to have at our disposal another derivation

$$\partial : \bigoplus_{m\in\mathbf{Z}} L^{\otimes m} \to \bigoplus_{m\in\mathbf{Z}} L^{\otimes m}$$

such that $\mathcal{D}$ and ∂ form a basis of the tangent sheaf $T_{X_1/R}$, for some Zariski open set $X_1 \subset X$ that meets the special fiber X_0. Consequently, by Proposition 3.49,

$$\partial_0, \mathcal{D}_0, ..., \partial_r, \mathcal{D}_r \tag{3.55}$$

form a basis of the $M_{X_1}^r$–module of R–derivations of $M_{X_1}^r$.

3.6.9. Commutation relation between ∂_j and Δ for flat bundles. Assume $\mathbf{X} = (X, \tilde{X}, \sigma_1, \sigma_2)$ is an étale correspondence smooth over R and $\mathbf{L}_r^{(b)} := \mathbf{F}(r, b)$ is a flat δ–line bundle. Consider the map

$$\Delta^b = \Delta^{\mathbf{L}} : H^0(X, \mathcal{O}^r) = \mathcal{O}^r(X) \to H^0(\tilde{X}, \mathcal{O}^r) = \mathcal{O}^r(\tilde{X})$$
$$f \mapsto \Delta^b f := \sigma_1^* f - b \cdot \sigma_2^* f$$

introduced in Equation 2.43.

PROPOSITION 3.59. *Let $\partial : \mathcal{O}_X \to \mathcal{O}_X$ be an R–derivation and $\tilde{\partial} := \sigma_1^* \partial : \mathcal{O}_{\tilde{X}} \to \mathcal{O}_{\tilde{X}}$ be the derivation induced by σ_1. Let $\partial_j : \mathcal{O}_X^\infty \to \mathcal{O}_X^\infty$, $\tilde{\partial}_j : \mathcal{O}_{\tilde{X}}^\infty \to \mathcal{O}_{\tilde{X}}^\infty$ be the induced conjugate operators with index j. Assume that*

$$\sigma_2^* \partial = a^{-1} \cdot \tilde{\partial}$$

for some $a \in R^\times$. Then we have an equality of maps $\mathcal{O}^r(X) \to \mathcal{O}^r(\tilde{X})$:

$$\tilde{\partial}_j \circ \Delta^b = \Delta^{a^{\phi^j} b} \circ \partial_j.$$

In particular

$$\partial_j(H^0(\mathbf{X}, \mathbf{L}_r^{(b)})) \subset H^0(\mathbf{X}, \mathbf{L}_r^{(a^{\phi^j} b)}).$$

PROOF. Using Equation 3.29 and Proposition 3.41 we have, for $f \in \mathcal{O}^\infty(X)$:

$$\begin{aligned}\tilde{\partial}_j \Delta^b f &= (\sigma_1^* \partial)_j (\sigma_1^* f - b \cdot \sigma_2^* f) = (\sigma_1^* \partial)_j (\sigma_1^* f) - b \cdot (a \cdot \sigma_2^* \partial)_j (\sigma_2^* f) \\ &= \sigma_1^*(\partial_j f) - b \cdot a^{\phi^j} \sigma_2^*(\partial_j f) = \Delta^{a^{\phi^j} b} \partial_j f.\end{aligned}$$

□

CHAPTER 4

Local theory

Our global $\delta-$ geometric approach to correspondences has a local counterpart. In particular one can formulate and prove results along local analogues of our Main Conjectures 2.54 and 2.55. On the other hand the local theory can be applied to study global situations by looking at fixed points and cycles of correspondences in the global case. The present Chapter is devoted to explaining this circle of ideas.

This Chapter is organized as follows. In Section 1 we introduce local (formal) analogues of correspondences, $\delta-$line bundles, $\delta-$sections, $\delta-$invariants, uniformization, etc., we formulate a local analogue of our Main Conjectures, and we prove results along the lines of the local conjectures. (The local analogue of correspondences will be power series in $R[[T]]$. The local analogue of $\delta-$sections will be power series in $R[[T]][T',...,T^{(r)}]\hat{}$. The local analogue of analytic uniformization will be "uniformization by a formal group".) In Section 2 we produce a useful upper bound for the rank of the module of $\delta-$invariants of a "formal correspondence". In Section 3 we completely determine the $\delta-$invariants for correspondences "uniformized" by the formal multiplicative group. In Section 4 we introduce and study $p-$jet spaces of formal group laws which will play a key role in the applications. In Section 5 we introduce the concepts of fixed points and cycles of a (global) correspondence and we explain the effect of the local theory on the global theory. As an application we will prove, in Section 6, a number of "converse results" saying that if $H^0(\mathbf{X},\mathbf{L}) \neq 0$ then $(\mathbf{X},\mathbf{L})$ possesses some remarkable properties. In the last 3 Chapters it will turn out that some of these properties are shared by uniformizable correspondences. In particular we shall prove, in Section 6, our converse Theorem 2.65.

4.1. Local analogue of the main conjectures

Let $R := \hat{\mathbf{Z}}_p^{ur}$, $K = R[1/p]$, and T be a variable.

DEFINITION 4.1. A *formal correspondence* is a series

$$\Sigma(T) = \sum_{n=1}^{\infty} \alpha_n T^n \in T \cdot R[[T]]^{\times}.$$

The element

$$a(\Sigma) := \frac{d\Sigma}{dT}(0) = \alpha_1 \in R^{\times}$$

is called the *multiplier* of Σ. The element

$$s(\Sigma) := \frac{2\frac{d\Sigma}{dT}\frac{d^3\Sigma}{dT^3} - 3\left(\frac{d^2\Sigma}{dT^2}\right)^2}{\left(\frac{d\Sigma}{dT}\right)^2}(0) = 3\frac{\alpha_1\alpha_3 - \alpha_2^2}{\alpha_1^2}$$

is called the *Schwarzian* of Σ.

For any two formal correspondences Σ_1 and Σ_2 we have

$$a(\Sigma_1 \circ \Sigma_2) = a(\Sigma_1)a(\Sigma_2),$$
$$s(\Sigma_1 \circ \Sigma_2) = a(\Sigma_2)^2 s(\Sigma_1) + s(\Sigma_2).$$

In particular, if $a(\Sigma)$ is not a root of unity then, for any $n \geq 1$, we have the following equality in K:

$$\frac{s(\Sigma^{\circ n})}{1 - a(\Sigma^{\circ n})^2} = \frac{s(\Sigma)}{1 - a(\Sigma)^2}. \tag{4.1}$$

If $a(\Sigma) \neq 1, -1$ we define

$$v(\Sigma) := \min\{v_p\left(\frac{s(\Sigma)}{1 - a(\Sigma)^2}\right), 0\} \in \mathbf{Z}_- := \{0, -1, -2, ...\}. \tag{4.2}$$

The number $v(\Sigma)$ will be viewed as a measure of the "non-linearity" of Σ; indeed this number will vanish if Σ is the series expansion of a linear fractional transformation.

DEFINITION 4.2. A *formal line bundle* is a series $B(T) \in R[[T]]^{\times}$. If Σ is a formal correspondence then its *canonical bundle* is the formal line bundle $K = K_{\Sigma} = d\Sigma/dT$. Two formal line bundles B_1 and B_2 are *isomorphic* if there exists $g \in R[[T]]^{\times}$ such that

$$B_2(T) = B_1(T) \cdot \frac{g(T)}{g(\Sigma(T))}.$$

A formal line bundle B is *flat* if it is isomorphic to a formal line bundle belonging to $R^{\times}$.

Hence the canonical bundle of Σ is flat if and only if there exists $g \in R[[T]]^{\times}$ such that

$$\frac{d\Sigma}{dT} = a \cdot \frac{g(T)}{g(\Sigma(T))} \tag{4.3}$$

where $a = a(\Sigma)$ is (necessarily) the multiplier of Γ.

DEFINITION 4.3. A formal correspondence $\Sigma(T) \in R[[T]]$ is *uniformized* by a formal group $\mathcal{F} \in R[[T_1, T_2]]$ if Σ is an endomorphism (hence an automorphism) of $\mathcal{F}$. We also say that $\mathcal{F}$ is a uniformization of Σ.

We view the notion above as a formal (local, p-adic) analogue of analytic uniformization (which is a global, complex analytic concept).

PROPOSITION 4.4. *If $\Sigma(T)$ admits a uniformization then the canonical bundle of $\Sigma(T)$ is flat.*

PROOF. If $g(T)dT$ is an R-basis for the invariant 1-forms of the formal group then, for some $a \in R^{\times}$ we have

$$a \cdot g(T)dT = \Sigma^*(g(T)dT) = g(\Sigma(T)) \cdot \frac{d\Sigma}{dT}dT.$$

□

PROPOSITION 4.5. *Let $\alpha \in R^{\times}$, $\gamma \in R$, and*

$$\Sigma(T) = \frac{\alpha T}{1 - \gamma T} \in R[[T]].$$

1) The multiplier and Schwarzian of Σ are given by $a(\Sigma) = \alpha$ and $s(\Sigma) = 0$ respectively.

2) If $\gamma \in (1-\alpha)R$ then $\Sigma(T)$ is uniformized by a formal group which is isomorphic to the additive formal group; in particular the canonical bundle of $\Sigma(T)$ is flat.

3) If $\gamma \notin (1-\alpha)R$ then the canonical bundle of $\Sigma(T)$ is not flat.

PROOF. Assertion 1 is trivial. To prove assertion 2 we may assume $\alpha \neq 1$ and set

$$\beta := \frac{\gamma}{1-\alpha} \in R, \quad L(T) := \frac{T}{1-\beta T} \in R[[T]].$$

Then it is trivial to check that

$$L(\Sigma(T)) = \alpha \cdot L(T), \tag{4.4}$$

and assertion 2 follows by considering the formal group law $L^{-1}(L(T_1) + L(T_2))$ plus Proposition 4.4. (Actually to see that $\Sigma(T)$ has a flat canonical bundle one can simply apply d/dT to Equation 4.4 and set $g(T) = dL/dT$.) To check assertion 3 assume Equation 4.3 holds for some $g = \sum c_n T^n \in R[[T]]^\times$. Equation 4.3 reads then as follows:

$$c_0 + c_1(\alpha T + \alpha\gamma T^2 + ...) + ... = (1 - 2\gamma T + \gamma^2 T^2)(c_0 + c_1 T + ...);$$

picking out the coefficient of T we get $2\gamma c_0 = c_1(1-\alpha)$, a contradiction. $\square$

DEFINITION 4.6. Let $\Sigma(T)$ be a formal correspondence. A *linearizing series* for $\Sigma(T)$ is a series $L = L(T) \in T \cdot K[[T]]$ with $(dL/dT)(0) = 1$ such that

$$L(\Sigma(T)) = a \cdot L(T) \tag{4.5}$$

where $a = a(\Sigma)$ is (necessarily) the multiplier of $\Sigma(T)$.

The following result is well known; cf. [**109**], p. 74 (Koenigs Linearization Theorem) for the holomorphic analogue.

PROPOSITION 4.7. *If a formal correspondence $\Sigma(T)$ has multiplier not a root of unity then $\Sigma(T)$ possesses a unique linearizing series.*

PROOF. If $\Sigma(T) = \sum_{n\geq 1} \alpha_n T^n$ and we seek L of the form $L(T) = \sum_{n\geq 1} \lambda_n T^n$, $\lambda_1 = 1$, $\lambda_n \in K$, then the equation 4.5 yields a (triangular, infinite) linear system of equations in $\lambda_2, \lambda_3, \lambda_4$,... with matrix having diagonal elements $\alpha_1 - \alpha_1^2$, $\alpha_1 - \alpha_1^3$, $\alpha_1 - \alpha_1^4$,... Since none of the latter vanishes, the system has a unique solution. For future reference the first λ's are given by

$$\lambda_2 = \frac{\alpha_2}{\alpha_1(1-\alpha_1)}, \quad \lambda_3 = \frac{\alpha_3\alpha_1 - \alpha_3\alpha_1^2 + 2\alpha_1\alpha_2^2}{\alpha_1^2(1-\alpha_1)^2(1+\alpha_1)}, \quad etc. \tag{4.6}$$

$\square$

REMARK 4.8. If Σ is an endomorphism of a formal group $\mathcal{F}$ and $a(\Sigma)$ is not a root of unity then the logarithm $l_{\mathcal{F}}(T)$ of $\mathcal{F}$ is, of course, the linearizing series of Σ.

We get the following interpretation of flatness of the canonical bundle as an integrality property for the derivative of the linearizing series.

PROPOSITION 4.9. *Assume the formal correspondence $\Sigma(T)$ has multiplier $a(\Sigma)$ not a root of unity and let L be its linearizing series.*

1) The canonical bundle of $\Sigma(T)$ is flat if and only if $dL/dT \in R[[T]]$.

2) If the canonical bundle of $\Sigma(T)$ is flat then $v(\Sigma) = 0$.

PROOF. Assertion 1 is clear. To check assertion 2 one computes, using Equation 4.6:

$$3(\lambda_3 - \lambda_2^2) = 3\frac{\alpha_1\alpha_3 - \alpha_2^2}{\alpha_1^2(1-\alpha_1^2)} = \frac{s(\Sigma)}{1-a(\Sigma)^2}. \tag{4.7}$$

By assertion 1, $3\lambda_3 \in R$; also $\lambda_2 \in R$ (because $p \neq 2$) and we are done. □

We will need a generalization of the concept of flatness of the canonical bundle. Assume we are given a formal correspondence $\Sigma(T)$ with multiplier not a root of unity and linearizing series L. Let m, n, r be integers, $m, r \geq 0$, and set

$$L_{m,n}(T) := L^m \cdot \left(\frac{dL}{dT}\right)^n \in K[[T]].$$

DEFINITION 4.10. The canonical bundle of Σ is $(m,n,r)-$*flat* if

$$p^{-rm}L_{m,n}(p^rT) \in R[[T]]. \tag{4.8}$$

The following clearly holds.

PROPOSITION 4.11. *Assume the multiplier of $\Sigma(T)$ is not a root of unity.*
1) The canonical bundle of Σ is flat if and only if it is $(0,1,0)-$ flat.
2) If the canonical bundle of Σ is flat then it is $(m,n,r)-$flat for all $r \geq 1$.

PROPOSITION 4.12. *If a formal correspondence $\Sigma(T)$ has multiplier not a root of unity and if $H(T) \in K((T))$ is such that $H(\Sigma(T)) = H(T)$ then $H \in K$.*

PROOF. Replacing, if necessary, H by H^{-1} we may assume $H \in K[[T]]$. Assume $H \notin K$. Replacing H by $H - H(0)$ we may assume $H = \sum_{n=n_0}^{\infty} c_nT^n$, $n_0 \geq 1$, $c_{n_0} \neq 0$. If a is the multiplier of Σ we get $a^{n_0} = 1$, a contradiction. □

We would like to extend the above discussion to "arbitrary order". So if T is a variable and $T', T'', ..., T^{(r)}, ...$ are new variables, one defines operators

$$\phi, \delta : R[[T]][T', ..., T^{(r)}]\hat{} \to R[[T]][T', ..., T^{(r+1)}]\hat{}$$

by the formulae 3.2. So ϕ are ring homomorphisms and δ are $p-$derivations. Consider the prolongation sequence

$$(R[[T]][T', ..., T^{(r)}]\hat{})$$

where the structure homomorphisms are the inclusions and the structure $p-$ derivations are the δ above.

DEFINITION 4.13. A *formal $\delta-$line bundle* (of order r) is a pair $\Lambda = (r, B)$ where

$$B = B(T, T', ..., T^{(r)}) \in (R[[T]][T', ..., T^{(r)}]\hat{})^{\times} = R[[T]]^{\times} + p \cdot R[[T]][T', ..., T^{(r)}]\hat{}.$$

(We sometimes denote $\Lambda = (r, B)$ by $\Lambda(r, B)$. If $\gamma \in R^{\times}$ we will sometimes abbreviate $\Lambda(r, \gamma^{-1})$ by either Λ_r or Λ_γ if γ or r are understood from the context respectively.) The constant coefficient

$$B(0, ..., 0) \in R^{\times}$$

will be called the *multiplier* of Λ. Two formal $\delta-$line bundles Λ_1 and Λ_2 of order r are called *isomorphic* if there exists a series

$$g \in (R[[T]][T', ..., T^{(r)}]\hat{})^{\times}$$

such that

$$B_2(T,T',...,T^{(r)}) = B_1(T,T',...,T^{(r)}) \cdot \frac{g(T,T',...,T^{(r)})}{g(\Sigma(T),\delta(\Sigma(T)),...,\delta^r(\Sigma(T)))}.$$

A $\delta-$line bundle of order r is called *flat* if it is isomorphic to a $\delta-$line bundle of the form $\Lambda(r,b)$ with $b \in R^\times$.

If $\Sigma(T)$ is a formal correspondence and $\Lambda = (r, B)$ is a formal $\delta-$line bundle then we define the $R-$linear endomorphism

$$\Delta^\Lambda = \Delta^B = \Delta \tag{4.9}$$

on $R[[T]][T',...,T^{(r)}]\hat{}$ by the formula

$$\Delta(F) := F - B \cdot F \circ \Sigma, \tag{4.10}$$

where, by abuse, we set

$$F \circ \Sigma := F(\Sigma(T),\delta(\Sigma(T)),...,\delta^r(\Sigma(T))). \tag{4.11}$$

Then define

$$H^0(\Sigma,\Lambda) = Ker\ \Delta, \quad H^1(\Sigma,\Lambda) = Coker\ \Delta. \tag{4.12}$$

DEFINITION 4.14. The elements of $H^0(\Sigma,\Lambda)$ are called *invariant $\delta-$sections* (or *$\delta-$invariants*) of Λ over Σ. So a series $F \in R[[T]][T',...,T^{(r)}]\hat{}$ is a *$\delta-$invariant* if and only if

$$F(\Sigma(T),\delta(\Sigma(T)),...,\delta^r(\Sigma(T))) \cdot B(T,T',...,T^{(r)}) = F(T,T',...,T^{(r)}). \tag{4.13}$$

If $r, d \geq 0$ we say that a series $F \in R[[T]][T',...,T^{(r)}]\hat{}$ is *$d-$regular* if

$$0 \neq F_*(T) := F(0,...,0,T) \in R \cdot T^d \cdot R[[T]]^\times.$$

We say that F is *$d-$perfect* if F is a $d-$ th power G^d of a non-zero series $G \in R[[T]][T',...,T^{(r)}]\hat{}$; of course any non-zero series is $1-$perfect and, if $d \not\equiv 0$ mod p, then any series in $(R[[T]][T',...,T^{(r)}]\hat{})^\times$ is $d-$perfect. A $d-$perfect series $F = G^d$ is called *$d-$coregular* if

$$G(T,0,...,0) \in T \cdot R[[T]]^\times.$$

If Λ_1 and Λ_2 are isomorphic formal $\delta-$line bundles then there is a natural isomorphism

$$H^0(\Sigma,\Lambda_1) \simeq H^0(\Sigma,\Lambda_2)$$

taking $d-$regular ($\delta-$ perfect, $d-$ perfect and $d-$ coregular) sections into $d-$regular ($\delta-$ perfect, $d-$ perfect and $d-$ coregular) sections, where d is any integer which, for the statements involving $d-$perfect sections, is assumed to be $\not\equiv 0$ mod p.

Consider any pair (r, w) with $w = \sum_{i=0}^r a_i\phi^i \in W$, $ord(w) \leq r$, $r \geq 1$. One can consider the formal $\delta-$line bundle "$w-$power of the canonical bundle":

$$\mathbf{K}_r^w := \left(r, \left(\frac{d\Sigma}{dT}\right)^w\right) \tag{4.14}$$

where, as usual,

$$\left(\frac{d\Sigma}{dT}\right)^w = \left(\frac{d\Sigma}{dT}\right)^{a_0} \cdots \left(\frac{d\Sigma^{(\phi^r)}}{dT}(T^{\phi^r})\right)^{a_r}.$$

If $r = ord(w)$ (i.e. if $a_r \neq 0$) we write

$$\mathbf{K}^w = \mathbf{K}_r^w. \tag{4.15}$$

We are in a position to state a local analogue of our Main Conjectures 2.54 and 2.55:

CONJECTURE 4.15. *Let $\Sigma(T)$ be a formal correspondence with multiplier not a root of unity. The following are equivalent:*
1) $\Sigma(T)$ admits a uniformization.
2) There exists $w \in W$, $deg(w) < 0$, such that $H^0(\Sigma, \mathbf{K}^w) \neq 0$.

We start by exploring the implication $2 \Rightarrow 1$ which, in the global case, is the more mysterious one.

PROPOSITION 4.16. *Assume the formal correspondence $\Sigma(T)$ has multiplier in $\mathbf{Z}_p^{\times}$, not a root of unity. Assume $H^0(\Sigma, \mathbf{K}_r^w)$ contains a $d-$ regular section for some $d > 0$. Then $d = -deg(w)$ (in particular $deg(w) < 0$) and the canonical bundle of Σ is $(d, a_r, r)-$flat.*

PROOF. It is trivial to see (by induction on r) that

$$T^{\phi^r} := \phi^r(T) = p^r T^{(r)} + P_{r-1}(T, T', ..., T^{(r-1)}), \tag{4.16}$$

where $P_{r-1} \in \mathbf{Z}[T, T', ..., T^{(r-1)}]$, $P_{r-1}(0, ..., 0) = 0$. This easily implies (induction on i) that

$$\delta^i(\Sigma(T))_{|T=T'=...=T^{(r-1)}=0} = 0, \quad for \quad i < r. \tag{4.17}$$

Equations 4.16 and 4.17 imply

$$\delta^r(\Sigma(T))_{|T=T'=...=T^{(r-1)}=0} = p^{-r}\Sigma^{(\phi^r)}(p^r T^{(r)}). \tag{4.18}$$

Recall that the superscript (ϕ^r) means "twisting coefficients by ϕ^r". Setting $T = T' = ... = T^{(r-1)} = 0$ in Equation 4.13 (where F is a regular section and $B = (d\Sigma/dT)^w$) and using equations 4.17 and 4.18 we get
(4.19)

$$F(0, ..., 0, p^{-r}\Sigma^{(\phi^r)}(p^r T^{(r)}))a^{deg(w)-a_r}\left(\frac{d\Sigma^{(\phi^r)}}{dT}(p^r T^{(r)})\right)^{a_r} = F(0, ..., 0, T^{(r)}),$$

where a is the multiplier of Σ. Set

$$\Sigma_r(T) := p^{-r}\Sigma^{(\phi^r)}(p^r T), \quad L_r(T) = p^{-r}L^{(\phi^r)}(p^r T),$$

where L is the linearizing series of Σ. Then Equation 4.19 reads

$$F_*(\Sigma_r(T))a^{deg(w)-a_r}\left(\frac{d\Sigma_r}{dT}\right)^{a_r} = F_*(T). \tag{4.20}$$

Since $L(\Sigma(T)) = aL(T)$ and $a \in \mathbf{Z}_p$ we get $L_r(\Sigma_r(T)) = aL_r(T)$ hence

$$\frac{dL_r}{dT}(\Sigma_r(T))\frac{d\Sigma_r}{dT} = a\frac{dL_r}{dT}. \tag{4.21}$$

Combining Equations 4.20 and 4.21 we get

$$\frac{F_*}{(dL_r/dT)^{a_r}}(\Sigma_r(T)) = a^{-deg(w)} \cdot \frac{F_*}{(dL_r/dT)^{a_r}}(T). \tag{4.22}$$

So if

$$H := \frac{F_*}{(dL_r/dT)^{a_r}} \cdot L_r^{deg(w)}$$

then $H(\Sigma_r(T)) = H(T)$. By Proposition 4.12 we get $H \in K$ and

$$L_r^{-deg(w)} \left(\frac{dL_r}{dT}\right)^{a_r} \in K \cdot F_* \in K \cdot T^d \cdot R[[T]]^\times.$$

Hence $d = -deg(w)$. Applying the operator $P \mapsto P^{(\phi^{-r})}$ to the last equation we get

$$L_{d,a_r}(p^r T) = \lambda \cdot T^d \cdot U(T),$$

where $\lambda \in K$, $U(T) \in R[[T]]^\times$. Since the left hand side of the above Equation has the form $p^{rd}T^d + ...$ we get $\lambda \in p^{rd} \cdot R^\times$ and we are done. □

COROLLARY 4.17. *Assume $\Sigma(T)$ has multiplier $a(\Sigma) \in \mathbf{Z}_p^\times$ not a root of unity. Assume $H^0(\Sigma, \mathbf{K}^{-1-\phi^r})$ contains a $d-$ regular section for some $d \geq 1$. Then $d = 2$ and $v(\Sigma) \geq -2r$.*

PROOF. Let $L(T) = \sum \lambda_n T^n$. By Proposition 4.16 $\Sigma(T)$ is $(2, -1, r)-$ flat; so there exists a series $\sum \beta_n T^n \in R[[T]]$ such that

$$(p^r T + \lambda_2 p^{2r} T^2 + \lambda_3 p^{3r} T^3 + ...)^2 = p^{2r}(1 + 2\lambda_2 p^r T + 3\lambda_3 p^{2r} T^2 + ...) \sum \beta_n T^n.$$

Picking out the coefficients of T^i for $i = 0, ..., 4$ we get $\beta_0 = \beta_1 = 0$, $\beta_2 = 1$, $\beta_3 = 0$, and

$$\beta_4 = p^{2r}(\lambda_2^2 - \lambda_3). \tag{4.23}$$

By Equation 4.7 we get $\frac{s(\Sigma)}{1-a(\Sigma)^2} \in p^{-2r}R$ and we are done. □

To proceed we need the following formal analogue of the theory of conjugate operators in the global case.

LEMMA 4.18. *Let $\partial : R[[T]] \to \bigcup_{r=0}^\infty R[[T]][T', ..., T^{(r)}]\hat{}$ be any $R-$derivation. Then, for each $j \geq 0$, there exists a unique derivation*

$$\partial_j : \bigcup_{r=0}^{\infty} R[[T]][T', ..., T^{(r)}]\hat{} \to \bigcup_{r=0}^{\infty} R[[T]][T', ..., T^{(r)}]\hat{}$$

satisfying the properties:

1) $\partial_j \circ \phi^s = 0$ on $R[[T]]$ for $s \neq j$,

2) $\partial_j \circ \phi^j = p^j \cdot \phi^j \circ \partial$ on $R[[T]]$.

Explicitly if

$$\partial := U \cdot \frac{d}{dT}, \quad U \in \bigcup_{r=0}^{\infty} R[[T]][T', ..., T^{(r)}]\hat{}$$

then

$$\partial_j := U^{\phi^j} \cdot \sum_{s=0}^{\infty} \Lambda_{sj}(T, T', ..., T^{(s-1)}) \cdot \frac{\partial}{\partial T^{(s)}}, \tag{4.24}$$

where Λ_{sj} are as in Lemma 3.34. Moreover, for any $\Sigma \in R[[T]]$ we have

$$\partial_j \delta^s \Sigma = (\Lambda_{sj} \circ \Sigma) \cdot U^{\phi^j} \cdot \left(\frac{d\Sigma}{dT}\right)^{\phi^j}. \tag{4.25}$$

The operators ∂_j and ∂_{j-1} are related by the following formula:

$$\partial_j \circ \phi = p \cdot \phi \circ \partial_{j-1}$$

on $\bigcup_{r=0}^{\infty} R[[T]][T',...,T^{(r)}]\hat{}$. *In particular, if* $\partial R[[T]] \subset R[[T]]$, *the restriction of* ∂_r *to* $R[[T]][T',...,T^{(r)}]\hat{}$ *is the unique* $R-$*derivation*

$$R[[T]][T',...,T^{(r)}]\hat{} \to R[[T]][T',...,T^{(r)}]\hat{}$$

satisfying the properties:

1) ∂_r *vanishes on* $R[[T]][T',...,T^{(r-1)}]\hat{}$,

2) $\partial_r \circ \delta^r = \phi^r \circ \partial$ *on* $R[[T]]$.

Explicitly if

$$\partial := U(T) \cdot \frac{d}{dT}, \quad U(T) \in R[[T]]$$

then

$$\partial_r := U(T)^{\phi^r} \cdot \frac{\partial}{\partial T^{(r)}} \quad on \quad R[[T]][T',...,T^{(r)}]\hat{}.$$

As in the global case the derivations $\partial_0, \partial_1, \partial_2, ...$ are called the *conjugates* of ∂.

PROOF. Uniqueness of ∂_j is clear. Now define ∂_j by Equation 4.24. By Lemma 3.34 we have

$$\partial_j T^{\phi^s} = p^j U^{\phi^j} \delta_{sj}.$$

Then one notes that the set of series in $R[[T]]$ on which $\partial_j \circ \phi^s$ and $p^j \cdot \delta_{sj} \cdot \phi^j \circ \partial$ coincide is a $T-$adically closed subring containing R and T. Therefore this set is the whole of $R[[T]]$. This proves the existence of ∂_j. Equation 4.25 holds for polynomials in $R[T]$ by Equation 3.31 hence it follows for any series in $R[[T]]$ by $T-$adic continuity of its right and left hand sides. The rest is clear. □

REMARK 4.19. The operators δ, ϕ have natural extensions

$$\delta, \phi : K[[T, T', ..., T^{(r-1)}]] \to K[[T, T', ..., T^{(r)}]]$$

and Lemma 4.18 is true if one replaces, everywhere in its statement, $R[[T]]$ by $K[[T]]$ and

$$\bigcup_{r=0}^{\infty} R[[T]][T',...,T^{(r)}]\hat{} \ \text{by} \ \bigcup_{r=0}^{\infty} K[[T,T',...,T^{(r)}]].$$

Moreover the derivation

$$\partial_{**} = \sum_{j=0}^{\infty} p^{-j}\partial_j : \bigcup_{r=0}^{\infty} K[[T,T',...,T^{(r)}]] \to \bigcup_{r=0}^{\infty} K[[T,T',...,T^{(r)}]]$$

satisfies, as in the global case, the equation

$$\partial_{**} \circ \phi = \phi \circ \partial_{**}. \tag{4.26}$$

DEFINITION 4.20. If $\partial : K[[T]] \to K[[T]]$ (respectively $\partial : K[[T]] \to A := \bigcup_{r=0}^{\infty} K[[T,T',...,T^{(r)}]]$) is a $K-$derivation and $\mathcal{L}$ is a $K-$linear subspace of A then ∂ is an *infinitesimal symmetry* (respectively a *generalized infinitesimal symmetry*) of $\mathcal{L}$ if

$$\partial_{**}\mathcal{L} \subset \mathcal{L}.$$

If $F \in A$ and $\partial_{**}F = 0$ we say that ∂ is a (generalized) *variational infinitesimal symmetry* of F.

There is an obvious analogue of Noether's Theorem in this context which we are not going to need so we will not state.

DEFINITION 4.21. If Σ is a flat formal correspondence with multiplier not a root of unity and linearizing series L then, by Corollary 4.9, $dL/dT \in R[[T]]$ so we may consider the following derivation on $R[[T]]$:

$$\partial = \partial^{can} := \left(\frac{dL}{dT}\right)^{-1} \cdot \frac{d}{dT}; \tag{4.27}$$

∂^{can} is called the *canonical derivation.*

LEMMA 4.22. *Let Σ be a flat formal correspondence with multiplier $a \in R^{\times}$, let $b \in R^{\times}$, let ∂ be the canonical derivation on $R[[T]]$, let $j, r \geq 0$, and let ∂_j be the conjugate derivation on $R[[T]][T', ..., T^{(r)}]\hat{}$ defined in Lemma 4.18. Then we have the following equality of endomorphisms of $R[[T]][T', ..., T^{(r)}]\hat{}$:*

$$\partial_j \circ \Delta^b = \Delta^{a^{\phi^j} b} \circ \partial_j.$$

PROOF. Since $L \circ \Sigma = a \cdot L(T)$ we get

$$\left(\frac{dL}{dT} \circ \Sigma\right) \cdot \frac{d\Sigma}{dT} = a \cdot \frac{dL}{dT}. \tag{4.28}$$

We have the following computation:

$$\partial_j \Delta^b G = \partial_j G - b \cdot \sum_{s=0}^{\infty} \left(\frac{\partial G}{\partial T^{(s)}} \circ \Sigma\right) \cdot \partial_j(\delta^s \Sigma)$$

$$= \partial_j G - b \cdot \sum_{s=0}^{\infty} \left(\frac{\partial G}{\partial T^{(s)}} \circ \Sigma\right) \cdot (\Lambda_{sj} \circ \Sigma) \cdot \left(\frac{dL}{dT}\right)^{-\phi^j} \cdot \left(\frac{d\Sigma}{dT}\right)^{\phi^j}, \quad \text{by Lemma 4.18}$$

$$= \partial_j G - b \cdot \sum_{s=0}^{\infty} \left(\frac{\partial G}{\partial T^{(s)}} \circ \Sigma\right) \cdot (\Lambda_{sj} \circ \Sigma) \cdot a^{\phi^j} \cdot \left(\frac{dL}{dT} \circ \Sigma\right)^{-\phi^j}, \quad \text{by Equation 4.28.}$$

On the other hand we have:

$$\Delta^{a^{\phi^j} b} \partial_j G = \partial_j G - a^{\phi^j} b \cdot (\partial_j G) \circ \Sigma$$

$$= \partial_j G - a^{\phi^j} b \cdot \sum_{s=0}^{\infty} \left(\frac{\partial G}{\partial T^{(s)}} \cdot \partial_j T^{(s)}\right) \circ \Sigma$$

$$= \partial_j G - a^{\phi^j} b \cdot \sum_{s=0}^{\infty} \left(\frac{\partial G}{\partial T^{(s)}} \cdot \left(\frac{dL}{dT}\right)^{-\phi^j} \Lambda_{sj}\right) \circ \Sigma, \quad \text{by Lemma 4.18}$$

$$= \partial_j \Delta^b G, \quad \text{by the previous computation.}$$

□

COROLLARY 4.23. *Let Σ be a flat formal correspondence with multiplier a, let $b \in R^{\times}$, and let ∂ be the canonical derivation on $R[[T]]$. Then, for any $j, r \geq 0$, the conjugate derivation ∂_j maps $H^0(\Sigma, \Lambda(r, b))$ into $H^0(\Sigma, \Lambda(r, a^{\phi^j} b))$.*

PROPOSITION 4.24. *Assume $\Sigma(T)$ is a formal correspondence with multiplier $a \in \mathbf{Z}_p^{\times}$, not a root of unity. Assume the canonical bundle of Σ is flat and let $w \in W$ be such that $d := -deg(w) > 0$, $d \not\equiv 0 \mod p$, $ord(w) \leq r$, $r \geq 1$.*

1) If $H^0(\Sigma, \mathbf{K}_r^w) \neq 0$ then $H^0(\Sigma, \mathbf{K}_r^w)$ contains a $d-$ perfect section.

2) If $H^0(\Sigma, \mathbf{K}_r^w)$ contains a $d-$ perfect, $d-$ coregular section then Σ admits a uniformization.

PROOF. Let us prove assertion 1. Set $\Lambda_r := \Lambda(r, a^{-1})$ so $\mathbf{K}_r^w$ is isomorphic to $\Lambda(r, a^{-d}) = \Lambda_r^d$. Hence we have $H^0(\Sigma, \mathbf{K}_r^w) \simeq H^0(\Sigma, \Lambda_r^d)$. So it is enough to show that if $H^0(\Sigma, \Lambda_r^d) \neq 0$ for some $d \geq 2$ then $H^0(\Sigma, \Lambda_r^{d-1}) \neq 0$. Let $0 \neq G \in H^0(\Sigma, \Lambda_s^d)$ and either $s = 0$ or $1 \leq s \leq r$ and $G \notin H^0(\Sigma, \Lambda_{s-1}^d)$. So either $s = 0$ or $G \notin R[[T]][T', ..., T^{(s-1)}]\hat{}$. By Corollary 4.23 $\partial_s G$ is a non-zero section of $H^0(\Sigma, \Lambda_s^{d-1})$, hence of $H^0(\Sigma, \Lambda_r^{d-1})$.

Let us prove assertion 2 of the Proposition. By the above discussion

$$H^0(\Sigma, \Lambda_r^d)$$

must contain a section of the form G^d where G is some series in $R[[T]][T', ..., T^{(r)}]\hat{}$, $d \not\equiv 0 \bmod p$, and $G(T, 0, ..., 0) \equiv T \bmod T^2$. Using the notation introduced in Equation 4.11 we have

$$G^d \circ \Sigma = a^d \cdot G^d.$$

Let $E(T) \in K[[T]]$ be the compositional inverse of $L(T)$. Hence $p^{-1}E(pT)$ is the compositional inverse of $p^{-1}L(pT)$. Since $dL/dT \in R[[T]]$ it follows that $p^{-1}L(pT) \in R[[T]]$ so $p^{-1}E(pT) \in R[[T]]$, hence $E(pT) \in pR[[T]]$, $E(pT) \equiv pT \bmod pT^2$. One can write

$$\begin{aligned} G^d \circ E(pT) \circ \frac{a}{p} L(T) &= G^d \circ E(pT) \circ \frac{1}{p} L(T) \circ \Sigma \\ &= G^d \circ \Sigma = a^d \cdot G^d = a^d \cdot G^d \circ E(pT) \circ \frac{1}{p} L(T). \end{aligned}$$

So if

$$H := G \circ E(pT) = G(E(pT), ..., \delta^r(E(pT))) \in R[[T]][T', ..., T^{(r)}]\hat{} \tag{4.29}$$

we get

$$H^d(aT, ..., \delta^r(aT)) = a^d \cdot H^d(T, ..., T^{(r)}).$$

Hence

$$H(aT, ..., \delta^r(aT)) = \zeta \cdot a \cdot H(T, ..., T^{(r)}), \tag{4.30}$$

where ζ is a $d-$th root of unity. View H as an element of $K[[T, T^{\phi}, ..., T^{\phi^r}]]$. Since a is not a root of unity Equation 4.30 implies that $\zeta = 1$ and $H = \sum_{i=0}^r \lambda_i T^{\phi^i}$, $\lambda_i \in K$. In particular

$$H \in K[T, T', ..., T^{(r)}] \cap R[[T, T', ..., T^{(r)}]] = R[T, T', ..., T^{(r)}].$$

Now it is easy to see that a polynomial in $R[T, T', ..., T^{(r)}]$ which can be written as a linear polynomial with $K-$coefficients in $T, T^{\phi}, ..., T^{\phi^r}$ is actually a linear polynomial with $R-$coefficients in $T, T^{\phi}, ..., T^{\phi^r}$ (cf. Lemma 5.4, to be proved later, for a more general statement); so $\lambda_0, ..., \lambda_r \in R$. Since $G(T, 0, ..., 0) \equiv T \bmod T^2$ we have

$$G \equiv \sum_{n \geq 1} \gamma_n T^n \ \ mod \ \ (T', ..., T^{(r)})$$

with $\gamma_n \in R$, $\gamma_1 = 1$. Hence

$$H \equiv \sum_{n \geq 1} \gamma_n (E(pT))^n \ \ mod \ \ (\delta(E(pT)), ..., \delta^r(E(pT))).$$

So $H(T, 0, ..., 0) \equiv pT \bmod T^2$ and so $\lambda_0 = p$. Consequently

$$L(T) + \frac{1}{p} \sum_{i=1}^r \lambda_i L^{(\phi^i)}(T^{\phi^i}) = H \circ \frac{1}{p} L(T) = G \in R[[T]][T', ..., T^{(r)}]\hat{}.$$

Setting $T' = T'' = ... = T^{(r)} = 0$ in the last equation we get

$$L(T) + \frac{1}{p}\sum_{i=1}^{r} \lambda_i L^{(\phi^i)}(T^{p^i}) \in R[[T]].$$

By Hazewinkel's functional equation lemma [**67**], p.74, the series $E(L(T_1)+L(T_2))$ belongs to $R[[T_1, T_2]]$; hence this series is a formal group law over R and, clearly, $\Sigma(T)$ is an endomorphism of this formal group law. □

REMARK 4.25. In view of Proposition 4.11 one sees that Propositions 4.16 and 4.24 can be viewed as partial results along the implication $2 \Rightarrow 1$ in Conjecture 4.15.

On the other hand we have the following partial result along the implication $1 \Rightarrow 2$ in Conjecture 4.15.

PROPOSITION 4.26. *Let $\mathcal{O}$ be a finite unramified extension of $\mathbf{Z}_p$. Assume $\Sigma(T) \in \mathcal{O}[[T]]$ is an endomorphism of a formal group $\mathcal{F} \in \mathcal{O}[[T_1, T_2]]$. Then there exists an integer $r_0 \geq 0$ such that for any $r \geq r_0$ and any $w \in W$ with $d := -deg(w) > 0$, $d \not\equiv 0$ mod p, $ord(w) \leq r$, the space $H^0(\Sigma, \mathbf{K}_r^w)$ contains a $d-$regular section. Also $H^0(\Sigma, \mathbf{K}_r^w)$ contains a $d-$ perfect, $d-$ coregular section at least if one assumes that $\mathcal{F}$ is either defined over $\mathbf{Z}_p$ or is a Lubin-Tate formal group attached to $\mathcal{O}$.*

PROOF. Write $\Sigma(T) = aT + ...$, $a \in \mathcal{O}$. Proposition 4.4 shows that the canonical bundle K_Σ is isomorphic to the formal line bundle a. So it is enough to find an r_0 such that for any $r \geq r_0$ and $d \geq 1$ there exist two series $F_1, F_2 \in R[[T]][T', ..., T^{(r)}]\hat{}$ such that F_1 is $d-$ regular, F_2 is $d-$ perfect and $d-$ coregular, and

$$F_i(\Sigma(T), ..., \delta^r(\Sigma(T))) = a^d F_i(T, ..., T^{(r)}), \quad i = 1, 2.$$

(We need F_2 only in case $\mathcal{F}$ is defined over $\mathbf{Z}_p$ or is a Lubin-Tate group.) Let $p^m := \sharp(\mathcal{O}/p\mathcal{O})$ and let $Fr_m(T) := T^{p^m}$ be the Frobenius endomorphism of the reduction mod p, $\bar{\mathcal{F}} \in (\mathcal{O}/p\mathcal{O})[[T_1, T_2]]$, of $\mathcal{F}$. If $\bar{\mathcal{F}}$ has infinite height (i.e. the multiplication by p in it is the zero series) then $\mathcal{F}$ is isomorphic to the additive group $\mathbf{G}_a$ over $\mathcal{O}$; cf. [**68**], p.329, Lemma A for the case $\mathcal{O} = \mathbf{Z}_p$ but the argument there works for general $\mathcal{O}$. So in this case we may assume $\mathcal{F} = \mathbf{G}_a$ and the conclusion of our Proposition is trivial: we may take $r_0 = 0$, $F_2 = T^d$, $F_1 = T^{d\phi^r}$. Assume now that $\bar{\mathcal{F}}$ has finite height h. By results of Dieudonné [**49**] it follows that Fr_m satisfies a monic polynomial of degree $s \leq h$ over $\mathbf{Z}_p$ i.e. there exist $a_1, ..., a_s \in \mathbf{Z}_p$ such that

$$T^{p^{ms}}[+]_{\bar{\mathcal{F}}}[a_1]_{\bar{\mathcal{F}}}(T^{p^{m(s-1)}})[+]_{\bar{\mathcal{F}}}...[+]_{\bar{\mathcal{F}}}[a_s]_{\bar{\mathcal{F}}}(T) = 0 \tag{4.31}$$

in $End_{\mathcal{O}/p\mathcal{O}}(\bar{\mathcal{F}})$, where $[a_i]_{\bar{\mathcal{F}}}(T)$ is the multiplication by a_i series and $[+]_{\bar{\mathcal{F}}}$ is the addition law on series defined by $\bar{\mathcal{F}}$. Set

$$P := P(T, T', ..., T^{(ms)}) := \phi^{ms}(T)[+]_{\mathcal{F}}[a_1]_{\mathcal{F}}(\phi^{m(s-1)}(T))[+]_{\mathcal{F}}...[+]_{\mathcal{F}}[a_s]_{\mathcal{F}}(T),$$

$$P \in R[[T]][T', ..., T^{(ms)}]\hat{}$$

where $\phi^i(T)$ is the $i-$th iterate of $\phi(T^{(j)}) = (T^{(j)})^p + pT^{(j+1)}$ and $[\ ]_{\mathcal{F}}$ denotes the corresponding operations induced by $\mathcal{F}$. By Equation 4.31 it follows that

$$P \in pR[[T]][T', ..., T^{(ms)}]\hat{}. \tag{4.32}$$

Let $l = l_{\mathcal{F}} = \sum c_n n^{-1} T^n$, $c_n \in \mathcal{O}$, be the logarithm of $\mathcal{F}$. From Equation 4.32 it follows that

$$\Psi(T, T', ..., T^{(ms)}) := \frac{1}{p} \cdot l(P(T, T', ..., T^{(ms)})) \in R[[T]][T', ..., T^{(ms)}]\hat{}.$$

On the other hand we have

$$\phi^{mi}(T) \circ \Sigma(T) = \Sigma^{(\phi^{mi})}(T) \circ \phi^{mi}(T) = \Sigma(T) \circ \phi^{mi}(T) \tag{4.33}$$

because the coefficients of $\Sigma(T)$ are ϕ^m-invariant. Also

$$[a_i]_{\mathcal{F}}(T) \circ \Sigma(T) = \Sigma(T) \circ [a_i]_{\mathcal{F}}(T)$$

because $[a_i]_{\mathcal{F}} \in End_{\mathcal{O}}(\mathcal{F})$. It follows that $P \circ \Sigma = \Sigma \circ P$ so

$$\Psi \circ \Sigma = \frac{1}{p} \cdot l \circ P \circ \Sigma = \frac{1}{p} \cdot l \circ \Sigma \circ P = \frac{1}{p} \cdot a \cdot l \circ P = a \cdot \Psi.$$

So we may take

$$r_0 = mh, \quad F_1 = \Psi^{d\phi^{r-mh}}.$$

Now note that if $\mathcal{F}$ is defined over $\mathbf{Z}_p$ then by [**68**],p. 329, Proposition C, the polynomial $x^s + a_1 x^{s-1} + ... + a_s$ is Eisenstein. The same is true if $\mathcal{F}$ is a Lubin-Tate formal group attached to $\mathcal{O}$ (in which case the latter polynomial equals $x - p$). So in this case we may take $F_2 = \Psi^d$ and note that

$$\Psi(T, 0, ..., 0) = \frac{1}{p}[l(T^{p^{ms}}) + a_1 l(T^{p^{m(s-1)}}) + ... + a_s T]$$

and the coefficient of T in the latter expression is $a_s/p \in \mathbf{Z}_p^\times$. □

REMARK 4.27. The implication $2 \Rightarrow 1$ in Conjecture 4.15 fails if the condition that the multiplier be not a root of unity is dropped. Indeed let

$$\Sigma(T) = \frac{T}{1-T} = T + T^2 + T^3 + ...$$

By Proposition 4.5 the canonical bundle of $\Sigma(T)$ is not flat; in particular, by Proposition 4.4 $\Sigma(T)$ is not an endomorphism of a formal group defined over R. On the other hand note that

$$\left(\left(\frac{T}{1-T}\right)^{\phi} - \frac{T}{1-T}\right)^2 \left(\frac{d}{dT}\left(\frac{T}{1-T}\right)\right)^{-1-\phi} = (T^{\phi} - T)^2,$$

hence $(T^{\phi} - T)^2$ is a (2− regular, 2−perfect, 2− coregular) section of $H^0(\Sigma, \mathbf{K}^{-1-\phi})$.

REMARK 4.28. Let Σ_1 and Σ be formal correspondences and let

$$\Sigma_2 := \Sigma^{\circ -1} \circ \Sigma_1 \circ \Sigma.$$

Then it is trivial to see that:

1) Σ_1 has a flat canonical bundle if and only if Σ_2 has a flat canonical bundle.
2) $v(\Sigma_1) = v(\Sigma_2)$.
3) There are natural isomorphisms

$$H^0(\Sigma_1, \mathbf{K}^w) \to H^0(\Sigma_2, \mathbf{K}^w)$$

sending $d-$ regular ($d-$ perfect, $d-$perfect and $d-$ coregular) sections into $d-$ regular ($d-$ perfect, $d-$perfect and $d-$ coregular) sections.

4.2. Bounding the rank of the module of δ–invariants

DEFINITION 4.29. An element $a \in R$ is called a *generalized root of unity* if there exists $0 \neq w \in W_+$ such that $a^w = 1$.

REMARK 4.30. For any positive integer λ set $R^{\phi^\lambda} := \{x \in R; x^{\phi^\lambda} = x\}$. So

$$\bigcup_\lambda R^{\phi^\lambda} = \mathbf{Z}_p^{ur} \subset \hat{\mathbf{Z}}_p^{ur} = R.$$

Then, for any $x \in R^{\phi^\lambda}$, one can consider the $R^{\phi^\lambda}/\mathbf{Z}_p$-*norm*,

$$Norm_\lambda(x) := x^{1+\phi+\ldots+\phi^{\lambda-1}} \in \mathbf{Z}_p.$$

Note that for any $x \in R^{\phi^\lambda}$ and $w \in W$ we have

$$Norm_\lambda(x^w) = Norm_\lambda(x)^{deg(w)};$$

in particular an element $a \in R^{\phi^\lambda}$ is a generalized root of unity if and only if $Norm_\lambda(a)$ is a root of unity.

In what follows we view W as an ordered set by setting $w \leq w'$ if and only if $w' - w \in W_+$; then, clearly, for any non-empty subset

$$S \subset \{w \in W_+ | ord(w) \leq r\}$$

the set S_{min} of minimal elements of S is finite.

We have the following interplay between linear dependence of $\delta-$ sections and multiplicative relations among $W-$powers of multipliers:

PROPOSITION 4.31. *Let Σ be a formal correspondence with multiplier $a = a(\Sigma)$ and let Λ be a $\delta-$line bundle with multiplier b. Let*

$$S := \{w \in W_+ | ord(w) \leq r, a^w = b^{-1}\}$$

Then the following hold:

1) $rank_R H^0(\Sigma, \Lambda) \leq \sharp\, S$.

2) If a is not a generalized root of unity then the $R-$module $H^0(\Sigma, \Lambda)$ is finitely generated and

$$rank_R H^0(\Sigma, \Lambda) \leq \sharp\, S_{min}.$$

We need the following trivial:

LEMMA 4.32. *If $a \in R$ and $H \in T^2R[[T]]$ then*

$$\delta^n(aT + H) - a^{\phi^n} T^{(n)} \in (T, T', \ldots, T^{(n)})^2.$$

PROOF. Induction on n. □

PROOF. (of Proposition 4.31.) Let $\Lambda = (r, B)$ and let us check assertion 1. Let $F_1, \ldots, F_d \in H^0(\Sigma, \Lambda)$ be $R-$linearly independent; it is enough to find d distinct elements in S. For any $v = \sum_{i=0}^r c_i \phi^i \in W_+$ set

$$T_v = T^{c_0}(T')^{c_1} \ldots (T^{(r)})^{c_r}.$$

We introduce a linear order on the set of monomials T_v as follows. If $v' = \sum_{i=0}^r c_i' \phi^i$, $v'' = \sum_{i=0}^r c_i'' \phi^i \in W_+$ we declare that $T_{v'} \leq_{lex} T_{v''}$ if and only if

$$(\sum c_i', c_0', \ldots, c_r') \leq_{lex} (\sum c_i'', c_0'', \ldots, c_r'')$$

in the lexicographical order. Let T_{v_i} be the smallest monomial present in F_i. Replacing, if necessary, $F_1, ..., F_d$ by linear combinations of them we may assume that $v_1, ..., v_d$ are distinct; this can be achieved, for instance by putting the $d \times \infty$ matrix of coefficients of $F_1, ..., F_d$ in echelon form. Each F_i satisfies a functional equation

$$F_i(\Sigma(T), ..., \delta^n(\Sigma(T))) \cdot B(T, ..., T^{(r)}) = F_i(T, ..., T^{(r)}). \tag{4.34}$$

By Lemma 4.32 we have

$$\delta^i(\Sigma(T)) = a^{\phi^i} T^{(i)} + M_i, \quad M_i \in (T, ..., T^{(i)})^2.$$

Picking out the coefficients of T_{v_i} in Equation 4.34 we get

$$a^{v_i} b = 1, \tag{4.35}$$

which ends the proof of assertion 1. To check assertion 2 assume that a is not a generalized root of unity. We may assume $H^0(\Sigma, \Lambda) \neq 0$. By assertion 1, S is not empty. Let $S_{min} := \{w_1, ..., w_s\}$ be the set of minimal elements of S. Since S is well ordered any element of S is greater than or equal to one of the minimal elements. Consider the R-linear map

$$H^0(\Sigma, \Lambda) \to R^s, \quad \sum c_v T_v \mapsto (c_{w_1}, ..., c_{w_s}).$$

To conclude it is enough to show that this map injective. Assume this is not the case so there exists $F = \sum c_v T_v \neq 0$ satisfying equation 4.34 and satisfying $c_{w_1} = ... = c_{w_s} = 0$. Let T_v be the smallest monomial present in F (with respect to the lexicographic order). Picking out the coefficient of T_v in Equation 4.34 we get $a^v = b^{-1}$ hence $v \in S$. So $v \geq w_i$ for some i. But $v \neq w_i$ because $c_v \neq 0$ while $c_{w_i} = 0$. Since $a^{w_i} = b^{-1}$ we get $a^{v-w_i} = 1$ with $v - w_i \geq 0$, $v - w_i \neq 0$, so a is a generalized root of unity, a contradiction. □

REMARK 4.33. Let $a, b \in \mathbf{Z}_p^\times$ be not roots of unity and let S be as in Proposition 4.31. Then two cases may occur: either there is no $m \in \mathbf{Z}_+$ with $a^m = b^{-1}$ in which case $S = \emptyset$ or there exists $m \in \mathbf{Z}_+$ with $a^m = b^{-1}$ in which case

$$S = S_{min} = \{w \in W_+ | ord(w) \leq r,\ deg(w) = m\}, \quad \sharp\ S = \sharp\ S_{min} = \frac{(m+r)!}{m!r!}.$$

REMARK 4.34. We will also need a variant "over K" of Proposition 4.31. Assume Σ is a formal correspondence and $\Lambda = (r, B)$ is a formal δ-line bundle. We set

$$H^0_K(\Sigma, \Lambda) := \{F \in K[[T, T', ..., T^{(r)}]] \mid F \circ \Sigma \cdot B = F\}. \tag{4.36}$$

Clearly we have

$$H^0(\Sigma, \Lambda) \otimes K \subset H^0_K(\Sigma, \Lambda)$$

but, as one can easily see, this inclusion may be strict (e.g. for Σ an endomorphism of the multiplicative formal group where the logarithm belongs to $H^0_K(\Sigma, \Lambda)$ but not to $H^0(\Sigma, \Lambda) \otimes K$.) Nevertheless, the same arguments as in the proof of Proposition 4.31 show that, under the assumptions of that Proposition, the following hold:

1) $dim_K H^0_K(\Sigma, \Lambda) \leq \sharp\ S$.

2) If a is not a generalized root of unity then $dim_K H^0_K(\Sigma, \Lambda) \leq \sharp\ S_{min}$.

In the case of flat bundles with multipliers powers of the multiplier of the correspondence the above upper bound over K is attained:

PROPOSITION 4.35. *Let Σ be a formal correspondence with multiplier $a \in \mathbf{Z}_p^\times$ not a root of unity and let $L \in K[[T]]$ be its linearizing series. Let $\Lambda_r := \Lambda(r, a^{-1})$. Then, for any integer $m \in \mathbf{Z}$, the K-linear space $H^0_K(\Sigma, \Lambda_r^m)$ consists of all the series of the form*

$$\Pi(L, L^\phi, ..., L^{\phi^r}) \tag{4.37}$$

where $\Pi \in K[x_0, ..., x_r]$ is a homogeneous polynomial of degree m.

PROOF. The space of all series of the form 4.37 is clearly contained in $H^0_K(\Sigma, \Lambda_r^m)$ and has dimension $\frac{(m+r)!}{m!r!}$. On the other hand $H^0_K(\Sigma, \Lambda_r^m)$ has dimension at most $\frac{(m+r)!}{m!r!}$ by Remarks 4.34 and 4.33. □

4.3. δ-invariants for the formal multiplicative group

For any $a \in \mathbf{Z}_p^\times$ consider the formal correspondence

$$\Sigma_a(T) = [a](T) = (1+T)^a - 1 = aT + \frac{a(a-1)}{2!}T^2 + ... \in \mathbf{Z}_p[[T]]; \tag{4.38}$$

this is of course the multiplication by a in the formal multiplicative group law. In what follows we shall fix n and set $\Lambda_a := \Lambda(n, a^{-1})$. For $m \in \mathbf{Z}$, the space of sections

$$H^0(\Sigma_a, \Lambda_a^m)$$

consists, by definition, of the series $F \in R[[T]][T', ..., T^{(n)}]\hat{}$ satisfying the functional equation:

$$G(\Sigma_a(T), \delta(\Sigma_a(T)), ..., \delta^n(\Sigma_a(T))) = a^m \cdot G(T, T', ..., T^{(n)}). \tag{4.39}$$

Consider the series:

$$\Psi = \Psi_{\mathbf{G}_m} := \frac{1}{p} log \frac{\phi(1+T)}{(1+T)^p} \in R[[T]][T']\hat{}, \tag{4.40}$$

where

$$log(1+T) := l_{\mathbf{G}_m}(T) = \sum_{n=1}^{\infty} (-1)^{n+1} \frac{T^n}{n}.$$

Our aim is to prove the following:

PROPOSITION 4.36. *For any series $G \in R[[T]][T', ..., T^{(n)}]\hat{}$ and any integer $m \geq 0$ the following are equivalent:*

1) $G \in H^0(\Sigma_a, \Lambda_a^m)$ for some $a \in \mathbf{Z}_p^\times$, a not a root of unity.

2) $G \in H^0(\Sigma_a, \Lambda_a^m)$ for all $a \in \mathbf{Z}_p^\times$.

3) There exists a homogeneous polynomial $\Phi \in K[x_1, ..., x_n]$ of degree m such that

$$G = \Phi(\Psi, \Psi^\phi, ..., \Psi^{\phi^{n-1}}),$$

For the proof of Proposition 4.36 we need some preliminaries. First note that, since $L(T) := log(1+T)$ is the linearizing series of Σ_γ and since

$$\Psi = \frac{1}{p}(L^\phi - pL),$$

Proposition 4.35 yields the following:

LEMMA 4.37. *If $a \in \mathbf{Z}_p^\times$ is not a root of unity the $K-$vector space $H^0_K(\Sigma_a, \Lambda_a^m)$ consists of all series of the form*

$$\Phi(log(1+T), \Psi, \Psi^\phi, ..., \Psi^{\phi^{n-1}}), \tag{4.41}$$

where $\Phi \in K[x_0, ..., x_n]$ is a homogeneous polynomial of degree m.

On the other hand we have

LEMMA 4.38. *Let $\Phi \in K[x_0, ..., x_n]$ be a homogeneous polynomial of degree $m \geq 1$ such that*

$$\Phi(log(1+T), \Psi, \Psi^\phi, ..., \Psi^{\phi^{n-1}}) \in R[[T]][T', ..., T^{(n)}]\hat{}.$$

Then

$$\frac{\partial \Phi}{\partial x_0} = 0.$$

PROOF. First note that the Lemma is true for $m = 1$ due to the fact that for any $\nu \in \mathbf{Z}_+$,

$$p^\nu \cdot log(1+T) \notin R[[T]].$$

Now we proceed by induction on $m+n$. If $m+n=1$ we must have $m=1$ and we are done. Assume now $m+n \geq 2$ and let Φ be as in the statement of the Lemma. If $m=1$ we are done so we may assume $m \geq 2$. If $n \geq 1$ write

$$\Phi = \Phi_0 + \Phi_1 x_n + ... + \Phi_s x_n^s,$$

with $\Phi_i \in K[x_0, ..., x_{n-1}]$. Consider the derivations

$$\partial = (1+T) \cdot \frac{d}{dT} : R[[T]] \to R[[T]],$$

$$\partial_n = (1+T^{\phi^n}) \cdot \frac{\partial}{\partial T^{(n)}} : R[[T]][T', ..., T^{(n)}]\hat{} \to R[[T]][T', ..., T^{(n)}]\hat{}.$$

Note that, by Lemma 4.18,

$$\partial_n(\Psi^{\phi^{n-1}}) = p^{n-1} \cdot \phi^{n-1}(\partial_1 \Psi) = p^{n-1}.$$

So we get

$$\frac{\partial \Phi}{\partial x_n}(log(1+T), \Psi, \Psi^\phi, ..., \Psi^{\phi^{n-1}}) \cdot p^{n-1} \in R[[T]][T', ..., T^{(n)}]\hat{}.$$

Since $\frac{\partial \Phi}{\partial x_n}$ has degree $m-1 \geq 1$ we get by induction that

$$\frac{\partial^2 \Phi}{\partial x_0 \partial x_n} = 0.$$

We conclude that

$$\frac{\partial \Phi_1}{\partial x_0} = ... = \frac{\partial \Phi_s}{\partial x_0} = 0$$

hence

$$(\Phi_1 x_n + ... + \Phi_s x_n^s)(log(1+T), \Psi, \Psi^\phi, ..., \Psi^{\phi^{n-1}}) \in R[[T]][T', ..., T^{(n)}]\hat{}$$

hence

$$\Phi_0(log(1+T), \Psi, \Psi^\phi, ..., \Psi^{\phi^{n-2}}) \in R[[T]][T', ..., T^{(n)}]\hat{}.$$

By induction we get

$$\frac{\partial \Phi_0}{\partial x_0} = 0$$

and we are done. If $n = 0$ then

$$\Phi = \lambda(log(1+T))^m,$$

$\lambda \in K$, $m \geq 2$. So

$$m\lambda(log(1+T))^{m-1} = \partial\Phi \in R[[T]].$$

By induction $\lambda = 0$ and we are done again. □

PROOF. (of Proposition 4.36.) Clearly assertion 3 implies assertion 2 and assertion 2 implies assertion 1. Now if assertion 1 holds then by Lemma 4.37 there exists a homogeneous polynomial $\Phi \in K[x_0, x_1, ..., x_n]$ of degree m such that

$$G = \Phi(log(1+T), \Psi, \Psi^{\phi}, ..., \Psi^{\phi^{n-1}}).$$

By Lemma 4.38, Φ does not depend on x_0 and we are done. □

4.4. p–jets of formal group laws

In this section we shall attach to any formal group law $\mathcal{F} \in R[[T_1, T_2]]$ in one variable and to any integer $r \geq 1$ a group

$$(\hat{\mathbf{A}}^r, [+])$$

in the category of formal schemes over R whose underlying formal scheme is the p–adic completion of the affine space over R; we will also produce, for each r a morphism from the above into the additive group

$$\hat{\mathbf{G}}_a := (\hat{\mathbf{A}}^1, +).$$

The idea is easy: we take the series $\delta\mathcal{F}, ..., \delta^r\mathcal{F}$ and we set $T_1 = T_2 = 0$ to get $[+]$ on $\hat{\mathbf{A}}^r$; also we apply $p^{-1}\phi^r$ to the logarithm of $\mathcal{F}$ and set $T = 0$ to obtain our homomorphism. Although the idea is simple, we will need some preparation to implement it.

We will actually need a relative version of the above so we start with a prolongation sequence

$$S^0 \xrightarrow{\delta} S^1 \xrightarrow{\delta} S^2 \xrightarrow{\delta} ...$$

of p–adically complete Noetherian rings which are flat over R. Consider a sequence of variables $T, T', T'', ..., T^{(r)}, ...$ and consider the ring homomorphisms

$$S^0[[T]] \xrightarrow{\phi} S^1[[T, T']] \xrightarrow{\phi} S^2[[T, T', T'']] \xrightarrow{\phi} ...$$

prolonging the ϕ homomorphisms in the sequence (S^r) and satisfying $\phi(T^{(r)}) = (T^{(r)})^p + pT^{(r+1)}$. Then define the prolongation sequence

$$S^0[[T]] \xrightarrow{\delta} S^1[[T, T']] \xrightarrow{\delta} S^2[[T, T', T'']] \xrightarrow{\delta} ... \tag{4.42}$$

by letting the structure p–derivations be given by the formula $\delta(f) := (\phi(f) - f^p)/p$ and by letting the structure homomorphisms be given by the inclusions.

Here is a useful notation: if

$$F_0, ..., F_s, g_0, ..., g_r \in S^r[[T, T', ..., T^{(r)}]]$$

are series and either $F_j \in S^r[T, ..., T^{(r)}]\hat{}$ or all $g_i \in (T, ..., T^{(r)})$ then we denote by

$$[F_0, ..., F_s] \circ [g_0, ..., g_r] \in S^r[[T, T', ..., T^{(r)}]]^{s+1} \tag{4.43}$$

the result obtained by replacing $T, T', ..., T^{(r)}$ in $F_0, ..., F_s$ by $g_0, ..., g_r$; in other words

$$[F_0, ..., F_s] \circ [g_0, ..., g_r] := [F_0(g_0, ..., g_r), ..., F_s(g_0, ..., g_r)].$$

With this convention and our previous conventions it is trivial to check, by induction, that for any $G \in S^0[[T]]$ we have

$$\phi^r(T) \circ G = [\phi^r(T)] \circ [G, \delta G, ..., \delta^r G] = \phi^r(G) \tag{4.44}$$

$$[G^{(\phi^r)}] \circ [\phi^r(T)] = \phi^r(G). \tag{4.45}$$

Also, by induction, one trivially checks that

$$\phi^r(T) - T^{p^r} - p(T')^{p^{r-1}} \in (pT, p^2);$$

consequently we have

$$[\phi^r(T)] \circ [0, T', ..., T^{(r)}] \equiv 0 \ \ mod \ \ (p) \tag{4.46}$$

$$G_r := \frac{1}{p}[\phi^r(T)] \circ [0, T', ..., T^{(r)}] \equiv (T')^{p^{r-1}} \ \ mod \ \ (p). \tag{4.47}$$

Next we claim that if $F \in S^0[[T]]$, $F(0) = 0$, then

$$[\delta^r F] \circ [0, T', ..., T^{(r)}] \in S^r[T', ..., T^{(r)}]\hat{}. \tag{4.48}$$

Indeed (as one can check by induction) $p^r \delta^r F$ belongs to the ring generated by $F, \phi(F), ..., \phi^r(F)$ so it is enough to check that

$$[\phi^r(F)] \circ [0, T', ..., T^{(r)}]$$

is a restricted power series. But

$$[\phi^r(F)] \circ [0, T', ..., T^{(r)}] = [F^{(\phi^r)}] \circ [\phi^r(T)] \circ [0, T', ..., T^{(r)}]$$

and we are done by Equation 4.46 above.

The above remarks and notation extend, in an obvious way, to the case when T is replaced by a tuple of variables and F is replaced by a tuple of series.

Now let $\mathcal{F}(T_1, T_2) \in S^0[[T_1, T_2]]$ be a formal group law over S^0 and consider the associated logarithm

$$l(T) = \sum_{n \geq 1} a_n T^n \in (S^0 \otimes K)[[T]].$$

We have

$$\mathcal{F}(T_1, T_2) = e(l(T_1) + l(T_2))$$

where $e(T) \in (S^0 \otimes K)[[T]]$ is the exponential, i.e. the compositional inverse of $l(T)$. Consider the variables

$$\mathbf{T} = (T_1, T_2), \ \mathbf{T}' = (T_1', T_2'), ..., \ \mathbf{T}^{(r)} = (T_1^{(r)}, T_2^{(r)})$$

PROPOSITION 4.39. *The $(r+1)$–tuple of formal series*

$$(\mathcal{F}, \delta\mathcal{F}, ..., \delta^r \mathcal{F}) \in S^r[[\mathbf{T}, \mathbf{T}', ..., \mathbf{T}^{(r)}]]^{r+1}$$

defines a (commutative) formal group law in $(r+1)$ variables.

PROOF. This is checked using an appropriate universality property of the prolongation sequence 4.42; we leave the details to the reader. □

Let us set $\mathbf{T} = 0$ in this tuple, in other words consider the r–tuple

$$([\delta\mathcal{F}] \circ [0, \mathbf{T}', ..., \mathbf{T}^{(r)}], ..., [\delta^r\mathcal{F}] \circ [0, \mathbf{T}', ..., \mathbf{T}^{(r)}]). \tag{4.49}$$

By the 2–variable analogue of Equation 4.48 the components of the latter tuple are restricted power series (rather than mere formal power series) i.e. belong to $S^r[\mathbf{T}', ..., \mathbf{T}^{(r)}]\hat{}$. For any p–adically complete ring S we may consider the p–adically completed affine space over S:

$$\hat{\mathbf{A}}^r_S := Spf\ S[T', ..., T^{(r)}]\hat{},$$

where Spf is taken with respect to the p–adic topology. Then, in particular, we get

COROLLARY 4.40. *The r–tuple 4.49 defines a group object $(\hat{\mathbf{A}}^r_{S^r}, [+])$ in the category of formal schemes over S^r.*

Now define, for each $r \geq 1$, the following series

$$L^r := \frac{1}{p}[\phi^r(l)] \circ [0, T', ..., T^{(r)}] = \frac{1}{p}[l^{(\phi^r)}] \circ [\phi^r(T)] \circ [0, T', ..., T^{(r)}], \tag{4.50}$$

viewed as an element of $(S^r \otimes K)[[T, T', ..., T^{(r)}]]$.

PROPOSITION 4.41. *We have $L^r \in S^r[T', ..., T^{(r)}]\hat{}$ and $L^r \equiv (T')^{p^{r-1}}$ mod (p). In particular $L^1 \equiv T'$ mod p.*

PROOF. By Equation 4.46 and 4.47 we may write

$$L^r = \sum_{n \geq 1} \phi^r(a_n) p^{n-1} G^n_r.$$

Since $na_n \in S^0$ and $p \geq 5$ we get that $p^{n-1}a_n$ belong to S^0, tend to 0, and are divisible by p if $n \geq 2$ which ends the proof due to Equation 4.47. □

PROPOSITION 4.42. *The series L^r defines a group homomorphism*

$$L^r : (\hat{\mathbf{A}}^r_{S^r}, [+]) \to (\hat{\mathbf{A}}^1_{S^r}, +) = \hat{\mathbf{G}}_{a,S^r}$$

in the category of formal schemes over S^r, in other words the following formula holds:

$$L^r((T'_1, ..., T^{(r)}_1)[+](T'_2, ..., T^{(r)}_2)) = L^r(T'_1, ..., T^{(r)}_1) + L^r(T'_2, ..., T^{(r)}_2).$$

Moreover, if K^r is the kernel of the natural projection

$$(\hat{\mathbf{A}}^r_{S^r}, [+]) \to (\hat{\mathbf{A}}^{r-1}_{S^r}, [+])$$

onto the first $r-1$ factors then the the induced homomorphism

$$L^r : K^r \to (\hat{\mathbf{A}}^1_{S^r}, +) = \hat{\mathbf{G}}_{a,S^r}$$

is p^{r-1} times an isomorphism; in particular $K^r \simeq \hat{\mathbf{G}}_{a,S^r}$. Also

$$L^1 : (\hat{\mathbf{A}}^1_{S^r}, [+]) \to (\hat{\mathbf{A}}^1_{S^r}, +) = \hat{\mathbf{G}}_{a,S^r}$$

is an isomorphism.

PROOF. Let us first check that L^r is a homomorphism. We have

$$L^r((T_1',...,T_1^{(r)})[+](T_2',...,T_2^{(r)}))$$

$$= L^r([\delta\mathcal{F}]\circ[0,\mathbf{T}',...,\mathbf{T}^{(r)}],...,[\delta^r\mathcal{F}]\circ[0,\mathbf{T}',...,\mathbf{T}^{(r)}])$$

$$= \frac{1}{p}[l^{(\phi^r)}]\circ[\phi^r(T)]\circ[0,[\delta\mathcal{F}]\circ[0,\mathbf{T}',...,\mathbf{T}^{(r)}],...,[\delta^r\mathcal{F}]\circ[0,\mathbf{T}',...,\mathbf{T}^{(r)}]]$$

$$= \frac{1}{p}[l^{(\phi^r)}]\circ[\phi^r(T)]\circ[\mathcal{F},\delta\mathcal{F},...,\delta^r\mathcal{F}]\circ[0,\mathbf{T}',...,\mathbf{T}^{(r)}]$$

$$= \frac{1}{p}[l^{(\phi^r)}]\circ[\phi^r(\mathcal{F})]\circ[0,\mathbf{T}',...,\mathbf{T}^{(r)}], \quad \text{by Equation 4.44}$$

$$= \frac{1}{p}[l^{(\phi^r)}]\circ[\mathcal{F}^{(\phi^r)}]\circ[\phi^r(\mathbf{T})]\circ[0,\mathbf{T}',...,\mathbf{T}^{(r)}], \quad \text{by Equation 4.45}$$

$$= \frac{1}{p}[T_1+T_2]\circ[l^{(\phi^r)}]\circ[\phi^r(\mathbf{T})]\circ[0,\mathbf{T}',...,\mathbf{T}^{(r)}]$$

$$= L_p^r(T_1',...,T_1^{(r)}) + L_p^r(T_2',...,T_2^{(r)}).$$

The second part of the Proposition follows because the restriction of L^r to K^r is given by the series

$$\frac{1}{p}[l^{(\phi^r)}]\circ[\phi^r(T)]\circ[0,0,...,0,T^{(r)}] = \frac{1}{p}\cdot l^{(\phi^r)}(p^rT^{(r)});$$

on the other hand the series

$$\frac{1}{p^r}\cdot l^{(\phi^r)}(p^rT^{(r)})$$

has compositional inverse

$$\frac{1}{p^r}\cdot e^{(\phi^r)}(p^rT^{(r)}) \in S^r[T^{(r)}]\hat{}$$

where e is the exponential of $\mathcal{F}$. □

4.5. Local versus global picture: fixed points and cycles

One of our main tools in the study of (global) correspondences is the local theory around a "fixed point", or more generally around an "$n-$cycle". In what follows we shall explain how this works.

4.5.1. Expansion map. We begin by comparing the global and the local pictures for a given scheme over R. Let X be a smooth curve over R with irreducible closed fiber $X\otimes k$, let P_0 be a closed point of $X\otimes k$ and assume we are given an isomorphism

$$\mathcal{O}_{X,P_0}^{for} \simeq R[[T]] \tag{4.51}$$

where we recall that the "for" superscript denotes completion with respect to the maximal ideal (and that the $\hat{}$ superscript is reserved to denote $p-$adic completion). By the universality property in Proposition 3.3 and by Proposition 3.19 we have a natural *expansion map*:

$$E:\mathcal{O}^r(X)\to R[[T]][T',...,T^{(r)}]\hat{}. \tag{4.52}$$

PROPOSITION 4.43. *The maps*

$$E : \mathcal{O}^r(X) \to R[[T]][T', ..., T^{(r)}]\hat{}, \quad \bar{E} : \mathcal{O}^r(X) \otimes k \to k[[T]][T', ..., T^{(r)}]$$

are injective. Moreover, for all $i < r$, we have

$$E^{-1}(R[[T]][T', ..., T^{(i)}]\hat{}) = \mathcal{O}^i(X).$$

PROOF. For the first assertion it is enough to check the injectivity of the second map. By Corollary 3.25 we may replace X by an open subset of it. So we may assume X is affine and we have an étale map $R[x] \subset \mathcal{O}(X)$ where x is an indeterminate. Let us view x as an element of $R[[T]]$ hence as a series $x = S(T) \in T \cdot R[[T]]^{\times}$. Let $S^{-1}(T)$ be the compositional inverse of $S(T)$. The automorphism of $R[[T]]$ mapping $T \mapsto S^{-1}(T)$ extends to automorphisms of $R[[T]][T', ..., T^{(r)}]\hat{}$ mapping

$$T^{(i)} \mapsto \delta^i(S^{-1}(T)).$$

Composing the expansion map E with these automorphisms we may assume $x = T$. By Propositions 3.13 and 3.19 we have

$$\mathcal{O}^i(X) \simeq \mathcal{O}(X)[T', ..., T^{(i)}]\hat{}, \quad \mathcal{O}^r(X) \otimes k \simeq \mathcal{O}(X \otimes k)[T', ..., T^{(r)}].$$

Since $X \otimes k$ is irreducible and reduced the map $\mathcal{O}(X \otimes k) \to k[[T]]$ is injective and we are done. To check the second assertion we are reduced to show that

$$\mathcal{O}(X)[T', ..., T^{(r)}]\hat{} \cap R[[T]][T', ..., T^{(i)}]\hat{} = \mathcal{O}(X)[T', ..., T^{(i)}]\hat{}$$

which is clear. □

PROPOSITION 4.44. *Let X be a smooth curve over R with irreducible closed fiber $X \otimes k$, let $P_0 \in X \otimes k$ be a closed point, and fix an isomorphism as in Equation 4.51. Let $\partial : \mathcal{O}_X \to \mathcal{O}_X$ be an $R-$derivation and $\partial : R[[T]] \to R[[T]]$ the induced derivation. Then, for any $j \geq 0$, the diagram*

$$\begin{array}{ccc} \mathcal{O}^r(X) & \stackrel{\partial_j}{\longrightarrow} & \mathcal{O}^r(X) \\ E \downarrow & & \downarrow E \\ R[[T]][T', ..., T^{(r)}]\hat{} & \stackrel{\partial_j}{\longrightarrow} & R[[T]][T', ..., T^{(r)}]\hat{} \end{array}$$

is commutative.

Here the maps ∂_j are the conjugates of ∂ defined in Proposition 3.43 and Lemma 4.18 while the vertical arrows E are the natural injective maps in Proposition 4.43.

PROOF. We may replace X by an open affine subset of it; cf. Proposition 3.25. So we may assume, upon changing T, that $T \in \mathcal{O}(X)$ and $R[T] \subset \mathcal{O}(X)$ is étale. To check the commutativity of the diagram it is enough to check that $E \circ \partial_j$ and $\partial_j \circ E$ have the same value on $T^{(s)}$ for any $s \geq 0$; this is however clear by Proposition 3.43 and Lemma 4.18: the common value is $\Lambda_{sj} \cdot (\partial T)^{\phi^j}$. □

PROPOSITION 4.45. *In the situation of Proposition 4.44 assume $X = G$ has a structure of group scheme over R with formal group law $\mathcal{F} \in R[[T_1, T_2]]$ and assume P_0 is the origin of G_0. Then the kernel of $J^r(G) \to G$ is isomorphic as a group formal scheme to the group formal scheme $(\hat{\mathbf{A}}^r, [+])$ attached to $\mathcal{F}$ via Proposition 4.40. In particular if $N^r := Ker(J^r(G) \to \hat{G})$ then $N^1 \simeq \hat{\mathbf{G}}_a$ and*

$$Ker(N^r \to N^{r-1}) \simeq \hat{\mathbf{G}}_a.$$

PROOF. This follows immediately from Propositions 3.13 and 4.42. □

We end this subsection by recording the following useful application of the expansion map.

PROPOSITION 4.46. *Let X be a smooth affine curve over R and let $\partial : \mathcal{O}_X \to \mathcal{O}_X$ be an $R-$derivation. Then, for any $j \geq 1$ and any $f, g \in \mathcal{O}(X)$, we have*

$$(\delta^j f) \cdot (\partial g)^{p^j} - (\delta^j g) \cdot (\partial f)^{p^j} \in p \cdot \mathcal{O}^j(X) + \mathcal{O}^{j-1}(X). \tag{4.53}$$

Equivalently

$$(\delta^j f) \cdot (\partial g)^{\phi^j} - (\delta^j g) \cdot (\partial f)^{\phi^j} \in p \cdot \mathcal{O}^j(X) + \mathcal{O}^{j-1}(X). \tag{4.54}$$

PROOF. For any $P \in X(R)$ choose an affine open set $U \subset X$, $P \in U(R)$, admitting an étale coordinate $T \in \mathcal{O}(U)$ such that the ideal $T \cdot \mathcal{O}(U)$ defines (the image of) P. Now we check that the left hand side, h_j, of Equation 4.53 belongs to

$$p \cdot \mathcal{O}^j(U) + \mathcal{O}^{j-1}(U).$$

We proceed by induction on j. Assume $j = 1$. We have an isomorphism $\mathcal{O}^{for}_{X,P_0} \simeq R[[T]]$ and we may consider the expansion map

$$E : \mathcal{O}^1(U) \to R[[T]][T']\hat{}$$

which we view here, for simplicity, as an inclusion. Then

$$f = E(f) = \sum_{n=0}^{\infty} a_n T^n, \quad g = E(g) = \sum_{n=0}^{\infty} b_n T^n,$$

and set $\partial T = H(T)$ for the induced derivation ∂ on $R[[T]]$. We have

$$h_1 \in \mathcal{O}^1(U) = \mathcal{O}(U)\hat{}[T']\hat{}.$$

Then, in $R[[T]][T']\hat{}$ we have

$$\begin{aligned} h_1 = & \frac{1}{p}[\sum_{n=0}^{\infty} a_n^{\phi}(T^p + pT')^n - (\sum_{n=0}^{\infty} a_n T^n)^p] \cdot [\sum_{n=0}^{\infty} n b_n T^{n-1}]^p \cdot H(T)^p \\ & -\frac{1}{p}[\sum_{n=0}^{\infty} b_n^{\phi}(T^p + pT')^n - (\sum_{n=0}^{\infty} b_n T^n)^p] \cdot [\sum_{n=0}^{\infty} n a_n T^{n-1}]^p \cdot H(T)^p \\ & \in \frac{1}{p}[\sum_{n=0}^{\infty} a_n^{\phi} n T^{p(n-1)} pT')] \cdot [\sum_{n=0}^{\infty} n b_n T^{n-1}]^p \cdot H(T)^p \\ & -\frac{1}{p}[\sum_{n=0}^{\infty} b_n^{\phi} n T^{p(n-1)} pT')] \cdot [\sum_{n=0}^{\infty} n a_n T^{n-1}]^p \cdot H(T)^p \\ & +pR[[T]][T']\hat{} + R[[T]] \subset pR[[T]][T']\hat{} + R[[T]]. \end{aligned}$$

Consequently

$$h_1 \in \mathcal{O}(U)\hat{}[T']\hat{} \cap (pR[[T]][T']\hat{} + R[[T]]) = p \cdot \mathcal{O}(U)\hat{}[T']\hat{} + \mathcal{O}(U)\hat{}$$

and the case $j = 1$ is checked. The induction step is trivial. Now, for each U, h_j can be written as $p\varphi_U + \psi_U$, $\varphi_U \in \mathcal{O}^j(U)$, $\psi_U \in \mathcal{O}^{j-1}(U)$. Clearly the images

$$\bar{\psi}_U \in \mathcal{O}^{j-1}(U) \otimes k = \mathcal{O}(J^{j-1}(U) \otimes k)$$

glue together to give an element $\bar{\psi} \in \mathcal{O}(J^{j-1}(X) \otimes k)$. Since X is affine $\bar{\psi}$ lifts to an element of $\mathcal{O}(J^{j-1}(X)) = \mathcal{O}^{j-1}(X)$ and we are done. $\square$

REMARK 4.47. For $\alpha, \beta \in \mathcal{O}^j(X)$ consider the $1-$form

$$\omega(\alpha, \beta) := \alpha d\beta - \beta d\alpha \in H^0(J^j(X), \Omega).$$

Then the left hand side of Equation 4.54 (which can be viewed as an arithmetic analogue of a Poisson bracket that mixes $p-$derivations with usual derivations) can be expressed as

$$(\delta^j f) \cdot \partial_j(\delta^j g) - (\delta^j g) \cdot \partial_j(\delta^j f) = \langle \partial_j, \omega(\delta^j f, \delta^j g) \rangle.$$

4.5.2. Germs at fixed points. Next we attach formal correspondences to fixed points of correspondences and compare the local and the global pictures. Let $\mathbf{X} = (X, \tilde{X}, \sigma_1, \sigma_2)$ be an étale, one dimensional correspondence of finite type and smooth over R, with irreducible base.

DEFINITION 4.48. A *fixed point* for $\mathbf{X}$ is a pair $(\tilde{P}, P)$ of points $\tilde{P} \in \tilde{X}(R)$, $P \in X(R)$ such that

$$\sigma_1(\tilde{P}) = \sigma_2(\tilde{P}) = P.$$

Let $\tilde{P}_0, P_0$ be the $k-$points obtained by reducing $\tilde{P}, P$ mod p. Assume we are given an isomorphism as in Equation 4.51 above. We shall always identify from now on $\mathcal{O}^{for}_{\tilde{X},\tilde{P}_0}$ with $\mathcal{O}^{for}_{X,P_0}$ via σ_1^*. So we have an induced isomorphism

$$\mathcal{O}^{for}_{\tilde{X},\tilde{P}_0} \simeq R[[T]], \tag{4.55}$$

which we view as an identification.

DEFINITION 4.49. The series

$$\Sigma(T) := \sigma_2^* T = a_1 T + a_2 T^2 + ... \in R[[T]]$$

is called the *germ* of $\mathbf{X}$ at the fixed point $(\tilde{P}, P)$.

The series Σ is then a formal correspondence. Its multiplier $a_1 \in R^\times$ does not depend on the isomorphism 4.51 and will be called the *multiplier* of $\mathbf{X}$ at the fixed point $(\tilde{P}, P)$; let's denote it by $a(\tilde{P}) = a(\mathbf{X}, \tilde{P})$. On the other hand if we are given a $\delta-$line bundle $\mathbf{L} = (L, \beta)$ of order r on $\mathbf{X}$, an isomorphism as in Equation 4.51, and a basis z_{for} of the pull-back L^{for} of L to *Spec* $R[[T]][T', ..., T^{(r)}]\hat{}$, the isomorphism β induces an isomorphism of line bundles

$$\beta : \sigma_2^* L^{for} \to \sigma_1^* L^{for}$$

on *Spec* $R[[T]][T, T', ..., T^{(r)}]\hat{}$. So we can write

$$\beta \sigma_2^* z_{for} = B \cdot \sigma_1^* z_{for}$$

for some series

$$B \in (R[[T]][T, T', ..., T^{(r)}]\hat{})^\times.$$

DEFINITION 4.50. The formal $\delta-$line bundle $\Lambda = \Lambda(r, B)$ is called the *germ* of $\mathbf{L}$ at $(\tilde{P}, P)$.

If we change the basis z_{for} the new formal $\delta-$line bundle will be isomorphic to Λ. Set $B(0, ..., 0) =: mult(\mathbf{L}, \tilde{P}) \in R^\times$; we call $mult(\mathbf{L}, \tilde{P})$ the *multiplier* of $\mathbf{L}$ at $(\tilde{P}, P)$ and we note that $mult(\mathbf{L}, \tilde{P})$ does not depend on either the isomorphism 4.51 or the choice of z_{for}. If $\mathbf{L} = \mathbf{F}(r, \lambda)$ is flat the multiplier at any fixed point equals the multiplier of $\mathbf{L}$; in our notation, $mult(\mathbf{L}, \tilde{P}) = \lambda$. On the other hand, clearly, the canonical bundle of the germ of $\mathbf{X}$ coincides with the germ of the canonical bundle of $\mathbf{X}$. Also the multiplier of the canonical line bundle $\mathbf{K}$ at any fixed point equals the multiplier of $\mathbf{X}$ at that point; in our notation,

$$mult(\mathbf{K},\tilde{P}) = a(\mathbf{X},\tilde{P}).$$

Note that, with the above notation, there is an induced *expansion* map (depending on the isomorphism 4.51 and on the trivialisation of L at P):

$$E : H^0(X,L) \to R[[T]][T',...,T^{(r)}]\hat{}. \tag{4.56}$$

PROPOSITION 4.51. *The maps*

$$E : H^0(X,L) \to R[[T]][T',...,T^{(r)}]\hat{}, \quad \bar{E} : H^0(X,L)\otimes k \to k[[T]][T',...,T^{(r)}]$$

are injective.

PROOF. By Corollary 3.25 we may replace X by an open subset of it. So we may assume X is affine and L trivial on X. So we are reduced to Proposition 4.43. □

PROPOSITION 4.52. *The following diagram is commutative*

$$\begin{array}{ccc} H^0(X,L) & \stackrel{E}{\to} & R[[T]][T',...,T^{(r)}]\hat{} \\ \Delta^{\mathbf{L}} \downarrow & & \downarrow \Delta^{\Lambda} \\ H^0(\tilde{X},\tilde{L}) & \stackrel{E}{\to} & R[[T]][T',...,T^{(r)}]\hat{} \end{array}.$$

In particular we have induced injective maps

$$E : H^0(\mathbf{X},\mathbf{L}) \to R[[T]][T',...,T^{(r)}]\hat{}, \tag{4.57}$$

$$\bar{E} : H^0(\mathbf{X},\mathbf{L})\otimes k \to k[[T]][T',...,T^{(r)}],$$

with $E(H^0(\mathbf{X},\mathbf{L})) \subset H^0(\Sigma,\Lambda)$. If in addition $\mathbf{X}$ is irreducible then

$$E^{-1}(H^0(\Sigma,\Lambda)) = H^0(\mathbf{X},\mathbf{L}).$$

PROOF. Follows directly from Proposition 4.51. □

DEFINITION 4.53. The image $E(f)$ of a section $f \in H^0(\mathbf{X},\mathbf{L})$ in $H^0(\Sigma,\Lambda)$ is called the *germ* of f (at the fixed point). We say f is *$d-$ regular* (*$d-$ perfect*, *$d-$ perfect and $d-$ coregular*) at the given fixed point if its germ is so.

By Remark 4.28 this definition is independent of the isomorphism 4.51 above. If $f \in H^0(\mathbf{X},\mathbf{L})$ is $d-$ regular ($d-$ perfect, $d-$ perfect and $d-$ coregular) at some fixed point and if $n \not\equiv 0$ mod p then $f^n \in H^0(\mathbf{X},\mathbf{L}^n)$ is $dn-$ regular ($dn-$ perfect, $dn-$ perfect and $dn-$ coregular) at that fixed point.

Let $\mathbf{X}$ be a correspondence with germ $\Sigma(T)$ at a fixed point $(\tilde{P},P)$ and $\mathbf{L}$ is a $\delta-$line bundle of order r on $\mathbf{X}$, with germ Λ at $(\tilde{P},P)$.

DEFINITION 4.54. $\mathbf{L}$ is *formally flat* at $(\tilde{P},P)$ if Λ is a flat formal $\delta-$line bundle on Σ.

This definition is independent of the isomorphism 4.51 above. Clearly, if $\mathbf{L}$ is flat then it is formally flat at any fixed point.

Propositions 4.44 and 4.52 show that, under quite general conditions, E commutes with both ∂_r and Δ. We already have looked at the commutation between ∂_r and Δ in Proposition 3.59. Here is a variant of that Proposition (where, as there, for a flat $\mathbf{L} = \mathbf{F}(r,b)$ we write Δ^b in place of $\Delta^{\mathbf{L}}$):

PROPOSITION 4.55. *Assume* $\mathbf{X}$ *is irreducible with multiplier* a *and let* $\mathbf{L} = \mathbf{F}(r, b)$. *Let* ∂ *be a derivation on* $\mathcal{O}_X$ *which induces the canonical derivation of* $R[[T]]$ *at the fixed point* $(\tilde{P}, P)$. *Let* $\tilde{\partial} = \sigma_1^* \partial$ *be the derivation on* $\mathcal{O}_{\tilde{X}}$ *induced by* ∂ *via* σ_1. *Then, for any* $j \geq 0$, *we have the following equality of maps* $\mathcal{O}^r(X) \to \mathcal{O}^r(\tilde{X})$:

$$\tilde{\partial}_j \circ \Delta^b = \Delta^{a^{\phi^j} b} \circ \partial_j.$$

PROOF. A trivial computation at the fixed point shows that

$$\sigma_2^* \partial = a^{-1} \cdot \sigma_1^* \partial,$$

so we may conclude by Proposition 3.59. Alternatively since the map $E : \mathcal{O}(\tilde{X}) \to R[[T]][T', ..., T^{(r)}]\hat{}$ is injective it is enough to check that

$$E \circ \partial_j \circ \Delta^b = E \circ \Delta^{a^{\phi^j} b} \circ \partial_j.$$

Since E commutes with both Δ and ∂_j the latter equality is true by Lemma 4.22. □

4.5.3. Germs at cycles. The concept of fixed point can be generalized as follows. We continue to consider an étale one dimensional correspondence $\mathbf{X}$ of finite type and smooth over R with irreducible base.

DEFINITION 4.56. If $n \geq 1$, an $n-$*cycle* (or a *cycle* of length n) on $\mathbf{X}$ is a tuple $C = (\tilde{P}_1, ..., \tilde{P}_n, P)$ where $\tilde{P}_1, ..., \tilde{P}_n \in \tilde{X}(R)$ and $P \in X(R)$ such that

$$\sigma_1(\tilde{P}_1) = \sigma_2(\tilde{P}_n) = P, \quad \sigma_2(\tilde{P}_i) = \sigma_1(\tilde{P}_{i+1}), \quad i = 1, ..., n-1.$$

A fixed point is, of course, the same thing as a $1-$cycle.

Recall that the $n-$th power of $\mathbf{X} = (X, \tilde{X}, \sigma_1, \sigma_2)$ is the correspondence

$$\mathbf{X}^{\circ n} := (X, \tilde{X}^{\circ n}, \sigma_1^{\circ n}, \sigma_2^{\circ n})$$

where

$$\tilde{X}^{\circ n} := \tilde{X} \times_X ... \times_X \tilde{X},$$

and $\sigma_1^{\circ n}, \sigma_2^{\circ n} : \tilde{X}^{\circ n} \to X$ are defined on $R-$points by

$$\sigma_1^{\circ n}(\tilde{P}_1, ..., \tilde{P}_n) = \sigma_1(\tilde{P}_1), \quad \sigma_2^{\circ n}(\tilde{P}_1, ..., \tilde{P}_n) = \sigma_2(\tilde{P}_n).$$

Any $n-$cycle on $\mathbf{X}$ gives rise to a fixed point on $\mathbf{X}^{\circ n}$. The *germ* of $\mathbf{X}$ at C will be, by definition, the germ of $\mathbf{X}^{\circ n}$ at the corresponding fixed point. One can define the multiplier and the Schwarzian

$$a(C), \ s(C) \tag{4.58}$$

of an $n-$cycle C on $\mathbf{X}$ as being the multiplier and the Schwarzian of the corresponding fixed point of $\mathbf{X}^{\circ n}$. We define

$$v(C) := \min\{v_p\left(\frac{s(C)}{1 - a(C)^2}\right), 0\}. \tag{4.59}$$

The $\delta-$line bundles on $\mathbf{X}$ and $\mathbf{X}^{\circ n}$ are related as follows. Assume $\mathbf{L} = (L, \beta)$ is a $\delta-$line bundle on $\mathbf{X}$. Then one defines a $\delta-$line bundle $\mathbf{L}^{\circ n}$ on $\mathbf{X}^{\circ n}$ by the formula $\mathbf{L}^{\circ n} := (L, \beta^{\circ n})$, where $\beta^{\circ n}$ is the obvious "$n-$fold product" of the various pull-backs of β. There is an obvious inclusion

$$H^0(\mathbf{X}, \mathbf{L}) \subset H^0(\mathbf{X}^{\circ n}, \mathbf{L}^{\circ n}).$$

Note that even if $\mathbf{X}$ is irreducible $\mathbf{X}^{\circ n}$ fails in general to be irreducible, although it remains, of course, étale. Now any point $\tilde{P} \in \tilde{X}^{\circ n}(R)$ has a Zariski open neighborhood $\tilde{X}_1^{\circ n} \subset \tilde{X}^{\circ n}$ with irreducible fibers over R. Indeed let Z be the connected component of $\tilde{X}^{\circ n}$ containing the point $\tilde{P} \otimes k$; since Z is regular and connected it is irreducible. Since $Z \otimes k$ is smooth the point $\tilde{P} \otimes k$ belongs to exactly one irreducible component of $Z \otimes k$ so one can take $\tilde{X}_1^{\circ n}$ to be the complement in Z of the union of all components of $Z \otimes k$ which do not contain $\tilde{P} \otimes k$. In particular any $n-$cycle on $\mathbf{X}$ gives rise to a fixed point of an irreducible open subcorrespondence

$$\mathbf{X}_1^{\circ n} := (X, \tilde{X}_1^{\circ n}, \sigma_1^{\circ n}, \sigma_2^{\circ n})$$

of $\mathbf{X}^{\circ n}$ with base X. Note that we have a natural inclusion

$$H^0(\mathbf{X}^{\circ n}, \mathbf{L}^{\circ n}) \subset H^0(\mathbf{X}_1^{\circ n}, \mathbf{L}^{\circ n}).$$

Recall that if $\mathbf{K} = \mathbf{K}_{\mathbf{X}}$ is the canonical bundle on $\mathbf{X}$ then $\mathbf{K}_{\mathbf{X}}^{\circ n} \simeq \mathbf{K}_{\mathbf{X}^{\circ n}}$. Also, for flat line bundles, $\mathbf{F}(\lambda^{-1})^{\circ n} \simeq \mathbf{F}(\lambda^{-n})$ for any λ.

We will say that a $\delta-$line bundle $\mathbf{L}$ of order r on $\mathbf{X}$ is *formally flat* at an $n-$cycle C of $\mathbf{X}$ if $\mathbf{L}^{\circ n}$ is formally flat at the corresponding fixed point of $\mathbf{X}^{\circ n}$. A section $f \in H^0(\mathbf{X}, \mathbf{L})$ is called $d-$regular ($d-$ perfect, $d-$ perfect and $d-$ coregular) at C if the image of f in $H^0(\mathbf{X}^{\circ n}, \mathbf{L}^{\circ n})$ has the corresponding property at the corresponding fixed point.

4.6. Some general global converse results

As an application of our local versus global machinery we will prove a number of general converse results, i.e. results showing that pairs $(\mathbf{X}, \mathbf{L})$ for which $H^0(\mathbf{X}, \mathbf{L})$ is sufficiently non-trivial must satisfy some remarkable properties (shared, as we shall later see, by correspondences that admit analytic uniformizations).

4.6.1. Generic triviality of $\mathbf{K}^{\nu}$ mod p. We start with a consequence of Proposition 4.46 which, in its turn, is a consequence of our theory of expansion maps (but does not use fixed points or cycles).

THEOREM 4.57. *Let $\mathbf{X}$ be an étale one dimensional irreducible correspondence of finite type and smooth over R and let $v \in W$. Assume*

$$\text{(4.60)} \qquad \textit{either} \quad H^0(\mathbf{X}, \mathbf{K}^v) \neq 0 \quad \textit{or} \quad H^1(\mathbf{X}, \mathbf{K}^v)[p] \neq 0.$$

Then there exists a non-empty open subcorrespondence $\mathbf{U} \subset \mathbf{X}$ and an integer $\nu \in \mathbf{Z}$, $\nu \equiv v(0)$ mod p, $\nu \geq v(p)$, such that the $\nu-$canonical bundle $\mathbf{K}_{\mathbf{U}\otimes k}^{\nu}$ of $\mathbf{U} \otimes k$ is trivial.

Note that condition 4.60 is equivalent to condition:

$$H^0(\mathbf{X}, \mathbf{K}^v; k) \neq 0.$$

PROOF. Let $f \in H^0(X, K^v)$ be such that its image $\bar{f} \in H^0(X, K^v) \otimes k$ is a non-zero element of $H^0(\mathbf{X}, \mathbf{K}^v; k)$. Let $U \subset X$ be an affine open set possessing an étale coordinate $T \in \mathcal{O}(U)$. Let $r = ord(v)$. Then we have an identification

$$\mathcal{O}^r(U) = \mathcal{O}(U)\hat{}[T', ..., T^{(r)}]\hat{}$$

so we may write

$$f = \left(\sum_{i_1,...,i_r} a_{i_1...i_r} (T')^{i_1} ... (T^{(r)})^{i_r} \right) \cdot (dT)^v,$$

with $a_{i_1\dots i_r} \in \mathcal{O}(U)\hat{}$, $a_{i_1\dots i_r} \to 0$ $p-$adically. Let $\bar{a}_{i_1\dots i_r} \in \mathcal{O}(U \otimes k)$ be the images of $a_{i_1\dots i_r}$. So not all $\bar{a}_{i_1\dots i_r}$ vanish. Let

$$\tilde{U} \subset \sigma_1^{-1}(U) \cap \sigma_2^{-1}(U)$$

be an open set with $\tilde{U} \otimes k \neq \emptyset$ and keep denoting $\sigma_1^* T \in \mathcal{O}(\tilde{U})$ by T. So we have an identification

$$\mathcal{O}^r(\tilde{U}) = \mathcal{O}(\tilde{U})\hat{}[T', \dots, T^{(r)}]\hat{}.$$

By Proposition 4.46 we have, for $j = 1, \dots, r$, an equality

$$\sigma_2^* T^{(j)} = \delta^j(\sigma_2^* T) = \kappa^{p^j} \cdot T^{(j)} + p \cdot \alpha_j + \beta_j, \tag{4.61}$$

where $\alpha_j \in \mathcal{O}^j(\tilde{U})$, $\beta_j \in \mathcal{O}^{j-1}(\tilde{U})$ and

$$\kappa := \frac{d(\sigma_2^* T)}{dT} \in \mathcal{O}(\tilde{U})^\times.$$

The equality

$$\sigma_2^* \bar{f} = \sigma_1^* \bar{f}$$

in $H^0(\tilde{X}, K^v) \otimes k$ yields an equality

$$\sum \sigma_2^* \bar{a}_{i_1\dots i_r} \cdot [\kappa^p T' + \bar{\beta}_1]^{i_1} \cdots [\kappa^{p^r} T^{(r)} + \bar{\beta}_r]^{i_r} \cdot (d\sigma_2^* T)^{v(p)}$$
$$= \sum \sigma_1^* \bar{a}_{i_1\dots i_r} \cdot (T')^{i_1} \cdots (T^{(r)})^{i_r} \cdot (dT)^{v(p)},$$

where

$$\bar{\beta}_j \in \mathcal{O}(\tilde{U} \otimes k)[T', \dots, T^{(j-1)}];$$

cf. Equation 4.61. Now introduce an order on the monomials

$$T_{i_1\dots i_r} := (T')^{i_1} \cdots (T^{(r)})^{i_r}$$

by declaring that $T_{i_1\dots i_r} \leq T_{j_1\dots j_r}$ if

$$(i_r, \dots, i_1) \leq_{lex} (j_r, \dots, j_1)$$

where $\leq_{lex}$ is the lexicographic order. Let

$$(l_r, \dots, l_1) = \max \{(i_r, \dots, i_1) \mid \bar{a}_{i_1\dots i_r} \neq 0\}.$$

Setting

$$\nu := pl_1 + \dots + p^r l_r + v(p)$$

and identifying the coefficients of $T_{l_1\dots l_r}$ we get

$$\sigma_2^* \bar{a}_{l_1\dots l_r} \cdot \bar{\kappa}^\nu = \sigma_1^* \bar{a}_{l_1\dots l_r}$$

where $\bar{\kappa} \in \mathcal{O}(\tilde{U} \otimes k)$ is the image of κ hence

$$\sigma_2^*(\bar{a}_{l_1\dots l_r} \cdot (dT)^\nu) = \sigma_2^*(\bar{a}_{l_1\dots l_r} \cdot (dT)^\nu). \tag{4.62}$$

After shrinking U and $\tilde{U}$ (without necessarily preserving the condition that $P \in U(R)$!) we may assume

$$\bar{a}_{l_1\dots l_r} \in \mathcal{O}(U \otimes k)^\times.$$

Set $\mathbf{U} := (U, \tilde{U}, \sigma_1, \sigma_2)$. Then Equation 4.62 implies that $\mathbf{K}^\nu_{\mathbf{U}\otimes k}$ is trivial. □

At this point we are ready to prove Theorem 2.65.

PROOF. Assertion 1 of the Theorem follows directly from Theorem 4.57. Assertion 2 follows from assertion 1 plus Theorem 1.52. □

4.6.2. Integrality. As a consequence of Propositions 4.16, 4.24, and Corollary 4.17 we get the following integrality properties for global correspondences:

COROLLARY 4.58. *Let* $\mathbf{X}$ *be an étale one dimensional correspondence of finite type and smooth over* R, *with irreducible base, and let* $\mathbf{L}$ *be a* $\delta-$*line bundle such that* $\mathbf{L}^m \simeq \mathbf{K}^w$ *for some positive* $m \in \mathbf{Z}$ *and some* $w \in W$, $ord(w) = r \geq 1$. *Assume* $d := -deg(w) = me$ *with* e *an integer. Let* C *be an* $n-$*cycle of* $\mathbf{X}$ *whose multiplier is in* $\mathbf{Z}_p^\times$, *not a root of unity.*

1) Assume $H^0(\mathbf{X}, \mathbf{L})$ *contains a section which is* $s-$ *regular at* C *for some* $s \geq 1$. *Then* $s = e$ *(hence* $deg(w) < 0$*) and the canonical bundle of the germ of* $\mathbf{X}^{\circ n}$ *at the fixed point corresponding to* C *is* $(d, a_r, r)-$*flat. If, in addition,* $w = -1 - \phi^r$ *then* $v(C) \geq -2r$.

2) Assume $d \not\equiv 0$ *mod* p *and* $d > 0$. *Moreover assume* $\mathbf{K}$ *is formally flat at* C, *and* $H^0(\mathbf{X}, \mathbf{L})$ *contains a section which is* $e-$*perfect and* $e-$ *coregular at* C. *Then the germ of* $\mathbf{X}$ *at* C *admits a uniformization.*

4.6.3. Accumulating multipliers and accumulating Schwarzians.

DEFINITION 4.59. Consider an étale one dimensional correspondence $\mathbf{X} = (X, \tilde{X}, \sigma_1, \sigma_2)$ of finite type and smooth over R. A sequence of pairwise disjoint cycles C_i, $i = 1, 2, ...$, of $\mathbf{X}$ is said to have $\mathbf{Z}_p-$ *accumulating multipliers* is all multipliers $a(C_i) \in R$ belong to $\mathbf{Z}_p$ and satisfy $1 \neq a(C_i) \to 1$ $p-$adically.

Here two cycles $(\tilde{P}_1, ..., \tilde{P}_n, P)$ and $(\tilde{Q}_1, ..., \tilde{Q}_m, Q)$ of $\mathbf{X}$ are said to be *disjoint* if

$$\{\sigma_1(\tilde{P}_1), ..., \sigma_1(\tilde{P}_n)\} \cap \{\sigma_1(\tilde{Q}_1), ..., \sigma_1(\tilde{Q}_m)\} = \emptyset$$

in $X(R)$.

The requirement that C_i be disjoint makes the above concept non-trivial. This assumption excludes the possibility that, for instance, all the cycles C_i are of the form $C_i = (\tilde{P}, ..., \tilde{P}, P)$ where $(\tilde{P}, P)$ is a fixed point. (If the latter is the case and, say, the multiplier $a = a(\tilde{P})$ is in $\mathbf{Z}_p^\times$ then $a(C_i) = a^{n(C_i)}$, where $n(C_i)$ is the length of C_i, and one can trivially find a sequence of integers $n(C_i) \to \infty$ such that $a(C_i) \to 1$, $p-$adically.)

A sequence (C_i) as above is said to have *accumulating Schwarzians* if $s(C_i) \to 0$ $p-$adically.

The significance of the above concepts comes from the following immediate consequence of Corollary 4.58.

COROLLARY 4.60. *Let* $\mathbf{X}$ *be an étale one dimensional correspondence of finite type and smooth over* R, *with irreducible base, and let* $\mathbf{L}$ *be a* $\delta-$*line bundle such that* $\mathbf{L}^m \simeq \mathbf{K}^{-1-\phi^r}$ *where* $m \in \{1, 2\}$ *and let* s *be an integer. Assume the following hold:*

1) $\mathbf{X}$ *has a sequence* (C_i) *of pairwise disjoint cycles with* $\mathbf{Z}_p-$*accumulating multipliers,*

2) $H^0(\mathbf{X}, \mathbf{L})$ *contains a section which is* $s-$*regular at all the cycles* C_i.

Then the sequence (C_i) *has accumulating Schwarzians.*

This Corollary should be viewed as a result in the spirit of Conjecture 2.55; indeed the conclusion about the Schwarzians should be viewed as a "local linearization" property. Note that this Corollary is *not* simply a consequence of the equality 4.1 because our cycles C_i are disjoint so, in particular, they are not obtained by iterating a given cycle.

Of course one can ask how natural the hypotheses of the Corollary are. In the body of this book we will show that a number of interesting classes of correspondences satisfy the above hypotheses; not surprisingly, these correspondences will turn out to also admit analytic uniformizations.

4.6.4. Rank of the space of δ–invariants. Here is another useful consequence of our analysis:

THEOREM 4.61. *Assume* $\mathbf{X}$ *is an étale one dimensional correspondence of finite type and smooth over* R*, with an irreducible base, let* C *be an* n*–cycle, and let* $\mathbf{L}$ *be a* δ*–line bundle on* $\mathbf{X}$ *of order* r *such that* $\mathbf{L}^m \simeq \mathbf{K}^w \otimes \mathbf{F}$*, where* $m \in \mathbf{Z}_+$*,* $m \neq 0$*,* $w \in W$*, and* $\mathbf{F} = \mathbf{F}(r, \lambda)$*,* $r \geq ord(w)$*. Let* $a = a(C)$ *be the multiplier of* $\mathbf{X}$ *at* C *and let*

$$S := \{v \in W_+ | ord(v) \leq r, \ a^{mv+w} = \lambda^{-n}\}.$$

Then $rank_R H^0(\mathbf{X}, \mathbf{L}) \leq \sharp\, S$*. If in addition that* $a = a(C)$ *is not a generalized root of unity the* R*–module* $H^0(\mathbf{X}, \mathbf{L})$ *is finitely generated.*

PROOF. Let $\mathbf{X}_1^{\circ n}$ be an irreducible open sub-correspondence of $\mathbf{X}^{\circ n}$ such that C gives rise to a fixed point $(\tilde{P}, P)$ of $\mathbf{X}_1^{\circ n}$. Take a trivialisation of L around P. Using this trivialisation and an isomorphism as in Equation 4.51 we consider the embedding

$$E : H^0(\mathbf{X}, \mathbf{L}) \to H^0(\Sigma, \Lambda) \subset R[[T]][T', ..., T^{(r)}]\hat{},$$

where $\Sigma(T)$ is the germ of $\mathbf{X}_1^{\circ n}$ at $(\tilde{P}, P)$ and Λ is the induced formal δ–line bundle. Let b be the multiplier of $\mathbf{L}^{\circ n}$ at $(\tilde{P}, P)$. By Proposition 4.31, $H^0(\Sigma, \Lambda)$ has rank bounded by the cardinality of the set

$$S' = \{v \in W_+ | ord(v) \leq r, \ a^v = b^{-1}\}$$

and it is finitely generated if a is not a generalized root of unity. By Equation 1.19 it follows that

$$(\mathbf{L}^{\circ n})^m \simeq (\mathbf{L}^m)^{\circ n} \simeq \mathbf{K}^w_{\mathbf{X}^{\circ n}} \otimes \mathbf{F}(r, \lambda^n)$$

hence $b^m = a^w \lambda^n$. This immediately implies that $S' \subset S$ and we are done. $\square$

4.6.5. Symmetry of the multiplier set. To state a Corollary of the above Theorem it is convenient to introduce some more terminology from [**29**].

DEFINITION 4.62. Assume $\mathcal{P}$ is a set equipped with two maps $a : \mathcal{P} \to R$ and $n : \mathcal{P} \to \mathbf{Z}_{>0}$. Let

$$W_{deg \neq 0} := \{w \in W; deg(w) \neq 0\} = W \backslash (\phi - 1)W. \tag{4.63}$$

Say that $\mathcal{P}$ is *symmetric of radius* $\rho \in \mathbf{Z}_p^\times$ if there exists a function $w : \mathcal{P} \to W_{deg \neq 0}$ such that

$$a(C)^{w(C)} = \rho^{n(C)}, \quad C \in \mathcal{P}.$$

Say $\mathcal{P}$ is *symmetric* if it is symmetric of some radius.

If $\mathcal{P}$ is not symmetric of radius 1 then there exists $C \in \mathcal{P}$ such that $a(C)$ is not a generalized root of unity. If $\mathbf{X}$ is an étale correspondence then any set $\mathcal{P}$ consisting of cycles on $\mathbf{X}$ is naturally equipped with two maps a and n as above: we let $a(C)$ be the multiplier of $\mathbf{X}$ at C and we let $n(C)$ be the length of C.

COROLLARY 4.63. *Assume* $\mathbf{X}$ *is an étale one dimensional correspondence of finite type and smooth over* R, *with an irreducible base, and let* $\mathbf{L}$ *be a* $\delta-$*line bundle on* $\mathbf{X}$ *of order* r *such that* $\mathbf{L}^m \simeq \mathbf{K}^w \otimes \mathbf{F}^{[r]}$, *where* $m \in \mathbf{Z}_+$, $m \neq 0$, $w \in W$, *and* $\mathbf{F} = \mathbf{F}(\lambda)$ *is a flat line bundle with multiplier* λ. *Let* $\mathcal{P}$ *be the set of all cycles on* $\mathbf{X}$ *whose multipliers belong to* $\mathbf{Z}_p^{ur}$. *Then the following hold:*

1) If λ *is not a root of unity and* $\mathcal{P}$ *is not symmetric then* $H^0(\mathbf{X}, \mathbf{L}^v) = 0$ *for all* $0 \neq v \in W_+$.

2) If λ *is a root of unity,* $deg(w) > 0$, *and* $\mathcal{P}$ *is not symmetric of radius* 1 *then* $H^0(\mathbf{X}, \mathbf{L}^v) = 0$ *for all* $0 \neq v \in W_+$.

3) If λ *is a root of unity,* $deg(w) = 0$, *and* $\mathcal{P}$ *is not symmetric of radius* 1 *then* $rank_R H^0(\mathbf{X}, \mathbf{L}^v) \leq 1$ *for all* $0 \neq v \in W_+$. *In particular* $H^0(\mathbf{X}, \mathbf{O}^{[r]}) = R$.
Consequently, in all 3 *cases above,* $(\mathbf{X}, \mathbf{L})$ *is* $\delta-$*trivial.*

The moral of the above Corollary is that if some positive $\mathbf{Z}-$power $\mathbf{L}^m$ of $\mathbf{L}$ is a $W-$linear combination of the canonical bundle and a flat bundle, and if $(\mathbf{X}, \mathbf{L})$ is $\delta-$non-trivial then either we have a remarkable symmetry of the multiplier set of $\mathbf{X}$ or $\mathbf{L}^m \simeq \mathbf{K}^w$, $deg(w) < 0$.

PROOF. To check assertion 1 assume λ is not a root of unity and $H^0(\mathbf{X}, \mathbf{L}^v) \neq 0$ for some $0 \neq v \in W_+$; we will show that $\mathcal{P}$ is symmetric. Note that

$$(\mathbf{L}^v)^m \simeq \mathbf{K}^{wv} \otimes \mathbf{F}(\lambda^{deg(v)})^{[r]}. \tag{4.64}$$

Also, if $a(C)$ is the multiplier of $\mathbf{L}$ at a cycle C then $a(C)^v$ is the multiplier of $\mathbf{L}^v$ at C. So, by Theorem 4.61, for each $C \in \mathcal{P}$ we can find $v_C \in W_+$ such that

$$(a(C)^v)^{mv_C + wv} = (\lambda^{deg(v)})^{-n(C)}. \tag{4.65}$$

We claim that $w(C) := v(mv_C + wv)$ has degree $\neq 0$; this will imply that $\mathcal{P}$ is symmetric with radius $\rho := \lambda^{-deg(v)}$ which will end the proof of assertion 1. Indeed, since $a(C) \in \mathbf{Z}_p^{ur}$ there exists a positive integer s such that $a(C)^{\phi^s} = a(C)$. Raising Equation 4.65 to the power $\omega_s := 1 + \phi + \phi^2 + ... + \phi^{s-1}$ we get

$$\rho^{n(C) \cdot s} = (a(C)^{\omega_s})^{w(C)} = a(C)^{\omega_s \cdot deg(w(C))};$$

in particular $deg(w(C)) \neq 0$ because otherwise ρ, and hence λ, would be a root of unity.

To check assertion 2 it is enough to assume $\lambda = 1$, $deg(w) > 0$, and $H^0(\mathbf{X}, \mathbf{L}^v) \neq 0$ for some $0 \neq v \in W_+$ and to show that $\mathcal{P}$ is symmetric of radius 1. Exactly as above, for each $C \in \mathcal{P}$ we can find $v_C \in W_+$ such that Equation 4.65 holds (with $\lambda = 1$). On the other hand, clearly, $deg(v(mv_C + wv)) > 0$.

Finally, to check assertion 3, it is enough to assume again $\lambda = 1$, $deg(w) = 0$, and $rank_R H^0(\mathbf{X}, \mathbf{L}) \geq 2$ for some $0 \neq v \in W_+$ and prove that $\mathcal{P}$ is symmetric of radius 1. By Theorem 4.61, for any $C \in \mathcal{P}$ one can find two distinct weights v'_C and v''_C in W_+ such that Equation 4.65 holds with $\lambda = 1$ and v_C replaced by v'_C and v''_C respectively. It is enough to check that at least one of the weights $mv'_C + wv$ and $mv''_C + wv$ has degree $\neq 0$. If both these weights have degree 0, since $deg(w) = 0$, it follows that $deg(v'_C) = deg(v''_C) = 0$ hence $v'_C = v''_C = 0$, hence v'_C and v''_C are not distinct, a contradiction. □

In applications it is important to have a criterion for the set $\mathcal{P}$ to be non-symmetric. In what follows we explain such a criterion by reproducing the discussion at the end of [**29**]. We need some preliminaries.

A polynomial $\sigma = \sigma(t) \in F[t]$ with coefficients in a field F will be called *dynamically separable* if there exists a positive integer M such that for all $m \equiv 1$ mod M the polynomial $\sigma^{\circ m}(t) - t$ has simple roots only (in the algebraic closure of F). We will only be interested in the above definition for F of positive characteristic.

EXAMPLE 4.64. Assume N is a positive integer, $N \not\equiv -1, 0, 1 \bmod p$. Let $\sigma_N(t) \in \mathbf{Z}[t]$ be either t^N or the Chebyshev polynomial $Cheb_N$ of degree N and let $\bar{\sigma}_N(t) \in \mathbf{F}_p[t]$ be the reduction mod p of $\sigma_N(t)$. Then $\bar{\sigma}_N(t)$ is dynamically separable. Indeed, if $\sigma_N(t) = t^N$, we have $\bar{\sigma}_N^{\circ m}(t) - t = t^{N^m} - t$; the roots of the latter are simple for $m \equiv 1 \bmod p-1$. The case when $\sigma_N(t)$ is the Chebyshev polynomial is an easy exercise; cf. [**29**].

Let $\sigma \in \mathbf{C}[t]$; we say that the *critical orbits of $\sigma(t)$ escape to infinity* if for any $\beta \in \mathbf{C}$ with $\sigma'(\beta) = 0$, we have $\sigma^{\circ n}(\beta) \to \infty$ as $n \to \infty$.

Here is our criterion for $\mathcal{P}$ to be non-symmetric.

PROPOSITION 4.65. *Let $\sigma(t) \in \mathbf{Z}[t]$ be a monic polynomial of degree N, satisfying the following properties:*

i) The image of $\sigma(t)$ in $\mathbf{F}_p[t]$ is dynamically separable,

ii) $\sigma'(t) \in N\mathbf{Z}[t]$,

iii) The critical orbits of $\sigma(t)$ escape to infinity.

Let $X \subset \mathbf{P}^1 := Proj\ R[x_0, x_1]$ be a Zariski open set such that the correspondence

$$\mathbf{X} := (X, X \cap \sigma^{-1}(X), \iota, \sigma)$$

is étale. Let $\mathcal{P}$ be the set of all periodic points of $\mathbf{X}$ whose multipliers belong to $\mathbf{Z}_p^{ur}$. Then $\mathcal{P}$ is not symmetric.

PROOF. Let us assume that $\mathcal{P}$ is symmetric of some radius $b \in \mathbf{Z}_p^{\times}$; we will derive a contradiction and this will end our proof.

Denote by $\beta_1, ..., \beta_{N-1} \in \mathbf{C}$ the roots of $\sigma'(t)$ taken with their corresponding multiplicities. Condition ii) in the statement of the Proposition implies that β_i belong to the ring of integers $\mathcal{O}_F$ of some number field F. For each $j = 1, ..., N-1$ set

$$c_j = \sigma(\beta_j) - \beta_j;$$

these quantities are non-zero because of condition iii) so we may consider the polynomials

$$\sigma_j(t) := \frac{\sigma(c_j t + \beta_j) - \beta_j}{c_j} \in F[t];$$

note that $\sigma_j(t)$ has actually coefficients in $\mathcal{O}_F$ and, of course, $\sigma_j(0) = 1$. Let us define, for all $m \geq 1$,

$$c_j^{(m)} := \sigma^{\circ m}(\beta_j) - \beta_j;$$

in particular $c_j^{(1)} = c_j$ and we have

$$c_j^{(m+1)}/c_j = \sigma_j(c_j^{(m)}/c_j).$$

It follows by induction that

$$c_j^{(m)}/c_j \in \mathcal{O}_F, \quad m \geq 1.$$

We claim that for any prime ideal Q in $\mathcal{O}_F$ there exists an integer $N_Q \geq 1$ such that

$$c_j^{(m)}/c_j \notin Q \ \textit{for all } j \textit{ and all } \ m \equiv 1 \bmod N_Q. \tag{4.66}$$

Indeed consider the maps $\bar{\sigma}_j : \mathcal{O}_F/Q \to \mathcal{O}_F/Q$ defined by $\bar{\sigma}_j(x) = \sigma_j(x)$ and consider the sequence

$$x_j^{(1)} = 1, x_j^{(m+1)} := \bar{\sigma}_j(x_j^{(m)}), \quad m = 1, 2, ... \tag{4.67}$$

So $x_j^{(m)}$ is the image in $\mathcal{O}_F/Q$ of $c_j^{(m)}/c_j$. If one of the terms of the sequence 4.67 is 0 then the next term is 1 so the sequence is periodic; in particular there exists a positive integer $N_{Q,j}$ such that

$$x_j^{(m)} \neq 0 \;\; for \;\; all \;\; m \equiv 1 \;\; mod \;\; N_{Q,j}. \tag{4.68}$$

If none of the terms of the sequence 4.67 is 0 then, again, condition 4.68 holds (for any choice of $N_{Q,j}$). Our claim is then proved by taking

$$N_Q := \prod_{j=1}^{N-1} N_{Q,j}.$$

Next note that there exists a finite set S of prime numbers with the following properties: 1) $p \in S$, 2) If $b^\lambda \in \mathbf{Z}$ for some $0 \neq \lambda \in \mathbf{Z}_+$ then any prime integer dividing b^λ belongs to S. We then let Σ be the set of all prime ideals Q of $\mathcal{O}_F$ lying over some prime q in S. By property iii) in the statement of the Proposition there exists a positive integer m_0 such that

$$\prod_{j=1}^{N-1} |c_j^{(m)}/c_j| > 1 \;\; for \;\; m \geq m_0. \tag{4.69}$$

By the dynamical separability condition i) in the statement of the Proposition, there is a positive integer M such that

$$\sigma^{\circ m}(t) - t \;\; has \;\; N^m \;\; distinct \;\; roots \;\; in \;\; k \;\; if \;\; m \equiv 1 \;\; mod \;\; M. \tag{4.70}$$

By Dirichlet's theorem on primes in arithmetic progressions there exists a prime $l \geq m_0$ such that

$$l \equiv 1 \;\; mod \;\; M \cdot \prod_{Q \in \Sigma} N_Q, \tag{4.71}$$

$$no \;\; point \;\; in \;\; \mathbf{P}^1(k) \backslash X(k) \;\; is \;\; periodic \;\; with \;\; period \;\; l. \tag{4.72}$$

By Hensel's Lemma, the N^l distinct roots of the reduction mod p of $\sigma^{\circ l}(t) - t$ in k (cf. condition 4.70) can be lifted to N^l distinct roots

$$\alpha_1, \alpha_2, ..., \alpha_{N^l} \in R$$

of $\sigma^{\circ l}(t) - t$; these are, of course, the fixed points of $\sigma^{\circ l}$. We may assume

$$\alpha_{N^l - N+1}, ..., \alpha_{N^l}$$

are the roots of the polynomial $\sigma(t) - t$, in other words that they are the fixed points of σ. Then σ acts on the set

$$\{\alpha_1, \alpha_2, ..., \alpha_{N^l - N}\}; \tag{4.73}$$

there are exactly $(N^l - N)/l$ orbits with l elements each. By condition 4.72 all the points 4.73 belong to $X(R)$; since all α_i's are integral over $\mathbf{Z}$ we have that $a(\alpha_i) \in \mathbf{Z}_p^{ur}$ hence all the points 4.73 belong to the set $\mathcal{P}$. We may arrange indices such that $\alpha_1, ..., \alpha_{(N^l-N)/l}$ are in distinct orbits. By the chain rule we have the following formula for the multiplier at α_i, $i = 1, ..., (N^l - N)/l$:

$$a(\alpha_i) = (\sigma^{\circ l})'(\alpha_i) = \sigma'(\sigma^{\circ(l-1)}(\alpha_i)) \cdots \sigma'(\alpha_i).$$

Taking the product over all $i = 1, ..., (N^l - N)/l$ and denoting by $Res(\ ,\ )$ the resultant of a pair of polynomials we get

$$\prod_{i=1}^{(N^l-N)/l} a(\alpha_i) = \frac{\prod_{i=1}^{N^l} \sigma'(\alpha_i)}{\prod_{i=N^l-N+1}^{N^l} \sigma'(\alpha_i)} = \frac{Res(\sigma^{\circ l}(t) - t, \sigma'(t))}{Res(\sigma(t) - t, \sigma'(t))} \tag{4.74}$$

$$= N^{N^l-N} \frac{\prod_{j=1}^{N-1}(\sigma^{\circ l}(\beta_j) - \beta_j)}{\prod_{j=1}^{N-1}(\sigma(\beta_j) - \beta_j)} = N^{N^l-N} \cdot s$$

where

$$s := \prod_{j=1}^{N-1} \frac{c_j^{(l)}}{c_j}. \tag{4.75}$$

The right hand side of the Equation 4.74 is in $\mathbf{Q}$ because it is a quotient of resultants of polynomials with integer coefficients. In particular

$$s \in \mathbf{Q} \cap \mathcal{O}_F = \mathbf{Z}. \tag{4.76}$$

By the symmetry property there exist $w_1, ..., w_{(N^l-N)/l} \in W_{deg \neq 0}$ such that

$$a(\alpha_i)^{w_i} = b^l.$$

There exists a positive integer λ such that $a(\alpha_i) \in R^{\phi^\lambda}$ for all i. Taking norms in the latter equation we get

$$Norm_\lambda(a(\alpha_i))^{\mu_i} = b^{\lambda l}, \tag{4.77}$$

where $\mu_i := deg(w_i) \neq 0$. Letting m be the product of the μ_i's, raising Equation 4.74 to the power m and taking norms we get an equality of the form

$$b^{\lambda l(m/\mu_1 + m/\mu_2 + ...)} = N^{\lambda m(N^l-N)} s^{\lambda m}. \tag{4.78}$$

By Equations 4.69, 4.75, 4.76, $|s|$ is an integer > 1; in particular the exponent of b in the left hand side of Equation 4.78 is non-zero and there exists a prime $q \in \mathbf{Z}$ dividing $|s|$. By Equation 4.78, $q \in S$. Let Q be a prime ideal in $\mathcal{O}_F$ that lies over q. Then $Q \in \Sigma$. On the other hand by Equation 4.75, since $s \in Q$, it follows that at least one of the factors $c_j^{(l)}/c_j$ must be contained in Q; this contradicts equation 4.66 which ends our proof. □

4.6.6. Quadratic ϕ−independence. Here is another consequence of Theorem 4.61 which is useful in applications. To state it we need the following:

DEFINITION 4.66. Let A be a ring equipped with a ring endomorphism ϕ. We say that an element $a \in A$ is *quadratically ϕ−independent to order* r if the following conditions hold:

1) a is not a generalized root of unity,
2) $a^{\phi^s} = a$ for some $s \geq 1$,
3) $a^{2\phi^i} \neq a^{\phi^j + \phi^r}$ for all $0 \leq i \leq r$ and all $0 \leq j \leq r - 1$.

COROLLARY 4.67. *Assume* $\mathbf{X}$ *is an étale one dimensional correspondence of finite type and smooth over* R, *with an irreducible base, let* C *be an* $n-$*cycle, and let* $\mathbf{L}$ *be a* $\delta-$*line bundle on* $\mathbf{X}$ *of order* r *such that* $\mathbf{L}^2 \simeq \mathbf{K}^{-\phi^\rho - \phi^r}$, $\rho < r$. *Assume the multiplier* $a = a(C)$ *of* $\mathbf{X}$ *at* C *is quadratically* $\phi-$*independent to order* r. *Then* $H^0(\mathbf{X}, \mathbf{L}) = 0$.

PROOF. Assume $H^0(\mathbf{X}, \mathbf{L}) \neq 0$. By Theorem 4.61 there exists $v \in W_+$ of order $\leq r$ such that

$$1 = a^{2v - \phi^\rho - \phi^r}. \tag{4.79}$$

Consider the "norm"

$$Norm_s(a) = \alpha := a^{\sum_{j=0}^{s-1} \phi^j}.$$

By condition 2 in Definition 4.66 $\alpha^\phi = \alpha$; so $\alpha \in \mathbf{Z}_p$. By condition 1 in Definition 4.66 α is not a root of unity. Taking "norms" $Norm_s$ in Equation 4.79 we get

$$1 = \alpha^{2v - \phi^\rho - \phi^r} = \alpha^{deg(2v - \phi^\rho - \phi^r)}$$

hence $deg(2v - \phi^\rho - \phi^r) = 0$, hence $deg(v) = 1$, hence $v = \phi^i$ for some $i \in \{0, ..., r\}$. But this plus Equation 4.79 contradict condition 3 in Definition 4.66. $\square$

CHAPTER 5

Birational theory

The "birational" properties of pairs $(\mathbf{X}, \mathbf{L})$ are captured by the local $\delta-p-$ring $R\langle \mathbf{X}, \mathbf{L}\rangle$ (and its residue field $k\langle \mathbf{X}, \mathbf{L}\rangle$). The ring $R\langle \mathbf{X}, \mathbf{L}\rangle$ is a ring of homogeneous fractions of the $\delta-$graded ring $R(\mathbf{X}, \mathbf{L})^*$. As we will show in later Chapters $R(\mathbf{X}, \mathbf{L})^*$ is often expressible in terms of certain graded rings $R\{x\}_*$, $R\{x\}^*$, $R\{x_*, y\}^*$; these rings will be introduced in Section 1 of this Chapter. Next we are going to investigate rings of homogeneous fractions of $R\{x\}_*$, $R\{x\}^*$, $R\{x_*, y\}^*$. This is done by investigating two interrelated problems: computing what we will call $\delta-$*Galois groups* of certain extensions of local $\delta-p-$rings of homogeneous fractions attached to our graded rings above and proving existence of $\delta-$bases for these local $\delta-p-$rings. In our applications this will lead to proofs of $\delta-$rationality of certain categorical quotients. In Section 2 we introduce $\delta-$*Galois groups*. In Section 3 we study a basic action of the group $PSL_2(\mathbf{Z}_p)$, and the induced actions of some of its subgroups, which play a key role in our computations of $\delta-$Galois groups and $\delta-$bases.

5.1. The basic graded rings

We will use the following general recipe to produce graded rings. Assume $R = \hat{\mathbf{Z}}_p^{ur}$, $K = R[1/p]$, and one is given a flat $R-$algebra B such that $B \otimes K$ has a structure of graded $K-$algebra,

$$B \otimes K = \bigoplus_{\gamma \in \Gamma} (B \otimes K)_\gamma,$$

where Γ is any Abelian semigroup. Then we can define

$$B_\gamma := (B \otimes K)_\gamma \cap B \subset B \otimes K,$$

and we can define the graded ring

$$B_* := \bigoplus_{\gamma \in \Gamma} B_\gamma. \tag{5.1}$$

Clearly the natural map $B_* \to B$ is injective and telescopic. Sometimes, in order to distinguish between various gradings on the same ring, we also use upper (as opposed to lower) indices: e.g. B^γ instead of B_γ, B^* instead of B_*, etc. As already done in Chapter 1, we adopt the convention that lower indices are associated to the case $\Gamma = \mathbf{Z}_+$ while upper indices will be associated to the case $\Gamma = W_+$.

In what follows we shall apply this recipe to a number of specific situations.

We start by letting $x, x', x'', ..., x^{(i)}, ...$ be $n-$tuples of variables, where $n \geq 1$. Recall that we have introduced, in Definition 3.1, the $\delta-$ring of $\delta-$polynomials

$$R\{x\} := R[x, x', x'', ...]. \tag{5.2}$$

For any r we denote by $x^{(\leq r)}$ the set of variables $x, x', ..., x^{(r)}$ and set

$$R[x^{(\leq r)}] := R[x, x', ..., x^{(r)}]. \tag{5.3}$$

Recall that we denoted by

$$\phi : R\{x\} \to R\{x\}$$

the unique ring homomorphism that extends $\phi : R \to R$ and sends

$$x^{(i)} \mapsto (x^{(i)})^p + px^{(i+1)}.$$

(Here the components of the n–tuple $(x^{(i)})^p$ are, as usual, the p–th powers of the components of the n–tuple $x^{(i)}$.) Write F^{ϕ^i} in place of $\phi^i(F)$; then, for any F, $\delta F = (F^\phi - F^p)/p$.

We will view the rings

$$R\{x\} \otimes K = K[x, x', x'', ...] = K[x, x^\phi, x^{\phi^2}, ...], \tag{5.4}$$

$$R[x^{(\leq r)}] \otimes K = K[x, x', ..., x^{(r)}] = K[x, x^\phi, ..., x^{\phi^r}] \tag{5.5}$$

as graded rings in two ways.

On way is to view the ring in Equation 5.4 as graded by $\mathbf{Z}_+$ such that the components $x_j^{\phi^i}$ of the n–tuples x^{ϕ^i} are viewed as homogeneous of degree 1:

$$deg(x_j^{\phi^i}) = 1. \tag{5.6}$$

We denote by $K[x, x^\phi, x^{\phi^2}, ...]_m$ the component of degree $m \in \mathbf{Z}_+$ in $R\{x\} \otimes K$. Then, according to our general recipe (with $B = R\{x\}$ and $B \otimes K$ graded by the condition 5.6), we may consider the graded ring

$$R\{x\}_* := \bigoplus_{m \in \mathbf{Z}_+} R\{x\}_m, \tag{5.7}$$

$$R\{x\}_m := R\{x\} \cap K[x, x^\phi, x^{\phi^2}, ...]_m.$$

Similarly one can consider a filtration of the above by

$$R[x^{(\leq r)}]_* := \bigoplus_{m \in \mathbf{Z}_+} R[x^{(\leq r)}]_m, \tag{5.8}$$

$$R[x^{(\leq r)}]_m := R[x^{(\leq r)}] \cap K[x, x^\phi, ..., x^{\phi^r}]_m.$$

Another way to view the ring in Equation 5.4 as a graded ring is as graded by the semigroup W_+ such that the components $x_j^{\phi^i}$ of x^{ϕ^i} have degree ϕ^i:

$$deg(x_j^{\phi^i}) = \phi^i. \tag{5.9}$$

We denote by $K[x, x^\phi, x^{\phi^2}, ...]^w$ the component of degree $w \in W_+$ in $R\{x\} \otimes K$. Then, according to our general recipe (with $B = R\{x\}$ and $B \otimes K$ graded by the condition 5.9), we may consider the graded ring

$$R\{x\}^* := \bigoplus_{w \in W_+} R\{x\}^w, \tag{5.10}$$

$$R\{x\}^w := R\{x\} \cap K[x, x^\phi, x^{\phi^2}, ...]^w.$$

By the way, clearly, $R\{x\}^w$ is contained in $R[x^{(\leq r)}]$, where $r = ord(w)$.

Finally we introduce one more ring which mixes features of both $R\{x\}_*$ and $R\{x\}^*$. We assume, as above, that x is a tuple of variables and y is one additional variable. Then we define the W_+–graded ring

$$R\{x_*, y\}^* := \bigoplus_{v \in W_+} R[x^{(\leq ord(v))}]_{deg(v)} \cdot y^v \subset R\{x, y\}. \tag{5.11}$$

The rings $R\{x\}^*$ and $R\{x_*, y\}^*$ defined above have natural structures of δ–graded rings in the sense of Definition 2.18. These rings (and some of their *Veronese rings* to be defined presently) will appear in some of our main results on the structure of the rings $R(\mathbf{X}, \mathbf{L})^*$. We will need 2 kinds of *Veronese rings*. Assume $A = \bigoplus_{\gamma \in \Gamma} A_\gamma$ is a ring graded by a semigroup Γ and let Γ' be a subsemigroup of Γ.

DEFINITION 5.1. The *Veronese ring* $A^{(\Gamma')}$ is the Γ–graded ring whose γ–components are defined as follows

$$(A^{(\Gamma')})_\gamma = A_\gamma \ \ or \ \ (A^{(\Gamma')})_\gamma = 0 \ \ according \ \ as \ \ \gamma \in \Gamma' \ \ or \ \ \gamma \notin \Gamma'. \tag{5.12}$$

If $\gamma_0 \in \Gamma$ and $\Gamma' = \gamma_0 \Gamma$ then we simply write $A^{(\gamma_0 \Gamma)} = A^{(\gamma_0)}$. The *Veronese* ring $A^{((\gamma_0))}$ is the Γ–graded ring whose γ–components are

$$(A^{((\gamma_0))})_\gamma = A_{\gamma_0 \gamma}. \tag{5.13}$$

The structure of the above rings is actually quite subtle. On the other hand, in low degrees, their structure is rather simple:

LEMMA 5.2. [**28**] *Let x be one indeterminate. For any $m \leq p$ the R–module $R\{x\}_m$ is free with basis*

$$\{x^w;\ w \in W_+,\ deg(w) = m\}.$$

REMARK 5.3. The above fails for $m = p + 1$, for instance, because

$$\frac{x^{p+\phi^2} - (x^\phi)^{p+1}}{p} \in R\{x\}_{p+1}.$$

PROOF. Any element $y \in R\{x\}_m$ can be written uniquely as

$$y = \sum \lambda_w x^w$$

with $\lambda_w \in K$ and $\lambda_w \neq 0$ only if $w \in W_+$ and $deg(w) = m$. We must show that $\lambda_w \in R$ for all w. Assume this is not the case. Let $-\nu < 0$ be the minimum of the p–adic valuations of the λ_w's and set $c_w := p^\nu \lambda_w$; then all the c_w's belong to R and not all their images $\bar{c}_w \in k$ in k are zero. On the other hand we have

$$\sum c_w x^w \in p \cdot R\{x\}.$$

Reducing mod p we get

$$\sum \bar{c}_w x^{w(p)} = 0 \in k[x, x', x'', ...].$$

To derive a contradiction it is enough to check that if a_i, b_i $(i = 0, 1, ...)$ are non-negative integers such that

$$\sum a_i p^i = \sum b_i p^i,$$
$$\sum a_i = \sum b_i \leq p,$$

then $a_i = b_i$ for all i; this is a trivial exercise, using the uniqueness of the expansion of integers to base p. $\square$

Using the same method one can prove the following:

LEMMA 5.4. [**28**] *Let x be an $n-$tuple of indeterminates $x_1, ..., x_n$. For any $w \in W_+$ with $deg(w) \leq p$ the $R-$module $R\{x\}^w$ is free with basis*

$$\{x_1^{w_1} ... x_n^{w_n};\ w_i \in W_+,\ w_1 + ... + w_n = w\}.$$

REMARK 5.5. The above fails for $deg(w) = p+1$, $n \geq 2$, for instance, because

$$\frac{x_i^p x_j^\phi - x_j^p x_i^\phi}{p} \in R\{x\}^{\phi+p}.$$

Note that the ring $R\{x\}$ is a global $p-$ring; in particular $R\{x\}_*$ and $R\{x\}^*$ are also global $p-$ rings because their inclusions in $R\{x\}$ are telescopic. Similarly $R\{x_*, y\}^*$ is a global $p-$ring because the map $R\{x_*, y\}^* \to R\{x\}_*$, $y \mapsto 1$, is telescopic. So we may consider the local $p-$rings $(R\{x\}_*)_{((p))}$ and $(R\{x\}^*)_{((p))}$; their residue fields

$$k\langle x\rangle_* \text{ and } k\langle x\rangle^* \tag{5.14}$$

are subfields in the residue field

$$k\langle x\rangle := R\{x\}_{(p)} \otimes k = k(x, x', x'', ...) \tag{5.15}$$

of $R\{x\}_{(p)}$. Similarly one can consider the local $p-$ring $(R\{x_*, y\}^*)_{((p))}$; this ring is, of course, isomorphic to $(R\{x\}_*)_{((p))}$. Recall that we have at our disposal a $p-$derivation

$$\delta : R\{x\}_{(p)} \to R\{x\}_{(p)}$$

defined by the formula

$$\delta\left(\frac{F}{G}\right) = \frac{\{G, F\}_\delta}{G^{\phi+p}},$$

where

$$\{G, F\}_\delta := \frac{G^p F^\phi - F^p G^\phi}{p} = G^p \delta F - F^p \delta G.$$

Note that δ sends each of the rings $(R\{x\}_*)_{((p))}$ and $(R\{x\}^*)_{((p))}$ into itself. So

$$R\{x\}_{(p)},\ (R\{x\}_*)_{((p))},\ (R\{x\}^*)_{((p))}$$

are local $\delta - p-$rings.

LEMMA 5.6. *Assume $x = \{x_0, ..., x_n\}$ and $y = \{y_1, ..., y_n\}$. There is an isomorphism*

$$R\{y\}_{(p)} \simeq R\{x\}^*_{((p))},\quad y^{(i)} \mapsto \delta^i(x_1/x_0); \tag{5.16}$$

in particular there is an induced isomorphism

$$k\langle y\rangle \simeq k\langle x\rangle^*.$$

PROOF. Clearly any $f \in R\{x\}^w$ can be written as

$$f = x_0^w \cdot F(\delta^i(x_j/x_0))$$

with F a polynomial with $R-$ coefficients. This shows immediately that the morphism in Equation 5.16 is well defined and bijective. □

5.2. δ–Galois groups

Our primary purpose in this Chapter is to show the existence of δ–bases for certain local $\delta - p$–rings arising from our graded rings introduced previously. It turns out that this problem is intimately related to computing invariants of certain profinite groups acting on certain fields; the latter depends on certain computations of δ– *Galois groups.*

DEFINITION 5.7. Let $A \subset B$ be an inclusion of local $\delta - p$–rings, with A a δ–subring of B, and let $k_A \subset k_B$ be the residue field extension. We denote by $Aut_\delta(B/A)$ the group of δ–ring A–isomorphisms of B and by $Aut(k_B/k_A)$ the group of k_A–automorphisms of k_B.

Morally the group $Aut_\delta(B/A)$ should be taken as the natural "Galois" group in our theory; the next Lemma shows that the group $Aut(k_B/k_A)$ can be interpreted as a natural extension of the former and, as we shall see later in our applications, $Aut(k_B/k_A)$ is often equal to $Aut_\delta(B/A)$.

LEMMA 5.8. *The natural homomorphism $Aut_\delta(B/A) \to Aut(k_B/k_A)$ is injective.*

PROOF. For $\sigma \in Aut_\delta(B/A)$ we denote by σ_n the induced automorphism of B/p^nB. Assume $\sigma_n = id$; we shall prove that $\sigma_{n+1} = id$. For $b \in B$ write $\sigma b = b + p^n b_1$, $b_1 \in B$. We have the following congruences in B:

$$\delta b \equiv \sigma\delta b = \delta\sigma b = \delta(b + p^n b_1) \equiv \delta b + p^{n-1}b_1^p \quad mod \quad p^n.$$

Hence $b_1 \equiv 0$ mod p and we are done. □

REMARK 5.9. The natural homomorphism $Aut(B/A) \to Aut(k_B/k_A)$ is not injective in general. E.g. take $A = R$, $B = R\{x\}_{(p)}$; then $\sigma \in Aut(B/A)$, $\sigma(x^{(i)}) = x^{(i)} + p$ maps to the identity in $Aut(k_B/k_A)$.

The following Lemma will play a role later.

LEMMA 5.10. *Assume A and B are complete and k_B/k_A is separable. Let $u \in B$ and let*

$$\bar{u}_0, \bar{u}_1, \bar{u}_2, ... \in k_B$$

be the images of

$$u, \delta u, \delta^2 u, ... \in B.$$

Assume

$$\bar{u}_0, \bar{u}_1, \bar{u}_2, ... \in k_A.$$

Then $u \in A$.

PROOF. It is enough to prove by induction on $n \geq 0$ that

$$u \in A + p^n B. \tag{5.17}$$

The case $n = 0$ follows from $\bar{u}_0 \in k_A$. Assume now

$$u = a + p^n c, \quad a \in A, \quad c \in B.$$

By Equation 3.11 we get

$$\delta^n u \equiv \delta^n a + c^{p^n} \quad mod \quad p.$$

If $\bar{c} \in k_B$ is the image of c we get

$$\bar{c}^{p^n} \in k_A.$$

Since k_B/k_A is separable we get

$$\bar{c} \in k_A$$

hence $u \in A + p^{n+1}B$ and we are done. $\square$

Assume in what follows that x is one variable. If $u \in (R\{x\}_{(p)})\hat{}$ we denote by

$$k\langle u \rangle := k(\bar{u}_0, \bar{u}_1, \bar{u}_2, ...) \tag{5.18}$$

the subfield of $k\langle x \rangle$ generated by the images

$$\bar{u}_0, \bar{u}_1, \bar{u}_2, ...$$

of the elements $u,\ \delta u,\ \delta^2 u, ...$ Recall that u is called $\delta-$independent if the images of the elements $u,\ \delta u,\ \delta^2 u, ...$ in $k\langle x \rangle$ are algebraically independent over k. In particular $u, \delta u, \delta^2 u, ...$ are algebraically independent over R and we set

$$R\{u\} := R[u, \delta u, \delta^2 u, ...] \subset R\{x\}.$$

Note that u is $\delta-$independent if and only if the $R-$algebra homomorphism

$$R\{x\} \to (R\{x\}_{(p)})\hat{},\quad x \mapsto u,\quad x' \mapsto \delta u, ...$$

extends to an endomorphism of $(R\{x\}_{(p)})\hat{}$. If this is the case then we have an induced inclusion of local $\delta - p-$rings

$$(R\{u\}_{(p)})\hat{} \subset (R\{x\}_{(p)})\hat{}$$

with residue field extension

$$k\langle u \rangle \subset k\langle x \rangle.$$

If u is $\delta-$independent then we call u^* the induced endomorphism of $(R\{x\}_{(p)})\hat{}$ and we call $\bar{u}^*$ the induced field endomorphism of $k\langle x \rangle$. Note that if u and v are $\delta-$independent we may consider the endomorphism $v^* \circ u^*$ of $(R\{x\}_{(p)})\hat{}$; the image of x via this endomorphism can be denoted by $u \circ v$ and should be interpreted as the "composition" of u and v. Clearly $u \circ v$ is still $\delta-$independent.

DEFINITION 5.11. A $\delta-$independent element u of $(R\{x\}_{(p)})\hat{}$ has $\delta-$ *slope zero* if the induced endomorphism $\bar{u}^*$ of $k\langle x \rangle$ sends $k(x, x', ..., x^{(n)})$ into itself for any $n \geq 0$. Given a $\delta-$independent element of $\delta-$slope zero its $\delta-$*degree* is defined as the sequence (d_n) where, for each $n \geq 0$, d_n is the degree of the field extension

$$k(\bar{u}_0, \bar{u}_1, ..., \bar{u}_n) \subset k(x, x', ..., x^{(n)}) \tag{5.19}$$

where $\bar{u}_i$ is the class of $\delta^i u$. Note that if u and v are $\delta-$independent of $\delta-$slope zero then so is $u \circ v$; if u has $\delta-$degree (d_n) and v has $\delta-$degree (e_n) then $u \circ v$ has $\delta-$degree $(d_n e_n)$. A $\delta-$independent element u of $\delta-$slope zero is $\delta-$*Galois* if all the extensions 5.19 are Galois; the sequence of Galois groups (G_n) of the extensions 5.19 is called then the $\delta-$ *Galois group* of u. If u is as above then, since $k(\bar{u}_{n+1}, \bar{u}_{n+2}, ...)$ and $k(x, x', ..., x^{(n)})$ are linearly disjoint over k, it follows that the extensions

$$k\langle u \rangle \subset k(x, x', ..., x^{(n)}, \bar{u}_{n+1}, \bar{u}_{n+2}, ...)$$

are (finite) Galois with Galois group G_n. Therefore the extension

$$k\langle u \rangle \subset k\langle x \rangle$$

is (algebraic) Galois with Galois group

$$G := \lim_{\leftarrow} G_n.$$

The elements of $k\langle u\rangle$ will be referred to as δ–*invariants* of G. (In applications they are sometimes related to our previously considered δ–invariants of correspondences.) Assume u is δ–Galois. We say u is *purely* δ–*Galois* if the injective homomorphism

$$Aut_\delta((R\{x\}_{(p)})\hat{}/(R\{u\}_{(p)})\hat{}) \to Aut(k\langle x\rangle/k\langle u\rangle)$$

is an isomorphism.

In most applications our u above will be an element of $R\{x\}_{(p)}$ rather than $(R\{x\}_{(p)})\hat{}$.

5.3. δ–invariants for subgroups of $PGL_2(\mathbf{Z}_p)$

PROPOSITION 5.12. *For any matrix*

$$\begin{pmatrix} a & b \\ c & d \end{pmatrix} \in GL_2(\mathbf{Z}_p) \tag{5.20}$$

the element

$$\frac{ax+b}{cx+d} \in R\{x\}_{(p)}$$

is δ–*independent of* δ–*slope zero and* δ–*degree* (d_n) *where* $d_n = 1$ *for all* $n \geq 0$.

PROOF. An easy exercise. □

For any ring S write

$$PGL_2(S) = GL_2(S)/S^\times \cdot id.$$

Let us remark that Proposition 5.12 gives rise to an action of $PGL_2(\mathbf{Z}_p)$ on the ring $R\{x\}_{(p)}$ hence on the field $k\langle x\rangle$. The purpose of this Section is to examine this action and the induced action of some remarkable subgroups of $PGL_2(\mathbf{Z}_p)$. We need some preliminaries.

LEMMA 5.13. *The group* $PGL_2(\mathbf{Z}/p^{n+1}\mathbf{Z})$ *acts faithfully on* $k(x, x', ..., x^{(n)})$.

PROOF. Assume matrix 5.20 acts as the identity on $k(x, x', ..., x^{(n)})$; we want to show that this matrix is congruent mod p^{n+1} to a scalar matrix. Since

$$\delta^i\left(\frac{ax+b}{cx+d}\right) \equiv \delta^i t \quad mod \quad p$$

in $R\{x\}_{(p)}$ for all $i \leq n$ it is easy to see that

$$\frac{ax+b}{cx+d} \equiv t \quad mod \quad p^{n+1}$$

in $R[x]_{(p)}$ which implies that p^{n+1} divides $a-d$, b, and c and our Lemma is proved. □

In order to study the PGL_2–action above it is convenient to start by studying the induced action of a maximal torus. We will then study the action of the full projective linear group, of its Borel subgroups and their unipotent radical, and of the special linear subgroup.

5.3.1. Maximal torus. For any ring S the natural homomorphism

$$S^{\times} = \mathbf{G}_m(S) \to PGL_2(S)$$

sending $a \in S^{\times}$ into the class of

$$\begin{bmatrix} a & 0 \\ 0 & 1 \end{bmatrix}$$

is injective so we can and will view $\mathbf{Z}_p^{\times}$ as a subgroup of $PGL_2(\mathbf{Z}_p)$ and we have an induced $\mathbf{Z}_p^{\times}-$action on $R\{x\}$ defined by

$$(\lambda, x^{(i)}) \mapsto \delta^i(\lambda x),$$

$\lambda \in \mathbf{Z}_p^{\times}$. We have an induced $\mathbf{Z}_p-$action on $k\langle x\rangle$. Let us denote, as usual, by $k\langle x\rangle^{\mathbf{Z}_p^{\times}}$ the field of elements of $k\langle x\rangle$ fixed by this action. Our main result on this $\mathbf{Z}_p^{\times}-$action is the following:

PROPOSITION 5.14. [**28**] *The element*

$$x^{\phi-1} = x^{\phi}/x \in R\{x\}_{(p)}$$

is $\delta-$independent of $\delta-$slope zero. It is purely $\delta-$Galois with $\delta-$Galois group

$$((\mathbf{Z}/p^{n+1}\mathbf{Z})^{\times}),$$

hence it has $\delta-$degree $((p-1)p^n)$. In particular the extension

$$k\langle x^{\phi}/x\rangle \subset k\langle x\rangle$$

is algebraic Galois with Galois group $\mathbf{Z}_p^{\times}$. So

$$k\langle x^{\phi}/x\rangle = k\langle x\rangle_* = k\langle x\rangle^{\mathbf{Z}_p^{\times}}.$$

The proof will be based on a series of Lemmas.

LEMMA 5.15. *The kernel of the homomorphism*

$$\rho_n : \mathbf{Z}_p^{\times} \to Aut(k(x, x', ..., x^{(n-1)})/k)$$

equals $1 + p^n\mathbf{Z}_p$. In particular each of the extensions

$$k(x, x', ..., x^{(n-1)})^{\mathbf{Z}_p^{\times}} \subset k(x, x', ..., x^{(n-1)})$$

is a Galois extension with Galois group $\mathbf{Z}_p^{\times}/(1 + p^n\mathbf{Z}_p) = (\mathbf{Z}/p^n\mathbf{Z})^{\times}$.

PROOF. This follows directly from Lemma 5.13. □

In what follows set

$$y_n := \delta^n(x^{\phi}/x) \in (R\{x\}_*)_{((p))}$$

and denote by $\eta_n \in k\langle x\rangle_*$ the image of y_n. Note that $\eta_0 = x^{p-1}$. Also set $\omega := \phi + p \in W$ and set

$$x^{\omega^n} \in R\{x\}_{(p+1)^n},$$

where ω^m is the $m-$th power of ω in the ring W. Define a sequence of elements $F_n \in R\{x\}$, $n \geq 0$, by letting $F_0 := x^{\phi} \in R\{x\}_1$ and letting

$$F_n := \{x^{\omega^{n-1}}, F_{n-1}\}_{\delta} \in R\{x\}_{(p+1)^n}$$

for $n \geq 1$. Then, of course, $y_n = F_n/x^{\omega^n}$. A direct computation needed later shows that

$$F_1 = \{x, x^{\phi}\}_{\delta} = x^p(x')^p + px^px'' - x'(x^p + px')^p. \tag{5.21}$$

LEMMA 5.16. *For any* $n \geq 1$ *we have*

$$x^{\phi^n} - p^n x^{(n)} \in R[x, x', ..., x^{(n-1)}].$$

Moreover

$$\delta(x^{\omega^n}) = x^{\phi\omega^n - \phi^{n+1}} p^n x^{(n+1)} + C_n,$$

where $C_n \in R[x, x', ..., x^{(n)}]$.

PROOF. The first assertion follows by induction. The second assertion follows directly from the first. □

Let us define a sequence of elements $w_n \in W_+$ by letting $w_1 = p$ and

$$w_n := \phi w_{n-1} + p\omega^{n-1}, \quad n \geq 2.$$

Note that $ord(w_n) \leq n - 1$ and

$$w_n(p) = (2^n - 1)p^n.$$

LEMMA 5.17. *For any* $n \geq 1$ *we have*

$$F_n = p \cdot x^{w_n} x^{(n+1)} + A_n + pB_n \tag{5.22}$$

with $A_n, B_n \in R[x, x', ..., x^{(n)}]$ *such that* A_n *has leading term*

$$\alpha_n \cdot (x^{(n)})^p,$$

with $\alpha_n \in R[x, x', ..., x^{(n-1)}]$ *and*

$$\alpha_n \equiv x^{w_n(p)} \mod p.$$

Here and later the *leading term* of a polynomial

$$\sum_{j=0}^{d} F_j(x, x', ..., x^{(n-1)})(x^{(n)})^j \in R[x, x', ..., x^{(n)}]$$

is the polynomial

$$F_d(x, x', ..., x^{(n-1)})(x^{(n)})^d$$

provided the latter is non-zero.

PROOF. We proceed by induction on n. For $n = 1$ we conclude by Equation 5.21. For the induction step assume Equation 5.22 is true for some $n \geq 1$ and note that

$$\delta F_n = \frac{1}{p}(F_n^\phi - F_n^p)$$

$$= \frac{1}{p}\{px^{\phi w_n}[(x^{(n+1)})^p + px^{(n+2)}] + A_n^\phi + pB_n^p + p^2\delta B_n - (px^{w_n}x^{(n+1)} + A_n + pB_n)^p\}$$

$$= p \cdot x^{\phi w_n} x^{(n+2)} + \tilde{A}_{n+1} + p\tilde{B}_{n+1},$$

where

$$\tilde{A}_{n+1} := x^{\phi w_n}(x^{(n+1)})^p + B_n^p + \frac{A_n^\phi - A_n^p}{p},$$

and $\tilde{B}_{n+1} \in R[x, ..., x^{(n+1)}]$. Note that the leading term of $\tilde{A}_{n+1}$ equals

$$(x^{\phi w_n} + \alpha_n^\phi p^{p-1}) \cdot (x^{(n+1)})^p.$$

Then

$$F_{n+1} = x^{p\omega^n} \cdot \delta F_n - F_n^p \cdot \delta(x^{\omega^n}) = p \cdot x^{\phi w_n + p\omega^n} x^{(n+2)} + A_{n+1} + pB_{n+1},$$

where, if C_n is as in Lemma 5.16, then

$$A_{n+1} = x^{p\omega^n}\tilde{A}_{n+1} - F_n^p C_n,$$

$$B_{n+1} = x^{p\omega^n}\tilde{B}_{n+1} - p^{n-1}F_n^p x^{\phi\omega^n - \phi^{n+1}} x^{(n+1)}.$$

Note that the leading term of A_{n+1} equals

$$(x^{p\omega^n + \phi w_n} + x^{p\omega^n}\alpha_n^\phi p^{p-1} - p^p x^{pw_n} C_n) \cdot (x^{(n+1)})^p$$

which ends our proof. □

Lemma 5.17 implies in particular that

$$\frac{\partial F_n}{\partial x^{(n+1)}} = p \cdot x^{w_n} \tag{5.23}$$

in $R\{x\}$. Also, formula 5.21 implies that

$$\frac{\partial F_1}{\partial x'} \equiv -x^{p^2} \quad mod \quad (p) \tag{5.24}$$

in $R\{x\}$. Let us define a sequence of integers $u_n \in \mathbf{Z}_+$ by $u_1 = p^2$,

$$u_{n+1} = pu_n + p\omega^n(p), \quad n \geq 1;$$

clearly,

$$u_n = (2^n + p - 2)p^n.$$

LEMMA 5.18. *For all $n \geq 1$ we have the following congruence in $R\{x\}$:*

$$\frac{\partial F_n}{\partial x^{(n)}} \equiv -x^{u_n} \quad mod \quad (p).$$

PROOF. We proceed by induction on n. The case $n = 1$ follows from Equation 5.24. Assume the congruence in our Lemma is true for some $n \geq 1$. Using the inductive definition of F_{n+1} we have

$$\frac{\partial F_{n+1}}{\partial x^{(n+1)}} = x^{p\omega^n}\frac{\partial \delta F_n}{\partial x^{(n+1)}} - \delta(x^{\omega^n}) \cdot p \cdot F_n^{p-1} \cdot \frac{\partial F_n}{\partial x^{(n+1)}} - F_n^p \cdot \frac{\partial(\delta(x^{\omega^n}))}{\partial x^{(n+1)}}. \tag{5.25}$$

By Lemma 5.16 the third term in the Equation 5.25 is $\equiv 0$ mod (p) in $R\{x\}$; the same is, of course, true for the second term in Equation 5.25. On the other hand the first term in the right hand side of the Equation 5.25 equals

$$x^{p\omega^n}\frac{\partial}{\partial x^{(n+1)}}\{\frac{1}{p}[F_n(x^p + px', ..., (x^{(n)})^p + px^{(n+1)}, (x^{(n+1)})^p + px^{(n+2)}) - F_n^p]\}$$

$$= \frac{1}{p}x^{p\omega^n}\left[\left(\frac{\partial F_n}{\partial x^{(n+1)}}\right)^\phi \cdot p \cdot (x^{(n+1)})^{p-1} + \left(\frac{\partial F_n}{\partial x^{(n)}}\right)^\phi \cdot p - p \cdot F_n^{p-1} \cdot \frac{\partial F_n}{\partial x^{(n+1)}}\right].$$

Using Equation 5.23 and the induction hypothesis the latter expression is congruent mod (p), in $R\{x\}$, to $-x^{u_{n+1}}$ which ends our proof. □

LEMMA 5.19. *For all $n \geq 1$ we have:*

$$\eta_n = x^{-p^n}(x^{(n)})^p - x^{p^{n+1}-2p^n}x^{(n)} + \gamma_n,$$

with $\gamma_n \in k[x, x^{-1}, x', ..., x^{(n-1)}]$.

PROOF. Combine Lemmas 5.17 and 5.18. □

PROOF. (of Proposition 5.14) Consider the extensions

$$k(\eta_0, \eta_1, ..., \eta_{n-1}) \overset{\alpha}{\subset} k(x, x', ..., x^{(n-1)})^{\mathbf{Z}_p^\times} \overset{\beta}{\subset} k(x, x', ..., x^{(n-1)}).$$

By Lemma 5.15 the degree of β equals $(p-1)p^{n-1}$. By Lemma 5.19 the extension $\beta \circ \alpha$ has degree at most $(p-1)p^{n-1}$. This forces α to be an equality. In particular $x^\phi/x \in R\{x\}_{(p)}$ is δ–independent of δ–slope zero and δ–Galois with δ–Galois group $((\mathbf{Z}/p^{n+1}\mathbf{Z})^\times)$. So the inclusion

$$k(\eta_0, \eta_1, \eta_2, ...) \subset k\langle x\rangle^{\mathbf{Z}_p^\times}$$

is an equality. But $k\langle x\rangle_*$ is an intermediate field in the latter inclusion. This forces the 3 fields in the statement of the Proposition to be equal. We are left to show that x^ϕ/x is purely δ–Galois (and not merely δ–Galois). So we need to show that the inclusion

$$Aut_\delta((R\{x\}_{(p)})\hat{}/(R\{x^\phi/x\}_{(p)})\hat{}) \subset Aut(k\langle x\rangle/k\langle x^\phi/x\rangle)$$

is an equality. This follows from the fact that $Aut(k\langle x\rangle/k\langle x^\phi/x\rangle)$ (which equals $\mathbf{Z}_p^\times$) acts on $R\{x\}_{(p)}$ by $R\{x^\phi/x\}_{(p)}$–automorphisms. $\square$

The following variant of Proposition 5.14 keeps track of denominators:

PROPOSITION 5.20. *The ring*

$$k[x, x^{-1}, x', ..., x^{(n-1)}]$$

is a free module over

$$k[\eta_0, \eta_0^{-1}, \eta_1, ..., \eta_{n-1}]$$

with basis

$$\{x^{i_0}(x')^{i_1}...(x^{(n-1)})^{i_{n-1}} | 0 \leq i_0 \leq p-2,\ 0 \leq i_1, ..., i_{n-1} \leq p-1\}. \tag{5.26}$$

Moreover we have an equality

$$k[x, x^{-1}, x', ..., x^{(n-1)}]^{\mathbf{Z}_p^\times} = k[\eta_0, \eta_0^{-1}, \eta_1, ..., \eta_{n-1}].$$

PROOF. The fact that the set in Equation 5.26 is a system of generators follows by induction from Lemma 5.19. The linear independence of the set in Equation 5.26 follows from the proof of Proposition 5.14. This proves the first assertion of the Proposition. By this first assertion the ring extension

$$k[\eta_0, \eta_0^{-1}, \eta_1, ..., \eta_{n-1}] \subset k[x, x^{-1}, x', ..., x^{(n-1)}]$$

is faithfully flat hence

$$k(\eta_0, \eta_0^{-1}, \eta_1, ..., \eta_{n-1}) \cap k[x, x^{-1}, x', ..., x^{(n-1)}] = k[\eta_0, \eta_0^{-1}, \eta_1, ..., \eta_{n-1}].$$

On the other hand, by the proof of Proposition 5.14, we have

$$k[x, x^{-1}, x', ..., x^{(n-1)}]^{\mathbf{Z}_p^\times} = k(\eta_0, \eta_0^{-1}, \eta_1, ..., \eta_{n-1}) \cap k[x, x^{-1}, x', ..., x^{(n-1)}]$$

which concludes the proof of the second assertion. $\square$

COROLLARY 5.21. *We have an equality*

$$(R[x, x^{-1}, x', x'', ...]\hat{})^{\mathbf{Z}_p^\times} = R\left[\frac{x^\phi}{x}, \delta\left(\frac{x^\phi}{x}\right), \delta^2\left(\frac{x^\phi}{x}\right), ...\right]^{cl}$$

where cl means p–adic closure in the ring

$$R[x, x^{-1}, x', x'', ...]\hat{}.$$

In particular

$$R\{x\}_* \subset R\left[\frac{x^\phi}{x}, \delta\left(\frac{x^\phi}{x}\right), \delta^2\left(\frac{x^\phi}{x}\right), \ldots\right]^{cl}[x].$$

PROOF. The first assertion follows directly from Proposition 5.20. The second assertion follows from the fact that for any $f \in R\{x\}_m$, the fraction f/x^m is $\mathbf{Z}_p^\times$-invariant. □

Proposition 5.14 can be slightly generalized as follows. Let d be a divisor of $p-1$ and consider the subgroups of d-th powers in $\mathbf{Z}_p^\times$:

$$\mathbf{Z}_p^{\times d} = \{a^d \mid a \in \mathbf{Z}_p^\times\} = \{\lambda \in \mathbf{Z}_p^\times \mid \lambda^{\frac{p-1}{d}} \equiv 1 \ \ mod \ \ p\}.$$

One can make a similar definition with $\mathbf{Z}/p^{n+1}\mathbf{Z}$ in place of $\mathbf{Z}_p$.

PROPOSITION 5.22. *The element*

$$v := x^{\frac{\phi-1}{d}} \in (R\{x\}_{(p)})\hat{}$$

is δ-independent of δ-slope zero. It is purely δ-Galois with δ-Galois group

$$((\mathbf{Z}/p^{n+1}\mathbf{Z})^{\times d}).$$

In particular the extension

$$k\langle v\rangle \subset k\langle x\rangle$$

is algebraic Galois with Galois group $\mathbf{Z}_p^{\times d}$.

PROOF. Note that $v^d = u$ where $u = x^{\phi-1} = x^\phi/x$. Let $\bar{u}_i, \bar{v}_i \in k\langle x\rangle$ be the images of $\delta^i u$ and $\delta^i v$ respectively. It is trivial to check that

$$\bar{v}_i \in k(x, x', \ldots, x^{(n)})^{(\mathbf{Z}/p^{n+1}\mathbf{Z})^{\times d}}.$$

So, by Proposition 5.14, we have a commutative diagram

$$\begin{array}{ccc} k(\bar{u}_0, \bar{u}_1, \ldots, \bar{u}_n) & = & k(x, x', \ldots, x^{(n)})^{(\mathbf{Z}/p^{n+1}\mathbf{Z})^\times} \\ \cap & & \cap \\ k(\bar{v}_0, \bar{v}_1, \ldots, \bar{v}_n) & \subset & k(x, x', \ldots, x^{(n)})^{(\mathbf{Z}/p^{n+1}\mathbf{Z})^{\times d}} \end{array}$$

We want to show that the bottom horizontal inclusion is an equality. Assume it is not. By Galois theory there exists an element

$$\lambda \in (\mathbf{Z}/p^{n+1}\mathbf{Z})^\times \backslash (\mathbf{Z}/p^{n+1}\mathbf{Z})^{\times d}$$

which fixes all elements of $k(\bar{v}_0, \bar{v}_1, \ldots, \bar{v}_n)$. But

$$\lambda^* v = (\lambda \cdot x)^{\frac{\phi-1}{d}} = \lambda^{\frac{\phi-1}{d}} \cdot v \equiv \lambda^{\frac{p-1}{d}} \cdot v \ \ mod \ \ p.$$

It follows that

$$\lambda^{\frac{p-1}{d}} \equiv 1 \quad mod \quad p,$$

a contradiction. □

PROPOSITION 5.23. [**28**] *Let $x = \{x_0, x_1\}$ be two indeterminates and t another indeterminate. Then the homomorphism $R\{x\}^* \to R\{t\}_*$ defined by $x_0 \mapsto t$, $x_1 \mapsto t^\phi$ is telescopic and the induced field extension*

$$k\langle x\rangle^* \to k\langle t\rangle_*$$

is an isomorphism.

PROOF. Any element in $f \in R\{x\}^w$ can be written as

$$x_0^w \cdot F(x_1/x_0, \delta(x_1/x_0), \delta^2(x_1/x_0), ...) \tag{5.27}$$

in the ring $R\{x_0, x_1\}[x_0^{-1}, (x_0^\phi)^{-1}, ...]$, where F is a polynomial with coefficients in R. To prove the telescopic property assume $f \notin p \cdot R\{x\}^*$ and let us check that the image of f in $R\{t\}_*$ is not in $pR\{t\}_*$. Note that not all coefficients of F are divisible by p; indeed this follows from the injectivity of the map

$$R\{x\}^w \otimes k \to R\{x\} \otimes k \to R\{x\}[x_0^{-1}, (x_0^\phi)^{-1}, ...] \otimes k.$$

But the image of the element in Equation 5.27 via the map

$$R\{x\}[x_0^{-1}, (x_0^\phi)^{-1}, ...] \to k\langle t\rangle$$

equals

$$t^{w(p)} \cdot F(\eta_0, \eta_1, \eta_2, ...)$$

and the latter is non-zero due to the algebraic independence of $\eta_0, \eta_1, \eta_2, ...$; cf. Proposition 5.14. In particular the image of f via the map

$$R\{x\}^w \to R\{t\}_{deg(w)} \otimes k$$

is non-zero which ends the proof that $R\{x\}^* \to R\{t\}_*$ is telescopic. Now note that, due to the representation in Equation 5.27, the image of the extension $k\langle x\rangle^* \to k\langle t\rangle_*$ equals the field

$$k(\eta_0, \eta_1, \eta_2, ...);$$

this field coincides with the field $k\langle x\rangle_*$ by Proposition 5.14 again and we are done. □

5.3.2. Projective linear group. In order to prove our main result on $\delta-$invariants of the full PGL_2 we will need to combine Proposition 5.14 with Propositions 5.24 and 5.25 below.

PROPOSITION 5.24. *Consider the element*

$$\Lambda = \sum_{i=0}^{r} \lambda_i x^{\phi^i} \in R\{x\}$$

where $\lambda_i \in R$, $r \in \{1, 2\}$, *and* $\lambda_r = 1$. *Then* Λ *is* $\delta-$*independent of* $\delta-$*slope zero and* $\delta-$*degree* (e_n), $e_n \leq p^{r(n+1)}$.

PROOF. We shall prove the statement for $r = 2$; the proof for $r = 1$ is similar (and easier). So we assume that

$$\Lambda(x) = \phi^2(x) + \lambda_1\phi(x) + \lambda_0 x,$$

where $\lambda_1, \lambda_0 \in R$. Our Proposition will follow if we prove the following:

Claim. For any $i \geq 0$ we have

$$\delta^i(\Lambda(x)) = p^2 x^{(i+2)} + pA_i + B_i,$$

where $A_i, B_i \in R\{x\}$ and

1) the leading term of A_i belongs to $(1 + pR) \cdot (x^{(i+1)})^p$,

2) the leading term of B_i belongs to $(1 + pR) \cdot (x^{(i)})^{p^2}$.

We prove the Claim by induction on i. For $i = 0$ note that

$$\Lambda(x) = (x^p + px')^p + p((x')^p + px'') + \lambda_1(x^p + px') + \lambda_0 x,$$

which clearly has the desired form. Assume the Claim holds for some i. Then

$$\delta^{i+1}(\Lambda(x)) = \frac{1}{p}\{p^2[(x^{(i+2)})^p + px^{(i+3)}] + p\phi(A_i) + \phi(B_i) - (p^2x^{(i+2)} + pA_i + B_i)^p\}$$
$$= p^2x^{(i+3)} + pA_{i+1} + B_{i+1},$$

where

$$A_{i+1} = (x^{(i+2)})^p + \frac{A_i^{(\phi)}(x^p + px', ..., (x^{(i+1)})^p + px^{(i+2)}) - A_i^{(\phi)}(x^p, ..., (x^{(i+1)})^p)}{p}$$
$$- \frac{(p^2x^{(i+2)} + pA_i + B_i)^p - (pA_i + B_i)^p}{p^2},$$
$$B_{i+1} = A_i^{(\phi)}(x^p, ..., (x^{(i+1)})^p) + \frac{B^{(\phi)}(x^p + px', ..., (x^{(i)})^p + px^{(i+1)}) - B_i^p}{p}$$
$$- \frac{(pA_i + B_i)^p - B_i^p}{p}.$$

If A_i has leading term $\alpha\cdot(x^{(i+1)})^p$ and B_i has leading term $\beta\cdot(x^{(i)})^{p^2}$ $(\alpha, \beta \in 1+pR)$ then it is trivial to check that A_{i+1} has leading term

$$(1 + \alpha^\phi p^{p-1} - p^{2p-2})\cdot(x^{(i+2)})^p$$

and B_{i+1} has leading term

$$(\alpha^\phi + \beta^\phi p^{p^2-1} - \alpha^p p^{p-1})\cdot(x^{(i+1)})^{p^2},$$

which concludes our proof. □

PROPOSITION 5.25. *The element $x(1-x^\phi) \in R\{x\}_{(p)}$ is $\delta-$independent of $\delta-$slope zero and $\delta-$degree (d_n) where $d_n \leq (p+1)p^n$ for all $n \geq 0$.*

From the proof of Proposition 5.26 below it will follow that actually $d_n = (p+1)p^n$ for all n.

PROOF. First one notes by induction that, if $y, y', y'', ...$ are new variables then, for any $n \geq 1$ we have the following equality in the ring $R[x^{(\leq n)}, y^{(\leq n)}]$:

$$\delta^n(xy) = x^{p^n}y^{(n)} + y^{p^n}x^{(n)} + pA_n + B_n, \tag{5.28}$$
$$A_n \in R[x^{(\leq n)}, y^{(\leq n)}], \quad B_n \in R[x^{(\leq n-1)}, y^{(\leq n-1)}].$$

On the other hand one checks by induction that

$$\delta^n(x-1) = x^{(n)} + C_n, \quad C_n \in R[x^{(\leq n-1)}].$$

Since δ and ϕ commute, we have

$$\delta^n(1-x^\phi) = (\delta^n(1-x))^\phi \equiv (\delta^n(1-x))^p \equiv -(x^{(n)})^p - C_n^p \ \ mod\ \ p.$$

Substituting $1-x^\phi$ for y in equation 5.28 we get, for any $n \geq 1$:

$$\delta^n(x(1-x^\phi)) \equiv x^{p^n}(-(x^{(n)})^p - C_n^p) + (1 - x^{p^{n+1}})x^{(n)}$$
$$+B_n(x, ..., x^{(n-1)}, -x^p + 1, ..., -(x^{(n-1)})^p - C_{n-1}^p) \ \ mod\ \ p.$$

This shows that if $n \geq 1$, $u = x(1-x^\phi)$, and $u_n \in k\langle x\rangle$ is the class of $\delta^n u$, then

$$[k(x, x', ..., x^{(n-1)}, x^{(n)}) : k(x, x', ..., x^{(n-1)}, u_n)] \leq p.$$

On the other hand

$$[k(x) : k(u)] = [k(x) : k(x(1-x^p))] = p+1.$$

Hence

$$[k(x, x', ..., x^{(n)}) : k(u, u_1, ..., u_n)] \leq (p+1)p^n;$$

we also get, of course, that u is δ–independent of δ–slope zero. □

We are ready to prove our main result about our $PGL_2(\mathbf{Z}_p)$–action.

PROPOSITION 5.26. *The element*

$$u := \frac{(x^{\phi^2} - x)^{\phi+1}}{(x^{\phi} - x)^{\phi^2+1}} = \frac{(x^{\phi^3} - x^{\phi})(x^{\phi^2} - x)}{(x^{\phi^3} - x^{\phi^2})(x^{\phi} - x)} \in R\{x\}_{(p)}$$

is δ–independent of δ–slope zero. It is purely δ–Galois with δ–Galois group

$$(PGL_2(\mathbf{Z}/p^{n+1}\mathbf{Z}))$$

and δ–degree $((p^2-1)p^{3n+1})$. In particular the extension

$$k\langle u\rangle \subset k\langle x\rangle$$

is algebraic Galois with Galois group $PGL_2(\mathbf{Z}_p)$. Moreover the extension

$$k(x) \cdot k\langle u\rangle \subset k\langle x\rangle$$

is algebraic Galois with Galois group

$$Ker(PGL_2(\mathbf{Z}_p) \rightarrow PGL_2(\mathbf{Z}/p\mathbf{Z})).$$

PROOF. Consider the following elements of $R\{x\}_{(p)}$:

$$f_1 := x^{\phi} - x, \quad f_2 := x^{\phi}/x, \quad f_3 := x+1,$$

$$f_4 := 1/x, \quad f_5 := x(1-x^{\phi}), \quad f_6 = f_4 = 1/x.$$

It is trivial to check that

$$u = f_6 \circ f_5 \circ f_4 \circ f_3 \circ f_2 \circ f_1.$$

By Propositions 5.14, 5.24, 5.25, 5.12 it follows that u is δ–independent of δ–slope zero and δ–degree (ϵ_n) where $\epsilon_n \leq (p^2-1)p^{3n+1}$. On the other hand, if $\bar{u}_i \in k\langle x\rangle$ is the image of $\delta^i u$, we have extensions

$$k(\bar{u}_0, \bar{u}_1, ..., \bar{u}_n) \overset{\alpha}{\subset} k(x, x', ..., x^{(n)})^{PGL_2(\mathbf{Z}/p^{n+1}\mathbf{Z})} \overset{\beta}{\subset} k(x, x', ..., x^{(n)}). \tag{5.29}$$

We know that $\beta \circ \alpha$ has degree at most

$$(p^2-1)p^{3n+1};$$

the latter number coincides with the order of the group

$$PGL_2(\mathbf{Z}/p^{n+1}\mathbf{Z}); \tag{5.30}$$

cf. [**122**], p. 22. By Lemma 5.13 the inclusion α in Equation 5.29 is an equality and the inclusion β is Galois with Galois group $PGL_2(\mathbf{Z}/p^{n+1}\mathbf{Z})$. Finally, let Γ_n be the kernel of the projection $PGL_2(\mathbf{Z}/p^{n+1}\mathbf{Z}) \rightarrow PGL_2(\mathbf{Z}/p\mathbf{Z})$ and consider the extensions

$$k(\bar{u}_0, \bar{u}_1, ..., \bar{u}_n) \subset k(x, \bar{u}_1, ..., \bar{u}_n) \subset k(x, x', ..., x^{(n)})^{\Gamma_n} \subset k(x, x', ..., x^{(n)}). \tag{5.31}$$

Since the extension $k(\bar{u}_0) \subset k(\bar{u}_0, \bar{u}_1, ..., \bar{u}_n)$ is purely transcendental and $k(\bar{u}_0) \subset k(x)$ has degree $(p^2-1)p$ it follows that the first extension in Equation 5.31 has degree $(p^2-1)p$. On the other hand the last extension in Equation 5.31 has degree p^{3n}. It follows that the middle extension in Equation 5.31 is an equality. So the extension

$$k(x, \bar{u}_0, \bar{u}_1, ...) = k(x) \cdot k\langle u\rangle \subset k\langle x\rangle$$

is Galois with Galois group the inverse limit of the groups Γ_n. We still have to show that u is purely $\delta-$Galois (and not merely $\delta-$Galois). This follows from the fact that $PGL_2(\mathbf{Z}_p)$ acts on $R\{x\}_{(p)}$ by $R\{u\}_{(p)}-$automorphisms. $\square$

5.3.3. Maximal unipotent subgroup. Note that for any ring S the natural homomorphism

$$S = \mathbf{G}_a(S) \to PGL_2(S),$$

sending $a \in S$ into the class of

$$\begin{bmatrix} 1 & a \\ 0 & 1 \end{bmatrix},$$

is injective so, in particular, $\mathbf{Z}_p$ acts on $R\{x\}_{(p)}$ by the rule

$$(\lambda, x^{(i)}) \mapsto \delta^i(x+\lambda).$$

Here is our main result about this action.

PROPOSITION 5.27. *The element*

$$\Lambda := x^{\phi} - x \in R\{x\}_{(p)}$$

is $\delta-$independent of $\delta-$slope zero. It is purely $\delta-$Galois with $\delta-$Galois group

$$(\mathbf{Z}/p^{n+1}\mathbf{Z})$$

and $\delta-$degree (p^{n+1}). In particular the extension

$$k\langle\Lambda\rangle \subset k\langle x\rangle$$

is algebraic Galois with Galois group $\mathbf{Z}_p$.

PROOF. If $\bar{\Lambda}_i \in k\langle x\rangle$ is the image of $\delta^i\Lambda$, we have extensions

$$k(\bar{\Lambda}_0, \bar{\Lambda}_1, ..., \bar{\Lambda}_n) \overset{\alpha}{\subset} k(x, x', ..., x^{(n)})^{\mathbf{Z}/p^{n+1}\mathbf{Z}} \overset{\beta}{\subset} k(x, x', ..., x^{(n)}). \tag{5.32}$$

By Proposition 5.24 $\beta \circ \alpha$ has degree at most p^{n+1}. By Lemma 5.13 the inclusion β in Equation 5.32 has degree p^{n+1} so α is an equality and we are done. $\square$

5.3.4. Special linear subgroup. For any ring S we write

$$PSL_2(S) := SL_2(S)/\mu_2(S) \cdot id,$$

where $\mu_2(S) := \{a \in S | a^2 = 1\}$. Then $PSL_2(S)$ is naturally a subgroup of $PGL_2(S)$. If S is $\mathbf{Z}_p$ or $\mathbf{Z}/p^{n+1}\mathbf{Z}$ then $PSL_2(S)$ has index 2 in $PGL_2(S)$; cf. [**122**], p. 22. Our main result about the $PSL_2(\mathbf{Z}_p)-$action on $k\langle x\rangle$ is the following:

PROPOSITION 5.28. *The element*

$$v := \frac{(x^{\phi^2} - x)^{\frac{\phi+1}{2}}}{(x^{\phi} - x)^{\frac{\phi^2+1}{2}}} \in (R\{x\}_{(p)})\hat{}$$

is purely $\delta-$Galois with $\delta-$Galois group

$$(PSL_2(\mathbf{Z}/p^{n+1}\mathbf{Z})).$$

In particular the extension

$$k\langle v\rangle \subset k\langle x\rangle$$

is algebraic Galois with Galois group $PSL_2(\mathbf{Z}_p)$.

PROOF. The argument is similar to the one in the proof of Proposition 5.22. Note that $v^2 = u$ where u is as in Proposition 5.26. Let $\bar{u}_i, \bar{v}_i \in k\langle x\rangle$ be the images of $\delta^i u$ and $\delta^i v$ respectively. It is trivial to check that

$$\bar{v}_i \in k(x, x', ..., x^{(n)})^{PSL_2(\mathbf{Z}/p^{n+1}\mathbf{Z})}.$$

So, by Proposition 5.26, we have a commutative diagram

$$\begin{array}{ccc} k(\bar{u}_0, \bar{u}_1, ..., \bar{u}_n) & = & k(x, x', ..., x^{(n)})^{PGL_2(\mathbf{Z}_p)} \\ \cap & & \cap \\ k(\bar{v}_0, \bar{v}_1, ..., \bar{v}_n) & \subset & k(x, x', ..., x^{(n)})^{PSL_2(\mathbf{Z}_p)} \end{array}$$

We want to show that the bottom horizontal inclusion is an equality. Since the right vertical inclusion is an extension of degree 2 it is enough to show that the left vertical inclusion is not an equality. Let $\epsilon \in \mathbf{Z}$ be such that

$$\left(\frac{\epsilon}{p}\right) = -1$$

and let $\gamma \in PGL_2(\mathbf{Z}/p^{n+1}\mathbf{Z})$ be the class of the matrix

$$\left[\begin{array}{cc} \epsilon & 0 \\ 0 & 1 \end{array}\right] \in GL_2(\mathbf{Z}).$$

Then we have

$$\gamma v = \epsilon^{\frac{\phi-\phi^2}{2}} \cdot v = \left(\frac{\epsilon}{p}\right) \cdot v = -v.$$

Hence

$$\bar{v}_0 \notin k(x, x', ..., x^{(n)})^{PGL_2(\mathbf{Z}_p)}$$

and we are done. □

5.3.5. Borel subgroup. For any ring S define the "affine group"

$$GA(S) := \{\left[\begin{array}{cc} a & b \\ 0 & 1 \end{array}\right]; a \in S^{\times},\ b \in S\} \subset GL_2(S).$$

The natural homomorphism

$$GA(S) \to PGL_2(S)$$

is clearly injective; so we may and will view $GA(S)$ as a subgroup of $PGL_2(S)$. In particular $GA(\mathbf{Z}_p)$ acts on $k\langle x\rangle$. Our main result about this actions is the following:

PROPOSITION 5.29. *The element*

$$y := \frac{x^{\phi^2} - x^{\phi}}{x^{\phi} - x} = (x^{\phi} - x)^{\phi - 1} \in R\{x\}_{(p)}$$

is purely $\delta-$Galois with $\delta-$Galois group

$$(GA(\mathbf{Z}/p^{n+1}\mathbf{Z})).$$

In particular the extension

$$k\langle y\rangle \subset k\langle x\rangle$$

is algebraic Galois with Galois group $GA(\mathbf{Z}_p)$.

PROOF. Consider the following elements of $R\{x\}_{(p)}$:

$$f_1 := x^{\phi} - x, \quad f_2 := x^{\phi}/x,$$

Clearly

$$y = f_2 \circ f_1.$$

By Propositions 5.14 and 5.27, it follows that y is $\delta-$independent of $\delta-$slope zero and $\delta-$degree

$$(e_n) := ((p-1)p^{2n+1}).$$

On the other hand, if $\bar{y}_i \in k\langle x\rangle$ is the image of $\delta^i y$, we have extensions

$$k(\bar{y}_0, \bar{y}_1, ..., \bar{y}_n) \overset{\alpha}{\subset} k(x, x', ..., x^{(n)})^{GA(\mathbf{Z}/p^{n+1}\mathbf{Z})} \overset{\beta}{\subset} k(x, x', ..., x^{(n)}). \tag{5.33}$$

We know that $\beta \circ \alpha$ has degree e_n; but the latter number coincides with the order of the group

$$GA(\mathbf{Z}/p^{n+1}\mathbf{Z}). \tag{5.34}$$

By Lemma 5.13 the inclusion β in Equation 5.33 has degree e_n so the inclusion α is an equality and we are done. □

For any ring S define the "special affine group"

$$SA(S) := GA(S) \cap PSL_2(S)$$

in $PGL_2(S)$. Clearly

$$SA(S) \simeq \{\begin{bmatrix} a^2 & b \\ 0 & 1 \end{bmatrix}; a \in S^{\times},\ b \in S\} \subset GL_2(S).$$

If $S = \mathbf{Z}_p$ and d is a divisor of $p-1$ we can define, more generally,

$$GA^{(d)}(\mathbf{Z}_p) = \{\begin{bmatrix} a^d & b \\ 0 & 1 \end{bmatrix}; a \in \mathbf{Z}_p^{\times},\ b \in \mathbf{Z}_p, \} \subset GL_2(\mathbf{Z}_p). \tag{5.35}$$

and one can make a similar definition with $\mathbf{Z}/p^{n+1}\mathbf{Z}$ instead of $\mathbf{Z}_p$. So

$$SA(\mathbf{Z}_p) \simeq GA^{(2)}(\mathbf{Z}_p)$$

and similarly for $\mathbf{Z}/p^{n+1}\mathbf{Z}$.

PROPOSITION 5.30. *The element*

$$z := (x^{\phi} - x)^{\frac{\phi-1}{d}} \in (R\{x\}_{(p)})\hat{}$$

is purely $\delta-$Galois with $\delta-$Galois group

$$(GA^{(d)}(\mathbf{Z}/p^{n+1}\mathbf{Z})).$$

In particular the extension

$$k\langle z\rangle \subset k\langle x\rangle$$

is algebraic Galois with Galois group $GA^{(d)}(\mathbf{Z}_p)$.

PROOF. The argument is similar to the one in the proof of Proposition 5.22. □

Part 3

Applications

CHAPTER 6

Spherical correspondences

We start by defining *spherical correspondences* over $R := \hat{\mathbf{Z}}_p^{ur}$; this is straightforward since the existence of an analytic uniformization is a purely algebraic property. Then we shall analyze spherical correspondences over R from the viewpoint of $\delta-$geometry. This strategy is what we earlier called a "vertical" strategy (in which we fix p and vary the other data); our results will then allow us to prove, among other things, our "horizontal" Main Theorem 2.59.

Here is the structure of this Chapter. In Section 1 we define spherical correspondences over R and we look at cycles on such correspondences. In Section 2 we compute the spaces of $\delta-$sections of $W-$powers of $\mathcal{O}(1)$ on the projective line; this computation is crucial for everything that follows. In Section 3 we analyze $\delta-$invariants and $\delta-$cohomology of (reducible) correspondences obtained by taking "disjoint unions" of correspondences of the form $(\mathbf{P}^1, \mathbf{P}^1, id, \tau_i)$, where τ_i are isomorphisms. This will allow us to study, in Section 4, (irreducible) spherical correspondences of the type encountered in our Theorem 2.59; we shall analyze $\delta-$invariants, $\delta-$cohomology, and $\delta-$fibers for such correspondences. In Section 5 we prove a "converse theorem" to the effect that the "general member" (in an arithmetic/geometric sense) of the "universal family of biquadratic correspondences" on the affine line over $\mathbf{Z}$ satisfies certain vanishing properties $H^0(\mathbf{X}, \mathbf{L}) = 0$. Many special members $\mathbf{X}$ of this family, however, are spherical and have $H^0(\mathbf{X}, \mathbf{L}) \neq 0$ for appropriate $\mathbf{L}$'s.

6.1. Spherical correspondences over R and their cycles

Start with a pair (Γ, τ) where Γ is a subgroup of $SL_2(\mathbf{Z}_p)$, and $\tau \in SL_2(\mathbf{Z}_p)$. We let $\mathbf{P}^1 := Proj\ R[x_0, x_1]$ and we let $SL_2(\mathbf{Z}_p)^{op}$ act on $R[x_0, x_1]$ by letting

$$\gamma = \begin{bmatrix} a & b \\ c & d \end{bmatrix}$$

act as

$$\gamma^* x_1 = a x_1 + b x_0, \quad \gamma^* x_0 = c x_1 + d x_0. \tag{6.1}$$

We let $\mathbf{P}^1/\Gamma$ be the corresponding quotient and $\pi : \mathbf{P}^1 \to \mathbf{P}^1/\Gamma$ the canonical projection. So we may consider the correspondence $(\mathbf{P}^1, \mathbf{P}^1/\Gamma, \pi, \pi \circ \tau)$ over R (i.e. in the category $\mathcal{C}_R$ of schemes over R).

DEFINITION 6.1. An irreducible correspondence $\mathbf{X} = (X, \tilde{X}, \sigma_1, \sigma_2)$ over R is *uniformized* by (Γ, τ) if $\mathbf{X}$ is smooth over R and there exists an open immersion of $\mathbf{X}$ into a correspondence of the form

$$(\mathbf{P}^1, \mathbf{P}^1/\Gamma, \pi, \pi \circ \tau)$$

such that $\pi : \mathbf{P}^1 \to \mathbf{P}^1/\Gamma$ is étale above X. An irreducible correspondence over R is called *spherical* if it admits a uniformization as above.

Any spherical correspondence over R is étale by definition. The natural "vertical" conjecture, in this setting, is the following:

CONJECTURE 6.2. *Let* $\mathbf{X}$ *be a spherical correspondence over* R. *Then*
1) $(\mathbf{X}, \mathbf{K}^{-1})$ *is* $\delta-$*non-trivial and* $\delta-$*rational,*
2) the extension $k\langle \mathbf{X}, \mathbf{K}^{-1}\rangle \subset k\langle X, K^{-1}\rangle$ *is algebraic.*

We will analyze, in this Chapter, this Conjecture for a number of quite general classes of examples; while doing so we will investigate more problems including $\delta-$invariants and $\delta-$cohomology of certain natural $\delta-$line bundles polarization equivalent to $\mathbf{K}^{-1}$, Galois properties of the above extension, infinitesimal symmetries, $\delta-$tangent maps, converse results, etc. We start by discussing cycles of spherical correspondences.

Let $\tilde{P} \in \tilde{X}(R)$ be such that

$$\tau(\tilde{P}) = \gamma(\tilde{P})$$

for some $\gamma \in \Gamma$. Let $P = \pi(\tilde{P})$. Then $(\tilde{P}, P)$ is a fixed point for $\mathbf{X}$.

LEMMA 6.3. *If* x_0, x_1 *are chosen such that* $\tilde{P}$ *has coordinates* $(1,0)$ *and if*

$$\gamma^{-1}\tau = \begin{bmatrix} a_{11} & a_{12} \\ a_{21} & a_{22} \end{bmatrix},$$

(hence $a_{12} = 0$*) then the germ* $\Sigma(T)$ *of* $\mathbf{X}$ *at* $(\tilde{P}, P)$ *with respect to* $T = x_1/x_0$ *is given by*

$$\Sigma(T) = \frac{a_{12}T}{a_{21}T + a_{22}}.$$

In particular:

1) If $a_{21} \in (a_{11} - a_{22})R$ *then* $\Sigma(T)$ *is uniformized by a formal group over* R *which is isomorphic to the additive formal group. In particular the canonical bundle* $\mathbf{K}$ *of* $\mathbf{X}$ *is formally flat at* $(\tilde{P}, P)$. *Moreover, setting*

$$\beta := \frac{a_{21}}{a_{11} - a_{22}},$$

the canonical derivation at $(\tilde{P}, P)$ *is given by*

$$\partial^{can} = (1 - \beta T)^2 \frac{d}{dT} \in Der_R(R[[T]], R[[T]]).$$

2) If $a_{21} \notin (a_{11} - a_{22})R$ *then the canonical bundle* $\mathbf{K}$ *of* $\mathbf{X}$ *is not formally flat at* $(\tilde{P}, P)$.

3) The expansion map

$$E = E_\beta : \bigoplus_{m \in \mathbf{Z}_+} H^0(\mathbf{P}^1, \mathcal{O}(m)) = R[x_0, x_1] \to R[[T]]$$

with respect to the local basis $\Delta := x_0 - \beta x_1$ *of* $\mathcal{O}(1)$ *at* P *is given by*

$$E(x_0) = \frac{1}{1 - \beta T}, \quad E(x_1) = \frac{T}{1 - \beta T}.$$

4) Consider the derivation

$$\partial := (x_0 - \beta x_1)\left[\beta \frac{\partial}{\partial x_0} + \frac{\partial}{\partial x_1}\right];$$

then $\partial\Delta = 0$ *and*

$$\partial^{can} \circ E = E \circ \partial$$

as maps $R[x_0, x_1] \to R[[T]]$.

PROOF. The statement about the shape of $\Sigma(T)$ follows directly from definitions. The statements 1 and 2 above follow from Proposition 4.5; the formula for ∂^{can} follows from the fact that the linearizing series for $\Sigma(T)$ is $\frac{T}{1-\beta T}$. Assertion 3 is a trivial computation. Assertion 4 follows from assertion 3. □

LEMMA 6.4. *Let* $\mathbf{X}$ *be a spherical correspondence with canonical bundle* $\mathbf{K}$. *Then there exists a non-empty open subcorrespondence* $\mathbf{U}$ *such that the line bundle*

$$\mathbf{K}_{\mathbf{U}\otimes k}^{(p+1)/2}$$

on $\mathbf{U} \otimes k$ *is trivial.*

The above can be proved directly rather easily. Since we are not going to use this Lemma in the sequel we postpone its proof; we shall deduce this Lemma from a more general argument later.

Spherical correspondences tend to have sequences of disjoint cycles with $\mathbf{Z}_p-$ accumulating multipliers. Here is a typical example:

PROPOSITION 6.5. *Assume* $\mathbf{X} = (X, \tilde{X}, \sigma_1, \sigma_2)$ *is a spherical correspondence over* R *uniformized by* (Γ, τ), *where* $\tilde{X} = Y \cap \tau^{-1}(Y)$, $Y \subset \mathbf{P}^1$ *is a* $\Gamma-$*invariant open subset and* $X = Y/\Gamma$. *Assume that:*

1) $\Gamma \subset SL_2(\mathbf{Z})$ *and* $\tau = \begin{bmatrix} 1 & 1 \\ 0 & 1 \end{bmatrix}$,

2) Γ *contains an element whose image in* $SL_2(\mathbf{F}_p)$ *is not upper triangular,*

3) none of the $k-$*points of* $\mathbf{P}^1 \backslash Y$ *is* $\mathbf{F}_p-$*rational.*

Then $\mathbf{X}$ *has a sequence of pairwise disjoint cycles with* $\mathbf{Z}_p-$ *accumulating multipliers.*

As an example one can take $\Gamma := \langle \gamma \rangle$,

$$\gamma = \begin{bmatrix} 0 & -1 \\ 1 & 0 \end{bmatrix},$$

$Y := \mathbf{P}^1 \backslash Z(x_0^2 + x_1^2)$, $p \equiv 3 \bmod 4$. (Cf Example 6.27 for more on this example.)

In order to prove this Proposition we need the following:

LEMMA 6.6. *Let* N *be an integer not divisible by two distinct odd primes* p *and* λ *and let* M *and* n *be any integers. Then one can find a positive integer*

$$\nu = \nu(N, M, p, \lambda, n)$$

such that if $N_\nu := N\nu + M$ *then the following hold:*

1) $v_\lambda(N_\nu^2 - 4) = 1$,

2) $v_p(N_\nu^2 - 4)$ *is even and* $\geq n$,

3) $\left(\frac{K}{p}\right) = 1$ *where* $N_\nu^2 - 4 = p^{v_p(N_\nu^2 - 4)} K$.

In the above statement v_λ and v_p are the corresponding normalized valuations and $\left(\frac{-}{p}\right)$ is the Legendre symbol.

PROOF. We may assume n is even and ≥ 2. Write $eN + fp\lambda = 1$ for some integers e and f. We may assume $e \geq 1$. Let $u = 4\lambda + p^s$ for some positive integer s; note that

$$\left(\frac{u}{p}\right) = \left(\frac{4\lambda}{p}\right);$$

Finally let $\nu := e\mu$ where

$$\mu := u\lambda p^n - (M-2)\sum_{i=0}^{n}(fp\lambda)^i;$$

by taking s large enough we obtain $\nu \geq 1$. We have the following computation in $\mathbf{Q}_p$:

$$v_p(N_\nu - 2) = v_p((1-fp\lambda)\mu + M - 2) = v_p\left(\mu + \frac{M-2}{1-fp\lambda}\right)$$

$$= v_p(u\lambda p^n + \sum_{i=n+1}^{\infty}(fp\lambda)^i) = n.$$

So $v_p(N_\nu + 2) = 0$ and hence $v_p(N_\nu^2 - 4) = n$. An identical computation in $\mathbf{Q}_\lambda$ yields $v_\lambda(N_\nu^2 - 4) = 1$. Finally we have the following computation in $\mathbf{Z}_p$:

$$K = \frac{N\nu + M - 2}{p^n}\cdot(N\nu + M + 2) = (\lambda u + \sum_{i=n+1}^{\infty} f^i\lambda^i p^{i-n})(4 + u\lambda p^n + \sum_{i=n+1}^{\infty} f^i\lambda^i p^i)$$

$$\equiv 4\lambda u \quad mod \quad p.$$

In particular

$$\left(\frac{K}{p}\right) = \left(\frac{4\lambda u}{p}\right) = 1.$$

□

PROOF. (of Proposition 6.5) Let $\gamma \in \Gamma$ have a reduction mod p which is not upper triangular. So we may write

$$\gamma^{-1} = \begin{bmatrix} a & b \\ c & d \end{bmatrix},$$

with $c \not\equiv 0 \bmod p$. Let $t = x_1/x_0$ be the standard affine coordinate on $\mathbf{P}^1$. For any positive integer ν a fixed point P_ν of $\gamma^{-1}\tau^\nu$ on $\mathbf{P}^1$ defined over the algebraic closure K^a of K has $t-$coordinate t_ν given by the equation $f_\nu(t_\nu) = t_\nu$ where

$$f_\nu(t) = \frac{at + a\nu + b}{ct + c\nu + d}.$$

Each P_ν gives rise to a $\nu-$cycle

$$C_\nu = (P_\nu, \tau(P_\nu), ..., \tau^{\nu-1}(P_\nu), \pi(P_\nu)).$$

on the correspondence

$$(\mathbf{P}^1, \mathbf{P}^1/\Gamma, \pi, \pi\circ\tau)$$

over K^a. Set $N_\nu = a + d + c\nu$. By a direct computation we get

$$t_\nu = \frac{a - c\nu - d + \sqrt{N_\nu^2 - 4}}{2c}.$$

Clearly, the multiplier of C_ν is given by

$$a(C_\nu) = \frac{df_\nu}{dt}(t_\nu) \in K^a.$$

A direct computation gives

$$a(C_\nu) - 1 = \frac{-2(N_\nu^2 - 4) + 2N_\nu\sqrt{N_\nu^2 - 4}}{N_\nu + \sqrt{N_\nu^2 - 4}}.$$

Next we construct inductively a sequence of odd primes $\lambda_0, \lambda_1, \lambda_2, ...$ and a sequence of positive integers $\nu_0, \nu_1, \nu_2, ...$ by taking λ_0 and ν_0 arbitrary and then choosing

$$\lambda_i > \max\{N_{\nu_0}^2 - 4, ..., N_{\nu_{i-1}}^2 - 4, c, p\}$$

and letting ν_i be such that

1) $v_{\lambda_i}(N_{\nu_i}^2 - 4) = 1$,
2) $v_p(N_{\nu_i}^2 - 4)$ is even and $\geq i$,
3) $\left(\frac{K_i}{p}\right) = 1$ where $N_{\nu_i}^2 - 4 = p^{v_p(N_{\nu_i}^2 - 4)}K_i$;

this is possible by Lemma 6.6. By conditions 2 and 3 above we get $\sqrt{N_{\nu_i}^2 - 4} \in p^{[i/2]}\mathbf{Z}_p$; in particular p does not divide N_{ν_i}. Hence

$$t_{\nu_i} \in \mathbf{Z}_p, \quad a(C_{\nu_i}) \in \mathbf{Z}_p,$$
$$v_p(a(C_{\nu_i}) - 1) = v_p(N_{\nu_i}^2 - 4)/2 \geq [i/2].$$

Note that

$$t_{\nu_i} \ mod \ p \ = \ \frac{a - c\nu_i - d}{c} \ mod \ p \ \in \ \mathbf{F}_p.$$

By hypothesis 3 we get that

$$\tau^j(P_{\nu_i}) \in \tilde{X}(R)$$

for all i and j hence C_{ν_i} are cycles on the correspondence $\mathbf{X}$. In order to conclude we need to show that the cycles C_{ν_i} are pairwise disjoint. Assume this is not the case, i.e. $\tau^r(t_{\nu_i}) = \gamma_1\tau^k(t_{\nu_j})$ for some $i < j$, some r, k, and some $\gamma_1 \in \Gamma$, and let us derive a contradiction. Consider the fields

$$\mathbf{Q}(t_{\nu_i}) = \mathbf{Q}(\sqrt{N_{\nu_i}^2 - 4})$$

and note that

$$\mathbf{Q}(t_{\nu_j}) = \mathbf{Q}(\gamma_1\tau^k(t_{\nu_j})) = \mathbf{Q}(\tau^r(t_{\nu_i})) = \mathbf{Q}(t_{\nu_i}).$$

Hence $\mathbf{Q}(t_{\nu_i}) = \mathbf{Q}(t_{\nu_j})$. Now λ_j divides the discriminant of $\mathbf{Q}(t_{\nu_j})$ hence the discriminant of $\mathbf{Q}(t_{\nu_i})$ hence it divides $N_{\nu_i}^2 - 4$, a contradiction. □

6.2. δ–sections of bundles on the projective line

The first step in analyzing spherical correspondences is, of course, to compute the spaces $H^0(\mathbf{P}^1, \mathcal{O}(1)^w)$. Here is the result:

THEOREM 6.7. *There are natural identifications*

$$H^0(\mathbf{P}^1, \mathcal{O}(1)^{w[s]}) = H^0(\mathbf{P}^1, \mathcal{O}(1)^w) = R\{x_0, x_1\}^w, \tag{6.2}$$

$$\mathcal{O}^r(\mathbf{V}^*(\mathcal{O}(1))) = R[x_0^{(\leq r)}, x_1^{(\leq r)}]\hat{}. \tag{6.3}$$

In particular

$$R(\mathbf{P}^1, \mathcal{O}(1))^* = R\{x_0, x_1\}^*, \quad \mathcal{O}^\infty(\mathbf{V}^*(\mathcal{O}(1)))\hat{} = R\{x_0, x_1\}\hat{}.$$

Furthermore the $R-$modules $H^0(\mathbf{P}^1, \mathcal{O}(1)^w)$ are finitely generated.

PROOF. Let $r = ord(w) + s$, $\mathbf{P}^1 = U_0 \cup U_1$, $U_0 = Spec\ R[x_1/x_0]$, $U_1 = Spec\ R[x_0/x_1]$. Hence we have identifications

$$\mathcal{O}^r(U_0) = R[x_1/x_0, \delta(x_1/x_0), ..., \delta^r(x_1/x_0)]\hat{},$$

$$\mathcal{O}^r(U_1) = R[x_0/x_1, \delta(x_0/x_1), ..., \delta^r(x_0/x_1)]\hat{}.$$

Let $C = R[x_0^{(\leq r)}, x_1^{(\leq r)}]$. Since $\mathbf{V}^*(\mathcal{O}(1)) = V_0 \cup V_1$, $V_i = Spec\ R[x_0, x_1, x_i^{-1}]$, it follows that

$$\mathcal{O}^r(\mathbf{V}^*(\mathcal{O}(1))) = C[1/x_0]\hat{} \cap C[1/x_1]\hat{} = \hat{C}$$

(the intersection being taken in $C[1/x_0x_1]\hat{}$). In particular Equation 6.3 follows. Now any section in $H^0(\mathbf{P}^1, \mathcal{O}(1)^{w[s]})$ gives rise to two elements

$$F_0 \cdot x_0^w \in R[x_1/x_0, \delta(x_1/x_0), ..., \delta^r(x_1/x_0)]\hat{} \cdot x_0^w \in C[1/x_0]\hat{}, \tag{6.4}$$

$$F_1 \cdot x_1^w \in R[x_0/x_1, \delta(x_0/x_1), ..., \delta^r(x_0/x_1)]\hat{} \cdot x_1^w \in C[1/x_1]\hat{}, \tag{6.5}$$

whose images in $C[1/x_0x_1]\hat{}$ coincide. So

$$f := F_0 \cdot x_0^w = F_1 \cdot x_1^w \in \hat{C}.$$

Also, clearly, if $\Lambda, \Lambda', ..., \Lambda^{(r)}$ are new variables, the equation

$$f(\Lambda x_0, \Lambda x_1, ..., \delta^r(\Lambda x_0), \delta^r(\Lambda x_1)) = \Lambda^w f(x_0, x_1, ..., x_0^{(r)}, x_1^{(r)}) \tag{6.6}$$

holds in the ring $C[\Lambda^{(\leq r)}]\hat{}$. Let us say that a monomial

$$x_0^{i_0} x_1^{j_0} (x_0')^{i_1} (x_1')^{j_1} ... (x_0^{(r)})^{i_r} (x_1^{(r)})^{j_r} \tag{6.7}$$

has *weak weight* N if

$$i_0 + j_0 + (i_1 + j_1)p + ... + (i_r + j_r)p^r = N.$$

Let $w = \sum a_i \phi^i$. We claim that all monomials in f have weak weight $w(p)$. To check the claim note that we have the following congruences:

$$\delta^i(\Lambda x_j) \equiv \Lambda^{p^i} x_j^{(i)} \quad mod \quad (\Lambda', ..., \Lambda^{(i)})$$

in the ring $C[\Lambda^{(\leq r)}]\hat{}$. For each $N \in \mathbf{Z}_+$ let f_N be the sum of all monomials in f having weak weight N and write $f = \sum_N f_N$. Moreover set $\Lambda' = ... = \Lambda^{(r)} = 0$ in Equation 6.6. Then we get

$$\sum_N \Lambda^N f_N = \Lambda^{w(p)} \sum_N f_N$$

in the ring $C[\Lambda]\hat{}$. We conclude that $f = f_{w(p)}$; this proves our claim. Since there are only finitely many monomials 6.7 of a given weak weight, our claim implies in particular that f is a polynomial, $f \in C$. We may write then $f \in K[x_0, x_1, x_0^\phi, x_1^\phi, ...]$. Equation 6.6 implies, again, that $f \in K[x_0, x_1, x_0^\phi, x_1^\phi, ...]^w$. So we conclude that $f \in R\{x_0, x_1\}^w$. Conversely, by reversing the above arguments, it is easy to see that any element of $R\{x_0, x_1\}^w$ defines a section in $H^0(\mathbf{P}^1, \mathcal{O}(1)^w)$. The finite generation of the latter module follows from the fact that, as previously noted, all monomials present in an element of $H^0(\mathbf{P}^1, \mathcal{O}(1)^w)$ have weak weight $w(p)$ hence $H^0(\mathbf{P}^1, \mathcal{O}(1)^w)$ is contained in a finitely generated $R-$submodule of C. □

COROLLARY 6.8. *$(\mathbf{P}^1, \mathcal{O}(1))$ is $\delta-$rational.*

PROOF. By Theorem 6.7 we have $R(\mathbf{P}^1, \mathcal{O}(1))^* = R\{x_0, x_1\}^*$ and we conclude by Lemma 5.6. □

REMARK 6.9. The results above about $\mathbf{P}^1$ have obvious generalizations to $\mathbf{P}^n$ for arbitrary n. In particular

$$R(\mathbf{P}^n, \mathcal{O}(1))^* \simeq R\{x_0, ..., x_n\}^*. \tag{6.8}$$

So $(\mathbf{P}^n, \mathcal{O}(1))$ is δ−rational. Equation 6.8 also has the following important consequence. Assume X is a smooth scheme over R, L is a δ−line bundle, and $s_0, ..., s_n \in H^0(X, L)$. Then there exists a unique morphism of δ−graded rings

$$R(\mathbf{P}^n, \mathcal{O}(1))^* \to R(X, L)^*, \tag{6.9}$$

sending x_i into s_i for all i.

6.3. Case Γ trivial: δ−invariants and δ−cohomology

For any $\tau \in SL_2(\mathbf{Z}_p)$ we will consider the correspondence

$$\mathbf{X}_\tau = (\mathbf{P}^1, \mathbf{P}^1, id, \tau); \tag{6.10}$$

clearly $\mathbf{X}_\tau$ is spherical, uniformized by $(\{id\}, \tau)$. Consider the line bundle

$$\mathbf{L} = \mathbf{L}(\tau) := (\mathcal{O}(1), can) \tag{6.11}$$

on $\mathbf{X}_\tau$ where

$$can : \tau^*\mathcal{O}(1) \to \mathcal{O}(1)$$

is the isomorphism corresponding to the isomorphism

$$\tau^* : \mathcal{O}(1) \to \tau_*\mathcal{O}(1)$$

defined at the level of global sections

$$\tau^* : H^0(\mathbf{P}^1, \mathcal{O}(1)) = Rx_0 + Rx_1 \to H^0(\mathbf{P}^1, \tau_*\mathcal{O}(1)) = H^0(\mathbf{P}^1, \mathcal{O}(1)) = Rx_0 + Rx_1$$

by the formulae 6.1.

LEMMA 6.10. $\mathbf{L}^2 \simeq \mathbf{K}^{-1}$.

PROOF. The isomorphism

$$can : \mathcal{O}(-2) \to K_{\mathbf{P}^1}$$

that sends

$$1/x_0^2 \mapsto d(x_1/x_0)$$

induces a commutative diagram:

$$\begin{array}{ccc} \tau^*\mathcal{O}(-2) & \stackrel{can}{\to} & \mathcal{O}(-2) \\ can \downarrow & & \downarrow can \\ \tau^*K_{\mathbf{P}^1} & \stackrel{can}{\to} & K_{\mathbf{P}^1} \end{array}$$

$$\begin{array}{ccccc} \tau^*(1/x_0^2) & \to & (cx_1 + dx_0)^{-2} & = & (c\frac{x_1}{x_0} + d)^{-2} \cdot (1/x_0^2) \\ \downarrow & & & & \downarrow \\ \tau^* d(x_1/x_0) & \to & d\left(\frac{ax_1 + bx_0}{cx_1 + dx_0}\right) & = & (ad - bc) \cdot (c\frac{x_1}{x_0} + d)^{-2} \cdot d(x_1/x_0) \end{array}$$

□

For any $w \in W$ define

$$\mathbf{L}_w = \mathbf{L}_w(\tau) := \mathbf{L}(\tau)^w; \tag{6.12}$$

in particular note that

$$\mathbf{L}_w^2 \simeq \mathbf{K}^{-w}. \tag{6.13}$$

More generally we consider a subset $\mathcal{T} \subset SL_2(\mathbf{Z}_p)$ and consider the disjoint union:

$$\mathbf{X}_{\mathcal{T}} := \coprod_{\tau \in \mathcal{T}} \mathbf{X}_\tau := (\mathbf{P}^1, \coprod_{\tau \in \mathcal{T}} \mathbf{P}^1, id, \coprod_{\tau \in \mathcal{T}} \tau). \tag{6.14}$$

For each $w \in W$ we may consider the naturally induced δ-line bundle

$$\mathbf{L}_w := \coprod_{\tau \in \mathcal{T}} \mathbf{L}_w(\tau)$$

on $\mathbf{X}_{\mathcal{T}}$. Again, Equation 6.13 holds; so if $w \in W_+ \backslash \phi W_+$ then $\mathbf{L}_w$ is polarization equivalent to $\mathbf{K}^{-1}$. We have

$$H^0(\mathbf{X}_{\mathcal{T}}, \mathbf{L}_w) = \bigcap_{\tau \in \mathcal{T}} H^0(\mathbf{X}_\tau, \mathbf{L}_w(\tau)).$$

We also consider $\mathbf{L}_{r,w} := \mathbf{L}_w^{[r-ord(w)]}$ for any $r \geq ord(w)$. For $0 \leq a < b \leq r$ consider the following sections:

$$f^{ab} := \det \begin{bmatrix} \phi^a x_0 & \phi^b x_0 \\ \phi^a x_1 & \phi^b x_1 \end{bmatrix} \in H^0(\mathbf{X}_{\mathcal{T}}, \mathbf{L}_{\phi^a + \phi^b}); \tag{6.15}$$

here we used the identification in Theorem 6.7. Set

$$f^r := f^{0r}; \tag{6.16}$$

the superscripts of f will never stand, therefore, for exponents. Note that we have the following relations:

$$(f^1)^\phi f^k - f^2 (f^{k-1})^\phi + f^1 (f^{k-2})^{\phi^2} = 0. \tag{6.17}$$

REMARK 6.11. It is easy to check that the sections $f^r \in H^0(\mathbf{X}_{\mathcal{T}}, \mathbf{L}_{1+\phi^r})$ are 1− regular and 1− coregular at all fixed points of $\mathbf{X}_{\mathcal{T}}$.

LEMMA 6.12. *With notation as in Lemma 6.3 we have*

$$E_\beta(f^{ab}) = \frac{T^{\phi^b} - T^{\phi^a}}{(1 - \beta T)^{\phi^a + \phi^b}}.$$

PROOF. A trivial computation. □

In particular we may consider the series

$$\Psi_{\mathbf{G}_a} := E_0(f^1) = T^\phi - T; \tag{6.18}$$

this series will play a role in the proof of Proposition 6.18, for instance, and should be viewed as an additive analogue of the series $\Psi_{\mathbf{G}_m}$ introduced in Equation 4.40.

Let $\beta_0, \beta_1 \in \mathbf{Z}_p$, not both in $p\mathbf{Z}_p$, let $\Delta := \beta_0 x_0 - \beta_1 x_1$, and let $\partial : R[x_0, x_1] \to R[x_0, x_1]$ be the derivation

$$\partial := -\frac{1}{2} \left[\frac{\partial(\Delta^2)}{\partial x_1} \cdot \frac{\partial}{\partial x_0} - \frac{\partial(\Delta^2)}{\partial x_0} \cdot \frac{\partial}{\partial x_1} \right] = (\beta_0 x_0 - \beta_1 x_1) \cdot \left[\beta_1 \frac{\partial}{\partial x_0} + \beta_0 \frac{\partial}{\partial x_1} \right];$$

so $\partial\Delta = 0$. Then ∂ induces a derivation

$$\partial : \bigoplus_{m\in\mathbf{Z}} \mathcal{O}(m) \to \bigoplus_{m\in\mathbf{Z}} \mathcal{O}(m),$$

such that $\partial(\mathcal{O}(m)) \subset \mathcal{O}(m)$ for all m. Let

$$\partial_0, \partial_1, \partial_2, ... : \mathcal{O}^\infty(\mathbf{V}^*(\mathcal{O}(1)))\hat{} = R\{x_0, x_1\}\hat{} \to R\{x_0, x_1\}\hat{}$$

be induced by the conjugate operators attached to ∂. Then a direct computation yields:

PROPOSITION 6.13.

$$\partial_j f^{ab} = 0 \quad for \quad j \neq a, b, \tag{6.19}$$

$$\partial_a f^{ab} = -\partial_a f^{ba} = -p^a \Delta^{\phi^a+\phi^b}. \tag{6.20}$$

In particular

$$\partial_i \partial_j f^{ab} = 0 \quad for\ all \quad i, j, a, b. \tag{6.21}$$

So if

$$V_{ab}^{ij} := \partial_i f^{ab} \cdot \partial_j - \partial_j f^{ab} \cdot \partial_i \in Der_R(R\{x_0, x_1\}\hat{}, R\{x_0, x_1\}\hat{})$$

is the $(i,j)-$ *Hamiltonian flow defined by* f^{ab} *and* ∂ *then any two Hamiltonian flows* V_{ab}^{ij} *and* V_{cd}^{mn} *commute. Moreover,*

$$(\alpha\partial)_* f^{ab} = \Delta^{\phi^a+\phi^b} \cdot (p^b\phi^b - p^a\phi^a)(\alpha), \quad \alpha \in R, \tag{6.22}$$

$$\partial_{**} f^{ab} = 0. \tag{6.23}$$

In particular ∂ *is a variational infinitesimal symmetry of each of the* f^{ab}*'s. Finally, the Euler-Lagrange equation* $\langle \partial, EL_1((f^1)^2)\rangle_1$ *attached to* $(f^1)^2$ *and* ∂ *equals*

$$2p\Delta^\phi[f^1\Delta - (f^1)^\phi \Delta^{\phi^2}].$$

The above (trivial) Proposition will not play a further role in our study of spherical correspondences; its interest lies in the fact that it possesses (quite non-trivial) flat and hyperbolic analogues.

REMARK 6.14. The f^{ab}'s also satisfy, of course, a set of differential equations with respect to the conjugates

$$\mathcal{D}_j : R\{x_0, x_1\}\hat{} \to R\{x_0, x_1\}\hat{}$$

of the Euler operator

$$\mathcal{D} = x_0 \frac{\partial}{\partial x_0} + x_1 \frac{\partial}{\partial x_1} : R[x_0, x_1] \to R[x_0, x_1],$$

which simply reflect the fact that f^{ab} have "weight $\phi^a + \phi^b$"; indeed, by Equation 3.54, we have

$$\mathcal{D}_j f^{ab} = 0 \quad for \quad j \neq a, b, \tag{6.24}$$

$$\mathcal{D}_a f^{ab} = p^a f^{ab}. \tag{6.25}$$

By the way note that ∂ and $\mathcal{D}$ form a basis of the tangent sheaf of the surface $\mathbf{V}^*(\mathcal{O}(1))$ on the locus $\mathbf{V}^*(\mathcal{O}(1))_\Delta$ where Δ is invertible. So, by Remark 3.58,

$$\partial_0, \mathcal{D}_0, ..., \partial_r, \mathcal{D}_r$$

form a basis of the tangent sheaf of $J^r(\mathbf{V}^*(\mathcal{O}(1))_\Delta)$. So Equations 6.24 and 6.25 together with Equations 6.19 and 6.20 provide a computation of the derivatives of each f^{ab} with respect to a basis of this tangent sheaf; this list of equations can therefore be interpreted as a Pfaffian system of equations satisfied by each of the f^{ab}'s as follows.

COROLLARY 6.15. *Let ω, η be a basis of 1–forms on $J^r(\mathbf{V}^*(\mathcal{O}(1))_\Delta)$ dual to $\partial, \mathcal{D}$ and let*

$$\omega_0, \eta_0, ..., \omega_r, \eta_r$$

be the conjugate 1–forms $\omega_i = p^{-i}(\phi^i)^*\omega$, $\eta_i = p^{-i}(\phi^i)^*\eta$. Then $Z = f^{ab}$ satisfies the equation

$$dZ = Z \cdot (p^a\eta_a - p^b\eta_b) + \Delta^{\phi^a+\phi^b} \cdot (p^b\omega_b - p^a\omega_a).$$

EXAMPLE 6.16. Proposition 6.13 can be used to compute δ–tangent maps; here is a typical example. Let $\mathbf{P}^1(R)_{f^1}$ be the set of all points $P \in \mathbf{P}^1(R)$ such that $f^1(P) \not\equiv 0$ mod p, let $f : \mathbf{P}^1(R)_{f^1} \to \mathbf{P}^1(R)$ be the map defined by the sections $(s_0, s_1) = ((f^1)^{1+\phi^2}, (f^2)^{1+\phi})$, let $P \in \mathbf{P}^1(R)_{f^1}$, and let $T_Pf : T_P\mathbf{P}^1 \to T_{f(P)}\mathbf{P}^1$ be the δ–tangent map at P; cf. Equation 3.44. Since $\partial\mathcal{O}(m) \subset \mathcal{O}(m)$, ∂ induces an R–derivation of $\mathcal{O}_{\mathbf{P}^1}$. Assume P is not a zero mod p of Δ; then the image, $\partial_P \in T_P\mathbf{P}^1$, of ∂ is an R–basis of $T_P\mathbf{P}^1$ and, for $\alpha \in R$, the δ–tangent map T_Pf acts as

$$(T_Pf)(\alpha\partial_P) = \left[(\alpha\partial)_*\left(\frac{(f^2)^{1+\phi}}{(f^1)^{1+\phi^2}}\right)\right](P).$$

It is a trivial exercise to explicitly compute the latter by using the formulae in Proposition 6.13. As a result, for instance, T_Pf turns out to be injective if none of the forms $\Delta, f^2, f^2 - \Delta^{\phi^2-\phi}f^1$ vanishes mod p at P.

PROPOSITION 6.17. *Assume the subgroup $\langle\mathcal{T}\rangle$ generated by $\mathcal{T}$ in $SL_2(\mathbf{Z}_p)$ is not solvable. Then, for any $w \in W$, $s := ord(w) \leq r$, the K–linear space*

$$H^0(\mathbf{X}_\mathcal{T}, \mathbf{L}_{r,w}) \otimes K$$

is generated by all monomials of the form

$$(f^1)^{w_1}(f^2)^{w_2}...(f^s)^{w_s}, \tag{6.26}$$

where $w_1, w_2, ..., w_s \in W_+$, $ord(w_i) \leq s - i$, and

$$(1+\phi)w_1 + (1+\phi^2)w_2 + ...(1+\phi^s)w_s = w.$$

PROOF. Clearly the monomials 6.26 are elements of $H^0(\mathbf{X}_\mathcal{T}, \mathbf{L}_{r,w})$. Conversely assume $f \in H^0(\mathbf{X}_\mathcal{T}, \mathbf{L}_{r,w})$ and write $x := \{x_0, x_1\}$. By Theorem 6.7, we have

$$f = f(x, x^\phi, ..., x^{\phi^s}) \in K[x, x^\phi, ..., x^{\phi^s}]^w,$$

$$f(\tau^*x, (\tau^*x)^\phi, ..., (\tau^*x)^{\phi^s}) = f(x, x^\phi, ..., x^{\phi^s}), \quad \tau \in \mathcal{T}. \tag{6.27}$$

Since the τ's have $\mathbf{Z}_p$–coefficients, we have $(\tau^*x)^{\phi^i} = \tau^*(x^{\phi^i})$ so Equation 6.27 is equivalent to

$$f(\tau^*x, \tau^*(x^\phi), ..., \tau^*(x^{\phi^s})) = f(x, x^\phi, ..., x^{\phi^s}) \tag{6.28}$$

for all $\tau \in \mathcal{T}$. Let K^a be the algebraic closure of K and let $G \subset SL_2(K^a)$ be the subgroup of all τ such that Equation 6.28 holds. Clearly G is an algebraic subgroup of $SL_2(K^a)$ and, by hypothesis, G is non-solvable. So $G = SL_2(K^a)$. In particular f belongs to the ring

$$K[x, x^{\phi}, ..., x^{\phi^s}]^{SL_2(K)} \tag{6.29}$$

of $SL_2(K)$–invariant polynomials in $K[x, x^{\phi}, ..., x^{\phi^s}]$. By the "Fundamental First Theorem" for SL_2 in classical invariant theory [**61**], p.183, the K–algebra 6.29 is generated by the elements f^{ab} with $0 \leq a < b \leq s$ which clearly implies the statement of our Proposition. □

In the next Proposition we let $u, u', u'', ...$ be new variables and we consider the ring $R\{u\}$. Moreover we denote by Ω the multiplicative system in $R\{x_0, x_1\}$ consisting of all elements of the form $(f^1)^{(1+\phi^2)v}$, $v \in W_+$ and we consider the ring of fractions $\Omega^{-1}R\{x_0, x_1\}$.

PROPOSITION 6.18. *Let* $\mathbf{X}_{\mathcal{T}}$ *be as in Proposition 6.17 and let* $w = 1+\phi+\phi^2+\phi^3$. *Then, for any* $f \in H^0(\mathbf{X}_{\mathcal{T}}, \mathbf{L}_w^v)$, $v \in W_+$, *there exists a polynomial* $P_f \in R\{u\}$ *such that the following equality holds in the ring* $\Omega^{-1}R\{x_0, x_1\}$:

$$\frac{f}{(f^1)^{(1+\phi^2)v}} = P_f(h, \delta h, ...), \quad h := \frac{(f^2)^{1+\phi}}{(f^1)^{1+\phi^2}}. \tag{6.30}$$

PROOF. By Proposition 6.17 and relations 6.17 it follows that for any $f \in H^0(\mathbf{X}_{\mathcal{T}}, \mathbf{L}_w^v)$ there exists $w_0 \in W_+$ such that $(f^1)^{w_0} f$ is a K–linear combination of monomials of the form

$$(f^1)^{w_1}(f^2)^{w_2} \tag{6.31}$$

with $w_1, w_2 \in W_+$. One can find $v_0 \in W_+$ such that $(1+\phi^2)v_0 \geq w_0$; so we may assume $w_0 = (1+\phi^2)v_0$. Then we may assume the only monomials 6.31 present are such that

$$w(v_0 + v) = (1+\phi)w_1 + (1+\phi^2)w_2.$$

Since $1+\phi$ and $1+\phi^2$ are prime elements in W it follows that $w_1 = (1+\phi^2)v_1$, $w_2 = (1+\phi)v_2$ so a relation like 6.30 holds for some $P_f \in K[u, u', u'', ...]$. We shall be done if we prove that P_f has coefficients in R. Assume this is not the case so there exists an integer $\nu \geq 1$ such that $p^{\nu}P_f \in R\{u\}$ but $p^{\nu-1}P_f \notin R\{u\}$. Consider the R–algebra homomorphism

$$E : \Omega^{-1}R\{x_0, x_1\} \to R\{t\}_{(p)}$$

$$E(x_0) = 1, \ E(x_0') = 0, \ E(x_0'') = 0, ...$$

$$E(x_1) = t, \ E(x_1') = t', \ E(x_1'') = t'', ...$$

(By the way this is induced by the expansion map E_0 of Lemma 6.3.) In particular note that $E(f^r) = t^{\phi^r} - t$ hence

$$E(h) = \frac{(t^{\phi^2} - t)(t^{\phi^3} - t^{\phi})}{(t^{\phi} - t)(t^{\phi^3} - t^{\phi^2})}. \tag{6.32}$$

So by Theorem 5.26, the images

$$\bar{h}_0, \bar{h}_1, \bar{h}_2, ... \in k\langle t \rangle \tag{6.33}$$

of $E(h)$, $\delta(E(h))$, $\delta^2(E(h))$,... are algebraically independent over k. Let $Q = p^{\nu}P_f$; so $Q \in R\{u\}$ and its image $\bar{Q} \in k[u, u', u'', ...]$ is non-zero. We have

$$Q(h, \delta h, \delta^2 h, ...) = p^{\nu}\frac{f}{(f^1)^{(1+\phi^2)v}} \in p^{\nu}\Omega^{-1}R\{x_0, x_1\}$$

so we get

$$Q(E(h), \delta(E(h)), \delta^2(E(h)), ...) \in pR\{t\}_{(p)};$$

this contradicts the algebraic independence of $\bar{h}_0, \bar{h}_1, \bar{h}_2, ...$ □

PROPOSITION 6.19. *Let $\mathbf{X}_{\mathcal{T}}$ be as in Proposition 6.17 and let $w = 1+\phi+\phi^2+\phi^3$. Then the map*

$$R(\mathbf{X}_{\mathcal{T}}, \mathbf{L}_w)^* \to R\{u\}, \quad f \mapsto P_f$$

in Proposition 6.18 is telescopic.

PROOF. Assume $f \in H^0(\mathbf{X}_{\mathcal{T}}, \mathbf{L}_w^v)$ is such that $P_f = pQ$ with $Q \in R\{u\}$. Then we have $f = p(f^1)^{(1+\phi^2)v}Q(h)$ in the ring $\Omega^{-1}R\{x_0, x_1\}$. It is enough to prove that $g := (f^1)^{(1+\phi^2)v}Q(h)$ belongs to $H^0(\mathbf{X}_{\mathcal{T}}, \mathbf{L}_w^v)$. Note that the morphism

$$R\{x_0, x_1\} \otimes k \to (\Omega^{-1}R\{x_0, x_1\}) \otimes k \tag{6.34}$$

is injective. Hence

$$pg \in (p\Omega^{-1}R\{x_0, x_1\}) \cap R\{x_0, x_1\} = pR\{x_0, x_1\}$$

so $g \in R\{x_0, x_1\}$. It is then clear that $g \in H^0(\mathbf{X}_{\mathcal{T}}, \mathbf{L}_w^v)$. □

THEOREM 6.20. *Let $w = 1+\phi+\phi^2+\phi^3$ and assume the subgroup $\langle \mathcal{T} \rangle$ generated by $\mathcal{T}$ in $SL_2(\mathbf{Z}_p)$ is not solvable. Then there is a natural isomorphism of $R-$algebras*

$$R\langle \mathbf{X}_{\mathcal{T}}, \mathbf{L}_w \rangle\hat{} \simeq (R\{u\}_{(p)})\hat{}, \quad \frac{(f^2)^{1+\phi}}{(f^1)^{1+\phi^2}} \mapsto u;$$

so, in particular, $(\mathbf{X}_{\mathcal{T}}, \mathbf{L}_w)$ is $\delta-$rational. In addition, for each integer $d \geq 0$, $H^0(\mathbf{X}_{\mathcal{T}}, \mathbf{L}_w^d)$ has an $R-$basis consisting of sections of the form $b_1^i b_2^j$, where $i+j = d$ and $b_1 = (f^1)^{1+\phi^2}$, $b_2 = (f^2)^{1+\phi}$.

PROOF. The telescopic map in Proposition 6.19 induces a morphism of local $\delta - p-$rings

$$R\langle \mathbf{X}_{\mathcal{T}}, \mathbf{L}_w \rangle \to R\{u\}_{(p)};$$

to get the ring isomorphism in the statement of the Theorem it is enough to show that the induced morphism

$$k\langle \mathbf{X}_{\mathcal{T}}, \mathbf{L}_w \rangle \to k\langle u \rangle := k(u, u', u'', ...)$$

between the residue fields is an isomorphism with u equal to the image of $\frac{(f^2)^{1+\phi}}{(f^1)^{1+\phi^2}}$; the latter is, however, clear. To prove the last part of the Theorem note that by Proposition 6.17 any section $f \in H^0(\mathbf{X}_{\mathcal{T}}, \mathbf{L}_w^d)$ is a $K-$linear combination of products of the form

$$\prod_{a<b} (f^{ab})^{m_{ab}}, \quad \textit{with} \ \sum_{a<b} m_{ab}(\phi^a + \phi^b) = d(1 + \phi + \phi^2 + \phi^3). \tag{6.35}$$

Equation 6.35 translates into the following system:

$$\begin{aligned} m_{01} + m_{02} + m_{03} &= d, \\ m_{01} + m_{12} + m_{13} &= d, \\ m_{02} + m_{12} + m_{23} &= d, \\ m_{03} + m_{13} + m_{23} &= d. \end{aligned}$$

The above equalities imply

$$m_{01} = m_{23}, \ m_{02} = m_{13}, \ m_{03} = m_{12},$$

which shows that the product in Equation 6.35 has the form

$$b_1^{m_{01}} b_2^{m_{02}} b_3^{m_{03}},$$

where $b_3 := f^{03} f^{12}$. But one can check that $b_3 = b_2 - b_1$; so f is a K−linear combination

$$f = \sum_{i+j=d} c_{ij} b_1^i b_2^j \in K[x_0, x_1, ..., x_0''', x_1'''], \quad c_{ij} \in K.$$

On the other hand, by Theorem 6.7

$$f \in R[x_0, x_1, ..., x_0''', x_1''']. \tag{6.36}$$

We claim that $c_{ij} \in R$ and this will end the proof. We proceed by induction on the highest $j \in \{-1, 0, ..., d\}$, call it j_0, such that $c_{ij} \neq 0$ (where $j_0 = -1$ if $f = 0$). For $j_0 = -1$ everything is clear. Assume $f \neq 0$, let q be an indeterminate, and consider the ring homomorphism

$$K[x_0, x_1, ..., x_0''', x_1'''] \to K[q].$$

$$x_0 \mapsto 1, x_1 \mapsto q, \quad x_0', ..., x_0''', x_1', ..., x_1''' \mapsto 0.$$

Under this homomorphism, the image of f equals

$$\sum_{i+j=d} c_{ij}(q^p - q)^i (q^{p^3} - q^{p^2})^i (q^{p^2} - q)^j (q^{p^3} - q^p)^j = \sum_{i+j=d} c_{ij}(q^{p^3 d + p^2 j + pi} + ...), \tag{6.37}$$

where the dots mean "lower powers of q". The coefficient of $q^{p^3 d + p^2 j_0 + p i_0}$ ($i_0 = d - j_0$) in Equation 6.37 equals $c_{i_0 j_0}$; by Equation 6.36 this coefficient belongs to R and we conclude by the induction hypothesis. □

Next we address the problem of computing δ−cohomology. We have the following result which, as we shall see later, is in deep contrast with the situation for flat and hyperbolic correspondences.

PROPOSITION 6.21. *$H^1(\mathbf{X}_{\mathcal{T}}, \mathbf{L}_{r,w})$ are finitely generated R−modules. In particular $H^1(\mathbf{X}_{\mathcal{T}}, \mathbf{L}_{r,w})[p]$ are finite dimensional k−linear spaces.*

PROOF. This follows directly from the fact that the R−modules $H^0(\mathbf{P}^1, \mathcal{O}(1)^v)$ are finitely generated; cf. Theorem 6.7. □

The "easiest possible" example of non-zero p−torsion elements in H^1 is given by the polynomial $x_0^p x_1 - x_1^p x_0$ which defines an element in

$$H^0(\mathbf{X}_{\mathcal{T}}, \mathbf{L}_{p+1}; k).$$

Since, by Proposition 6.17, $H^0(\mathbf{X}_{\mathcal{T}}, \mathbf{L}_{p+1}) = 0$ it follows that $x_0^p x_1 - x_1^p x_0$ defines a non-zero element in $H^1(\mathbf{X}_{\mathcal{T}}, \mathbf{L}_{p+1})$; cf. the exact sequence in Equation 2.52.

6.4. Case Γ non-trivial: δ−invariants, δ−cohomology and δ−fiber

Let $\mathbf{X} = (X, \tilde{X}, \sigma_1, \sigma_2)$ be a spherical correspondence as in Definition 6.1. Recall that $\mathbf{X}$ is obtained as follows. Consider a finite subgroup $\Gamma = \{\gamma_1, ..., \gamma_m\} \subset SL_2(\mathbf{Z}_p)$ and an element $\tau \in SL_2(\mathbf{Z}_p)$. Let $Y \subset \mathbf{P}^1$ be a Γ−invariant affine open subset such that $X := Y/\Gamma$ is smooth and $\pi : Y \to X$ is étale. Let

$$\tilde{X} \subset Y \cap \tau^{-1}(Y).$$

Then

$$\mathbf{X} = \mathbf{X}^{\Gamma,\tau} := (X, \tilde{X}, \pi, \pi \circ \tau). \tag{6.38}$$

We intend to study the pair $(\mathbf{X}^{\Gamma,\tau},\mathbf{K}^{-w})$ via the pair $(\mathbf{X}_{\tau\circ\Gamma},\mathbf{L}_w)$ where

$$\mathbf{X}_{\tau\circ\Gamma} := \coprod_{i=1}^{m} \mathbf{X}_{\tau\circ\gamma_i},\ \ \mathbf{L}_w := \coprod_{i=1}^{m} \mathbf{L}_w(\tau\circ\gamma_i).$$

DEFINITION 6.22. A subgroup G of $SL_2(\mathbf{Z}_p)$ is *full* in $SL_2(\mathbf{Z}_p)$ if for all $n\geq 0$ the homomorphism

$$G \to PSL_2(\mathbf{Z}/p^{n+1}\mathbf{Z})$$

is surjective.

EXAMPLE 6.23. The group $SL_2(\mathbf{Z})$ is full in $SL_2(\mathbf{Z}_p)$ because

$$SL_2(\mathbf{Z}) \to SL_2(\mathbf{Z}/p^{n+1}\mathbf{Z})$$

is surjective. For a slightly more general example see Example 6.27.

THEOREM 6.24. *There is a natural telescopic map*

$$R(\mathbf{X}_{\tau\circ\Gamma},\mathbf{L}_w^2)^* \to R(\mathbf{X}^{\Gamma,\tau},\mathbf{K}^{-w})^*. \tag{6.39}$$

If, in addition, $w=1+\phi+\phi^2+\phi^3$ *the induced field extensions*

$$k\langle \mathbf{X}_{\tau\circ\Gamma},\mathbf{L}_w^2\rangle \subset k\langle \mathbf{X}^{\Gamma,\tau},\mathbf{K}^{-w}\rangle \subset k\langle X,K^{-w}\rangle \tag{6.40}$$

are algebraic. If in addition $\langle\Gamma,\tau\rangle$ *is full in* $SL_2(\mathbf{Z}_p)$ *then* $(\mathbf{X}^{\Gamma,\tau},\mathbf{K}^{-w})$ *is* $\delta-$ *rational and* $k\langle X,K^{-w}\rangle$ *can be embedded into a Galois extension of the field* $k\langle \mathbf{X}^{\Gamma,\tau},\mathbf{K}^{-w}\rangle$ *with Galois group* $PSL_2(\mathbf{Z}_p)$ *or* $PGL_2(\mathbf{Z}_p)$.

PROOF. Let $f\in H^0(\mathbf{X}_{\tau\circ\Gamma},\mathbf{L}_w^{2v})$ so in particular

$$f\in H^0(\mathbf{P}^1,\mathcal{O}_{\mathbf{P}^1}(1)^{2wv}) = R\{x_0,x_1\}^{2wv}$$

and f is $\langle\Gamma,\tau\rangle-$invariant. We shall naturally attach to f a section in

$$H^0(\mathbf{X}^{\Gamma,\tau},\mathbf{K}^{-wv})$$

as follows. Let s be a local basis of K_X^{-1}. Then $(\pi^*s)^{wv}$ is a local basis of K_Y^{-wv}. Recall that the canonical isomorphism

$$can: K_Y^{-1}\simeq \mathcal{O}(2),\ \ can((d(x_1/x_0)^{-1}) = x_0^2$$

is $SL_2(\mathbf{Z}_p)-$equivariant (cf. Lemma 6.10) hence so are the induced isomorphisms

$$can: K_Y^{-wv}\simeq \mathcal{O}(2)^{wv} = \mathcal{O}(1)^{2wv}.$$

Of course $can(\pi^*s)^{wv}$ is a local basis of $\mathcal{O}(1)^{2wv}$ so we may locally write

$$f=\varphi\cdot can(\pi^*s)^{wv},\ \ \ \varphi\in\mathcal{O}^r(Y),\ \ \ r=ord(wv).$$

By $SL_2(\mathbf{Z}_p)-$equivariance we get that, for any $\gamma\in\Gamma$, γ^* commutes with "can" hence:

$$f=\gamma^*f=\gamma^*\varphi\cdot(\gamma^*(can(\pi^*s)))^{wv} = \gamma^*\varphi\cdot(can(\gamma^*(\pi^*s)))^{wv} = \gamma^*\varphi\cdot can(\pi^*s)^{wv}$$

hence

$$\gamma^*\varphi=\varphi.$$

By Proposition 3.27, since $Y\to X$ a finite Galois étale map, we have

$$\varphi=\pi^*\psi,\ \ \ \psi\in\mathcal{O}^r(X).$$

Then

$$\psi\cdot s^{wv}$$

is a local section of K^{-wv}. These local sections clearly glue to give a global section in

$$H^0(X, K^{-wv}).$$

We claim this section belongs to

$$H^0(\mathbf{X}^{\Gamma,\tau}, \mathbf{K}^{-wv}).$$

Indeed

$$\begin{aligned} can(\tau^*\pi^*(\psi \cdot s^{wv})) &= \tau^*(can(\pi^*(\psi \cdot s^{wv}))) \\ = \tau^*(\varphi \cdot can(\pi^* s)^{wv}) &= \tau^* f = f = can(\pi^*(\psi \cdot s^{wv})); \end{aligned}$$

in particular

$$\tau^*\pi^*(\psi \cdot s^{wv}) = \pi^*(\psi \cdot s^{wv})$$

and our claim is proved. We have constructed the desired map 6.39 in the statement of our Theorem. If

$$h = \frac{(f^2)^{\phi+1}}{(f^1)^{\phi^2+1}}$$

is as in the statement of Proposition 6.18 and if $t = x_1/x_0$ then the image of h in $k\langle t\rangle$ equals

$$h = \frac{(t^{\phi^2} - t)^{\phi+1}}{(t^{\phi} - t)^{\phi^2+1}}.$$

So the classes mod p of h, δh, $\delta^2 h, ...$ are elements

$$\bar{h}_0, \bar{h}_1, \bar{h}_2, ... \in k\langle \mathbf{X}_{\tau\circ\Gamma}, \mathbf{L}_w\rangle.$$

To prove the rest of the assertions consider the field extensions

$$L_1 \subset L_2 \subset L_3 \subset L_4 \subset L_5 \subset L_6,$$

$$L_1 := k\langle h\rangle = k(\bar{h}_0, \bar{h}_1, \bar{h}_2, ...), \quad L_2 := k\langle \mathbf{X}_{\tau\circ\Gamma}, \mathbf{L}_w^2\rangle, \quad L_3 := k\langle \mathbf{X}^{\Gamma,\tau}, \mathbf{K}^{-w}\rangle,$$

$$L_4 := k\langle X, K^{-w}\rangle, \quad L_5 := k\langle Y, \mathcal{O}_{\mathbf{P}^1}(2)^w\rangle, \quad L_6 := k\langle t\rangle := k(t, t', t'', ...).$$

By Proposition 5.26 the extension $L_1 \subset L_6$ is algebraic so the extension $L_2 \subset L_4$ is algebraic. Assume in addition that $\langle \Gamma, \tau\rangle$ is full in $SL_2(\mathbf{Z}_p)$. Then

$$L_3 \subset L_6^{PSL_2(\mathbf{Z}_p)}.$$

On the other hand, by Proposition 5.26, $L_1 \subset L_6$ is Galois with Galois group $PGL_2(\mathbf{Z}_p)$, in particular

$$L_1 = L_6^{PGL_2(\mathbf{Z}_p)}.$$

Therefore the extension $L_1 \subset L_3$ has degree 2. There are 2 possibilities:

Case 1: $L_3 = L_6^{PSL_2(\mathbf{Z}_p)}$. In this case, by Proposition 5.28,

$$L_3 = k\langle v\rangle, \quad v = \frac{(t^{\phi^2} - t)^{\frac{\phi+1}{2}}}{(t^{\phi} - t)^{\frac{\phi^2+1}{2}}} \in (R\{t\}_{(p)})\hat{}.$$

Since L_6/L_3 is separable and the images

$$\bar{v}_0, \bar{v}_1, \bar{v}_2, ... \in L_6$$

of $v, \delta v, \delta^2 v, ...$ belong to L_3 it follows by Lemma 5.10 that

$$v \in R\langle \mathbf{X}^{\Gamma,\tau}, \mathbf{K}^{-w}\rangle\hat{}$$

hence, in particular, $(\mathbf{X}^{\Gamma,\tau}, \mathbf{K}^{-w})$ is δ−rational. Also L_4 is embedded into the Galois extension L_6 of L_3 with Galois group $PSL_2(\mathbf{Z}_p)$ and we are done.

Case 2: $L_3 = L_6^{PGL_2(\mathbf{Z}_p)}$. In this case $L_3 = k\langle h\rangle$ where

$$h \in R\langle \mathbf{X}^{\Gamma,\tau}, \mathbf{K}^{-w}\rangle^{\wedge}$$

and L_4 is embedded into the Galois extension L_6 of L_3 with Galois group $PGL_2(\mathbf{Z}_p)$ and we are done again. □

REMARK 6.25. Theorem 6.24 clearly implies Theorem 2.59. One only needs to note that if Y is a smooth affine scheme over a ring of $S-$integers $\mathcal{O}$ in a number field and Γ is a finite group of automorphisms of Y such that $Y \to Y/\Gamma$ is flat and if $R := \hat{\mathcal{O}}_{\wp}^{ur}$ for a place $\wp$ then

$$(Y \otimes R)/\Gamma \simeq (Y/\Gamma) \otimes R.$$

REMARK 6.26. The argument in the proof above can be used to prove Lemma 6.4. Indeed one can consider the section

$$\bar{f} := x_0^p x_1 - x_1^p x_0 \in H^0(\mathbf{P}^1_k, \mathcal{O}(p+1))$$

and the isomorphism

$$can : K_Y^{-(p+1)/2} \simeq \mathcal{O}(p+1)$$

as in the proof above. Exactly as in that proof one can show that

$$\bar{f} = can(\pi^*\bar{\eta}), \quad \bar{\eta} \in H^0(X \otimes k, K^{-(p+1)/2}), \quad \tau^*\pi^*\bar{\eta} = \pi^*\bar{\eta}.$$

We conclude that $\mathbf{K}_{\mathbf{U}\otimes k}^{-(p+1)/2}$ is trivial (hence so is its inverse) on a suitable non-empty open subcorrespondence $\mathbf{U}$ of $\mathbf{X}$.

EXAMPLE 6.27. Let

$$\gamma = \begin{bmatrix} 0 & -1 \\ 1 & 0 \end{bmatrix}, \quad \tau = \begin{bmatrix} 1 & b \\ 0 & 1 \end{bmatrix},$$

where $b \in \mathbf{Z}$, $b \not\equiv 0 \bmod p$. Set $\Gamma = \langle\gamma\rangle$. Since γ and τ have no common complex eigenvector, $\langle \Gamma \circ \tau\rangle = \langle \Gamma, \tau\rangle$ is not solvable. So, in particular, by Theorem 6.20, $(\mathbf{X}_{\Gamma\circ\tau}, \mathbf{L}_w)$ is $\delta-$rational, provided $w = 1 + \phi + \phi^2 + \phi^3$. Note that γ and τ act as $\gamma^* x_0 = x_1$, $\gamma^* x_1 = -x_0$, $\sigma^* x_0 = x_0$, $\sigma^* x_1 = bx_0 + x_1$; so if $t := x_1/x_0$ then

$$\gamma^*(t) = -t^{-1}, \quad \tau^*(t) = t + b.$$

The invariant ring of Γ is

$$R[x_0, x_1]^{\Gamma} = R[x_0^2 + x_1^2, x_0x_1(x_0^2 - x_1^2), x_0^2x_1^2].$$

The sum of the pieces of degree divisible by 4 in the above ring is the same as the sum of the pieces of degree divisible by 4 in the ring $R[x_0x_1, x_0^2 - x_1^2]$. So $\mathbf{P}^1/\Gamma$ is naturally isomorphic to a projective line $Proj\ R[\xi_0, \xi_1]$ where $\pi : \mathbf{P}^1 \to \mathbf{P}^1/\Gamma$ is defined by $\pi^*\xi_0 = g_0 := x_0x_1$, $\pi^*\xi_1 = g_1 := x_0^2 - x_1^2$. Note that $\gamma^* g_0 = -g_0$ and $\gamma^* g_1 = -g_1$. If $s = \xi_1/\xi_0$ then $\pi^*(s) = t - t^{-1}$. The fixed points of γ are the solutions to the equation $t^2 + 1 = 0$. Take, in what follows,

$$Y := \mathbf{P}^1 \backslash Z(\Phi)$$

where $Z(\Phi)$ is the closed subscheme defined by the form $\Phi := x_0^2 + x_1^2$. The maps $\sigma_1 = \pi, \sigma_2 = \pi \circ \tau : \tilde{X} \to X$ are given by

$$\sigma_1^*(s) = S_1(t) := t - t^{-1}, \quad \sigma_2^*(s) = S_2(t) := (t+b) - (t+b)^{-1}.$$

One readily sees that the closure of the image of the map

$$\sigma_1 \times \sigma_2 : \tilde{X} \to X \times X \subset \mathbf{P}^1 \times \mathbf{P}^1,$$

base changed to K, has affine equation

$$s_1(s_1 - s_2 + 2b)(bs_2 - b^2) = (bs_2 - b^2)^2 - (s_1 - s_2 + 2b)^2,$$

where s_1, s_2 are the rational functions on $\mathbf{P}^1 \times \mathbf{P}^1$ arising from s.

Note now that $\langle \gamma, \tau \rangle$ is actually full. Indeed, for any $n \geq 0$, there exists $m_n \in \mathbf{Z}$ such that $m_n b \equiv 1 \bmod p^{n+1}$; so

$$\tau^{m_n} \equiv \begin{bmatrix} 1 & 1 \\ 0 & 1 \end{bmatrix} \quad mod \quad p^{n+1}$$

so the image of $\langle \gamma, \tau \rangle$ in $PSL_2(\mathbf{Z}/p^{n+1}\mathbf{Z})$ contains the images of

$$\begin{bmatrix} 0 & -1 \\ 1 & 0 \end{bmatrix}, \begin{bmatrix} 1 & 1 \\ 0 & 1 \end{bmatrix}.$$

But the latter images generate $PSL_2(\mathbf{Z}/p^{n+1}\mathbf{Z})$ because the homomorphism

$$SL_2(\mathbf{Z}) \to SL_2(\mathbf{Z}/p^{n+1}\mathbf{Z})$$

is surjective. This proves our claim. Consequently, by Theorem 6.24, $(\mathbf{X}^{\Gamma,\tau}, \mathbf{K}^{-w})$ is $\delta-$rational and $k\langle X, K^{-w} \rangle$ can be embedded into a Galois extension of the field $k\langle \mathbf{X}^{\Gamma,\tau}, \mathbf{K}^{-w} \rangle$ with Galois group $PSL_2(\mathbf{Z}_p)$ or $PGL_2(\mathbf{Z}_p)$.

Assume, in addition, that

$$b^2 \neq 1, 4, \quad \left(\frac{b^2 - 4}{p} \right) = 1.$$

(It is easy to find infinitely many examples when this happens; for instance one can take $p \equiv 1 \bmod 20$ and $b \equiv 3 \bmod p$.) By our Legendre symbol assumption, the equation $t^2 + bt + 1 = 0$ has at least a solution $\theta \in \mathbf{Z}_p$. Let $\tilde{P} \in \mathbf{P}^1(R)$ be the point with affine coordinate $t = \theta$. Since $b \not\equiv 0 \bmod p$, it follows that $\bar{\Phi}$, the reduction mod p of Φ, does not vanish at the reduction mod p, $(1 : \bar{\theta}) \in \mathbf{P}^1(k)$ of $\tilde{P}$; hence $\tilde{P} \in Y(R)$. Similarly $\bar{\Phi}$ does not vanish at the reduction mod p, $(1 : \bar{\theta} + \bar{b}) \in \mathbf{P}^1(k)$, of $\tau(\tilde{P})$; hence $\tilde{P} \in (\tau^{-1}Y)(R)$. So, assuming in addition that $\tilde{X} = Y \cap \tau^{-1}(Y)$, we have $\tilde{P} \in \tilde{X}(R)$. Let $P := \pi(\tilde{P}) \in X(R)$. In particular Since $\tau(\theta) = \gamma(\theta)$, $(\tilde{P}, P)$ is a fixed point of the correspondence $\mathbf{X}^{\Gamma,\tau}$. The multiplier of this correspondence at this fixed point is

$$\left(\frac{dS_2}{dt}(\theta) \right) \left(\frac{dS_1}{dt}(\theta) \right)^{-1} = \left(1 + \frac{1}{(\theta + b)^2} \right) \left(1 + \frac{1}{\theta^2} \right)^{-1} = \theta^2.$$

Since $b \neq -2, -1, 0, 1, 2$, it follows that θ is not a root of unity. By Theorem 4.61 the $R-$modules $H^0(\mathbf{X}^{\Gamma,\tau}, \mathbf{L})$ are finitely generated for any $\delta-$line bundle $\mathbf{L}$. On the other hand, if $w = 1 + \phi + \phi^2 + \phi^3$, the $R-$module $H^0(\mathbf{X}^{\Gamma,\tau}, \mathbf{K}^{-w})$ is quite non-trivial: indeed, by Theorem 6.24, this module contains $H^0(\mathbf{X}_{\tau \circ \Gamma}, \mathbf{L}_w^2)$ and the latter contains non-trivial elements like

$$(f^1)^{i(1+\phi^2)}(f^2)^{j(1+\phi)}, \quad i + j = 2, \quad i, j \geq 0.$$

Next we address the problem of computing $\delta-$cohomology. We have the following result which, as we shall see, contrasts with the situation in the flat and hyperbolic cases.

PROPOSITION 6.28. *Assume* $Y = \mathbf{P}^1 \backslash Z(\Phi)$, $\Phi \in \mathcal{O}[x_0, x_1]$ *a form of degree* $d < p$, $\Phi(0, x_1) \not\equiv 0$ *mod* p, $\tau = \begin{bmatrix} 1 & 1 \\ 0 & 1 \end{bmatrix}$, *and let* m *be an even integer. Then the* $k-$*linear space* $H^1(\mathbf{X}^{\Gamma,\tau}, \mathbf{K}^{-m})[p]$ *is finite dimensional.*

PROOF. We view $H^0(X, K^{-m}) \otimes k$ as embedded into $H^0(Y, \mathcal{O}(2m)) \otimes k$ via π^*. The latter space identifies with the space of all rational fractions

$$\frac{\bar{G}(x_0, x_1)}{\bar{\Phi}(x_0, x_1)^n} = \frac{\bar{G}(1,t)}{\bar{\Phi}(1,t)^n} \cdot x_0^{2m} \tag{6.41}$$

where $\bar{G}(x_0, x_1) \in k[x_0, x_1]$ is homogeneous of degree $2m + nd$, $\bar{\Phi}$ is the reduction of Φ mod p, and $t = x_1/x_0$. So the space $H^0(\mathbf{X}^{\Gamma,\tau}, \mathbf{K}^{-m}; k)$ can be embedded into the space of all fractions in Equation 6.41 which are τ–invariant. Since $\tau x_0 = x_0$, any τ–invariant fraction 6.41 yields an element in

$$k(t)^\tau \cdot x_0^m = k(t^p - t) \cdot x_0^m,$$

hence we can write

$$\frac{\bar{G}(1,t)}{\bar{\Phi}(1,t)^n} = c \cdot \frac{\prod_{i=1}^{\nu}(t^p - t - \alpha_i)}{\prod_{j=1}^{\mu}(t^p - t - \beta_j)}, \tag{6.42}$$

for some $c, \alpha_i, \beta_j \in k$, $\nu, \mu \geq 0$, $\alpha_i \neq \beta_j$ (all i, j). We claim that $\mu = 0$. Indeed, if $\mu \geq 1$, the polynomial $t^p - t - \beta_1$ has p distinct roots in k and all these roots must also be roots of $\bar{\Phi}(1,t)$ contradicting the fact that Φ has degree $< p$; this proves our claim. By the claim, the right hand side of Equation 6.42 is a polynomial of degree $N = p\nu$. If $N > 2m$ the right hand side of Equation 6.41 has the form $P(x_0, x_1)/x_0^{N-2m}$ where P is a polynomial not divisible by x_0; then Equation 6.41 implies that x_0 divides $\bar{\Phi}(x_0, x_1)$, a contradiction. So $N \leq 2m$, hence the space of all τ–invariant fractions in Equation 6.41 has dimension $\leq 2m + 1$. This shows that the k–linear space $H^0(\mathbf{X}^{\Gamma,\tau}, \mathbf{K}^{-m}; k)$ and hence the space $H^1(\mathbf{X}^{\Gamma,\tau}, \mathbf{K}^{-m})[p]$ is finite dimensional. $\square$

6.5. Case $\langle \Gamma, \tau \rangle$ solvable

In the last Section we concentrated on the case when our group $\langle \Gamma, \tau \rangle$ is full, in particular not solvable. Here we would like to consider the solvable case. Although the solvable case should in principle be easier we will see that an extra complication arises from the role of "Wieferich primes" (which played no role in the "full" case).

We continue to investigate the correspondences $X^{\Gamma,\tau}$ introduced in the previous Section. We assume in addition that

$$Y \subset \mathbf{A}^1 = Spec\ R[t],\ t = x_1/x_0,$$
$$\Gamma \subset SA(\mathbf{Z}_p),\quad \tau \in SA(\mathbf{Z}_p).$$

We view $SA(\mathbf{Z}_p) \subset PSL_2(\mathbf{Z}_p)$ as acting, in the natural way on the affine line $\mathbf{A}^1$.

DEFINITION 6.29. 1) A subgroup G of $SA(\mathbf{Z}_p)$ is *fully Borel* in $SA(\mathbf{Z}_p)$ if there exists an even integer d dividing $p - 1$ such that for all $n \geq 0$ the image of the homomorphisms

$$G \to SA(\mathbf{Z}/p^{n+1}\mathbf{Z})$$

coincides with the group

$$GA^{(d)}(\mathbf{Z}/p^{n+1}\mathbf{Z}).$$

2) A subgroup G of $SA(\mathbf{Z}_p)$ is *fully toral* in $SA(\mathbf{Z}_p)$ if there exists an even integer d dividing $p - 1$ such that for all $n \geq 0$ the image of the homomorphisms

$$G \to SA(\mathbf{Z}/p^{n+1}\mathbf{Z})$$

coincides with the group

$$\mathbf{G}_m(\mathbf{Z}/p^{n+1}\mathbf{Z})^d.$$

3) A subgroup G of $SA(\mathbf{Z}_p)$ is *fully unipotent* in $SA(\mathbf{Z}_p)$ if for all $n\geq 0$ the image of the homomorphisms

$$G\to SA(\mathbf{Z}/p^{n+1}\mathbf{Z})$$

coincides with the group

$$\mathbf{G}_a(\mathbf{Z}/p^{n+1}\mathbf{Z}).$$

THEOREM 6.30. *Set $w=1+\phi^2$. The field extension*

$$k\langle\mathbf{X}^{\Gamma,\tau},\mathbf{K}^{-w}\rangle\subset k\langle X,K^{-w}\rangle \tag{6.43}$$

is algebraic. If in addition $\langle\Gamma,\tau\rangle$ is fully Borel in $SL_2(\mathbf{Z}_p)$ then $(\mathbf{X}^{\Gamma,\tau},\mathbf{K}^{-w})$ is $\delta-$rational and $k\langle X,K^{-w}\rangle$ can be embedded into a Galois extension of the field $k\langle\mathbf{X}^{\Gamma,\tau},\mathbf{K}^{-w}\rangle$ with Galois group $GA^{(d)}(\mathbf{Z}_p)$ where d divides $p-1$.

PROOF. Consider the sections

$$f_1:=(t^\phi-t)^2(dt)^{-w},\quad f_2:=(t^\phi-t)^{\phi+1}(dt)^{-w}\in H^0(Y,K^{-w}).$$

Clearly

$$g^*f_i=f_i,\quad i=1,2,$$

for all $g\in SA(\mathbf{Z}_p)$. We shall naturally attach to f_i sections

$$\omega_i\in H^0(\mathbf{X}^{\Gamma,F,\tau},\mathbf{K}^{-w})$$

as follows. Let s be a local basis of K_X^{-1}. Then $(\pi^*s)^w$ is a local basis of K_Y^{-w}. So we may locally write

$$f_i=\varphi_i\cdot(\pi^*s)^w,\quad \varphi\in\mathcal{O}^2(Y).$$

For any $\gamma\in\Gamma$ we have

$$f_i=\gamma^*f_i=\gamma^*\varphi_i\cdot(\gamma^*(\pi^*s))^w=\gamma^*\varphi_i\cdot(\pi^*s)^{wv}$$

hence

$$\gamma^*\varphi_i=\varphi_i.$$

By Proposition 3.27, since $Y\to X$ is a finite Galois étale map, we have

$$\varphi_i=\pi^*\psi_i,\quad \psi_i\in\mathcal{O}^2(X).$$

Then

$$\psi_i\cdot s^w$$

is a local section of K^{-w}. These local sections clearly glue to give global sections

$$\omega_i\in H^0(X,K^{-w}).$$

We claim these sections belong to

$$H^0(\mathbf{X}^{\Gamma,\tau},\mathbf{K}^{-w}).$$

Indeed

$$\tau^*\pi^*(\psi\cdot s^{wv})=\tau^*(\varphi\cdot(\pi^*s)^{wv})=\tau^*f=f=\pi^*(\psi\cdot s^{wv})$$

and our claim is proved. The image of

$$h=\frac{f_2}{f_1}$$

in $k\langle t\rangle$ equals

$$h=(t^\phi-t)^{\phi-1}.$$

So the classes mod p of h, δh, $\delta^2 h$, ... are elements

$$\bar{h}_0,\bar{h}_1,\bar{h}_2,...\in k\langle\mathbf{X}^{\Gamma,\tau},\mathbf{K}^{-w}\rangle.$$

Consider the field extensions

$$L_1 \subset L_3 \subset L_4 \subset L_5 \subset L_6,$$

$$L_1 := k\langle h\rangle = k(\bar{h}_0, \bar{h}_1, \bar{h}_2, ...), \quad L_3 := k\langle \mathbf{X}^{\Gamma,\tau}, \mathbf{K}^{-w}\rangle,$$

$$L_4 := k\langle X, K^{-w}\rangle, \quad L_5 := k\langle Y, K^{-w}\rangle, \quad L_6 := k\langle t\rangle := k(t, t', t'', ...).$$

By Proposition 5.29 the extension $L_1 \subset L_6$ is algebraic so the extension $L_3 \subset L_4$ is algebraic. Assume in addition that $\langle \Gamma, \tau\rangle$ is fully Borel. Then

$$L_3 \subset L_6^{GA_2^{(e)}(\mathbf{Z}_p)}$$

for some e dividing $p-1$. On the other hand, by Proposition 5.29, $L_1 \subset L_6$ is Galois with Galois group $GA_2(\mathbf{Z}_p)$. By Galois theory there exists a divisor d of e such that

$$L_3 = L_6^{GA^{(d)}(\mathbf{Z}_p)}.$$

By Proposition 5.30 we have

$$L_3 = k\langle v\rangle, \quad v = (t^{\phi} - t)^{\frac{\phi-1}{d}} \in (R\{t\}_{(p)})\hat{}.$$

Since L_6/L_3 is separable and the images

$$\bar{v}_0, \bar{v}_1, \bar{v}_2, ... \in L_6$$

of $v, \delta v, \delta^2 v, ...$ belong to L_3 it follows by Lemma 5.10 that

$$v \in R\langle \mathbf{X}^{\Gamma,\tau}, \mathbf{K}^{-w}\rangle\hat{}$$

hence, in particular, $(\mathbf{X}^{\Gamma,F,\tau}, \mathbf{K}^{-w})$ is δ–rational. Also L_4 is embedded into the Galois extension L_6 of L_3 with Galois group $GA^{(d)}(\mathbf{Z}_p)$ and we are done. □

Similarly one proves:

THEOREM 6.31. *Set $w = 1 + \phi$ and assume $\langle \Gamma, \tau\rangle$ is fully toral respectively fully unipotent in $SL_2(\mathbf{Z}_p)$. Then $(\mathbf{X}^{\Gamma,\tau}, \mathbf{K}^{-w})$ is δ–rational and $k\langle X, K^{-w}\rangle$ can be embedded into a Galois extension of the field $k\langle \mathbf{X}^{\Gamma,\tau}, \mathbf{K}^{-w}\rangle$ with Galois group $\mathbf{Z}_p^{\times d}$ (d dividing $p-1$) respectively $\mathbf{Z}_p$. In addition $BS(\mathbf{X}^{\Gamma,\tau}, \mathbf{K}^{-1})$ has at most one point (in the toral case) respectively is empty (in the unipotent case).*

PROOF. Same argument as for the previous Theorem; instead of the f_1 and f_2 there one needs to use

$$f_1 = t^2(dt)^{-w}, \quad f_2 = t^{\phi+1}(dt)^{-w}$$

and

$$f_1 = (dt)^{-w}, \quad f_2 = (t^{\phi} - t)(dt)^{-w}$$

respectively. □

EXAMPLE 6.32. Here is a typical example when $\langle \Gamma, \tau\rangle$ is a fully Borel subgroup of $SA(\mathbf{Z}_p)$. Let $\Gamma := \langle \gamma\rangle$,

$$\gamma = \begin{bmatrix} -1 & 1 \\ 0 & 1 \end{bmatrix}, \quad \tau = \begin{bmatrix} a & 1 \\ 0 & 1 \end{bmatrix},$$

where $a \in \mathbf{Z}$ satisfies the following conditions

$$\left(\frac{a}{p}\right) = \left(\frac{-1}{p}\right) = 1 \tag{6.44}$$

$$a \not\equiv -1 \quad mod \quad p \tag{6.45}$$

$$a^{p-1} \not\equiv 1 \quad mod \quad p^2. \tag{6.46}$$

Condition 6.46 is usually referred to as saying that p is not a *Wieferich prime* to base a; recall that it is unknown whether for a given $a \neq 1, -1$ there are always infinitely many non-Wieferich primes to base a. Due to condition 6.44, $\gamma, \tau \in SA(\mathbf{Z}_p)$. Of course $\gamma^2 = id$.

We claim that $\langle \gamma, \tau \rangle$ is fully Borel. Start by noting that

$$\tau \cdot \gamma \cdot \tau^{-1} \cdot \gamma^{-1} = \begin{bmatrix} 1 & a+1 \\ 0 & 1 \end{bmatrix}.$$

By Condition 6.45 it follows that for any n the image of $\langle \gamma, \tau \rangle$ in $SA(\mathbf{Z}/p^{n+1}\mathbf{Z})$ contains the group

$$\mathbf{G}_a(\mathbf{Z}/p^{n+1}\mathbf{Z}).$$

Let $(p-1)/d$ be the order of a in $(\mathbf{Z}/p\mathbf{Z})^{\times}$. To conclude the proof of our claim it is enough to show that the group generated by a in $(\mathbf{Z}/p^{n+1}\mathbf{Z})^{\times}$ coincides with $(\mathbf{Z}/p^{n+1}\mathbf{Z})^{\times d}$. This is an easy exercise using condition 6.46.

Similar examples can be given of fully toral and fully unipotent subgroups.

6.6. A converse theorem: biquadratic correspondences

In this section we will analyze the "universal family of polynomial biquadratic correspondences" on the affine line over $\mathbf{Z}$. We will show that for a "generic" such correspondence we have certain vanishings $H^0(\mathbf{X}, \mathbf{L}) = 0$. On the other hand there is a whole subfamily of such correspondences which are spherical and for which the corresponding vanishing doesn't hold.

Let $a_i, b_i, c_i \in \mathbf{Z}$ be integers, where $i = 1, 2$, and consider the quadratic polynomials $Q_i = a_i t^2 + b_i t + c_i \in \mathbf{Z}[t]$. Set $\Delta_i = b_i^2 - 4a_i c_i$, $i = 1, 2$, and define

$$\Delta_{12} := (b_1 - b_2)^2 - 4(a_1 - a_2)(c_1 - c_2). \tag{6.47}$$

Consider the maps $\sigma_i : \mathbf{A}^1_{\mathbf{Z}} = Spec\ \mathbf{Z}[t] \to \mathbf{A}^1_{\mathbf{Z}}$, $i = 1, 2$, given by the homomorphisms $\sigma_i^* : \mathbf{Z}[t] \to \mathbf{Z}[t]$, $\sigma_i^* t = Q_i$. Set

$$X_i := Spec\ \mathbf{Z}[t, (4a_i t + \Delta_i)^{-1}] \subset \mathbf{A}^1_{\mathbf{Z}},$$

$$\tilde{X}_i := \sigma_i^{-1}(X_i) = Spec\ \mathbf{Z}[t, (2a_i t + b_i)^{-1}] \subset \mathbf{A}^1_{\mathbf{Z}},$$

$$X := X_1 \cap X_2, \quad \tilde{X} := \sigma_1^{-1}(X) \cap \sigma_2^{-1}(X).$$

Consider the correspondence

$$\mathbf{X}_{Q_1,Q_2} := (X, \tilde{X}, \sigma_1, \sigma_2) \tag{6.48}$$

and, for any rational prime p, let $\mathbf{X}_{Q_1,Q_2,p}$ be the induced correspondence over $R_p := \hat{\mathbf{Z}}_p^{ur}$. Note that for any p not dividing $a_1 a_2$ the correspondence $\mathbf{X}_{Q_1,Q_2,p}$ is étale. It is trivial to check that the canonical bundle $\mathbf{K}$ of $\mathbf{X}_{Q_1,Q_2,p}$ is isomorphic to $(\mathcal{O}, \beta)$ where β is the composition

$$\beta : \sigma^* \mathcal{O}_X \overset{can}{\simeq} \mathcal{O}_{\tilde{X}} \overset{B}{\to} \mathcal{O}_{\tilde{X}} \overset{can}{\simeq} \iota^* \mathcal{O}_X,$$

and the middle arrow is the multiplication by B map,

$$B := \frac{2a_2 t + b_2}{2a_1 t + b_1}.$$

We simply write $\mathbf{K} \simeq (\mathcal{O}, B)$. Let $w \in W$ with $deg(w) = w(1) = 2m$ even. We can define a "square root" of $\mathbf{K}^w$ by the formula

$$\mathbf{L}_w := (\mathcal{O}, B^{w/2}); \tag{6.49}$$

cf. Remark 3.23. Clearly

$$\mathbf{L}_w^2 \simeq \mathbf{K}^w; \tag{6.50}$$

so $\mathbf{L}_w$ is polarization equivalent to $\mathbf{K}^{-1}$ if $-w \in W_+ \backslash \phi W_+$.

THEOREM 6.33. *There exists a non-zero polynomial H in 6 variables, with $\mathbf{Z}$–coefficients, such that the following holds. Assume*

1) $H(a_1, b_1, c_1, a_2, b_2, c_2) \neq 0$,

2) Δ_{12} is not a perfect square.

Then there exist infinitely many primes p such that

$$H^0(\mathbf{X}_{Q_1,Q_2,p}, \mathbf{L}_{-1-\phi}) = 0.$$

REMARK 6.34. Note that if Q_1, Q_2 are such that

$$Q_2(t) = Q_1(mt + n) \tag{6.51}$$

for some $m, n \in \mathbf{Z}$ then

$$\Delta_{12} = (b_1 - mb_1 + 2a_1 n)^2,$$

in particular Δ_{12} is a perfect square. So the above Theorem does not apply to such a situation. By the way, if the condition in Equation 6.51 holds, the correspondence $\mathbf{X}_{Q_1,Q_2,p}$ is spherical. We will examine this case later.

PROOF. As we will see our determination of H will be entirely effective. Define the polynomials

$$Q = (a_1 - a_2)t^2 + (b_1 - b_2)t + (c_1 - c_2),$$
$$Q_{12} = \sigma_1^*(4a_2 t + \Delta_2) = 4a_1 a_2 t^2 + 4a_2 b_1 t + b_2^2 + 4a_2(c_1 - c_2),$$
$$Q_{21} = \sigma_2^*(4a_1 t + \Delta_1) = 4a_1 a_2 t^2 + 4a_1 b_2 t + b_1^2 + 4a_1(c_2 - c_1).$$

Let R_{12} be the resultant of Q and Q_{12}; similarly let R_{21} be the resultant of Q and Q_{21}. Clearly R_{12} and R_{21} are polynomial expressions in $a_1, b_1, c_1, a_2, b_2, c_2$ and they are not identically zero. (Indeed R_{12} does not vanish for $a_2 = 1$, $b_2 = 0$, $c_2 = 0$, $c_1 = 1$.) The resultant of Q and $2a_1 t + b_1$ is, up to a constant,

$$C_1 = (a_1 - a_2)b_1^2 - 2a_1 b_1(b_1 - b_2) + 4a_1^2(c_1 - c_2).$$

Similarly the resultant of Q and $2a_2 t + b_2$ is, up to a constant,

$$C_2 = (a_1 - a_2)b_2^2 - 2a_2 b_2(b_1 - b_2) + 4a_2^2(c_1 - c_2).$$

Define now

$$H(a_1, b_1, c_1, a_2, b_2, c_2) = a_1 a_2(a_1 - a_2)(a_1 b_2 - a_2 b_1)C_1 C_2(C_1^2 - C_2^2)R_{12}R_{21},$$

assume the above expression is not 0 and assume Δ_{12} is not a perfect square. Then there exist infinitely many primes p such that

$$\left(\frac{\Delta_{12}}{p}\right) = -1. \tag{6.52}$$

We will show that for any sufficiently large p satisfying condition 6.52, $\mathbf{X}_{Q_1,Q_2,p}$ possesses a fixed point P with multiplier $a = a(P) \in R_p^\times$ with $a^{\phi^2} = a$, $a^\phi \neq a$, and $aa^\phi \in \mathbf{Z}_p$, aa^ϕ not a root of unity. This will close the proof by Corollary 4.67. Indeed let $\theta \in R_p$ be a root of Q. Due to the non-vanishing mod p of our resultants,

θ defines a fixed point P of $\mathbf{X}_{Q_1,Q_2,p}$. By condition 6.52 we have $\theta^\phi \neq \theta$ and, of course, $\theta^{\phi^2} = \theta$. The multiplier $a = a(P)$ equals

$$a = \frac{2a_2\theta + b_2}{2a_1\theta + b_1}.$$

Then the conditions $a_1b_2 - a_2b_1 \neq 0$, $\theta \neq \theta^\phi$ imply $a \neq a^\phi$. Finally an easy computation yields

$$aa^\phi = \frac{C_2}{C_1} \in \mathbf{Q};$$

so if aa^ϕ were a root of unity then it would equal 1 or -1 which would contradict the condition $C_1^2 - C_2^2 \neq 0$. □

The conditions 1 and 2 in the Theorem above are crucial as shown by the following:

PROPOSITION 6.35. *Assume Q_1, Q_2 are such that $Q_2(t) = Q_1(mt+n)$ with $m \in \{1,-1\}$ and $n \in \mathbf{Z}$. Then, for any prime p not dividing a_1, the R module*

$$H^0(\mathbf{X}_{Q_1,Q_2,p}, \mathbf{L}_{-1-\phi})$$

has infinite rank.

PROOF. Choose any integer $l \geq 0$ such that $\frac{p+1}{2} + l$ is even. We claim that

$$m^{\frac{1+\phi}{2}} = m^l. \tag{6.53}$$

Indeed, this is clear if $m = 1$; if $m = -1$ we get

$$(-1)^{\frac{1+\phi}{2}} = (-1)\cdot\left(\frac{-1}{p}\right) = (-1)^{1+\frac{p-1}{2}} = (-1)^l.$$

Define

$$g := (2a_1t + b_1)^{\frac{1+\phi}{2}}(t^\phi - t)^l \in R_p[t, (2a_1t+b_1)^{-1}, t']\hat{} \in \mathcal{O}(J^1(X_1)).$$

The map $\tilde{X}_1 \to X_1$ pulled back to R_p is a finite étale Galois cover with Galois groups generated by the involution γ acting as $\gamma^* t = -a_1^{-1}b_1 - t$. Note that g is $\gamma-$ invariant; indeed:

$$\gamma^* g = (-2a_1t - b_1)^{\frac{1+\phi}{2}}(-t^\phi + t)^l = (-1)^{\frac{1+\phi}{2}}\cdot(-1)^l\cdot g = g.$$

By Proposition 3.27, there exists $f \in \mathcal{O}(J^1(\tilde{X}_1))$ such that $\sigma_1^* f = g$. We claim that f defines a section in $H^0(\mathbf{X}_{Q_1,Q_2,p}, \mathbf{L}_{-1-\phi})$ which will of course end the proof. To check the claim set $L(t) = mt+n$; then $\sigma_2^* = L^* \circ \sigma_1^*$. Also $a_2 = a_1m^2$, $b_2 = 2a_1mn + mb_1$. Hence

$$\left(\frac{2a_2t + b_2}{2a_1t+b_1}\right)^{\frac{-1-\phi}{2}}\cdot\sigma_2^* f = \left(\frac{2a_1m^2t + 2a_1mn + mb_1}{2a_1t+b_1}\right)^{\frac{-1-\phi}{2}}\cdot L^* g$$

$$= m^{\frac{-1-\phi}{2}}\cdot\left(\frac{2a_1(mt+n)+b_1}{2a_1t+b_1}\right)^{\frac{-1-\phi}{2}}\cdot(2a_1(mt+n)+b_1)^{\frac{1+\phi}{2}}\cdot((mt+n)^\phi - (mt+n))^l$$

$$= m^{\frac{-1-\phi}{2}}\cdot m^l\cdot(2a_1t+b_1)^{\frac{1+\phi}{2}}\cdot(t^\phi - t)^l = g = \sigma_1^* f.$$

□

CHAPTER 7

Flat correspondences

We will adopt again a "vertical" viewpoint by developing the theory for a fixed p. Since, by Proposition 1.27, the existence of a flat analytic uniformization is a purely algebraic property it will be straightforward to define a notion of *flat correspondence* over $R := \hat{\mathbf{Z}}_p^{ur}$. Then we shall analyze flat correspondences over R from the viewpoint of δ–geometry. We start by following [**17**] and [**29**] where this work was initiated; then we shall push the theory forward by proving a series of new results. In particular this will allow us to prove our "horizontal" Main Theorem 2.60.

Here is the structure of this Chapter. In Section 1 we define *flat correspondences* over R and we examine line bundles and cycles on them. In Section 2 we review and complement the theory of δ–characters [**17**] which is crucial for everything that follows. A δ–character is simply a δ–function on the R–points of a group scheme which is a homomorphism to the additive group of R; as such it is in particular a δ–section of an appropriate flat δ–line bundle. In Section 3 we will essentially show that all δ–invariants of flat correspondences are δ–polynomials in δ–characters; in doing so a crucial role will be played by our theory of conjugate operators introduced previously. In Sections 4 and 5 we examine δ–cohomology and δ–fibers respectively. In Section 6 we prove some "converse theorems". In particular we show that, among the correspondences arising from quadratic dynamical systems $t \to t^2 + c$ with $c \in \mathbf{Z}$, the correspondences with $c = 0$ and $c = -2$ are essentially the only ones that possess δ–invariants.

7.1. Flat correspondences over R: δ–line bundles and cycles

Start with a tuple $(G, \Gamma, \tilde{\Gamma}, \tau_1, \tau_2)$ where G is either an elliptic curve over R or isomorphic to $\mathbf{G}_m$ over R, Γ and $\tilde{\Gamma}$ are finite groups of group scheme automorphisms of G and $\tau_1, \tau_2 : G \to G$ are finite étale maps of schemes such that $\tau_1 \tilde{\Gamma} \subset \Gamma \tau_1$ and $\tau_2 \tilde{\Gamma} \subset \Gamma \tau_2$. Of course we necessarily have that Γ and $\tilde{\Gamma}$ are cyclic generated, say, by γ_0 and $\tilde{\gamma}_0$. If $P_i := \tau_i(0)$ and $\tau_{i0}(P) := \tau_i(P) - P_i$ then the conditions $\tau_1 \tilde{\Gamma} \subset \Gamma \tau_1$ and $\tau_2 \tilde{\Gamma} \subset \Gamma \tau_2$ are equivalent to the conditions

$$\tau_{i0} \tilde{\gamma}_0 = \gamma_0^{\nu_i} \tau_{i0}, \quad \gamma_0^{\nu_i}(P_i) = P_i, \quad i = 1, 2,$$

for suitable integers ν_i. Given such a tuple there are induced morphisms of schemes $\bar{\tau}_1, \bar{\tau}_2 : G/\tilde{\Gamma} \to G/\Gamma$. So we may consider the correspondence $(G/\Gamma, G/\tilde{\Gamma}, \bar{\tau}_1, \bar{\tau}_2)$ over R (i.e. in the category $\mathcal{C}_R$ of schemes over R).

In view of Proposition 1.27 it is reasonable to introduce the following:

DEFINITION 7.1. A correspondence $\mathbf{X} = (X, \tilde{X}, \sigma_1, \sigma_2)$ over R is *uniformized* by $(G, \Gamma, \tilde{\Gamma}, \tau_1, \tau_2)$ if $\mathbf{X}$ is smooth over R and there exists an open immersion of $\mathbf{X}$

into a correspondence of the form

$$(G/\Gamma, G/\tilde{\Gamma}, \bar{\tau}_1, \bar{\tau}_2)$$

above such that the natural projections $\pi : G \to G/\Gamma$ and $\tilde{\pi} : G \to G/\tilde{\Gamma}$ are étale above X and $\tilde{X}$ respectively. A correspondence over R is called *flat* if it admits a uniformization as above.

The natural "vertical" conjecture, in this setting, is the following:

CONJECTURE 7.2. *Let $\mathbf{X}$ be a flat correspondence over R. Assume (for simplicity) X is either affine or projective. Then:*
1) $(\mathbf{X}, \mathbf{K}^{-1})$ is $\delta-$non-trivial and $\delta-$rational;
2) the extension $k(X) \cdot k\langle \mathbf{X}, \mathbf{K}^{-1}\rangle \subset k\langle X, K^{-1}\rangle$ is algebraic.

We will analyze, in this Chapter, this Conjecture in a special (but "generic") case; see below. While doing so we will investigate more problems including $\delta-$invariants and $\delta-$cohomology of certain natural $\delta-$line bundles polarization equivalent to $\mathbf{K}^{-1}$, Galois properties of the above extension, infinitesimal symmetries, $\delta-$tangent maps, converse results, etc.

The special case we will look at is the following. We start, as above, with G either $\mathbf{G}_m$ or E, an elliptic curve. For any integer $N \in \mathbf{Z}$ we denote by $[N] : G \to G$ the multiplication by N endomorphism. Now we fix an integer $N \in \mathbf{Z}$, $N \notin \{1, -1\}$, $N \not\equiv 0 \bmod p$, and we fix $\epsilon \in \{1, -1\}$. We set

$$\Gamma = \langle [\epsilon] \rangle,$$

the group generated by $[\epsilon] : G \to G$. The endomorphism $[N] : G \to G$ induces a morphism

$$\sigma : G/\Gamma \to G/\Gamma.$$

We let $X \subset G/\Gamma$ be an open set such that the canonical projection $\pi : G \to G/\Gamma$ is étale above X and we consider the correspondence

$$\mathbf{X} = \mathbf{X}_{G,N,\epsilon} = (X, \tilde{X} := X \cap \sigma^{-1}(X), \iota, \sigma) \tag{7.1}$$

where ι is the inclusion. Then clearly $\mathbf{X}$ is flat, uniformized by

$$(G, \Gamma, \Gamma, [1], [N]).$$

(The results below can be appropriately generalized to more general flat correspondences although the statements become somewhat technical.)

REMARK 7.3. There are 4 possibilities according as ϵ is 1 or -1 and G is

$$\mathbf{G}_m = Spec\ R[x, x^{-1}]$$

or an elliptic curve

$$E = Proj\ R[x_0, x_1, x_2]/(x_2^2 x_0 - x_1^3 - a x_1 x_0^2 - b x_0^3),$$

$a, b \in R$, $4a^3 + 27b^2 \in R^\times$. Here is a brief description of these 4 situations.

1) If $\epsilon = 1$, $G = \mathbf{G}_m$ then $\sigma(x) = x^N$.

2) If $\epsilon = 1$ and $G = E$ then $\sigma = [N]$.

3) If $\epsilon = -1$ and $G = \mathbf{G}_m$ then $G/\Gamma = \mathbf{A}^1 = Spec\ R[t]$, the map $\mathbf{G}_m \to \mathbf{G}_m/\Gamma$ is given by $t \mapsto x + x^{-1}$, and the map $\sigma : G/\Gamma \to G/\Gamma$ is given by the Chebyshev polynomial $\sigma_N = Cheb_N \in \mathbf{Z}[t]$; recall that the latter is defined as the unique polynomial satisfying

$$x^N + x^{-N} = Cheb_N(x + x^{-1}).$$

(So, for instance $\sigma_2(t) = t^2 - 2, \sigma_3(t) = t^3 - 3t, ...$) Hence we have a commutative diagram

$$\begin{array}{ccc} \mathbf{G}_m & \stackrel{x^N}{\to} & \mathbf{G}_m \\ x+x^{-1} \downarrow & & \downarrow x+x^{-1} \\ \mathbf{A}^1 & \stackrel{\sigma_N}{\to} & \mathbf{A}^1 \end{array}$$

The locus in $\mathbf{A}^1$ above which the projection $\mathbf{G}_m \to \mathbf{A}^1$ is étale is

$$Spec\ R\left[t, \frac{1}{t^2-4}\right].$$

4) If $\epsilon = -1$ and $G = E$ then

$$G/\Gamma = \mathbf{P}^1 = Proj\ R[x_0, x_1]$$

and the map $\sigma = \sigma_N : \mathbf{P}^1 \to \mathbf{P}^1$ is a special case of a Lattès map Lat_{N^2} of degree N^2; cf. Example 1.29. We have a commutative diagram

$$\begin{array}{ccc} E & \stackrel{[N]}{\to} & E \\ \downarrow & & \downarrow \\ \mathbf{P}^1 & \stackrel{\sigma_N}{\to} & \mathbf{P}^1 \end{array}$$

The locus in $\mathbf{P}^1$ above which the projection $E \to \mathbf{P}^1$ is étale is

$$Spec\ R\left[x, \frac{1}{x^3+ax+b}\right]$$

where $x := x_1/x_0$.

In the rest of this Section we discuss δ–line bundles and cycles on the correspondences $\mathbf{X} = \mathbf{X}_{G,N,\epsilon}$.

For any $s \in \mathbf{Z}_+$ we can equip $\mathbf{X}$ with the flat δ–line bundle

$$\mathbf{L}_s := \mathbf{F}(s, N^{-1}) \tag{7.2}$$

of order s and multiplier N^{-1}. Let e be the order of $[\epsilon]$, i.e. e is 1 or 2 according as ϵ is 1 or -1.

LEMMA 7.4. *We have $\mathbf{L}_s^2 \simeq \mathbf{K}^{-2\phi^s} \simeq \mathbf{K}^{-1-\phi^s}$ for all $s \geq 0$. In particular $\mathbf{L}_s$ is polarization equivalent to $\mathbf{K}^{-1}$, $\mathbf{K}^2$ is flat, and $\mathbf{K}^{p-1}_{\mathbf{X}\otimes k}$ is trivial.*

PROOF. Let ω be a translation invariant 1–form on G. Then

$$[N]^*\omega = N\omega$$

which shows that if $e = 1$, $\mathbf{K} \simeq \mathbf{F}(0, N)$; cf. Lemma 2.28. So if $e = 1$ we get $\mathbf{K} \simeq \mathbf{L}_0^{-1}$ and the conclusion follows. Assume from now on that $e = 2$. Since X is an open set of the affine line its canonical bundle is trivial; let α be a nowhere vanishing 1–form on X. Since ω^2 is Γ–invariant we have $\omega^2 = \pi^*\eta$ for some $\eta = f \cdot \alpha^2$ where $f \in \mathcal{O}(X)^\times$. Since $[N]^*\omega^2 = N^2\omega^2$ it follows that $\sigma_2^*\eta = N^2\sigma_1^*\eta$ hence $\mathbf{L}_0^2 \simeq \mathbf{K}^{-2}$ hence $\mathbf{L}_s^2 \simeq \mathbf{K}^{-2\phi^s}$ for all $s \geq 0$. Now set $w = -1 - \phi^s$ and consider the section

$$\theta := f^{w/2} \cdot \alpha^w;$$

recall that $f^{w/2} \in \mathcal{O}^s(X)^\times$ is well defined because $deg(w)$ is even. Then θ is a basis of K_X^w. We have $\theta^2 = f^w \cdot \alpha^{2w} = \eta^w$ hence

$$\sigma_2^*\theta^2 = (\sigma_2^*\eta)^w = (N^2\sigma_1^*\eta)^w = N^{-4}\sigma_1^*\eta^w = N^{-4}\sigma_1^*\theta^2,$$

hence $\sigma_2^*\theta = c \cdot N^{-2}\sigma_1^*\theta$ where $c \in \{1, -1\}$. If $c = 1$ we are done by Lemma 2.28. Assume $c = -1$ and let us derive a contradiction. We have the following congruences mod p in the rings $\mathcal{O}^s(\mathbf{V}^*(K_X))$ and $\mathcal{O}^s(\mathbf{V}^*(K_{\tilde{X}}))$:

$$\theta \equiv f^{-\frac{p+1}{2}} \cdot \alpha^{-p-1} \equiv (f \cdot \alpha^2)^{-\frac{p+1}{2}} \equiv \eta^{-\frac{p+1}{2}}.$$

Consequently

$$\sigma_2^*\theta \equiv (\sigma_2^*\eta)^{-\frac{p+1}{2}} \equiv N^{-p-1}(\sigma_1^*\eta)^{-\frac{p+1}{2}} \equiv N^{-2}(\sigma_1^*\eta)^{-\frac{p+1}{2}}.$$

On the other hand

$$\sigma_2^*\theta \equiv -N^{-2}\sigma_1^*\theta \equiv -N^{-2}(\sigma_1^*\eta)^{-\frac{p+1}{2}}.$$

So $N^2 \equiv -N^2$ mod p, a contradiction.

Since $\mathbf{K}^2$ is flat with multiplier a square mod p it follows that $\mathbf{K}_{\mathbf{X}\otimes k}^{p-1}$ is trivial. $\square$

REMARK 7.5. Lemma 7.4 shows that $\mathbf{K}^2$ is flat. Let us note that $\mathbf{K}$ itself need not be flat. Indeed consider the case when $G = \mathbf{G}_m$, $e = 2$, $N = 2$. So, by the condition that $\mathbf{G}_m \to \mathbf{G}_m/\langle[-1]\rangle$ be étale above X we get

$$X \subset Spec\ R[t, \frac{1}{t^2-4}].$$

As noticed before $\sigma(t) = t^2 - 2$ is the quadratic Chebyshev polynomial. By the way, one can see directly that $\mathbf{K}^2$ is flat with multiplier 4 by taking

$$\eta = \frac{(dt)^2}{t^2-4}$$

and noting that

$$\sigma^*\eta = 4 \cdot \eta.$$

On the other hand, if $\mathbf{K}$ were flat with multiplier λ then $\lambda^2 = 4$ and one could find a section

$$\theta := a(t) \cdot dt, \quad a(t) \in \mathcal{O}(X) \subset K(t),$$

such that

$$\sigma^*\theta = \lambda \cdot \theta.$$

In particular we would get

$$\sigma^*\theta^2 = 4 \cdot \theta^2$$

hence $\theta^2/\eta = a^2(t) \cdot (t^2 - 4)$ would be a $\sigma-$invariant element of $K(t)$, in particular it would belong to K. Hence $t^2 - 4$ would be a constant in K times a square in $K(t)$, a contradiction because $t^2 - 4$ has simple roots.

By the way, Lemma 7.4 also shows for instance that $\mathbf{K}^{-1-\phi}$ is flat. One can see this directly, in the example above, by setting

$$\theta := (t^2-4)^{\frac{\phi+1}{2}}(dt)^{-1-\phi}$$

and noting that

$$\sigma^*\theta = (t^4 - 4t^2)^{\frac{\phi+1}{2}}(2tdt)^{-1-\phi} = t^{\phi+1} \cdot (t^2-4)^{\frac{\phi+1}{2}} \cdot 2^{-2} \cdot t^{-\phi-1} \cdot (dt)^{-1-\phi} = \frac{1}{4} \cdot \theta.$$

LEMMA 7.6. *Let $A \in G(R)$ be such that $[N](A) = A$. Assume A projects, via $G \to G/\langle[\epsilon]\rangle$, to a point $P \in \tilde{X}(R)$ and consider the induced fixed point (P, P) of* $\mathbf{X}$. *Fix an isomorphism*

$$\mathcal{O}_{G,0}^{for} \simeq R[[T]] \tag{7.3}$$

and consider the isomorphism

$$\mathcal{O}_{X,P}^{for} \simeq R[[T]] \tag{7.4}$$

induced by the isomorphism 7.3 and the morphism

$$G \stackrel{+A}{\to} G \to G/\langle[\epsilon]\rangle.$$

Then the germ $\Sigma(T)$ of $\mathbf{X}$ *at (P, P) (with respect to the isomorphism 7.4) is given by*

$$\Sigma(T) = [N](T),$$

the multiplication by $N \in \mathbf{Z}$ in the formal group law of G (corresponding to the isomorphism 7.3). In particular thc canonical bundlc $\mathbf{K}$ *of* $\mathbf{X}$ *is formally flat at (P, P).*

PROOF. The case $A = 0$ is clear. The general case reduces to the case $A = 0$ via the commutative diagram

$$\begin{array}{ccccc} G & \stackrel{[1]}{\leftarrow} & G & \stackrel{[N]}{\to} & G \\ +A\downarrow & & \downarrow +A & & \downarrow +A \\ G & \stackrel{[1]}{\leftarrow} & G & \stackrel{[N]}{\to} & G \end{array}$$

The claim about formal flatness follows from Proposition 4.4. □

The following Proposition addresses the existence of sequences of disjoint cycles with $\mathbf{Z}_p-$ accumulating multipliers on flat correspondences.

PROPOSITION 7.7. *Let* $\mathbf{X}$ *be a flat correspondence as in Equation 7.1. Then* $\mathbf{X}$ *has a sequence of pairwise disjoint cycles with $\mathbf{Z}_p-$ accumulating multipliers.*

We need a technical Lemma. For the purposes of our proof of Proposition 7.7 a less precise Lemma is necessary; but we will need later the full power of this Lemma.

LEMMA 7.8. *Let N be an integer not divisible by two distinct odd primes p and λ and let n be an integer. Then one can find an even positive integer*

$$\nu = \nu(N, p, \lambda, n)$$

such that the following hold:

1) $v_\lambda(1 - N^\nu)$ is odd,

2) $v_p(1 - N^\nu)$ is even and $\geq n$,

3) $\left(\frac{K}{p}\right) = 1$ where $1 - N^\nu = p^{v_p(1-N^\nu)}K$.

PROOF. We may assume $n \geq 2$. Choose an even integer m such that $N^m \equiv 1 \bmod p^n\lambda^n$; write

$$N^m = 1 + p^a s = 1 + \lambda^b t$$

where s is not divisible by p, t is not divisible by λ and $a, b \geq n$. For any $\mu \geq 1$ we have

$$N^{m\mu} = 1 + \mu p^a s + p^{2a} s_0 = 1 + \mu \lambda^b t + \lambda^{2b} t_0,$$

for some integers s_0, t_0. So if $v_p(\mu) \leq 1$ and $v_\lambda(\mu) \leq 1$ we have

$$v_p(N^{m\mu} - 1) = a + v_p(\mu),$$

$$v_\lambda(N^{m\mu} - 1) = b + v_\lambda(\mu),$$

$$\frac{N^{m\mu} - 1}{p^{v_p(N^{m\mu}-1)}} \equiv s \cdot \frac{\mu}{p^{v_p(\mu)}} \quad mod \ p.$$

Choose $\mu = \lambda^d p^c \mu_1$ such that the following hold:
i) c is either 1 or 0 according as a is odd or even,
ii) d is either 0 or 1 according as b is odd or even,
iii) $\mu_1 \geq 1$ is not divisible by p or λ and

$$\left(\frac{\mu_1}{p}\right) = \left(\frac{-\lambda^d s}{p}\right).$$

If $\nu = m\mu$ then conditions 1, 2, 3 in the statement of the Lemma obviously hold. □

PROOF. (of Proposition 7.7) Let $\lambda_1, \lambda_2, ...$ be a sequence of odd distinct primes not dividing N such that λ_i is larger than all the orders of all points in $G(k)\backslash\tilde{X}(k)$. By Lemma 7.8 one can find a sequence of positive numbers $\nu_1, \nu_2, ...$ such that $v_p(N^{\nu_i} - 1) \geq i$ and $v_{\lambda_i}(N^{\nu_i} - 1) \geq 1$. Since $\lambda_i \neq p$, $G(R)$ contains a point P_i of exact order λ_i and clearly $[N^{\nu_i}](P_i) = P_i$. Since the map from the λ_i-torsion of $G(R)$ to the λ_i-torsion of $G(k)$ is injective it follows that the image of P_i in $G(k)$ has exact order λ_i hence it belongs to $\tilde{X}(k)$. But then P_i must belong to $\tilde{X}(R)$. Consequently we may consider the following ν_i-cycles

$$C_i := (P_i, [N](P_i), ..., [N^{\nu_i - 1}](P_i), \pi(P_i))$$

on the correspondence $\mathbf{X}$. The multiplier of C_i is given by

$$a(C_i) = N^{\nu_i},$$

as one can see from the argument of Lemma 7.6. To conclude it is enough to note that the cycles C_i are pairwise disjoint. But if we assume the contrary then one can find $i \neq j$ and integers r, s such that $[N^r](P_i)$ equals either $[N^s](P_j)$ or $[-N^s](P_j)$. This is however impossible because $[N^r](P_i)$ has exact order λ_i while $[N^s](P_j)$ and $[-N^s](P_j)$ have exact order λ_j. □

The interest in Proposition 7.7 comes, of course, from our discussion on $\mathbf{Z}_p-$accumulating multipliers vs accumulating Schwarzians; but the only way we will apply Proposition 7.7 in what follows will be through its consequence that any flat correspondence as in Equation 7.1 has at least a cycle; this can be proved directly rather easily.

7.2. $\delta-$characters

The key ingredient in the construction of $\delta-$invariants for flat correspondences is given by the following:

DEFINITION 7.9. [**17**] Let G be a smooth group scheme over R. A *$\delta-$character* of G (of order $\leq r$) is a $\delta-$function $\psi : G(R) \to R$ (of order $\leq r$) which is a group homomorphism into the additive group of R.

In what follows we quickly review the theory of δ–characters following both [**17**] and [**26**]. To give a δ–character of order $\leq r$ is the same as to give a morphism of group formal schemes

$$J^r(G) \to \hat{\mathbf{G}}_a$$

into the additive group formal scheme $\hat{\mathbf{G}}_a := (Spf\ R[T]\hat{}, +)$; this follows from Proposition 3.19 (applied to both G and $G \times G$).

EXAMPLE 7.10. If $\mathbf{G}_m = (Spec\ R[T, T^{-1}], \cdot)$ is the multiplicative group scheme over R then the function

(7.5)

$$\psi_1 : \mathbf{G}_m(R) = R^{\times} \to R, \quad \psi_1(x) = \frac{1}{p} log\left(\frac{\phi(x)}{x^p}\right) = \sum_{n \geq 1} (-1)^{n-1} \frac{p^{n-1}}{n} (\delta x)^n x^{-pn}$$

is a δ–character which we call the *basic* δ–character on $\mathbf{G}_m$ of order 1. This δ–character should be viewed as an arithmetic analogue of the *logarithmic derivative* in Ritt-Kolchin theory.

Our aim in what follows is to construct an analogue of the δ–character 7.5 for elliptic curves E over R. In what follows we need some preparatory discussion. Set

$$N^r := Ker(J^r(E) \to \hat{E}). \tag{7.6}$$

Exactly as in the theory of algebraic group extensions in the last Chapter of [**119**] we have an exact sequence

$$0 \to Hom(J^r(E), \hat{\mathbf{G}}_a) \to Hom(N^r, \hat{\mathbf{G}}_a) \xrightarrow{\partial} H^1(E, \mathcal{O}). \tag{7.7}$$

Since our context is that of group formal schemes rather than algebraic groups and since we need the explicit shape of the map ∂ above, we quickly review the construction of ∂ and the exactness of the sequence. First note that the projection $\pi_r : J^r(E) \to \hat{E}$ has a natural structure of a principal homogeneous space under $\hat{E} \times N^r \to \hat{E}$. Recall that, by Proposition 3.13, π_r has sections locally in the Zariski topology; let

$$E = \bigcup U_i, \quad s_i : \hat{U}_i \to \pi_r^{-1}(\hat{U}_i) \tag{7.8}$$

be a Zariski open covering and a corresponding family of sections of π_r. We may and will assume there is an index i_0 such that $0 \in U_{i_0}(R)$ and $s_{i_0}(0) = 0$. Then the morphism

$$\tau_i : \hat{U}_i \times N^r \to \pi_r^{-1}(\hat{U}_i) \tag{7.9}$$

which at the level of R–points is given by

$$(A, B) \mapsto s_i(A) + B$$

is an isomorphism of principal homogeneous spaces under $\hat{U}_i \times N^r \to \hat{U}_i$.

REMARK 7.11. Note that the isomorphism τ_{i_0} induces the identity

$$0 \times N^r \to \pi_r^{-1}(0) = N^r.$$

Moreover, if in addition $T \in \mathcal{O}(U_{i_0})$ is an étale coordinate on U_{i_0} then we have an induced $\mathcal{O}(U_{i_0})$–isomorphism

(7.10)

$$\tau_{i_0}^* : \mathcal{O}(\hat{U}_{i_0})[T', ..., T^{(r)}]\hat{} \overset{can}{\simeq} \mathcal{O}^r(\pi_r^{-1}(\hat{U}_{i_0})) \overset{\circ\tau_{i_0}}{\to} \mathcal{O}^r(\hat{U}_{i_0} \times N^r) = \mathcal{O}^r(\hat{U}_{i_0})[T', ..., T^{(r)}]\hat{}$$

where *can* is the unique isomorphism sending $T^{(i)}$ into $\delta^i T$. Moreover $\tau_{i_0}^*$ is the identity modulo T.

The sections in Equation 7.8 induce maps

$$s_i - s_j : \hat{U}_{ij} = \hat{U}_i \cap \hat{U}_j \to J^r(E)$$

that obviously factor through maps

$$s_{ij} : \hat{U}_{ij} \to N^r. \tag{7.11}$$

In particular, at the level of points

$$(\tau_j^{-1} \circ \tau_i)(A, B) = (A, s_{ij}(A) + B).$$

We define the map ∂ in the sequence 7.7 by attaching to any $\Theta : N^r \to \hat{\mathbf{G}}_a$ the cohomology class

$$\varphi := \partial(\Theta) \in H^1(\hat{E}, \mathcal{O}) = H^1(E, \mathcal{O})$$

represented by the cocycle of functions

$$\varphi_{ij} := \Theta \circ s_{ij} \in \mathcal{O}(\hat{U}_{ij}).$$

PROPOSITION 7.12. *The sequence 7.7 is exact.*

PROOF. The only less obvious part is to show that if $\partial(\Theta) = 0$ then Θ lifts to a homomorphism $J^r(E) \to \hat{\mathbf{G}}_a$. If $\partial(\Theta) = 0$ we may write

$$\varphi_{ij} = \varphi_i - \varphi_j$$

with $\varphi_i \in \mathcal{O}(\hat{U}_i) = \mathcal{O}(U_i)\hat{}$ and set

$$\psi_i := \Theta + \varphi_i \in \mathcal{O}(\hat{U}_i \times N^r).$$

By adding the same constant in R to all φ_i we may arrange that there is an index i_0 such that $0 \in U_{i_0}(R)$ and $\varphi_{i_0}(0) = 0$. Note that ψ_i is N^r-equivariant with respect to the action of N^r on $\hat{U}_i \times N^r$ via addition on the second component and the action of N^r on $\hat{\mathbf{G}}_a$ defined by

$$N^r \times \hat{\mathbf{G}}_a \overset{\Theta \times id}{\to} \hat{\mathbf{G}}_a \times \hat{\mathbf{G}}_a \overset{mult}{\to} \hat{\mathbf{G}}_a;$$

indeed, for $A \in \hat{U}_i(R)$, $B, C \in N^r(R)$ we have

$$(\Theta + \varphi_i)(A, B + C) = \Theta(B + C) + \varphi_i(A)$$

$$= \Theta(B) + \Theta(C) + \varphi_i(A) = \Theta(B) + (\Theta + \varphi_i)(A, C).$$

Moreover, with the notation in Equation 7.9, the functions

$$\psi_i \circ \tau_i^{-1} \in \pi_r^{-1}(\hat{U}_i)$$

glue together to give a function $\psi \in \mathcal{O}(J^r(E))$; indeed, if $(A, B) \in \hat{U}_{ij}(R) \times N^r(R)$, then

$$(\Theta + \varphi_j)(\tau_j^{-1}(\tau_i(A, B))) = \Theta(B) + \Theta(s_{ij}(A)) + \varphi_j(A)$$

$$= \Theta(B) + \varphi_i(A) = (\Theta + \varphi_i)(A, B)$$

Now ψ is a homomorphism due to the following Lemma 7.13. □

LEMMA 7.13. *Let $f : J^r(E) \to \hat{\mathbf{G}}_a$ be a morphism of formal schemes such that*
1) f takes the zero element of $J^r(E)$ into the zero element of $\hat{\mathbf{G}}_a$;
2) The restriction of f to N^r is a homomorphism $N^r \to \hat{\mathbf{G}}_a$;
3) f is N^r−equivariant with respect to the actions

$$N^r \times J^r(E) \overset{can \times id}{\longrightarrow} J^r(E) \times J^r(E) \overset{mult}{\to} J^r(E),$$

$$N^r \times \hat{\mathbf{G}}_a \overset{f \times id}{\longrightarrow} \hat{\mathbf{G}}_a \times \hat{\mathbf{G}}_a \overset{mult}{\to} \hat{\mathbf{G}}_a.$$

Then f is a homomorphism.

PROOF. Fix an R−point $P \in J^r(E)(R)$ and consider the function $g_P \in \mathcal{O}(J^r(E))$, $g_P : J^r(E) \to \hat{\mathbf{G}}_a$ which on R−points $Q \in J^r(E)(R)$ satisfies

$$g_P(Q) = f(P+Q) - f(P) - f(Q).$$

Note that for any point $A \in N^r(R)$ we have

$$\begin{aligned} g_P(A+Q) &= f(P+A+Q) - f(P) - f(A+Q) \\ &= (f(A) + f(P+Q)) - f(P) - (f(A) + f(Q)) \quad \text{by condition 3} \\ &= f(P+Q) - f(P) - f(Q) = g_P(Q). \end{aligned}$$

Pulling back via the isomorphism in Equation 7.9 we get

$$g_P \circ \tau_i \in \mathcal{O}(\hat{U}_i \times N^r)^{N^r(R)} = \mathcal{O}(\hat{U}_i).$$

Since this is true for all i we get $g_P \in \mathcal{O}(\hat{E}) = R$. Since, by condition 1, $g_P(0) = f(P+0) - f(P) - f(0) = 0$ we get that $g_P = 0 \in R$. This implies that f is a homomorphism at the level of R−points hence is a homomorphism. $\square$

In what follows we fix an invertible 1−form ω on E and an isomorphism $\mathcal{O}_{E,0}^{for} \simeq R[[T]]$. We will define a sequence of elements

$$f_{jet}^1(E,\omega), f_{jet}^2(E,\omega), f_{jet}^3(E,\omega), \ldots \in R$$

that play a key role in the theory. (It is easy to check that $f_{jet}^r(E,\omega)$ does not depend on the isomorphism $\mathcal{O}_{E,0}^{for} \simeq R[[T]]$ and in the next Chapter we will investigate the dependence of the $f_{jet}^r(A,\omega)$'s on A and ω but at this point we do not need to do that; we will also introduce there a sequence $f_{crys}^1(E,\omega), f_{crys}^2(E,\omega), f_{crys}^3(E,\omega), \ldots$ which will be compared to the sequence above.) By Proposition 4.45 one has a natural isomorphism

$$N^r \simeq (\hat{\mathbf{A}}^r, [+]),$$

$\hat{\mathbf{A}}^r = Spf\ R[T', \ldots, T^{(r)}]\hat{}$, which we view, in what follows as an identification. So we have an identification

$$s_{ij} = (\alpha_{ij}^1, \ldots, \alpha_{ij}^r) \in \mathcal{O}(\hat{U}_{ij})^r,$$

$$\alpha_{ij}^k := T^{(k)} \circ s_{ij}.$$

Recall from Proposition 4.42 that we have, at our disposal, series $L^r \in R[T', \ldots, T^{(r)}]\hat{}$ that define homomorphisms

$$L^r : N^r \simeq (\hat{\mathbf{A}}^r, [+]) \to \hat{\mathbf{G}}_a = (\hat{\mathbf{A}}^1, +).$$

In particular

$$L^2 = \frac{1}{p}(\sum c_i^{\phi^2}(p(T')^p + p^p(T')^p + p^2T'')^i),\ L^1 = \frac{1}{p}(\sum c_i^{\phi}(pT')^i), \tag{7.12}$$

where $l_E(T) = \sum c_n T^n \in K[[T]]$ is the logarithm of the formal group of E. Define

$$\varphi_{ij}^r := L^r(\alpha_{ij}^1, ..., \alpha_{ij}^r) \in \mathcal{O}(\hat{U}_{ij}).$$

The collection of these sections, for varying i, j, defines a cohomology class

$$\varphi^r \in H^1(\hat{E}, \mathcal{O}_{\hat{E}}) = H^1(E, \mathcal{O}_E)$$

which is, of course, the class $\partial(L^r)$ defined in the general setting. Denote by

$$\langle\ ,\ \rangle : H^1(E, \mathcal{O}) \times H^0(E, \Omega^1) \to R$$

the Serre duality pairing and define

$$f_{jet}^r(E, \omega) := \langle \varphi^r, \omega \rangle \in R. \tag{7.13}$$

We have the following relation with the Kodaira-Spencer class.

LEMMA 7.14. *The condition $f_{jet}^1(E, \omega) \equiv 0$ mod p holds if and only if $KS(E) =$ 0.*

PROOF. Since, by Proposition 4.41, $L^1(T') \equiv T'$ mod p we have $f_{jet}^1(E, \omega) \equiv 0$ mod p if and only if the $1-$cocycle (α_{ij}^1 mod p) with values in $\mathcal{O}_{E_0}$ is a coboundary. This happens if and only if $J^1(E)_0 \to E_0$ has a section in the category of varieties which holds if and only if $KS(E) = 0$, by the very definition of the latter. □

PROPOSITION 7.15. *The following conditions are equivalent:*
1) $f_{jet}^1(E, \omega) = 0$,
2) E has a lift of Frobenius,
3) E has a non-zero $\delta-$character of order ≤ 1.

PROOF. If 1) holds then $\partial(L^1) = \varphi^1 = 0$ so by the exact sequence 7.7 $L^1 : N^1 \to \hat{\mathbf{G}}_a$ lifts to a unique homomorphism $\psi_1 : J^1(E) \to \hat{\mathbf{G}}_a$ hence 3) holds; by the way, by the proof of Proposition 7.12, ψ_1 is obtained by gluing functions

$$\psi_{1,i} \circ \tau_i^{-1} \in \mathcal{O}(\pi_1^{-1}(\hat{U}_i)) \tag{7.14}$$

$$\psi_{1,i} := L^1 + \varphi_i \in \mathcal{O}(U_i)[T']\hat{} = \mathcal{O}(U_i \times N^1), \tag{7.15}$$

with $\varphi_i \in \mathcal{O}(\hat{U}_i) = \mathcal{O}(U_i)\hat{}$.

Assume 3) holds. By the isomorphism $L^1 : N^1 \simeq \hat{\mathbf{G}}_a$ in Proposition 4.45 we see that the middle term in the exact sequence 7.7 is isomorphic to R so the connecting homomorphism ∂ in that sequence must be zero and 1) follows.

On the other hand 2) is equivalent to $\hat{E}$ having a lift of Frobenius (Grothendieck's existence theorem in formal geometry) and the latter is equivalent to the projection $J^1(E) \to \hat{E}$ having a section (by the universality property in Proposition 3.3). The latter is clearly equivalent to the cocycle $s_{ij} : \hat{U}_{ij} \to N^1$ being a coboundary which is equivalent to 1). □

DEFINITION 7.16. If E has a lift of Frobenius the $\delta-$character ψ_1 constructed in the proof of Lemma 7.15 will be called the *basic* $\delta-$character on E of order 1.

PROPOSITION 7.17. [**17**] *Any elliptic curve E over R has a non-zero $\delta-$character of order ≤ 2.*

PROOF. Consider the homomorphism $\Theta : N^2 \to \hat{\mathbf{G}}_a$ given by

$$\Theta := f^1_{jet}(E,\omega)L^2 - f^2_{jet}(E,\omega)L^1 \in R[T',T'']\hat{}.$$

We claim that $\partial(\Theta)=0$. Indeed $\partial(\Theta)$ is the class $\gamma \in H^1(\hat{E},\mathcal{O})$ of the cocycle

$$\gamma_{ij} := f^1_{jet}(E,\omega)\cdot\varphi^2_{ij} - f^2_{jet}(E,\omega)\cdot\varphi^1_{ij} \in \mathcal{O}(\hat{U}_{ij}).$$

But $\gamma = 0$ because

$$\langle\gamma,\omega\rangle = f^1_{jet}(E,\omega)\cdot\langle\varphi^2,\omega\rangle - f^2_{jet}(E,\omega)\cdot\langle\varphi^1,\omega\rangle = 0.$$

By the exactness of the sequence 7.7 Θ lifts to a unique homomorphism $\psi_2 : J^2(E) \to \hat{\mathbf{G}}_a$; by the way, by the proof of Proposition 7.12, ψ_2 is obtained by gluing functions

$$\psi_{2,i}\circ\tau_i^{-1} \in \mathcal{O}(\pi_2^{-1}(\hat{U}_i)) \tag{7.16}$$

$$\psi_{2,i} := f^1_{jet}(E,\omega)\cdot L^2 - f^2_{jet}(E,\omega)\cdot L^1 + \gamma_i \in \mathcal{O}(U_i)[T',T'']\hat{}, \tag{7.17}$$

with $\gamma_i \in \mathcal{O}(\hat{U}_i)$. If $f^1_{jet}(E,\omega)\neq 0$ then $\psi_2 \neq 0$ and we are done. If $f^1_{jet}(E,\omega)=0$ we are done by Proposition 7.15. □

DEFINITION 7.18. The δ–character ψ_2 constructed in the proof of Proposition 7.17 will be called the *basic* δ–character on E of order ≤ 2. If E doesn't have a lift of Frobenius (i.e. if $f^1_{jet}(E,\omega)\neq 0$) then $\psi_2\neq 0$.

REMARK 7.19. It was proved in [**17**] that if E is an elliptic curve with ordinary reduction E_0 then $f^1_{jet}(E,\omega)\not\equiv 0$ mod p if and only if the Serre-Tate parameter $q(E)$ of E (with respect to some or equivalently any parameter in the moduli space) is not in $1+p^2R$; we will not need so we will not check this here.

In the next Proposition we let G be either $\mathbf{G}_m$ or an elliptic curve E and we fix an invertible 1–form ω on G and an isomorphism $\mathcal{O}^{for}_{G,0}\simeq R[[T]]$; if $G = \mathbf{G}_m = Spec\ R[x,x^{-1}]$ we let $T = x-1$ and $\omega = dx/x$. If $G=\mathbf{G}_m$ or $G=E$ and E has a lift of Frobenius we let ψ_1 be the basic δ–character of order 1. If $G=E$ we let ψ_2 be the basic δ–character of order ≤ 2. For $r=1,2$ we let $E:\mathcal{O}^r(G)\to R[[T]][T',\dots,T^{(r)}]\hat{}$ be the expansion map at a torsion point $A\in G(R)$ corresponding to the isomorphism $\mathcal{O}_{G,A}\simeq R[[T]]$ induced by the translation by A map $+A : G\to G$. Finally let $l_G(T)$ be the logarithm of the formal group law of G.

PROPOSITION 7.20. *The following formulae hold:*
1) If $G=\mathbf{G}_m$ *then*

$$E(\psi_1) = \frac{1}{p}(l_G(T)^{\phi} - p l_G(T)).$$

2) If $G=E$ *and* E *has a lift of Frobenius then*

$$E(\psi_1) = \frac{1}{p}(l_G(T)^{\phi} + \beta_1 l_G(T)),\quad \beta_1\in R.$$

3) If $G=E$ *then*

$$E(\psi_2) = \frac{1}{p}(f^1_{jet}(E,\omega)l_G(T)^{\phi^2} - f^2_{jet}(E,\omega)l_G(T)^{\phi} + \beta_2 l_G(T)),\ \beta_2\in R.$$

REMARK 7.21. Looking at the coefficient of T in $E(\psi_1)$ and $E(\psi_2)$ above we see that $\beta_1, \beta_2 \in pR$; in particular we may define

$$f^{1,2}_{jet}(E,\omega) := \frac{\beta_2}{p} \in R.$$

Later we will show that $f^{1,2}_{jet}(E,\omega)$ and $f^1_{jet}(E,\omega)$ are related by a simple formula; cf. Corollary 8.86. Also if $G = E$ and E has a lift of Frobenius then we set

$$f_{lift}(E,\omega) := \frac{\beta_1}{p} \in R. \tag{7.18}$$

It is also convenient to extend the above notation by setting

$$f_{lift}(\mathbf{G}_m, dx/x) = -1. \tag{7.19}$$

Finally note that, if $G = \mathbf{G}_m$, then $E(\psi_1)$ coincides with the series $\Psi_{\mathbf{G}_m}$ introduced in Equation 4.40.

PROOF. Let $r = 1$ in assertions 1) and 2) and let $r = 2$ in assertion 3). We have $\psi_r(A) = 0$ because A is torsion. So ψ_r is invariant under translation by A. So we may assume $A = 0$. If $G = \mathbf{G}_m$ we have $l_G(T) = log(1+T)$ and our claim is trivial. Let us prove assertion 3); assertion 2) is proved similarly. Recall that ψ_2 is obtained by gluing functions $\psi_{2,i} \circ \tau_i^{-1}$ where $\psi_{2,i}$ are as in in Equation 7.17. Since the isomorphism $\tau^*_{i_0}$ in Equation 7.10 is congruent to the identity mod T it follows that $E(\psi_2)$ is congruent mod T with an expression of the form

$$f^1_{jet} \cdot L^2 - f^2_{jet} \cdot L^1 + \gamma_{i_0} \in \mathcal{O}(\hat{U}_{i_0})[T', T'']\hat{} \subset R[[T]][T', T'']\hat{},$$

with $\gamma_{i_0} \in \mathcal{O}(\hat{U}_{i_0})$; cf. Equation 7.17. Consider the ring homomorphism

$$R[[T]][T', T'']\hat{} \to R[[T]][T', T'']\hat{},$$

$$F(T, T', T'') \mapsto F \circ e(pT) := F(e(pT), \delta(e(pT)), \delta^2(e(pT))),$$

where $e(T)$ is the exponential of the formal group law of E, inverse of $l(T) = l_G(T)$. Then we have that $\Lambda(T, T', T'') := E(\psi_2) \circ e(pT)$ is congruent mod T to

$$\begin{aligned}\frac{1}{p}\{f^1_{jet}[l^{(\phi^2)}] \circ [\phi^2(T)] \circ [0, T', T''] \circ [\delta(e(pT)), \delta^2(e(pT))] \\ -f^2_{jet}[l^{(\phi)}] \circ [\phi(T)] \circ [0, T'] \circ [\delta(e(pT))]\} + u(T),\end{aligned} \tag{7.20}$$

with $u(T) \in TR[[T]]$. We want to set $T = 0$ in 7.20. Note that

$$\begin{aligned}&[l^{(\phi^2)}] \circ [\phi^2(T)] \circ [0, T', T''] \circ [\delta(e(pT)), \delta^2(e(pT))] \circ [0, T', T''] \\ &= [l^{(\phi^2)}] \circ [\phi^2(T)] \circ [e(pT), \delta(e(pT)), \delta^2(e(pT))] \circ [0, T', T''] \\ &= [l^{(\phi^2)}] \circ [e^{\phi^2}] \circ [p\phi^2(T)] \circ [0, T', T''] \\ &= p[\phi^2(T)] \circ [0, T', T'']\end{aligned}$$

Similarly

$$[l^{(\phi)}] \circ [\phi^1(T)] \circ [0, T'] \circ [\delta(e(pT))] \circ [0, T'] = p[\phi(T)] \circ [0, T'].$$

So $\Lambda(T, T', T'')$ is congruent mod T with

$$f^1_{jet}(E,\omega)\phi^2(T) - f^2_{jet}(E,\omega)\phi(T).$$

On the other hand $\Lambda(T, T', T'')$ is clearly additive in the sense that

$$\Lambda(T_1 + T_2, \delta(T_1 + T_2), \delta^2(T_1 + T_2)) = \Lambda(T_1, T_1', T_1'') + \Lambda(T_2, T_2', T_2'')$$

so we must have

$$\Lambda = c_2\phi^2(T) + c_1\phi(T) + c_0 T,$$

with $c_i \in R$. Taking Λ mod T we conclude that $c_2 = f^1_{jet}(E,\omega)$ and $c_1 = -f^2_{jet}(E,\omega)$ which ends the proof. $\square$

Here is the main result of this section:

THEOREM 7.22. *Assume E is an elliptic curve that doesn't have a lift of Frobenius. Then*

$$f^2_{jet}(E,\omega), f^{1,2}_{jet}(E,\omega) \in f^1_{jet}(E,\omega)\cdot R$$

and E has a δ–character $\psi_{2,nor}$ of order ≤ 2 whose expansion is

$$E(\psi_{2,nor}) = \frac{1}{p}(l_E(T)^{\phi^2} - \frac{f^2_{jet}(E,\omega)}{f^1_{jet}(E,\omega)} l_E(T)^{\phi} + p\frac{f^{1,2}_{jet}(E,\omega)}{f^1_{jet}(E,\omega)} l_E(T)).$$

Moreover under the canonical identification $\mathcal{O}^2(U) = \mathcal{O}(U)[T',T'']\hat{}$ (for $U \subset E$ an affine containing the origin, on which T is an étale coordinate, $T' = \delta T$, $T'' = \delta^2 T$) we have

$$\psi_{2,nor} = L^2 - \frac{f^2_{jet}(E,\omega)}{f^1_{jet}(E,\omega)} L^1 + T(\Omega_1 + p\Omega_2)$$

with $\Omega_1 \in \mathcal{O}(\hat{U})[T']\hat{}$, $\Omega_2 \in \mathcal{O}(\hat{U})[T',T'']\hat{}$.

The idea of the proof is to change the coordinates on N^2 so that the group law acquires a particularly simple form. That form will be then used to explicitly construct our δ–character $\psi_{2,nor}$. The rest of the claims will follow by comparing $\psi_{2,nor}$ with our previously constructed ψ_2; by the way, it will be shown in the proof that

$$\psi_2 = f^1_{jet}(E,\omega)\cdot\psi_{2,nor}. \tag{7.21}$$

We need some preliminaries.

Choose a section $s : N^1 \to N^2$ in the category of formal schemes; there are plenty of such sections due to Proposition 4.45. We then have an isomorphism of formal schemes

$$u : N^2 \to N^1 \times K^2$$

which on the level of points is given by

$$u(B) = (\pi(B), B - s(\pi(B))),$$

for $B \in N^2(R)$, where $\pi : N^2 \to N^1$ is the canonical projection. The map $\Gamma : N^1 \times N^1 \to N^2$ given on points by

$$\Gamma(A_1, A_2) = s(A_1) + s(A_2) - s(A_1 + A_2) \tag{7.22}$$

factors through a map

$$\Gamma : N^1 \times N^1 \to K^2.$$

Transporting the group law on N^2 via u we get a group law $\oplus$ on $N^1 \times K^2$ given, on points, by

$$(A_1, B_1) \oplus (A_2, B_2) = (A_1 + A_2, B_1 + B_2 + \Gamma(A_1, A_2)). \tag{7.23}$$

Choose isomorphisms $N^1 \simeq K^2 \simeq \hat{\mathbf{G}}_a$ (cf. Proposition 4.45) which we view as identifications. In particular we view Γ as a map $\hat{\mathbf{A}}^2 \to \hat{\mathbf{A}}^1$ and we view $(N^1 \times K^2, \oplus)$ identified with a group formal scheme $(\hat{\mathbf{A}}^2, \oplus)$ where $\oplus$ is still defined by

Equation 7.23. We denote by y, z the coordinates on $\hat{\mathbf{A}}^2$. In particular Γ can be viewed as an element of the ring $R[y_1, y_2]\hat{}$ where y_1, y_2 is a pair of variables "corresponding" to y; Γ is a 2−cocycle of $\hat{\mathbf{G}}_a$ with values in $\hat{\mathbf{G}}_a$ in the sense that

$$\Gamma(y_2, y_3) - \Gamma(y_1 + y_2, y_3) + \Gamma(y_1, y_2 + y_3) - \Gamma(y_1, y_2) = 0 \tag{7.24}$$

for any variable y_3. Write Γ as a series; then each of the homogeneous components of Γ satisfies Equation 7.24. A basic Theorem of Lazard [**90**] says that any homogeneous polynomial of degree d with R−coefficients that satisfies Equation 7.24 is either of the form

$$r(y_1 + y_2)^d - ry_1^d - ry_2^d,$$

with $r \in R$, if d is not a power of p or of the form

$$r\frac{(y_1 + y_2)^{p^n} - y_1^{p^n} - y_2^{p^n}}{p}$$

if $d = p^n$. So we get that Γ is of the form

$$s'(y_1 + y_2) - s'(y_1) - s'(y_2) + \sum_n r_n \cdot \frac{(y_1 + y_2)^{p^n} - y_1^{p^n} - y_2^{p^n}}{p}, \tag{7.25}$$

for some $s' \in R[y]\hat{}$ and some $r_n \in R$, $r_n \to 0$. Changing s (by adding s' to its second component) we may assume $s' = 0$. The situation is summarized by the following diagram of group formal schemes:

$$\begin{array}{ccc} Spf\ R[T', T'']\hat{} = N^2 & \overset{u}{\simeq} & (\hat{\mathbf{A}}^2, \oplus) = Spf\ R[y, z]\hat{} \\ \downarrow \pi & & \downarrow pr_1 \\ Spf\ R[T']\hat{} = N^1 & \simeq & \hat{\mathbf{G}}_a = Spf\ R[y]\hat{} \end{array}$$

Let $s_{ij} : \hat{U}_{ij} \to N^2 \simeq (\hat{\mathbf{A}}^2, \oplus)$ be as in Equation 7.11 and view s_{ij} as pairs

$$s_{ij} = (a_{ij}, b_{ij}) \in \mathcal{O}(\hat{U}_{ij})^2$$

satisfying the Cech 1−cocycle condition with respect to $\oplus$. Consider the series

$$\Theta_0(y, z) := \sum r_n y^{p^n} - pz \in R[y, z]\hat{};$$

it is trivial to see that this series defines a homomorphism

$$\Theta_0 : (\hat{\mathbf{A}}^2, \oplus) \to \hat{\mathbf{G}}_a.$$

Therefore the collection

$$c_{ij} := \Theta_0(a_{ij}, b_{ij}) = \sum r_n a_{ij}^{p^n} - pb_{ij}$$

is a Cech 1−cocycle with respect to usual addition. Let $[c_{ij}], [a_{ij}] \in H^1(\hat{E}, \mathcal{O}) = H^1(E, \mathcal{O})$ be the cohomology classes of the Cech cocycles c_{ij} and a_{ij}.

We have the following crucial:

LEMMA 7.23. *$[c_{ij}] \in R \cdot [a_{ij}]$ in $H^1(E, \mathcal{O})$.*

PROOF. Write $[a_{ij}] = p^m[e_{ij}]$ where e_{ij} is a cocycle in $\mathcal{O}_{\hat{E}}$ such that $[e_{ij}]$ is a basis of $H := H^1(\hat{E}, \mathcal{O})$. It is of course enough to check that $[c_{ij}] \in p^m H$. If $m = 0$ there is nothing to prove. Assume $m = 1$. Then we may write $a_{ij} \equiv a_i - a_j \bmod p$ with $a_i \in \mathcal{O}(\hat{U}_i)$. We get

$$a_{ij}^{p^n} \equiv a_i^{p^n} - a_j^{p^n} \mod p$$

hence $[c_{ij}] \equiv 0$ mod (p) and our claim is proved. Assume finally that $m \geq 2$ and that

$$[a_{ij}] \equiv 0 \mod p^m. \tag{7.26}$$

In particular we may write

$$a_{ij} + a_j \equiv a_i \quad mod \quad p^m \tag{7.27}$$

for some $a_i \in \mathcal{O}(\hat{U}_i)$. Condition 7.26 implies that

$$J^1(E) \otimes R/p^m R \to E \otimes R/p^m R \tag{7.28}$$

has a section in the category of schemes. By the universality property in Proposition 3.3 applied to the prolongation sequence

$$S^i := \mathcal{O}(J^{i+1}(U)) \otimes R/p^{m-i}R, \ U \subset E$$

there exists an induced section of

$$J^2(E) \otimes R/p^{m-1}R \to E \otimes R/p^{m-1}R$$

in the category of schemes, compatible with the section of the map 7.28 that we already have. This means that the cocycle

$$(a_{ij}, b_{ij}) \ \ mod \ \ p^{m-1}$$

is the coboundary, with respect to $\oplus$, of a family of the form

$$(a_i, b_i) \ \ mod \ \ p^{m-1}$$

with $b_i \in \mathcal{O}(\hat{U}_i)$. In particular

$$b_{ij} + b_j + \Gamma(a_{ij}, a_j) \equiv b_i \quad mod \quad p^{m-1}. \tag{7.29}$$

Multiplying Equation 7.29 by p and using Equation 7.26 we get

$$pb_{ij} + pb_j + \sum r_n(a_i^{p^n} - a_j^{p^n} - a_{ij}^{p^n}) \equiv pb_i \mod p^m \tag{7.30}$$

which implies that

$$[c_{ij}] \equiv 0 \mod p^m.$$

□

We are ready to prove Theorem 7.22:

PROOF. We start by constructing $\psi = \psi_{2,nor}$. By Lemma 7.23 we may write $[c_{ij}] = m[a_{ij}]$ in $H^1(E, \mathcal{O})$ with $m \in R$. Consider the series

$$\Theta(y, z) := \Theta_0(y, z) - my = \sum r_n y^{p^n} - my - pz \in R[y, z]\hat{}.$$

This series defines a homomorphism

$$\Theta : (\hat{\mathbf{A}}^2, \oplus) \to \hat{\mathbf{G}}_a$$

and clearly $\partial(\Theta) = 0$ so by Proposition 7.12 Θ lifts to a homomorphism $\psi : J^2(E) \to \hat{E}$; by the way, by the proof of Proposition 7.12, $\psi = \psi_{i_0} \circ \tau_{i_0}^{-1}$ where

$$\psi_{i_0} := \sum r_n y^{p^n} - my - pz + g \in \mathcal{O}(\hat{U})[y, z]\hat{} \subset R[[T]][T', T'']\hat{} \tag{7.31}$$

where $g \in \mathcal{O}(\hat{U})$ and U contains the origin. We may assume, of course, that $g \in TR[[T]]$ so

$$g(T) = \lambda T + (\text{higher degree terms in } T).$$

By the identifications $R[y]\hat{} = R[T']\hat{}$ and $R[y,z]\hat{} = R[T',T'']\hat{}$ we may express y as a restricted series in T' and z as a restricted series in T', T'' such that

$$y = \alpha T' + (\text{higher degree terms in } T'),$$

$$z = \beta T' + \gamma T'' + (\text{higher degree terms in } T', T''),$$

with $\alpha, \gamma \in R^\times$, $\beta \in R$. Exactly as in the proof of Proposition 7.20 we have

$$E(\psi) \circ e(pT) = \lambda_2 \phi^2(T) + \lambda_1 \phi(T) + \lambda_0 T$$

with $\lambda_i \in R$ and $E(\psi) \circ e(pT)$ is congruent mod T with

$$\psi_{i_0}(e(pT), \delta(e(pT)), \delta^2(e(pT))).$$

Now we have the following congruences in the ring $K[[T, T', T'']]$:

$$\psi_{i_0} \equiv -my - pz + \lambda T \equiv -m\alpha T' - p(\beta T' + \gamma T'') + \lambda T,$$

$$\psi_{i_0}(e(pT), \delta(e(pT)), \delta^2(e(pT))) \equiv \psi_{i_0}(pT, pT', pT'')$$

$$\equiv -m\alpha p T' - p^2(\beta T' + \gamma T'') + \lambda p T \quad mod \quad (T, T', T'')^2.$$

On the other hand

$$\lambda_2 \phi^2(T) + \lambda_1 \phi(T) + \lambda_0 T \equiv \lambda_2 p^2 T'' + \lambda_1 p T' + \lambda_0 T \quad mod \quad (T, T', T'')^2.$$

So we get

$$\lambda_2 p^2 T'' + \lambda_1 p T' + \lambda_0 T \equiv -p^2 \gamma T'' - (m\alpha p + p^2 \beta) T' + \lambda p T \quad mod \quad (T) + (T', T'')^2,$$

hence

$$\lambda_2 p^2 T'' + \lambda_1 p T' = -p^2 \gamma T'' - (m\alpha p + p^2\beta) T'.$$

We get that $\lambda_2 p^2 = -\gamma p^2$ hence $\lambda_2 = -\gamma \in R^\times$. Dividing ψ by λ_2 we may assume $\lambda_2 = 1$.

Next we claim that $\psi_2 = f^1_{jet}(E,\omega) \cdot \psi$. Indeed,

$$E(\psi_2 - f^1_{jet}(E,\omega) \cdot \psi) \in R[[T]][T']\hat{}$$

so by Proposition 4.44 $\psi_2 - f^1_{jet}(E,\omega) \cdot \psi \in \mathcal{O}^1(E)$. So $\psi_2 - f^1_{jet}(E,\omega) \cdot \psi$ is a $\delta-$character of order ≤ 1 on E. By Proposition 7.15 this $\delta-$character must vanish which proves our claim.

By Proposition 7.20 we get

$$f^1_{jet}(E,\omega)\phi^2(T) - f^2_{jet}(E,\omega)\phi(T) + p f^{1,2}_{jet}(E,\omega) T = E(\psi_2) \circ e(pT)$$

$$= f^1_{jet}(E,\omega) \cdot E(\psi) \circ e(pT)$$

$$= f^1_{jet}(E,\omega)\phi^2(T) + f^1_{jet}(E,\omega)\lambda_1 \phi(T) + f^1_{jet}(E,\omega)\lambda_0 T.$$

We get that

$$-\frac{f^2_{jet}(E,\omega)}{f^1_{jet}(E,\omega)} = \lambda_1 \in R, \quad \frac{f^{1,2}_{jet}(E,\omega)}{f^1_{jet}(E,\omega)} = \lambda_0 \in R.$$

Now recall that, by the very construction of ψ_2, we have

$$\tau^*_{i_0}\psi_2 = f^1_{jet}(E,\omega) L^2 - f^2_{jet}(E,\omega) L^1 + \gamma, \quad \gamma \in \mathcal{O}(\hat{U}_{i_0});$$

cf. Equation 7.17. Since $f^1_{jet}(E,\omega)$ divides ψ_2 it also divides $\tau^*_{i_0}\psi_2$ in $\mathcal{O}(\hat{U}_{i_0})[T', T'']\hat{}$. Since $f^1_{jet}(E,\omega)$ divides $f^1_{jet}(E,\omega)L^2 - f^2_{jet}(E,\omega)L^1$ in $\mathcal{O}(\hat{U}_{i_0})[T', T'']\hat{}$ it follows that $f^1_{jet}(E,\omega)$ also divides γ in $\mathcal{O}(\hat{U}_{i_0})[T', T'']\hat{}$, hence in $\mathcal{O}(\hat{U}_{i_0})$; so $\gamma = f^1_{jet}(E,\omega) \cdot \eta$

for some $\eta \in \mathcal{O}(\hat{U}_{i_0})$. On the other hand γ (and hence η) are $\equiv 0$ mod T and $\tau_{i_0}^* \psi \equiv \psi$ mod T. This implies that we have an equality of the form

$$\psi = L^2 - \frac{f^2_{jet}(E,\omega)}{f^1_{jet}(E,\omega)} L^1 + T \cdot \Omega, \quad \Omega \in \mathcal{O}(\hat{U})[T', T'']\hat{}. \tag{7.32}$$

Next we claim that

$$\psi = \psi^* + p\psi^{**}, \quad \psi^* \in \mathcal{O}(\hat{U})[T']\hat{}, \quad \psi^{**} \in \mathcal{O}(\hat{U})[T', T'']\hat{}. \tag{7.33}$$

Indeed, by the first assertion of our Theorem, the coefficient of $(T'')^n$ in $E(\psi)$ equals

$$p^{2n-1} \sum_{i=n}^{\infty} c_i^{\phi^2} \cdot \frac{i!}{(i-n)!n!} \cdot [(T^p + pT')^p + p(T')^p]^{i-n},$$

where $l_E(T) = \sum_{i=1}^{\infty} c_i T^i$. Recall from [**124**], p. 123, that $ic_i \in R$. Also $p^{n-1}/n! \in R$ for $n \geq 1$. We get that

$$p^{2n-1} \cdot c_i^{\phi^2} \cdot \frac{i!}{(i-n)!n!} = p^n \frac{p^{n-1}}{n!} (ic_i)^{\phi^2} \frac{(i-1)!}{(i-n)!} \in p^n R$$

for $n \geq 1$. It follows that the image of $E(\psi)$ in $k[[T]][T', T'']$ belongs to $k[[T]][T']$. Since

$$k[[T]][T'] \cap \mathcal{O}(U \otimes k)[T', T''] = \mathcal{O}(U \otimes k)[T']$$

it follows that the image of ψ in $\mathcal{O}(U \otimes k)[T', T'']$ belongs to $\mathcal{O}(U \otimes k)[T']$. This proves Claim 7.33. Since

$$L^2 \equiv (T')^p \quad mod \quad p,$$
$$L^1 \equiv T' \quad mod \quad p,$$

in $\mathcal{O}(\hat{U})[T', T'']\hat{}$ it follows that

$$(T')^p - \frac{f^2_{jet}(E,\omega)}{f^1_{jet}(E,\omega)} T' + T \cdot \Omega \equiv \psi \equiv \psi^* \quad mod \quad p.$$

It follows that the image $\bar{\Omega}$ of Ω in $\mathcal{O}(U \otimes k)[T', T'']$ is such that $T \cdot \bar{\Omega} \in \mathcal{O}(U \otimes k)[T']$ so $\bar{\Omega} \in \mathcal{O}(U \otimes k)[T']$ so

$$\Omega = \Omega_1 + p\Omega_2, \quad \Omega_1 \in \mathcal{O}(\hat{U})[T']\hat{}, \quad \Omega_2 \in \mathcal{O}(\hat{U})[T', T'']\hat{}. \tag{7.34}$$

The last assertion of the Theorem follows now directly from Equations 7.34 and 7.32. □

DEFINITION 7.24. Let G be $\mathbf{G}_m$ or an elliptic curve E and let $\omega = dx/x$ if $G = \mathbf{G}_m$ or any invertible 1–form if $G = E$. The *canonical* $\delta-$ character ψ_{can} on G is defined as follows:

1) $\psi_{can} = \psi_1$, the basic $\delta-$character of order 1, if $G = \mathbf{G}_m$ or if $G = E$ and E has a lift of Frobenius;

2) $\psi_{can} = \psi_{2,nor}$ if $G = E$ and E doesn't have a lift of Frobenius.

One can show that ψ_{can} depends on the pair (E, ω) only. Since $E(\psi_{can}) = \Psi_{\mathbf{G}_m}$ in case $G = \mathbf{G}_m$ it is reasonable to extend this notation by letting

$$\Psi_E = \Psi_{E,\omega} := E(\psi_{can}) \tag{7.35}$$

for $G = E$.

Let $\partial : \mathcal{O}_G \to \mathcal{O}_G$ be the R-derivation inducing the canonical derivation

$$\partial = \partial^{can} = \left(\frac{dl}{dT}\right)^{-1} \cdot \frac{d}{dT}$$

on $R[[T]]$, $l := l_G(T)$. In particular $\partial(l(T)) = 1$. We also denote, as usual, by $e(T) = e_G(T)$ the compositional inverse of $l(T)$. Let

$$\partial_j : \mathcal{O}_G^{\infty} \to \mathcal{O}_G^{\infty},$$

$j \geq 0$, be the conjugate operators of ∂. By Propositions 3.42, 3.43, 3.45, 4.44, Lemma 4.18, and Remark 4.19 we have

$$\partial_s \circ \delta^s = \phi^s \circ \partial, \quad \partial_s \circ \phi^s = p^s \cdot \phi^s \circ \partial, \quad \text{on} \quad \mathcal{O}_G, \tag{7.36}$$

$$\partial_j \circ \phi^s = 0 \quad \text{on} \quad \mathcal{O}_G \quad \text{for} \quad j \neq s;$$

moreover

$$\partial_j \circ \phi = p \cdot \phi \circ \partial_{j-1} \quad \text{on} \quad \mathcal{O}_G^{\infty}$$

and ∂_j are compatible (via the expansion map) with derivations (still denoted by)

$$\partial_j : \bigcup_{s=0}^{\infty} K[[T, ..., T^{(s)}]] \to \bigcup_{s=0}^{\infty} K[[T, ..., T^{(s)}]] \tag{7.37}$$

satisfying the conditions 7.36 on $K[[T]]$.

PROPOSITION 7.25. *Assume $G = E$ is an elliptic curve that doesn't have a lift of Frobenius and set $\psi = \psi_{can} = \psi_{2,nor}$. Then*

$$\partial_2\psi = p, \quad \partial_1\psi = -\frac{f^2_{jet}(E,\omega)}{f^1_{jet}(E,\omega)}, \quad \partial_0\psi = \frac{f^{1,2}_{jet}(E,\omega)}{f^1_{jet}(E,\omega)}.$$

PROOF. By Theorem 7.22, Proposition 4.44, and formula 7.36 we get

$$E(\partial_2\psi) = \partial_2(E(\psi)) = \partial_2(\frac{1}{p}\phi^2(l(T))) = \frac{1}{p} \cdot p^2 \cdot \phi^s(\partial(l(T))) = p.$$

Similarly we obtain

$$E(\partial_1\psi) = -\frac{f^2_{jet}(E,\omega)}{f^1_{jet}(E,\omega)}, \quad E(\partial_0\psi) = \frac{f^{1,2}_{jet}(E,\omega)}{f^1_{jet}(E,\omega)},$$

and we are done by the injectivity of the expansion map E. $\square$

Similarly one proves:

PROPOSITION 7.26. *Assume $G = \mathbf{G}_m$ or $G = E$ is an elliptic curve that has a lift of Frobenius and set $\psi = \psi_{can} = \psi_1$. Then*

$$\partial_1\psi = 1, \quad \partial_0\psi = f_{lift}(G,\omega).$$

REMARK 7.27. Proposition 7.25 provides a computation of the derivatives of $\psi = \psi_{can}$ with respect to

$$\partial_0, \partial_1, \partial_2$$

which form a basis of the tangent sheaf of $J^2(E)$; therefore the Equations in Proposition 7.25 can be viewed as a system of Pfaffian equations satisfied by ψ. A similar remark can be made about Proposition 7.26. Here is the result for the case of Proposition 7.25; a similar result holds in the case of Proposition 7.26.

COROLLARY 7.28. *Assume the hypotheses and notation of Proposition 7.25, let ω be the invertible 1–form on E dual to ∂, and let $\omega_0, \omega_1, \omega_2$ be the conjugate 1–forms on $J^2(E)$, $\omega_i = p^{-i}(\phi^i)^*\omega$. Then $Z := \psi$ satisfies the equation*

$$dZ = \frac{f^{1,2}_{jet}(E,\omega)}{f^1_{jet}(E,\omega)} \cdot \omega_0 - \frac{f^2_{jet}(E,\omega)}{f^1_{jet}(E,\omega)} \cdot \omega_1 + p \cdot \omega_2.$$

In particular the 1–form in the right hand side of the above Equation is exact.

The next result can be viewed as an analogue of Proposition 6.13:

COROLLARY 7.29. *Let G be either $\mathbf{G}_m$ or an elliptic curve E and let $\psi = \psi_{can}$ be the canonical δ–character on G. Then*

$$\partial_i \partial_j \psi^{\phi^r} = 0 \quad \text{for all} \quad i, j, r. \tag{7.38}$$

In particular if

$$V^{ij}_r := \partial_i \psi^{\phi^r} \cdot \partial_j - \partial_j \psi^{\phi^r} \cdot \partial_i \in Der_R(\mathcal{O}^\infty_E, \mathcal{O}^\infty_E)$$

is the (i,j)–Hamiltonian flow defined by ψ^{ϕ^r} and ∂ then any two Hamiltonian flows V^{ij}_r and V^{mn}_s commute.

Next we address the existence problem for infinitesimal symmetries and shape of tangent maps:

PROPOSITION 7.30. *Let G be $\mathbf{G}_m$ or an elliptic curve E and let $\psi = \psi_{can}$ be the canonical δ–character on G corresponding to the invertible 1–form ω.*

1) The following holds in $K[[T, T', T'']]$:

$$(l_G(T) \cdot \partial)_{**} E(\psi) = E(\psi); \tag{7.39}$$

in particular $l_G(T) \cdot \partial$ is an infinitesimal symmetry of $K \cdot E(\psi)$.

2) Assume $G = \mathbf{G}_m$ or $G = E$ and E has a lift of Frobenius. Then, for any $\alpha \in p\mathcal{O}(G)$, we have

$$(\alpha\partial)_* \psi = (\phi + f_{lift}(G,\omega))(\alpha); \tag{7.40}$$

$$(\alpha\partial)_{**} \psi = \frac{1}{p}[\phi + p \cdot f_{lift}(G,\omega)](\alpha) = E(\psi)(e_G(\alpha)). \tag{7.41}$$

Also, the Euler-Lagrange equation, $\langle \partial, EL_1(\psi^2)\rangle_1$, attached to ψ^2 and ∂ equals

$$2[\psi + p(f_{lift}(G,\omega))^\phi \psi^\phi].$$

3) If $G = E$ and E doesn't have a lift of Frobenius then

$$(\alpha\partial)_* \psi = \left[p\phi^2 - \frac{f^2_{jet}(E,\omega)}{f^1_{jet}(E,\omega)} \cdot \phi + \frac{f^{1,2}_{jet}(E,\omega)}{f^1_{jet}(E,\omega)}\right](\alpha), \tag{7.42}$$

$$(\alpha\partial)_{**} \psi = \frac{1}{p}\left[\phi^2 - \frac{f^2_{jet}(E,\omega)}{f^1_{jet}(E,\omega)}\phi + p\frac{f^{1,2}_{jet}(E,\omega)}{f^1_{jet}(E,\omega)}\right](\alpha) = E(\psi)(e_G(\alpha)). \tag{7.43}$$

Also, the Euler-Lagrange equation, $\langle \partial, EL_2(\psi^2)\rangle_2$, attached to ψ^2 and ∂ equals

$$2p\left[\psi - \left(\frac{f^2_{jet}(E,\omega)}{f^1_{jet}(E,\omega)}\right)^\phi \psi^\phi + p\left(\frac{f^{1,2}_{jet}(E,\omega)}{f^1_{jet}(E,\omega)}\right)^{\phi^2} \psi^{\phi^2}\right].$$

REMARK 7.31. The following conditions are easily seen to be equivalent:

1) There exists $\theta \in Der_R(\mathcal{O}_G, \mathcal{O}_G)$, $\theta \neq 0$, with $\theta_{**}\psi = 0$ (i.e. θ is a variational infinitesimal symmetry of ψ).

2) $G = E$ and

$$E \textit{ doesn't have a lift of Frobenius and } \frac{f^2_{jet}(E,\omega)}{f^1_{jet}(E,\omega)} \not\equiv 0 \textit{ mod } p. \tag{7.44}$$

It follows from results in the next Chapter that condition 7.44 is implied, for instance, by the condition

$$E \textit{ has ordinary reduction and Kodaira-Spencer class } KS(E) \neq 0. \tag{7.45}$$

Indeed, by Lemma 7.14 the condition $KS(E) \neq 0$ implies that

$$f^1_{jet}(E,\omega) \not\equiv 0 \textit{ mod } p.$$

On the other hand, it will be proved in Proposition 8.55 that the reduction mod p of $f^2_{jet}(E,\omega)$ equals the reduction mod p of $(f^1_{jet}(E,\omega))^p$ times the value of the Hasse invariant at (E,ω) which proves our claim.

In a similar way one sees, for instance, that if $G = E$ and condition 7.44 holds then there exists $\alpha \in R$, $\alpha \neq 0$, with $(\alpha\partial)_*\psi = 0$; here one needs to use Corollary 8.84.

PROOF. (of Proposition 7.30) Assume first that either $G = \mathbf{G}_m$ or $G = E$ and E has a lift of Frobenius. By Proposition 7.20,

$$(\alpha\partial)_{**}\psi = \frac{1}{p}[\phi + p \cdot f_{lift}(G,\omega)](\alpha) = \frac{1}{p}[\phi + p \cdot f_{lift}(G,\omega)](l_G(e_G(\alpha))) = E(\psi)(e_G(\alpha)).$$

Assume now $G = E$ is an elliptic curve without a lift of Frobenius. Then, by Proposition 7.20, we have, for $\alpha \in \mathcal{O}(E) = R$:

$$\begin{aligned}(\alpha\partial)_{**}\psi &= \alpha\partial_0\psi + p^{-1}\alpha^{\phi}\partial_1\psi + p^{-2}\alpha^{\phi^2}\partial_2\psi \\ &= \alpha\frac{f^{1,2}_{jet}(E,\omega)}{f^1_{jet}(E,\omega)} - p^{-1}\alpha^{\phi}\frac{f^2_{jet}(E,\omega)}{f^1_{jet}(E,\omega)} + p^{-1}\alpha^{\phi^2}\end{aligned}$$

$$= \frac{1}{p}\left[\phi^2 - \frac{f^2_{jet}(E,\omega)}{f^1_{jet}(E,\omega)}\phi + p\frac{f^{1,2}_{jet}(E,\omega)}{f^1_{jet}(E,\omega)}\right](l_G(e_G(\alpha))) = E(\psi)(e_G(\alpha)).$$

Equations 7.40 and 7.42 follow in a similar way. Equation 7.39 follows directly from Equation 7.41. The statements about the Euler-Lagrange equations are left to reader. □

EXAMPLE 7.32. Proposition 7.30 can be used to explicitly compute $\delta-$tangent maps. Here is an example. Let

$$f : G(R)_{\psi} \to \mathbf{P}^1(R), \ P \mapsto f(P) := (\psi(P) : \psi^{\phi}(P))$$

be the map defined by the pair of sections $(s_0, s_1) = (\psi, \psi^{\phi})$, where $G(R)_{\psi}$ is the set of all $P \in G(R)$ such that $\psi(P) \not\equiv 0$ mod p. Then, for any $P \in G(R)_{\psi}$, one can consider the $\delta-$tangent map

$$T_P f : T_P G \to T_{f(P)}\mathbf{P}^1,$$

as in Equation 3.44. For $\alpha \in R$, a trivial computation yields:

$$(T_P f)(\alpha\partial_P) = \psi(P)^{\phi-1}\left[p\cdot\left(\frac{(\alpha\partial)_*\psi}{\psi}\right)^{\phi} - \left(\frac{(\alpha\partial)_*\psi}{\psi}\right)\right](P).$$

The above quantity vanishes if and only if the quantity $(\alpha\partial)_*\psi$ vanishes. The latter is explicitly given in Equation 7.42. So, in particular, $rank_{\mathbf{Z}_p} Ker(T_P f)$ is independent of P and is > 0 if $G = \mathbf{G}_m$ or if $G = E$ and condition 7.45 holds.

The following Lemma will be useful later; it is a formal scheme version of Weil's principle according to which "generically defined homomorphisms of algebraic groups are everywhere defined"; this Lemma will be useful for constructing δ–characters. Since here we are dealing with formal schemes rather than varieties, it might be helpful to express Weil's argument in a form that applies to formal schemes.

LEMMA 7.33. *Set $U := J^s(X)$, $U^2 := U\times U$ and let $U^{(2)} := U^2 \cap \mu^{-1}(U)$ where $\mu : J^s(E)\times J^s(E) \to J^s(E)$ is the addition map. Let $\mu, p_1, p_2 : U^{(2)} \to U$ be the maps induced by μ and the two projections and let $\mu^*, p_1^*, p_2^* : \mathcal{O}(U) \to \mathcal{O}(U^{(2)})$ be the corresponding algebra maps. Assume $f \in \mathcal{O}(U)$ satisfies*

$$\mu^* f = p_1^* f + p_2^* f$$

in $\mathcal{O}(U^{(2)})$. Then f is the restriction of a homomorphism $J^s(E) \to \hat{\mathbf{G}}_a$, hence comes from a δ–character of E.

PROOF. One considers the morphism

$$U^2 \times_{J^s(E)} U^2 = (U^2,\mu)\times_{J^s(E)} (U^2,\mu) \to \hat{\mathbf{G}}_a$$

defined on points by

$$(B_1, C_1, B_2, C_2) \mapsto f(B_1) + f(C_1) - f(B_2) - f(C_2).$$

This morphism is constant (equal to 0) on the open formal subscheme

$$U^{(2)} \times_{J^s(E)} U^{(2)}$$

so it must be constant (equal to 0). So the morphism

$$U^2 \to \hat{\mathbf{G}}_a,\quad (B, C) \mapsto f(B) + f(C)$$

is "constant on the fibers of" $\mu : U^2 \to J^s(E)$ (in the "functorial sense"). Since $\mu : U^2 \to J^s(E)$ is surjective and is an open immersion composed with the locally trivial fibration $J^s(E)^2 \to J^s(E)$, it is easy to check that $U^2 \to \hat{\mathbf{G}}_a$ descends to a morphism $J^s(E) \to \hat{\mathbf{G}}_a$; then the latter is easily seen to be a homomorphism. □

7.3. δ–invariants

THEOREM 7.34. [**29**]. *Assume we are given a flat correspondence $\mathbf{X} = \mathbf{X}_{G,N,\epsilon}$ as in Equation 7.1. Let r be 1 or 2 according as G has a lift of Frobenius or not and let $e = \sharp\Gamma$, i.e. e is 1 or 2 according as Γ is trivial or not. Then, for any $m, s \in \mathbf{Z}_+$ we have*

$$H^0(\mathbf{X}, \mathbf{L}^m_{r+s}) = R[x^{(\leq s)}]_m \quad or \quad H^0(\mathbf{X}, \mathbf{L}^m_{r+s}) = 0$$

according as m is divisible by e or not. Consequently

$$R(\mathbf{X}, \mathbf{L}_r)^* = (R\{x_*, y\}^*)^{(W_{+,e})}$$

$$R(\mathbf{X}, \mathbf{L}_r)_* \simeq R\{x\}^{(e)}_*, \quad R\langle\mathbf{X}, \mathbf{L}_r\rangle \simeq (R\{x\}_*)_{((p))};$$

in particular $(\mathbf{X}, \mathbf{L}_r)$ *is* $\delta-$*rational (cf. Proposition 5.14).*

In the above statement y is, of course, an additional variable, $W_{+,e}$ is the subsemigroup of W_+ consisting of weights of degree divisible by e, the superscript $(W_{+,e})$ means, as usual, the sum of all pieces of weight belonging to $W_{+,e}$, and the superscript (e) means, as usual, the sum of all pieces of degree divisible by e. Now there are 3 cases in the above Theorem:

1) $G = \mathbf{G}_m$,
2) $G = E$ an elliptic curve that has a lift of Frobenius,
3) $G = E$ an elliptic curve that doesn't have a lift of Frobenius.

Each of the 3 cases above has 2 sub-cases corresponding to $\epsilon = 1$ (Γ trivial) and $\epsilon = -1$ (Γ non-trivial) respectively. We will only address the (apparently) most complicated case 3) here; we will first look at the sub-case $\epsilon = 1$ and then we shall reduce the sub-case $\epsilon = -1$ to the sub-case $\epsilon = 1$. The cases 1 and 2 can be treated similarly.

PROPOSITION 7.35. *Theorem 7.34 holds in case* $G = E$ *doesn't have a lift of Frobenius and* Γ *is trivial.*

In this case

$$\mathbf{X} = (X, \tilde{X} = X \cap [N]^{-1}(X), \iota, [N]).$$

So, in particular, for any positive integer n we have

$$\mathbf{X}^{\circ n} = (X, \tilde{X}^{\circ n} = X \cap [N]^{-1}(X) \cap ... \cap [N^n]^{-1}(X), \iota, [N^n]).$$

We need some preliminaries.

By Proposition 7.7 $\mathbf{X}$ has an $n-$cycle for some n; hence $\mathbf{X}^{\circ n}$ has a fixed point (A, A) where $A \in X(R)$, $[N^i](A) \in X(R)$ for $1 \leq i \leq n-1$, and $[N^n](A) = A$. Note that A is torsion. Consider the injective expansion maps at this point

$$E : \mathcal{O}^s(X) \to R[[T]][T', ..., T^{(s)}]\hat{}, E : \mathcal{O}^s(\tilde{X}^{\circ n}) \to R[[T]][T', ..., T^{(s)}]\hat{}$$

corresponding to the choice of an étale coordinate T at the origin and the translation by A; cf. Proposition 4.43 and Lemma 7.6. By that Lemma the germ of $\mathbf{X}^{\circ n}$ at (A, A) is $\Sigma(T) = [N^n](T)$, the multiplication by N in the formal group of E. Also, clearly, we have

$$\mathbf{L}_s^{\circ n} = \mathbf{L}_s^n$$

so the germ of $\mathbf{L}_s^{\circ n}$ at (A, A) is the flat formal $\delta-$line bundle

$$\Lambda_s^n := \Lambda(s, N^{-n}). \tag{7.46}$$

(Here, according to our standard notation, we write $\Lambda_s := \Lambda(s, N^{-1})$.) By Proposition 4.43, for any $f \in \mathcal{O}^s(X)$ and any $i < s$ we have

$$f \in \mathcal{O}^i(X) \text{ if and only if } E(f) \in R[[T]][T', ..., T^{(i)}]\hat{}. \tag{7.47}$$

We have the maps

$$\Delta_{nm} := \Delta^{\mathbf{L}_s^{mn}} : \mathcal{O}^s(X) \to \mathcal{O}^s(\tilde{X}^{\circ n}), \tag{7.48}$$

$$\Delta_{nm} := \Delta^{\Lambda_s^{nm}} : R[[T]][T', ..., T^{(s)}]\hat{} \to R[[T]][T', ..., T^{(s)}]\hat{},$$

both defined by

$$\Delta_{nm} f := f - N^{-mn} \cdot f \circ [N^n],$$

with kernels

$$H^0(\mathbf{X}^{\circ n}, \mathbf{L}_s^{nm}), \quad H^0(\Sigma, \Lambda_s^{nm}).$$

By Proposition 4.52 we have

$$\Delta_{nm} \circ E = E \circ \Delta_{nm}, \tag{7.49}$$

hence, for any $f \in \mathcal{O}^s(X)$,

$$f \in H^0(\mathbf{X}^{\circ n}, \mathbf{L}_s^{nm}) \text{ if and only if } E(f) \in H^0(\Sigma, \Lambda_s^{nm}). \tag{7.50}$$

By Proposition 4.35 we have

(7.51)

$$E(H^0(\mathbf{X}^{\circ n}, \mathbf{L}_s^{nm})) \subset \{\Pi(l, l^\phi, ..., l^{\phi^s}) | \Pi \in K[x_0, ..., x_s] \text{ homogeneous of degree } m\},$$

where $l = l(T) \in K[[T]]$ is the logarithm of the formal group law of E.

Let $\psi = \psi_{2,nor}$ be the δ–character on E defined by Theorem 7.22. Since ψ is a homomorphism we have $\psi \circ [N] = N \cdot \psi$ so for all $j \geq 1$ we have

$$\Delta_{nm}\psi^j = (1 - N^{n(j-m)}) \cdot \psi^j. \tag{7.52}$$

In particular

$$\psi^j \in H^0(\mathbf{X}^{\circ n}, \mathbf{L}_2^{nj}). \tag{7.53}$$

Also, by Proposition 7.25, we have

$$\partial_s \psi^{\phi^{s-2}} = p^{s-1}. \tag{7.54}$$

Recall from Proposition 4.55 that

$$\partial_s \circ \Delta_{nm} = \Delta_{n(m-1)} \circ \partial_s \tag{7.55}$$

so in particular

$$\partial_s(H^0(\mathbf{X}^{\circ n}, \mathbf{L}_s^{nm})) \subset H^0(\mathbf{X}^{\circ n}, \mathbf{L}_s^{n(m-1)}). \tag{7.56}$$

The main step in the proof of Proposition 7.35 is to prove a statement of that type in which we do not take torsion into account; we will deal with torsion phenomena later.

LEMMA 7.36. *For any integer $m \in \mathbf{Z}_+$ and any $s \geq 2$ the natural map*

$$K[x, x^\phi, ..., x^{\phi^{s-2}}]_m = R[x, x', ..., x^{(s-2)}]_m \otimes_R K \to H^0(\mathbf{X}^{\circ n}, \mathbf{L}_s^{nm}) \otimes_R K$$

defined by sending x^{ϕ^i} into ψ^{ϕ^i} is an isomorphism. Moreover, for $m \geq 1$,

$$H^0(\mathbf{X}^{\circ n}, \mathbf{L}_1^{nm}) = 0.$$

PROOF. *Step 0.* We check the case $m = 0$ of the Lemma, i.e. we check that for $m = 0$ and $s \geq 2$ we have that any element of $H^0(\mathbf{X}^{\circ n}, \mathbf{L}_s^0) \otimes K$ comes from K. This statement is, however, a direct consequence of Equation 7.51.

Step 1. We prove the case $m = 1$, $s = 1$ of our Lemma; in other words we show that $H^0(\mathbf{X}^{\circ n}, \mathbf{L}_1^n) = 0$. Let $g \in H^0(\mathbf{X}^{\circ n}, \mathbf{L}_1^n)$. By Equation 7.51 $E(g)$ is a K–linear combination of $l(z)$ and $l(z)^\phi$ so we have

$$\mu^* E(g) = p_1^* E(g) + p_2^* E(g)$$

where μ^*, p_1^*, p_2^* are the obvious maps on series induced by comultiplication and the two projections. It follows that

$$\mu^* g = p_1^* g + p_2^* g$$

in $\mathcal{O}(U^{(2)})$, where $U^{(2)}$ is as in Lemma 7.33. By Lemma 7.33 g comes from a δ–character of order ≤ 1 on E; but by Proposition 7.15 any such δ–character vanishes so $g = 0$.

Step 2. We prove by induction on $m+s$ that for any $f \in H^0(\mathbf{X}^{\circ n}, \mathbf{L}_s^{nm})$ the following hold:

1) If $m \geq 1$, $s \geq 2$ then f has an $R-$multiple which is a homogeneous polynomial, of degree m, with $R-$coefficients in $\psi, \psi^{\phi}, ..., \psi^{\phi^{s-2}}$;

2) If $m \geq 1$ and $s \in \{0,1\}$ then $f=0$.

This will, of course, complete the proof of our Lemma. The case $m+s=1$ (equivalently $m=1$ and $s=0$) follows from Step 1. To perform the induction step, assume that $f \in H^0(\mathbf{X}^{\circ n}, \mathbf{L}_s^{nm}) \subset \mathcal{O}(J^s(X))$ where $m+s \geq 2$. We treat the cases $m=1$ and $m \geq 2$ separately.

Case $m=1$. If $s=1$ we are done again by Step 1 so we may assume $s \geq 2$. By Equation 7.51 we may write

$$E(f) = \sum_{i=0}^{s} a_i l(z)^{\phi^i}$$

with $a_i \in K$. Let $b \in R$ be such that $ba_s \in R$ and set

$$g := b \cdot f - ba_s p \cdot \psi^{\phi^{s-2}} \in H^0(\mathbf{X}^{\circ n}, \mathbf{L}_s^n).$$

So $E(g) \in R[[T, T', ..., T^{(s-1)}]]$. By Equation 7.47 $g \in \mathcal{O}(J^{s-1}(X))$ hence

$$g \in H^0(\mathbf{X}^{\circ n}, \mathbf{L}_{s-1}^n).$$

By the induction hypothesis g is a $K-$linear combination of $\psi, \psi^{\phi}, ..., \psi^{\phi^{s-3}}$ if $s \geq 3$ or $g=0$ if $s=2$. Hence f is a $K-$linear combination of $\psi, \psi^{\phi}, ..., \psi^{\phi^{s-2}}$ which settles the case $m=1$.

Case $m \geq 2$. By Equation 7.56 $\partial_s f \in H^0(\mathbf{X}^{\circ n}, \mathbf{L}_s^{n(m-1)})$. By induction $\partial_s f = 0$ if $s \in \{0,1\}$ or, in case $s \geq 2$, $\partial_s f$ is a homogeneous polynomial, of degree $m-1$, with $K-$coefficients, in $\psi, \psi^{\phi}, ..., \psi^{\phi^{s-2}}$. If $s=0$ then $f \in R$ and, since $m \geq 1$, we must have $f=0$. If $s=1$ then $\partial_1(E(f))=0$ hence $E(f) \in R[[T]]$ hence, by Equation 7.47 $f \in \mathcal{O}(X)\hat{}$ hence $f \in H^0(\mathbf{X}^{\circ n}, \mathbf{L}_0^{nm})$ hence, by induction, $f=0$. Assume $s \geq 2$. Then write

$$\partial_s f = \sum_{i=0}^{m-1} c_i A_i(\psi, \psi^{\phi}, ..., \psi^{\phi^{s-3}}) \cdot (\psi^{\phi^{s-2}})^i$$

where $c_i \in K$ and A_i are homogeneous polynomials of degree $m-1-i$, with $R-$coefficients. Let $b \in R$, $b \neq 0$, such that $bc_i \in R$ for all i and define

$$g := p^{s-1} \cdot m! \cdot b \cdot f - m! \cdot b \cdot \sum_{i=0}^{m-1} c_i A_i(\psi, \psi^{\phi}, ..., \psi^{\phi^{s-3}}) \frac{(\psi^{\phi^{s-2}})^{i+1}}{i+1}.$$

By Equation 7.54 it follows that $\partial_s g = 0$. Hence $\partial_s(E(g))=0$ hence

$$E(g) \in R[[T]][T', ..., T^{(s-1)}]\hat{}$$

hence, by Equation 7.47 $g \in \mathcal{O}(J^{s-1}(X))$ hence

$$g \in H^0(\mathbf{X}^{\circ n}, \mathbf{L}_{s-1}^{nm}).$$

By induction g is a homogeneous polynomial of degree m, with $K-$coefficients, in $\psi, \psi^{\phi}, ..., \psi^{\phi^{s-3}}$ if $s \geq 3$ or $g=0$ if $s=2$. Then f itself is a homogeneous polynomial of degree m, with $K-$coefficients, in $\psi, \psi^{\phi}, ..., \psi^{\phi^{s-2}}$ and we are done. □

LEMMA 7.37. *For any $s \geq 2$ the natural map*

$$R[x, x', ..., x^{(s-2)}] \otimes k \to \mathcal{O}(J^s(X)) \otimes k,$$

$$x^{(i)} \mapsto \delta^i \psi,$$

is injective. In particular, if $f \in \mathcal{O}(J^s(X))$ is such that $p^\nu f$ belongs to the image of

$$R[x, x', ..., x^{(s-2)}] \to \mathcal{O}(J^s(X)),$$

then f itself belongs to this image.

PROOF. It is enough to show that the composition

$$R[x, x', ..., x^{(s-2)}] \otimes k \to \mathcal{O}(J^s(X)) \otimes k \stackrel{E}{\to} R[[T, T', ..., T^{(s)}]] \otimes k \to$$

$$\to R[T, T', ..., T^{(s)}]^{\hat{}} \otimes k = k[T, T', ..., T^{(s)}]$$

is injective where the last map above is induced by the map

$$R[[T, T', ..., T^{(s)}]] \to R[T, T', ..., T^{(s)}]^{\hat{}},$$

$$T^{(i)} \mapsto \delta^i(e(pT));$$

here $e(z) \in K[[T]]$ is the exponential of the formal group of E. By Theorem 7.22 the image of $x^{(i)}$ under the above composition equals the image of $\delta^i(\Lambda(T)) \in R[T, T', ..., T^{(s)}]$ in $k[T, T', ..., T^{(s)}]$, where

$$\Lambda(T) = \phi^2(T) + \lambda_1 \phi(T) + \lambda_0 T,$$

$\lambda_1, \lambda_0 \in R$. So it is enough to show that the images of

$$\Lambda(T), \delta(\Lambda(T)), ..., \delta^{s-2}(\Lambda(T))$$

in $k[T, T', ..., T^{(s)}]$ are algebraically independent. This is a consequence of Proposition 5.24. This concludes our proof. □

We are ready to conclude the proof of Proposition 7.35:

PROOF. Note that, by Lemma 7.36 the map

$$K[x, x^\phi, ..., x^{\phi^{s-2}}]_m \to H^0(\mathbf{X}^{\circ n}, \mathbf{L}_s^{nm}) \otimes K$$

is an isomorphism. By Lemma 7.37 the composition

$$R[x, x', ..., x^{(s-2)}]_m \to H^0(\mathbf{X}, \mathbf{L}_s^m) \to H^0(\mathbf{X}^{\circ n}, \mathbf{L}_s^{nm})$$

is an isomorphism. Since both maps above are injective, the first one must be an isomorphism. □

PROPOSITION 7.38. *Theorem 7.34 holds in case $G = E$ doesn't have a lift of Frobenius and Γ is non-trivial.*

PROOF. Let $\pi : E \to E/\langle [\epsilon] \rangle$ be the canonical projection, let $Y = \pi^{-1}(X)$ and consider, in addition to our correspondence $\mathbf{X}$, the correspondence

$$\mathbf{Y} = (Y, Y \cap [N]^{-1}(Y), \iota, [N]).$$

Let $\psi = \psi_{2,can}$. By Corollary 3.30 we have

$$H^0(\mathbf{X}, \mathbf{L}_s^m) \simeq H^0(\mathbf{Y}, \mathbf{L}_s^m)^\Gamma.$$

By Proposition 7.35

$$H^0(\mathbf{Y}, \mathbf{L}_s^m) \simeq (R[x, x', ..., x^{(s-2)}]_m)^\Gamma.$$

Since $[-1]^*\psi = -\psi$ the action of Γ is easily seen to be given by

$$[-1]^* x^{(i)} = -x^{(i)}.$$

So $(R[x, x', ..., x^{(s-2)}]_m)^\Gamma$ is $R[x, x', ..., x^{(s-2)}]_m$ or 0 according as m is even or odd. This ends our proof. □

REMARK 7.39. Note that $R(\mathbf{X}, \mathbf{L}_s)^*$ in Theorem 7.34 depends on X, G, and ϵ but not on N. Therefore Theorem 7.34 remains true for the correspondences of the form

$$\mathbf{X}_{G,\mathcal{N},\epsilon} := \coprod_{N \in \mathcal{N}} \mathbf{X}_{G,N,\epsilon}$$

equipped with the δ–line bundle

$$\mathbf{L}_s := \coprod_{N \in \mathcal{N}} \mathbf{F}(s, N^{-1}),$$

where X, G, ϵ are fixed and N runs through the set $\mathcal{N}$ of integers such that $N \neq 1, -1$ and $N \not\equiv 0 \bmod p$.

7.4. δ–base loci

We begin with the following result (cf. [**17**] + [**19**]) that should be interpreted as an arithmetic analogue of Manin's Theorem of the Kernel [**97**]. For a more precise result (including the supersingular case) and a higher dimensional version we refer to [**17**], [**19**].

THEOREM 7.40. *([**17**] + [**19**]) Let E be an elliptic curve over $R = \hat{\mathbf{Z}}_p^{ur}$. Let $\psi = \psi_{can} : E(R) \to R$ be the canonical δ–character and consider the group*

$$p^\infty E(R) := \bigcap p^n E(R) \subset Ker\ \psi.$$

Then the group

$$\frac{Ker\ \psi}{p^\infty E(R)}$$

is torsion. Assume furthermore that E is defined over $R' := \mathbf{Z}_p^{ur}$. Then

$$E(R') \cap (Ker\ \psi) = E(R')_{tors}.$$

We will only present the proof in case E doesn't have a lift of Frobenius, hence $\psi = \psi_{2,nor}$; the case when E has a lift of Frobenius is easier and is left to the reader. For the proof we need some preliminaries. First we need the following two facts due to Tate:

LEMMA 7.41. *Assume E is an elliptic curve over $R = \hat{\mathbf{Z}}_p^{ur}$ with ordinary reduction and no lift of Frobenius and let $E(pR) := Ker(E(R) \to E(k))$. Then*

$$E(pR) \cap p^\infty E(R) \simeq \mathbf{Z}_p.$$

PROOF. This is implicit in [**129**]; see also [**17**], p. 331. □

LEMMA 7.42. *Assume E is an elliptic curve over $R = \hat{\mathbf{Z}}_p^{ur}$ with ordinary reduction and assume E is defined over $R' = \mathbf{Z}_p^{ur}$. Then*

$$E(R') \cap p^\infty E(R) \subset E(R')_{tors}.$$

PROOF. This is implicit in [**129**]; see also [**19**], p. 209. □

We also need the following

LEMMA 7.43. *Let $\lambda_1 \in R, \lambda_0 \in R$. Then the $\mathbf{Z}_p-$module*

$$\{x \in R \mid x^{\phi^2} + \lambda_1 x^{\phi} + p\lambda_0 x = 0\}$$

has rank ≤ 1. If $\lambda_1 \in pR$ then the above module vanishes.

PROOF. An easy exercise; see also [**17**], p. 338. $\square$

PROOF. (of Theorem 7.40) Consider the following groups

$$E(pR) \cap p^{\infty}E(R) \subset E(pR) \cap (Ker\ \psi) \subset Ker\ \psi.$$

The group

$$\frac{Ker\ \psi}{E(pR) \cap (Ker\ \psi)}$$

injects into the torsion group $E(k)$ so it is torsion. We claim that the group

$$\frac{E(pR) \cap (Ker\ \psi)}{E(pR) \cap p^{\infty}E(R)}$$

is also torsion; this will imply that the group

$$\frac{Ker\ \psi}{p^{\infty}E(R)}$$

is torsion. To check our Claim note that, by Theorem 7.22 we have

$$E(pR) \cap Ker\ \psi \simeq \{x \in R \mid x^{\phi^2} + \lambda_1 x^{\phi} + p\lambda_0 x = 0\}$$

where

$$\lambda_1 = -\frac{f^2_{jet}(E,\omega)}{f^1_{jet}(E,\omega)}, \quad \lambda_0 = \frac{f^{1,2}_{jet}(E,\omega)}{f^1_{jet}(E,\omega)}.$$

If E has supersingular reduction then, by Corollary 8.89 to be proved later, $\lambda_1 \in pR$ hence, by Lemma 7.43, $E(pR) \cap Ker\ \psi = 0$ and our claim is proved. Assume now E has ordinary reduction. Then, by Lemma 7.43, the $\mathbf{Z}_p-$module $E(pR) \cap Ker\ \psi$ has rank ≤ 1. On the other hand, by Lemma 7.41, the $\mathbf{Z}_p-$module $E(pR) \cap p^{\infty}E(R)$ has rank 1. Again, our Claim follows.

Assume now E is defined over R'. By what we already proved, any point $P \in E(R') \cap Ker\ \psi$ has a multiple in $E(R') \cap p^{\infty}E(R)$. If E has ordinary reduction the latter group is contained in $E(R')_{tors}$, by Lemma 7.42. So $P \in E(R')_{tors}$ and we are done. If E has supersingular reduction, a multiple of P belongs to $E(pR) \cap Ker\ \psi$; but, by what we already proved, the latter group vanishes hence, again, $P \in E(R')_{tors}$. $\square$

The elliptic Theorem 7.40 has a multiplicative analogue whose proof is trivial and will be left to the reader:

THEOREM 7.44. *Let $\psi = \psi_{can} : \mathbf{G}_m(R) = R^{\times} \to R$ be the canonical $\delta-$character of $\mathbf{G}_m$. Then*

$$Ker\ \psi = \mathbf{G}_m(R)_{tors}.$$

In what follows we prove a finiteness result for the strict $\delta-$base locus of flat correspondences over number fields. We recall the context. Let $\mathcal{O} \subset \mathbf{C}$ be the ring of $S-$integers in a number field F, S a saturated finite set of finite places containing all the ramified places of F and all the places above 2 and 3. Let G be either the multiplicative group $\mathbf{G}_m$ or an elliptic curve E over $\mathcal{O}$, let $N \in \mathbf{Z}$ be invertible in $\mathcal{O}$ with $N \notin \{1, -1\}$, let $[N] : G \to G$ be the multiplication by N endomorphism, and let $\epsilon \in \{1, -1\}$. Then $[N]$ induces a morphism of schemes $\sigma : G/\langle[\epsilon]\rangle \to G/\langle[\epsilon]\rangle$. Let $X \subset G/\langle[\epsilon]\rangle$ be an affine Zariski open set such that the natural projection $\pi : G \to G/\langle[\epsilon]\rangle$ is étale above X, let $\tilde{X} = X \cap \sigma^{-1}(X)$ and consider the étale irreducible correspondence over $\mathcal{O}$,

$$\mathbf{X}_{\mathcal{O}} = \mathbf{X} = (X, \tilde{X}, \iota, \sigma),$$

smooth over $\mathcal{O}$, where ι is the inclusion. With the above notation we have:

THEOREM 7.45. *The $\delta-$base locus $\overline{BS}(\mathbf{X}, \mathbf{K}^{-1})$ is finite. If $\epsilon = 1$, $G = \mathbf{G}_m$, $F = \mathbf{Q}$, then the $\delta-$base locus $\underline{B}(\mathbf{X}, \mathbf{K}^{-1})$ is contained in the Wieferich set*

$$\mathcal{W} := \{a \in \mathbf{Q} \mid a^p \equiv a \text{ mod } p^2 \text{ in } \mathbf{Z}_{(p)} \text{ for almost all } p\}.$$

For the proof, in the elliptic case, we will need Merel's deep result on the torsion of elliptic curves:

THEOREM 7.46. [**106**] *For any integer d there exists an integer $N(d)$ such that*

$$\sharp E(L)_{tors} \leq N(d)$$

for all pairs (L, E) where L is a number field of degree d and E is an elliptic curve over L.

Of course the above statement is still true is one replaces E by $\mathbf{G}_m$.

We are ready to prove Theorem 7.45.

PROOF. Let $e = \sharp\Gamma$ and let d be the degree of F. Let $2, 3 \notin \wp \notin S$ and $P \in X(\mathcal{O}_\wp) \cap BS(\mathbf{X}_\wp, \mathbf{K}_\wp^{-1})$. Since π is étale above X there exists $Q \in G(R)$ such that $\pi(Q) = P$. Then $Q \in G(F')$ where F' is an extension of F of degree e. On the other hand if ψ is a canonical $\delta-$character of G then we must have $\psi(Q) = 0$ (because otherwise a suitable W_+-power of ψ would yield a section of an appropriate W_+-power of $\mathbf{K}_\wp^{-1}$ not vanishing at P). By Theorem 7.40 it follows that $Q \in G(F')_{tors}$. By Merel's theorem (resp. its $\mathbf{G}_m-$analogue) it follows that Q is annihilated by $(N(de))!$ hence there are only finitely many possibilities for P.

If $\epsilon = 1$, $G = \mathbf{G}_m$, $F = \mathbf{Q}$, the $\delta-$base locus $\underline{B}(\mathbf{X}, \mathbf{K}^{-1})$ is contained in the set of all $a \in \mathbf{Q}$ such that $\psi(a) \equiv 0$ mod p for almost all p, where $\psi = \psi_1$ is the canonical $\delta-$character in Equation 7.5. By Equation 7.5 we have

$$\psi(a) \equiv \frac{\delta a}{a^p} \quad mod \quad p \quad in \quad \mathbf{Z}_{(p)}$$

so if $\psi(a) \equiv 0$ mod p then $a \equiv a^p$ mod p^2 which ends the proof of the last assertion in the Theorem. $\square$

7.5. $\delta-$cohomology

In this section we present a number of cohomological computations for correspondences with flat uniformizations. We do not attempt to perform these computations in full generality: we will make a series of simplifying assumptions so that the results acquire a nicer form. It should be clear, however, that our computations can be generalized. Our point here is to show that our H^1 invariants are, in some sense, quite computable. Here are our main results. (In what follows, if G is a commutative group scheme over R then $G[N]$ will denote the kernel of $[N] : G \to G$.)

THEOREM 7.47. *Let* $\mathbf{X} = \mathbf{X}_{\mathbf{G}_m,N,1}$ *be a correspondence as in Equation 7.1. Assume* $X = \mathbf{G}_m \backslash \mathbf{G}_m[N']$ *where* $N' \geq 1$ *is an integer. Let* $\mathbf{L}_1 = \mathbf{F}(1, N^{-1})$, *let* $\psi = \psi_{can} = \psi_1$ *be the canonical* $\delta-$*character on* $\mathbf{G}_m$ *and fix an integer* $m \geq 1$. *For any non-negative integer* $j \neq m$ *with* $N^{j-m} \equiv 1$ *mod* p *in* $\mathbf{Z}_p$ *let* $N_{j,m} := (1 - N^{j-m})/p \in \mathbf{Z}_p$. *Then the following hold:*

1) The family

$$\{[N_{j,m}\psi^j] \in H^1(\mathbf{X}, \mathbf{L}_1^m) \mid j \geq 0, \quad j \neq m, \ N^{j-m} \equiv 1 \quad mod \quad p\}$$

is a basis of the $k-$*vector space* $H^1(\mathbf{X}, \mathbf{L}_1^m)[p]$. *Equivalently, the family*

$$\{\psi^j \mid j \geq 0, \ N^{j-m} = 1 \ mod \ p\}$$

is a basis of the $k-$*vector space* $H^0(\mathbf{X}, \mathbf{L}_1^m; k)$.

2) The element $[\psi^m]$ *has infinite order in the group* $H^1(\mathbf{X}, \mathbf{L}_1^m)$.

THEOREM 7.48. *Consider a correspondence* $\mathbf{X} = \mathbf{X}_{E,N,1}$ *as in Equation 7.1 where* E *is an elliptic curve which has a lifting of Frobenius,* $N \geq 3$ *is an integer such that* $N \not\equiv 0, 1$ *mod* p, *and* $X = E \backslash E[N']$ *where* $N' \geq 1$ *is an integer coprime with* $N - 1$. *Let* $\mathbf{L}_1 = \mathbf{F}(1, N^{-1})$, *let* $\psi = \psi_{can} = \psi_1$ *be a canonical* $\delta-$*character on* E *and fix an integer* $m \geq 1$. *For any non-negative integer* $j \neq m$ *with* $N^{j-m} \equiv 1$ *mod* p *in* $\mathbf{Z}_p$ *let* $N_{j,m} := (1 - N^{j-m})/p \in \mathbf{Z}_p$. *Then the following hold:*

1) The family

$$\{[N_{j,m}\psi^j] \in H^1(\mathbf{X}, \mathbf{L}_1^m) \mid j \geq 0, \quad j \neq m, \ N^{j-m} \equiv 1 \quad mod \quad p\}$$

is a basis of the $k-$*vector space* $H^1(\mathbf{X}, \mathbf{L}_1^m)[p]$. *Equivalently, the family*

$$\{\psi^j \mid j \geq 0, \ N^{j-m} \equiv 1 \ mod \ p\}$$

is a basis of the $k-$*vector space* $H^0(\mathbf{X}, \mathbf{L}_1^m; k)$.

2) The element $[\psi^m]$ *has infinite order in the group* $H^1(\mathbf{X}, \mathbf{L}_1^m)$.

Note that, by Theorem 7.34, the ring $A := \bigoplus_{n \geq 0} H^0(\mathbf{X}, \mathbf{L}_1^n)$ is isomorphic to a ring of polynomials $R[\psi]$. Consequently the assertion 2 in each of the Theorems 7.47 and 7.48 is equivalent to saying that the canonical map

$$\cup[1] : H^0(\mathbf{X}, \mathbf{L}_1^m) \otimes K \to H^1(\mathbf{X}, \mathbf{L}_1^m) \otimes K$$

is injective. Moreover if M is the $A-$module $\bigoplus_{m \geq 1} H^1(\mathbf{X}, \mathbf{L}_1^m)$ then its $p-$torsion $M[p]$ is a $A/pA = k[\psi]-$module. It follows from assertion 1) of the above Theorems that the cup product maps

$$\psi \cup : H^1(\mathbf{X}, \mathbf{L}_1^m)[p] \to H^1(\mathbf{X}, \mathbf{L}_1^{m+1})[p]$$

are injective, with cokernel of dimension 1 or 0 over k, according as $N^{m+1} = 1$ mod p or not. In particular, $M[p]$ is a torsion free $A/pA-$module.

THEOREM 7.49. *Consider the correspondence* $\mathbf{X} = \mathbf{X}_{E,N,1}$ *as in Equation 7.1 where* E *is an elliptic curve,* $N \geq 3$ *is an integer such that* $N \not\equiv 0, 1$ *mod* p, *and* $X = E \backslash E[N']$ *where* $N' \geq 1$ *is an integer coprime with* $N - 1$. *Assume* E *has Kodaira-Spencer class* $KS(E) \neq 0$. *Let* $\mathbf{L}_1 = \mathbf{F}(1, N^{-1})$, $\mathbf{L}_2 = \mathbf{F}(2, N^{-1})$, *let*

$\psi = \psi_{can} = \psi_{2,nor}$ be a canonical $\delta-$character on E and fix an integer $m \geq 1$. Then the following hold.

1) There exists $\psi_1 \in \mathcal{O}^1(X)$ such that $\psi \equiv \psi_1 \bmod p$ in $\mathcal{O}^2(X)$. Moreover there exist elements $\beta_{j,m} \in \mathcal{O}^1([N]^{-1}(X))$ such that

$$\Delta_m \psi_1^j = p \cdot \beta_{j,m}$$

and the family

$$\{[\beta_{j,m}] \in H^1(\mathbf{X}, \mathbf{L}_1^m) \mid j \geq 0,\ N^{j-m} \equiv 1 \quad mod \quad p\}$$

is a basis for $H^1(\mathbf{X}, \mathbf{L}_1^m)[p]$. Also the family

$$\{\psi_1^j \mid j \geq,\ N^{j-m} \equiv 1 \ mod \ p\}$$

is a basis for $H^0(\mathbf{X}, \mathbf{L}_1^m; k)$.

2) The element $[\psi^m]$ has infinite order in the group $H^1(\mathbf{X}, \mathbf{L}_2^m)$.

Recall that, under the assumptions of Theorem 7.49 above ψ^m is a basis for $H^0(\mathbf{X}, \mathbf{L}_2^m)$; cf. Theorem 7.34. So we get that the assertion 2 in Theorem 7.49 is equivalent to saying that the canonical map

$$\cup[1] : H^0(\mathbf{X}, \mathbf{L}_2^m) \otimes K \to H^1(\mathbf{X}, \mathbf{L}_2^m) \otimes K$$

is injective. Also, the ring $\bigoplus_{n \geq 0} H^0(\mathbf{X}, \mathbf{L}_1^n)$ is isomorphic to R.

Our Theorems above can be used to compute the cohomology of certain correspondences $\mathbf{X} = \mathbf{X}_{G,N,-1}$ as in Equation 7.1. Assume for simplicity that $X = Y/\Gamma$ where $Y = G \backslash G[2]$; this is, by the way, the classical context of the Chebyshev and Lattès dynamical systems [**109**]. Consider the correspondence

$$\mathbf{Y} = (Y, [N]^{-1}(Y), \iota, [N])$$

and refer to it as the *covering correspondence* of $\mathbf{X}$. Let us assume for simplicity that N has an even order modulo p; so for any integer $j \geq 0$ such that $N^{j-m} \equiv 1 \bmod p$ we must have $j \equiv m \bmod 2$. Also assume, in the case $G = E$, that $N \geq 3$ and either E has a lifting of Frobenius (in which case we set $r = 1$) or E has Kodaira-Spencer class $KS(E) \neq 0$ (in which case we set $r = 2$). We also set $r = 1$ in case $G = \mathbf{G}_m$. Set $\mathbf{L}_i = \mathbf{F}(i, N^{-1})$. Then we will prove

COROLLARY 7.50. *Assume $\mathbf{X}$ with covering correspondence $\mathbf{Y}$ satisfies the above conditions. Then, for any even $m \in \mathbf{Z}_+$, the map*

$$H^1(\mathbf{X}, \mathbf{L}_1^m)[p] \to H^1(\mathbf{Y}, \mathbf{L}_1^m)[p]$$

is an isomorphism. On the other hand, for any odd $m \in \mathbf{Z}_+$, we have

$$H^1(\mathbf{X}, \mathbf{L}_1^m)[p] = 0.$$

Finally, for any $m \in \mathbf{Z}_+$, the map

$$\cup[1] : H^0(\mathbf{X}, \mathbf{L}_r^m) \otimes K \to H^1(\mathbf{X}, \mathbf{L}_r^m) \otimes K$$

is injective.

On the other hand the spaces $H^1(\mathbf{Y}, \mathbf{L}_1^m)[p]$ were computed in Theorems 7.47, 7.48, 7.49 above. Moreover, recall from Theorem 7.34 that, if $r = 1$, then $H^0(\mathbf{X}, \mathbf{L}_1^2)$ is free of rank one with basis ψ^2. Then by the last Corollary, the map

$$\psi^2 \cup : H^1(\mathbf{X}, \mathbf{L}_1^m)[p] \to H^1(\mathbf{X}, \mathbf{L}_1^{m+2})[p]$$

is injective with cokernel of dimension 0 or 1 over k according as $N^{m+1} \not\equiv 1$ and $N^{m+2} \not\equiv 1 \bmod p$ or otherwise.

We embark now on our proof of Theorems 7.49 and 7.47 and Corollary 7.50. The proof of Theorem 7.48, which shall be omitted, is entirely similar to that of Theorem 7.47 once one uses the concepts that appear in the proof of Theorem 7.49.

Note that we have natural inclusions

$$\mathcal{O}^r(X) \otimes k \subset \mathcal{O}(J^r(X) \otimes k), \tag{7.57}$$
$$\mathcal{O}^r(\sigma^{-1}(X)) \otimes k \subset \mathcal{O}(J^r(\sigma^{-1}(X)) \otimes k);$$

if X is affine then the inclusions above are equalities.

PROOF. (of Theorem 7.47.) By Corollary 3.22

$$\mathcal{O}^1(X) = R[T, T', T^{-1}, g^{-1}]\hat{}, \quad g = 1 - T^{N'}$$

The canonical δ–character ψ on $\mathbf{G}_m$ corresponds to the series

$$\psi := \frac{1}{p} log \left(1 + p\frac{T'}{T^p} \right) \in \mathcal{O}^1(\mathbf{G}_m) = R[T, T', T^{-1}]\hat{}.$$

Now we claim that any element $f \in \mathcal{O}^1(X)$ can be written uniquely as a restricted power series

$$f = \sum_{i=0}^{\infty} a_i(T)\psi^i \tag{7.58}$$

where the coefficients $a_i(T)$ satisfy $a_i(T) \in R[T, T^{-1}, g^{-1}]\hat{}$ and

$$a_i(T) \to 0$$

p–adically as $i \to \infty$. To prove uniqueness in our claim assume $f = 0$ in Equation 7.58 and assume that not all the a_i's vanish. We may assume not all a_i's are divisible by p; then we obtain a contradiction by looking at the reduction mod p of Equation 7.58. The existence part in our claim follows by noting that

$$T' = \frac{T^p(exp(p\psi) - 1)}{p}.$$

So we may write

$$\mathcal{O}^1(X) = S[\psi]\hat{}, \quad S = R[T, T^{-1}, g^{-1}]\hat{};$$

we shall view S as embedded into the ring $R((T))\hat{}$ of all series $\sum_{i=-\infty}^{\infty} \lambda_i T^i$ with $\lambda_i \to 0$, p–adically, as $i \to -\infty$. Consequently we can embed

$$\mathcal{O}^1(X) \subset R((T))\hat{}[\psi]\hat{}.$$

There is a natural R–module endomorphism

$$[N] : R((T))\hat{}[\psi]\hat{} \to R((T))\hat{}[\psi]\hat{} \tag{7.59}$$
$$f = \sum_j a_j(T)\psi^j \mapsto f \circ [N] = \sum_j a_j(T^N)N^j\psi^j$$

inducing, for any $m \in \mathbf{Z}$, endomorphisms

$$\Delta_m : R((T))\hat{}[\psi]\hat{} \to R((T))\hat{}[\psi]\hat{}, \ F \mapsto F - N^{-m} \cdot F \circ [N],$$
$$\Delta_m \otimes k = \bar{\Delta}_m : k((T))[\psi] \to k((T))[\psi].$$

To check assertion 2 of the Theorem assume

$$\Delta_m(f) = \lambda\psi^m$$

for some $\lambda \in R$ and some $f \in \mathcal{O}^1(X)$, $f = \sum_j a_j(T)\psi^j$. Picking out the coefficient of ψ^m we get

$$a_m(T) - a_m(T^N) = \lambda$$

hence $\lambda = 0$.

To check assertion 1 of the Theorem note that, by Lemma 2.32, it is enough to show that the space

$$H^0(\mathbf{X}, \mathbf{L}_1^m; k) = Ker(\bar{\Delta}_m : k[T, T^{-1}, g^{-1}][\psi] \to k[T, T^{-1}, g^{-1}][\psi])$$

has a basis consisting of all ψ^j with $N^{j-m} \equiv 1 \bmod p$. Clearly the family of these ψ^j is linearly independent. To show it is spanning the above kernel assume $\bar{\Delta}_m(\bar{f}) = 0$ where $\bar{f} \in k[T, T^{-1}, g^{-1}][\psi] \subset k((T))[\psi]$, $\bar{f} = \sum_j \bar{a}_j(T)\psi^j$. We get

$$\bar{a}_j(T) = N^{j-m}\bar{a}_j(x^N) \in k((T)),$$

which easily implies that $\bar{a}_j(T) \in k$ for all j and $\bar{a}_j(T) = 0$ if $N^{j-m} \not\equiv 1 \bmod p$; this ends the proof. $\square$

Next we prove Theorem 7.49 so we will assume from now the hypothesis and notation of Theorem 7.49. Also we will freely use our discussion in the proof of Theorem 7.34. Due to the fact that N' and $N-1$ in Theorem 7.49 are relatively prime and $N \geq 3$, $N \not\equiv 1 \bmod p$, it follows easily that there exists a point $A \in X(R)$ such that $[N](A) = A$; in particular the integer n in our proof of Theorem 7.34 can be taken to be $n = 1$.

LEMMA 7.51. *If one has an equality of the form*

$$\Delta_m f = \lambda \cdot \psi^m \tag{7.60}$$

with $f \in \mathcal{O}^2(X)$, $m \geq 0$, $\lambda \in R$, then necessarily $\lambda = 0$. In particular $[\psi^m]$ is not a torsion element in $H^1(\mathbf{X}, \mathbf{L}_2^m)$.

PROOF. We proceed by induction on m. If $m = 0$ then we have

$$f - f \circ [N] = \lambda.$$

Taking the image of this equality via the expansion map E and setting $F = E(f)$ we get

$$F(T, T', T'') - F([N](T), \delta([N](T)), \delta^2([N](T))) = \lambda. \tag{7.61}$$

The series $\delta^i([N](T))$ are without constant term. Picking out the constant terms in Equation 7.61 we get $\lambda = 0$. To perform the induction step assume $m \geq 1$. Applying ∂_2 to Equation 7.60 and using Equations 7.55 and 7.54 we get

$$\Delta_{m-1}\partial_2 f = \partial_2\Delta_m f = \lambda \cdot m \cdot p \cdot \psi^{m-1}.$$

We conclude by the induction hypothesis. $\square$

In what follows we need a general:

LEMMA 7.52. *Let $\mathbf{T}$ be a tuple of variables, let $F \in R[[\mathbf{T}]]$ be such that $F(0) = 0$, and let L and Q be the homogeneous components of F of degree 1 and 2 respectively. Then we have the following congruences mod p in $R[[\mathbf{T}, \mathbf{T}']]$ and $R[[\mathbf{T}, \mathbf{T}', \mathbf{T}'']]$ respectively:*

$$(\delta F)(0, \mathbf{T}') \equiv L^{(\phi)}(\mathbf{T}')$$

$$(\delta^2 F)(0, \mathbf{T}', \mathbf{T}'') \equiv L^{(\phi^2)}(\mathbf{T}'') + Q^{(\phi^2)}((\mathbf{T}')^p) + \frac{L^{(\phi^2)}((\mathbf{T}')^p) - (L^{(\phi)}(\mathbf{T}'))^p}{p}.$$

PROOF. A direct computation. □

Recall from [**119**], p. 173, that if $\mathbf{G}_a$ denotes the additive group over k then $Ext(\mathbf{G}_a, \mathbf{G}_a)$ is a k–linear space with basis α_i, $(i = 0, 1, 2, ...)$, where $\alpha = \alpha_0$ is the class corresponding to the polynomial

$$C_p(X, Y) = \frac{X^p + Y^p - (X+Y)^p}{p}$$

and α_i, for $i \geq 1$, correspond to the polynomials $C_p(X^{p^i}, Y^{p^i})$. Let now E/R be an elliptic curve as in our Theorem 7.49 and consider the algebraic k–groups

$$E_i := J^i(E) \otimes k, \ \ N_i = N^i \otimes k := Ker(E_i \to E_0).$$

Recall from Proposition 4.45 that for each i, $Ker(E_{i+1} \to E_i) \simeq \mathbf{G}_a$; in particular $N_1 \simeq \mathbf{G}_a$ and N_2 is an extension of $\mathbf{G}_a$ by $\mathbf{G}_a$.

LEMMA 7.53. *The group N_2, viewed as an extension of $\mathbf{G}_a$ by $\mathbf{G}_a$ corresponds to the class α.*

PROOF. By Proposition 4.45, N_2 is isomorphic to the affine plane equipped with the group law defined by the pair of polynomials in

$$k[T_1', T_2', T_1'', T_2'']$$

obtained by reducing mod p the pair of formal series

$$((\delta F)(0, 0, T_1', T_2'), (\delta^2 F)(0, 0, T_1', T_2', T_1'', T_2'')), \tag{7.62}$$

where $F \in R[[T_1, T_2]]$ is the formal group law of E/R. (By loc. cit. the formal series 7.62 are actually restricted formal series so their reductions mod p are indeed polynomials.) It is a general fact about formal group laws (cf., say, [**124**], p.115) that one can arrange that

$$F = T_1 + T_2 + (\text{terms of degree} \geq 3).$$

By Lemma 7.52 we get congruences mod p:

$$(\delta F)(0, 0, T_1, T_2'') \equiv T_1' + T_2'$$

$$(\delta^2 F)(0, 0, T_1', T_2', T_1'', T_2'') = T_1'' + T_2'' + \frac{(T_1')^p + (T_2')^p - (T_1' + T_2')^p}{p},$$

and we are done. □

REMARK 7.54. As a corollary of the proof above note that, if T is a parameter at the origin of E and T', T'' have the meaning above then $\mathcal{O}(N_1)$ identifies with $k[T']$ and the group law on N_1 is given by the comultiplication $T' \mapsto T_1' + T_2'$. Also we have an identification $\mathcal{O}(N_2) = k[T', T'']$ with comultiplication given by the right hand side of the formulae at the end of the preceding proof.

REMARK 7.55. So far we have not used, in our proofs, the hypothesis that

$$KS(E) \neq 0 \tag{7.63}$$

so all our Lemmas above hold without hypothesis 7.63. Starting with Lemma 7.56 below we will need hypothesis 7.63.

Let us consider the extensions

$$0 \to N_1 \to E_1 \to E_0 \to 0 \tag{7.64}$$

$$0 \to N_2 \to E_2 \to E_0 \to 0 \tag{7.65}$$

and the induced commutative diagram

$$\begin{array}{ccccccc} 0 & \to & Hom(E_1, \mathbf{G}_a) & \to & Hom(N_1, \mathbf{G}_a) & \stackrel{s}{\to} & Ext(E_0, \mathbf{G}_a) \\ & & u \downarrow & & v \downarrow & & id \downarrow \\ 0 & \to & Hom(E_2, \mathbf{G}_a) & \to & Hom(N_2, \mathbf{G}_a) & \to & Ext(E_0, \mathbf{G}_a) \end{array} \tag{7.66}$$

LEMMA 7.56. *Assume $KS(E) \neq 0$. Then the morphism v is an isomorphism. Moreover the map s is surjective and u is an isomorphism.*

PROOF. Consider the Hom/Ext exact sequence of the extension

$$0 \to \mathbf{G}_a \to N_2 \to N_1 = \mathbf{G}_a \to 0.$$

To prove that v is an isomorphism it is enough to check that the connecting homomorphism

$$\partial = \partial_{N_2} : Hom(\mathbf{G}_a, \mathbf{G}_a) \to Ext(\mathbf{G}_a, \mathbf{G}_a)$$

is an isomorphism. Recall that $Hom(\mathbf{G}_a, \mathbf{G}_a)$ is isomorphic to the (non-commutative) ring $k[\pi]$ where π is the Frobenius endomorphism hence it has a $k-$basis consisting of $1, \pi, \pi^2, \ldots$. By definition of ∂ and by Lemma 7.53 we have $\partial(1) = \alpha$. It is then easy to check that $\partial(\pi^i) = \alpha_i$ for $i \geq 1$. This ends the proof that v is an isomorphism. Now the image of the isomorphism $N_1 \simeq \mathbf{G}_a$ via s is the class of the extension 7.64. By our hypothesis that $KS(E) \neq 0$ this class is non-trivial. By [**119**], p. 185, Theorem 7, $Ext(E_0, \mathbf{G}_a)$ is one dimensional. So s is surjective. We conclude that u is an isomorphism. □

REMARK 7.57. Let

$$X_0 = X \otimes k \subset E_0,$$
$$X_1 = J^1(X) \otimes k = X_0 \times_{E_0} E_1 \subset E_1.$$

Since the map u in Lemma 7.56 is an isomorphism the image,

$$\bar{\psi} \in Hom(E_2, \mathbf{G}_a),$$

of the homomorphism

$$\psi \in Hom(J^2(E), \hat{\mathbf{G}}_a)$$

comes from an element

$$\bar{\psi}_1 \in Hom(E_1, \mathbf{G}_a) \subset \mathcal{O}(E_1) \subset \mathcal{O}(X_1) = \mathcal{O}(J^1(X)) \otimes k.$$

Let $\psi_1 \in \mathcal{O}(J^1(X)) = \mathcal{O}^1(X)$ be *any* lifting of $\bar{\psi}_1$. Then, of course, $\psi_1 \equiv \psi$ mod p in $\mathcal{O}^2(X)$.

LEMMA 7.58. *The image of $\psi \in \mathcal{O}(J^2(E))$ via the map*

$$\mathcal{O}(J^2(E)) \mapsto \mathcal{O}(J^2(E) \otimes k) = \mathcal{O}(E_2) \to \mathcal{O}(N_2) = k[T', T'']$$

has the form $\Lambda := (T')^p - cT'$ with $c \in k^\times$. In particular the image of $\bar{\psi}_1$ via the map

$$\mathcal{O}(E_1) \to \mathcal{O}(N_1) = k[T']$$

equals Λ.

PROOF. This is a direct consequence of Equation 7.17 and of the equality $f^1_{jet}(E,\omega)\psi_{2,nor} = \psi_2$. (We use the fact that the isomorphism $\tau^*_{i_0}$ in Equation 7.10 is congruent to the identity mod T and that T vanishes on N_2.) □

LEMMA 7.59. *Let $L_1 = Spec\ \mathcal{O}(E_1)$. Then the composition $N_1 \subset E_1 \to L_1$ is an isogeny of degree p and identifies $\mathcal{O}(L_1) = \mathcal{O}(E_1)$ with $k[\Lambda]$ inside $\mathcal{O}(N_1) = k[T']$. In particular, by Lemma 7.58, we have $\mathcal{O}(L_1) = \mathcal{O}(E_1) = k[\bar{\psi}_1]$.*

PROOF. We use the notation of Remark 7.57. By Lemma 7.58 the image of $\bar{\psi}_1$ in $\mathcal{O}(N_1) = k[T']$ equals Λ; since $\bar{\psi}_1 \in \mathcal{O}(E_1) = \mathcal{O}(L_1)$ we have

$$k[\Lambda] \subset \mathcal{O}(L_1) \subset \mathcal{O}(N_1) = k[T'].$$

Since the composition of isogenies

$$Spec\ k[T'] \to Spec\ \mathcal{O}(L_1) \to Spec\ k[\Lambda]$$

has degree p two situations can occur. Either $k[\Lambda] = \mathcal{O}(L_1)$ in which case we are done or $\mathcal{O}(L_1) = \mathcal{O}(N_1)$ in which case the map $N_1 \to L_1$ is an isomorphism. The latter implies that the natural map from $Ker(E_1 \to L_1)$ to E_0 is an isomorphism; this implies that the extension 7.64 is split, contradicting the surjectivity of s in Lemma 7.56. □

PROOF. (of Theorem 7.49.) Assertion 2 of the Theorem was proved in Lemma 7.51. Let us prove assertion 1. By Remark 7.57 we have at our disposal an element $\psi_1 \in \mathcal{O}^1(X)$ such that $\psi_1 \equiv \psi$ mod p in $\mathcal{O}^2(X)$. So we may write $\psi_1^j = \psi^j + p\gamma_j$ with $\gamma_j \in \mathcal{O}^2(X)$. By Equation 7.52, for j such that $N^{j-m} \equiv 1$ mod p, we have $\Delta_m \psi_1^j = p \cdot \beta_{j,m}$ where

$$\beta_{j,m} = N_{j,m}\psi^j + \Delta_m \gamma_j.$$

Since

$$p\beta_{j,m} \in \mathcal{O}^1([N]^{-1}(X)) \cap p\mathcal{O}^2([N]^{-1}(X)) = p\mathcal{O}^1([N]^{-1}(X))$$

it follows that $\beta_{j,m} \in \mathcal{O}^1([N]^{-1}(X))$. By Theorem 7.34 we have $H^0(\mathbf{X}, \mathbf{L}_1^m) = 0$; hence, by Lemma 2.32, assertion 1 of Theorem 7.49 will be proved if we prove that the k-linear space

$$H^0(\mathbf{X}, \mathbf{L}_1^m; k) = Ker(\bar{\Delta}_m : \mathcal{O}(X_1) \to \mathcal{O}([N]^{-1}(X_1))),$$

where $\bar{\Delta}_m = \Delta_m \otimes k$, has basis

$$\{\bar{\psi}_1^j \mid j \in \mathbf{Z}_+,\ N^{j-m} \equiv 1 \ \ mod \ \ p\}.$$

So let $f \in \mathcal{O}(X_1)$ be such that $\Delta_m f = 0 \in \mathcal{O}([N]^{-1}(X_1))$; we will prove the following

Claim 1. $f \in k[\bar{\psi}_1]$.

Note that Claim 1 is enough to conclude the proof because if $f = \sum_{j \geq 0} c_j \bar{\psi}_1^j$, $c_j \in k$, then we will have

$$N^m \cdot f = f \circ [N] = \sum_{j \geq 0} c_j N^j \bar{\psi}_1^j$$

which implies $(1 - N^{j-m})c_j = 0$ for all $j \in \mathbf{Z}_+$, hence $c_j = 0$ whenever $N^{j-m} \not\equiv 1$ mod p. To prove Claim 1 we need a preparation. Consider the map $[p] : E_1 \to E_1$. This map factors through a homomorphism $E_0 \to E_1$; let $\tilde{E}_0 \subset E_1$ be the image of this homomorphism and set

$$\tilde{E}_1 = E_1 \times_{E_0} \tilde{E}_0,$$

$$\tilde{X}_0 = X_0 \times_{E_0} \tilde{E}_0,$$
$$\tilde{X}_1 = \tilde{X}_0 \times_{X_0} X_1.$$

Note that

$$\tilde{E}_1 \simeq \tilde{E}_0 \times N_1,$$
$$\tilde{X}_1 \simeq \tilde{X}_0 \times N_1.$$

Consider the diagram

$$\begin{array}{ccccccc} \tilde{X}_1 & \subset & \tilde{E}_1 & \rightarrow & Spec\ \mathcal{O}(\tilde{E}_1) & = & N_1 \\ \downarrow & & \downarrow & & & & \downarrow \\ X_1 & \subset & E_1 & \rightarrow & Spec\ \mathcal{O}(E_1) & = & L_1 \end{array}$$

and let us view all the rings of regular functions on the varieties appearing in this diagram as subrings of the field, $k(\tilde{E}_1)$, of rational functions on $\tilde{E}_1$.

Claim 2. $f \in \mathcal{O}(N_1)$.

Assuming for a moment that Claim 2 is true let us show how Claim 1 follows. By Claim 2 we get $f \in k(N_1)$. On the other hand $f \in \mathcal{O}(X_1) \subset k(X_1)$. So $f \in k(X_1) \cap k(N_1)$. By Lemma 7.59, the composition

$$k(L_1) \subset k(X_1) \cap k(N_1) \subset k(N_1)$$

has degree p. So one of the following must happen:

(i) $k(L_1) = k(X_1) \cap k(N_1)$ or

(ii) $k(N_1) \subset k(X_1)$.

If *(i)* is true then $f \in k(L_1)$. Since the extension $\mathcal{O}(L_1) \subset \mathcal{O}(N_1)$ is faithfully flat we have $\mathcal{O}(N_1) \cap k(L_1) = \mathcal{O}(L_1)$. By Claim 2 again we get $f \in \mathcal{O}(L_1)$. Then Claim 1 follows from Lemma 7.59. If *(ii)* is true then there is a rational map $E_1 \rightarrow N_1$ which is generically a homomorphism, hence a homorphisim, such that the composition $N_1 \rightarrow E_1 \rightarrow N_1$ is the identity. As in the proof of Lemma 7.59 this forces the extension 7.64 to split which is a contradiction.

We are left to check Claim 2. Note that

$$f \in \mathcal{O}(\tilde{X}_1) = \mathcal{O}(\tilde{X}_0) \otimes_k \mathcal{O}(N_1) = \mathcal{O}(\tilde{X}_0) \otimes_k k[T'] = \mathcal{O}(\tilde{X}_0)[T'].$$

Hence we have a representation

$$f = \sum_{j \geq 0} f_j \cdot (T')^j, \quad f_j \in \mathcal{O}(\tilde{X}_0).$$

We know that

$$N^m \cdot f = f \circ [N] = \sum_{j \geq 0} (f_j \circ [N]) \cdot (NT')^j.$$

We conclude that

$$f_j \circ [N] = N^{m-j} \cdot f_j \tag{7.67}$$

in $k(\tilde{E}_0)$ for all $j \in \mathbf{Z}_+$. In particular, if $D_j = div(f_j) \in Div(\tilde{E}_0)$ is the principal divisor of f_j then

$$[N]^* D_j = D_j. \tag{7.68}$$

Assume there is a j such that $f_j \notin k$. Then, for that j, we have $Supp(D_j) \neq \emptyset$ and for any point $P \in Supp(D_j)$, we have $[N]^{-1}(P) \subset Supp(D_j)$. So $[N^n]^{-1}(P) \subset Supp(D_j)$ for any $n \in \mathbf{Z}_+$. But the cardinality of $[N^n]^{-1}(P)$ goes to infinity as $n \rightarrow \infty$, a contradiction. We conclude that $f_j \in k$ for all j which ends the proof of Claim 2 and hence of the Theorem. □

Finally we note that Corollary 7.50 follows from Theorems 7.47, 7.48, 7.49 plus Corollary 3.30.

7.6. The relative generic $\delta-$fiber

THEOREM 7.60. *Assume we are given a flat correspondence* $\mathbf{X} = \mathbf{X}_{G,N,\epsilon}$ *as in Equation 7.1. Let* r *be* 1 *or* 2 *according as* G *has a lift of Frobenius or not and set* $\mathbf{L} = \mathbf{L}_r = \mathbf{F}(r, N^{-1})$, $L = \mathcal{O}^r$. *Then the field extension*

$$k(X) \cdot k\langle \mathbf{X}, \mathbf{L}\rangle \subset k\langle X, L\rangle \tag{7.69}$$

is algebraic. If $\epsilon = 1$ *and* $G = \mathbf{G}_m$ *then the extension 7.69 is Galois with Galois group* $\mathbf{Z}_p^\times$. *If* $\epsilon = 1$ *and* $G = E$ *is an elliptic curve with ordinary reduction and no lift of Frobenius then there exists an intermediate field* M *in the extension 7.69 such that:*

a) The extension $k(X) \cdot k\langle \mathbf{X}, \mathbf{L}\rangle \subset M$ *is Galois with Galois group* $\mathbf{Z}_p^\times$ *and*
b) The automorphism group of the extension $M \subset k\langle X, L\rangle$ *contains* $\mathbf{Z}_p$.

PROOF. The case $\epsilon = -1$ reduces to the case $\epsilon = 1$ so we assume $\epsilon = 1$. We assume now $G = E$.

We treat first the case when E doesn't have a lift of Frobenius. Let $U \subset G$ be an affine open set containing the zero section and possessing an étale coordinate $T \in \mathcal{O}(U)$. Then, by Proposition 3.13 we have identifications

$$\mathcal{O}(J^r(U)) = \mathcal{O}(U)[T', T'', ..., T^{(r)}]^{\hat{}}.$$

inducing an identification

$$k\langle X, L\rangle = k(X)(T', T'', ...). \tag{7.70}$$

Consider the element $x \in H^0(\mathbf{X}, \mathbf{L}_2)$ in Theorem 7.34; note that x comes from the $\delta-$character $\psi_{2,nor}$ in Theorem 7.22. By Proposition 5.14 we have an identification

$$k\langle \mathbf{X}, \mathbf{L}\rangle = k(\eta_0, \eta_1, \eta_2, ...), \tag{7.71}$$

where η_r is the image of $\delta^r(x^\phi/x)$. So the field extension 7.69 can be identified with the extension

$$k(X)(\eta_0, \eta_1, \eta_2, ...) \subset k(X)(T', T'', ...). \tag{7.72}$$

We claim that, for all $r \geq 0$, we have

$$\begin{aligned} x^{(r)} = & (T^{(r+1)})^p + \lambda^{\phi^r} T^{(r+1)} + pT^{(r+2)} \\ & + p^2 O(r+2) + pO(r+1) + O(r) \\ & + \sum_{i=0}^{r} T^{(i)} \cdot O(r+1) + p\sum_{i=0}^{r} T^{(i)} \cdot O(r+2), \end{aligned} \tag{7.73}$$

where $O(i)$ denote arbitrary elements in $\mathcal{O}(U)[T', ..., T^{(i)}]^{\hat{}}$ and

$$\lambda = -\frac{f_{jet}^2(E, \omega)}{f_{jet}^1(E, \omega)} \in R.$$

For $r = 0$ this follows directly from the last assertion of Theorem 7.22 plus Equation 7.12. The general case follows easily by induction. We claim that the extension

$$k(X)(T', T'', ..., T^{(r)}, x^{(r)}) \subset k(X)(T', T'', ..., T^{(r)}, T^{(r+1)}) \tag{7.74}$$

is algebraic. Indeed this extension is generated by $T^{(r+1)}$ and, by Equation 7.73, $T^{(r+1)}$ is a root of a polynomial

$$F(t) = \sum u_n t^n \tag{7.75}$$

with coefficients

$$u_n \in k(X)(T', T'', ..., T^{(r)}, x^{(r)})$$

such that

$$u_p = 1 + \sum_{i=0}^{r} T^{(i)} \cdot O(r) \neq 0.$$

In particular $F \neq 0$. We deduce that the extensions

$$k(X)(x, x', x'', ..., x^{(r)}) \subset k(X)(T', T'', ..., T^{(r+1)}) \tag{7.76}$$

are algebraic. On the other hand, by Proposition 5.14, the extension

$$k(\eta_0, \eta_1, ..., \eta_r) \subset k(x, x', ..., x^{(r)})$$

is algebraic. Hence the extension 7.69 is algebraic. Now set

$$M := k(X)(x, x', x'', ...).$$

Since the fields $k(X)$ and $k(x, x', x'', ...)$ are linearly disjoint over k and the extension

$$k(\eta_0, \eta_1, \eta_2, ...) \subset k(x, x', x'', ...)$$

is Galois with Galois group $\mathbf{Z}_p^\times$ (cf. Proposition 5.14) it follows that the extension

$$k(X)(\eta_0, \eta_1, \eta_2, ...) \subset k(X)(x, x', x'', ...)$$

is Galois with Galois group $\mathbf{Z}_p^\times$. Assume now E has ordinary reduction and no lift of Frobenius. By Lemma 7.41 we have

$$E(pR) \cap p^\infty E(R) \simeq \mathbf{Z}_p.$$

Then $\mathbf{Z}_p$ acts (via translations) on $E(R)$ and hence on $k\langle X, L\rangle$. The induced homomorphism,

$$\rho : \mathbf{Z}_p \to Aut\ k\langle X, L\rangle,$$

is injective by Proposition 3.18. On the other hand $\rho(\mathbf{Z}_p)$ consists of automorphisms that fix the elements of $k(X)$ (because translation by points in $E(pR)$ induces the identity on $k(X)$) and also fix the elements of $k(\eta_0, \eta_1, \eta_2, ...)$ (because $\psi_{2,nor}$ vanishes on $\cap_{n=1}^\infty p^n E(R)$). Hence $\rho(\mathbf{Z}_p)$ consists of $M-$automorphisms.

The case when E has a lift of Frobenius can be treated similarly. In that case the element x comes from ψ_1 hence identifies with an element of the form

$$x = L^1 + T \cdot \Omega,$$

$\Omega \in \mathcal{O}(\hat{U})[T']\hat{}$. Then one can check by induction that for all $r \geq 0$,

$$x^{(r)} = T^{(r+1)} + p \cdot O(r+1) + O(r) + \sum_{i=0}^{r} T^{(i)} \cdot O(r+1)$$

and the argument proceeds as in the case that E doesn't have a lift of Frobenius.

If $G = \mathbf{G}_m = Spec\ R[T, T^{-1}]$ and $\epsilon = 1$ then $k(X) = k(T)$ and

$$k\langle X, L\rangle = k(T, T', T'', ...)$$

and Equation 7.71 still holds, with $x = \psi = \psi_1 \in R[T, T', T^{-1}]\hat{}$. One checks by induction that

$$x^{(r)} = \frac{T^{(r+1)}}{T^{p^{r+1}}} + pO(r+1) + O(r).$$

Therefore

$$k(T, x, x', x'', ...) = k(T, T', T'', T''', ...)$$

and we may conclude by Proposition 5.14. □

REMARK 7.61. Clearly, Theorems 7.34, 7.45 and 7.60 imply Theorem 2.60.

REMARK 7.62. Assume the hypotheses of Theorem 7.34 and let (P, P) be a fixed point as in Lemma 7.6. We claim that $H^0(\mathbf{X}, \mathbf{L}_r^e)$ contains a section which is $e-$ regular at $(\tilde{P}, P)$. Let us give the argument for the case $r = 2$, $e = 1$; all other cases are similar. By the diagram in the proof of Lemma 7.6 it is enough to find a section of $H^0(\mathbf{X}, \mathbf{L}_2)$ whose germ at the origin is $1-$ regular. But the section x in the proof of Theorem 7.60 has germ $E(\psi)$ at the origin and, clearly,

$$E(\psi)(0, 0, T) = \sum_{i \geq 1} c_i^{\phi^2} p^{2i-1} T^i \in p \cdot T \cdot R[[T]]^{\times}.$$

If, in case G is an elliptic curve, one assumes in addition that G is defined over $\mathbf{Z}_p$ then we claim that $H^0(\mathbf{X}, \mathbf{L}_r^e)$ contains a section which is $e-$coregular and $e-$perfect at (P, P). In the elliptic case this follows from [**19**], Theorem 1.10; in the multiplicative case the claim above is an easy exercise.

REMARK 7.63. As in Remark 7.39 note that Theorem 7.60 remains true for correspondences of the form

$$\mathbf{X}_{G, \mathcal{N}, \epsilon} := \coprod_{N \in \mathcal{N}} \mathbf{X}_{G, N, \epsilon}$$

equipped with the $\delta-$line bundle

$$\mathbf{L}_r := \coprod_{N \in \mathcal{N}} \mathbf{F}(r, N^{-1}).$$

7.7. Converse theorems: quadratic dynamical systems

The aim of this section is to look at correspondences attached to quadratic dynamical systems $\sigma(t) = t^2 + c$, where c is in a number field. We would like to understand for what values of c the correspondence has a non-trivial categorical quotient in $\delta-$geometry.

Start with a ring $\mathcal{O}$ of $S-$integers in a number field F where S is a saturated finite set of finite places containing all the ramified places and all places above 2 and 3; so $\mathcal{O} := \mathcal{O}_F[1/m]$ where m is an integer divisible by 6. Consider the correspondence

$$\mathbf{X}_{\mathcal{O}}^* := (\mathbf{A}^1, \mathbf{A}^1, id, \sigma)$$

where $\mathbf{A}^1 = Spec\ \mathcal{O}[t]$ is the affine line over $\mathcal{O}$, id is the identity, and $\sigma : \mathbf{A}^1 \to \mathbf{A}^1$ is the map defined by the polynomial $\sigma(t) = t^2 + c$, $c \in \mathcal{O}$. This correspondence is not étale; to get an étale correspondence consider a polynomial $g(t) \in t\mathcal{O}[t]$ and set

$$X_g := Spec\ \mathcal{O}[t, g(t-c)^{-1}] \subset \mathbf{A}^1,$$

$$\tilde{X}_g := Spec\ \mathcal{O}[t, g(t-c)^{-1}, g(t^2)^{-1}] = X_g \cap \sigma^{-1}(X_g);$$

then the induced correspondence

$$\mathbf{X}_{t^2+c,g} = \mathbf{X}_{t^2+c,g,\mathcal{O}} = (X_g, \tilde{X}_g, \iota, \sigma) \tag{7.77}$$

is étale because $g(t^2)$ is divisible by $\sigma'(t) = 2t$. As usual, for any place $\wp \notin S$ we denote by $\mathbf{X}_{t^2+c,g,\wp}$ the induced correspondence over $R_\wp := \hat{\mathbf{Z}}_p^{ur}$. We shall refer to $\mathbf{X}_{t^2+c,g}$, $\mathbf{X}_{t^2+c,g,\wp}$ as *quadratic correspondences.* Let us fix a quadratic correspondence

$$\mathbf{X}_{t^2+c,g,\wp}. \tag{7.78}$$

It is trivial to check that its canonical bundle $\mathbf{K}$ is isomorphic to $(\mathcal{O}, \beta)$ where β is the composition

$$\beta : \sigma^*\mathcal{O}_{X_g} \overset{can}{\simeq} \mathcal{O}_{\tilde{X}_g} \overset{2t}{\to} \mathcal{O}_{\tilde{X}_g} \overset{can}{\simeq} \iota^*\mathcal{O}_{X_g}, \tag{7.79}$$

where the middle arrow is the multiplication by $2t$ map.

The cases $c = 0$ and $c = -2$ (corresponding to the multiplicative polynomial $\sigma(t) = t^2$ and the Chebyshev polynomial $\sigma(t) = t^2 - 2$ respectively) fit into the frame of both Theorems 7.34 and 7.45 (where $G = \mathbf{G}_m$, $N = 2$, and $\epsilon = 1$ or $\epsilon = -1$ respectively). Hence we have, for instance:

COROLLARY 7.64. *Assume one of the following holds:*
i) $c = 0$ *and* $g(t) = t$,
ii) $c = -2$ *and* $g(t+2) = t^2 - 4$.
Then the following hold:

1) The $\delta-$*base locus* $\underline{BS}(\mathbf{X}_{t^2+c,g,F'}, \mathbf{K}_{F'}^{-1})$ *is finite for all finite extensions* F' *of* F;

2) $H^0(\mathbf{X}_{t^2+c,g,\wp}, \mathbf{K}^{-1-\phi}) \neq 0$ *for almost all* $\wp$.

On the other hand we have the following result in the converse direction:

THEOREM 7.65. *Assume the* $\delta-$*base locus* $\underline{BS}(\mathbf{X}_{t^2+c,g,F'}, \mathbf{K}_{F'}^{-1})$ *is finite for all finite extensions* F' *of* F. *Then the postcritical locus* P_{t^2+c} *is finite hence, in case* $F = \mathbf{Q}$, $c \in \{-2, -1, 0\}$.

PROOF. By Theorem 2.64 the postcritical set P_{t^2+c} is finite. We conclude by Lemma 1.39. □

The cases $c = -2$ and $c = 0$ in Theorem 7.65 really occur, as shown by Corollary 7.64. We conjecture that the case $c = -1$ in Theorem 7.65 does not occur. The next result addresses, in particular, the case $c = -1$.

THEOREM 7.66. *Let* $c \not\equiv 0, -2$ *mod* $\wp$ *and let* $v \in W$ *satisfy either* $v(0) \not\equiv 0$ *mod* p *or* $v(p) > 0$. *Let* $g \in t\mathcal{O}[t]$, $g \not\equiv 0$ *mod* $\wp$. *Then*

$$H^0(\mathbf{X}_{t^2+c,g,\wp}, \mathbf{K}^v; k_\wp) = 0;$$

in particular $H^0(\mathbf{X}_{t^2+c,g,\wp}, \mathbf{K}^v) = 0$.

PROOF. Recall that $k_\wp := R_\wp/pR_\wp$. Assume $H^0(\mathbf{X}_{t^2+c,g,\wp}, \mathbf{K}^v; k_\wp) \neq 0$. By Theorem 4.57 there exists an integer $\nu \in \mathbf{Z}$, $\nu \equiv v(0)$ mod p, $\nu \geq v(p)$, and an open non-empty subcorrespondence $\mathbf{U} \subset \mathbf{X}$ such that $\mathbf{K}^\nu_{\mathbf{U}\otimes k_\wp}$ is trivial. Note in particular that $\nu \neq 0$. It is an easy exercise then to show that $c = 0$ or -2 in $k_\wp$, a contradiction. □

The next Theorem addresses the case $c \equiv 0$ or -2 mod $\wp$.

THEOREM 7.67. *Let* $\mathbf{X}_{t^2+c,g,p}$ *be a quadratic correspondence with* $c \in \mathbf{Z}$ *and* $g \in t\mathbf{Z}[t]$ *not divisible by* p. *Assume one of the following holds:*

i) $c \equiv 0 \bmod p$ *and* $c \neq 0$;

ii) $c \equiv -2 \bmod p$ *and* $c \neq -2$.

Then the following hold:

1) $H^0(\mathbf{X}_{t^2+c,g,p}, \mathbf{K}^w) = 0$ *for all* $w \in W$ *with* $deg(w) > 0$;

2) $H^0(\mathbf{X}_{t^2+c,g,p}, \mathbf{K}^w \otimes \mathbf{F}(\lambda)) = 0$ *for all* $w \in W$ *and all* $\lambda \in \mathbf{Z}_p^\times$ *not a root of unity.*

PROOF. By Proposition 4.65 and Example 4.64 the set $\mathcal{P}$ of all cycles on $\mathbf{X}_{t^2+c,g,p}$ whose multipliers lie in $\mathbf{Z}_p^{ur}$ is not symmetric. We conclude by Corollary 4.63. $\square$

CHAPTER 8

Hyperbolic correspondences

We adopt again the vertical viewpoint by developing the theory for a fixed prime p. Recall that dense correspondences over $\mathbf{C}$ that admit a hyperbolic analytic uniformization can be described in purely algebraic / arithmetic terms via quaternion algebras over totally real fields E. For $E = \mathbf{Q}$ we get (up to commensurability) a description of these correspondences in terms of moduli spaces. Cf. Proposition 1.32, Proposition 1.34 and Remark 1.35. This will allow us to define a class of correspondences over $R := \hat{\mathbf{Z}}_p^{ur}$ (which we call *Hecke correspondences*) that (up to commensurability) are an algebraic / arithmetic analogue of correspondences over $\mathbf{C}$ that possess hyperbolic, $\mathbf{Q}-$arithmetic uniformizations. We will then treat this class from the viewpoint of $\delta-$geometry. We start by following [**21**], [**2**], [**3**], [**23**], [**24**], where this theory was initiated. Then we shall push the theory forward and prove a series of new results. In particular we will prove our "horizontal" Main Theorem 2.62.

There will be 2 cases in our theory according as the quaternion $\mathbf{Q}-$algebra in question is split or not. The split case leads to (usual, non-compact) modular curves (parameterizing elliptic curves) whereas the non-split case leads to (compact) Shimura curves (parameterizing false elliptic curves). This Chapter assumes some familiarity with the modular curves as found, for instance, in Katz' papers [**80**] and [**81**] and with Shimura curves [**48**], [**11**].

Here is the structure of this Chapter. In Section 1 we review some basic concepts related to Abelian schemes and their crystals. In Section 2 we define *Hecke correspondences* and analyze certain $\delta-$line bundles and cycles on them. Section 3 is mainly concerned with understanding what our general concepts of *expansion maps* and *conjugate operators* become in the case of Hecke correspondences; we are lead to what we shall call *$\delta-$Serre-Tate expansions* and *$\delta-$Serre operators*. We are going to spell out some of the basic commutation relations, between these operators, that play a key role later. Section 4 is devoted to the construction of $\delta-$invariants for Hecke correspondences. We do this by constructing a "stacky" version of these $\delta-$invariants, called here *smooth isogeny covariant differential modular forms*; they are a "smooth" version of the objects studied in [**21**], [**2**], [**3**], [**23**], [**24**]. By the way we will not *use* these papers but redo and expand (part of) the theory therein. There are essentially two types of constructions of $\delta-$invariants. One is based on a fine analysis of $p-$jet spaces of Abelian schemes. The other is based on crystalline cohomology. The interplay between the two types of $\delta-$invariants is a central part of the theory; for instance, one of the main results will imply that the two sets of $\delta-$invariants are proportional. Sections 5 and 6 are devoted to the determination of $\delta-$invariants for Hecke correspondences in the ordinary and non-ordinary case respectively. The main effort is to show that "all $\delta-$invariants come from the crystalline ones". A key technical point is the use of the $\delta-$Serre operators, a technique initiated by M. Barcau in [**2**]. In Sections 7 and 8 we deal with δ-cohomology and δ-fibers respectively. All the above theory is developed for Hecke correspondences with a "Hecke" fixed point. In Section 9 we show how part of the theory generalizes to the case of Hecke correspondences with a "Hecke" cycle; here we shall conclude the proof of Theorem 2.62. Finally, in Section 10, we prove a "converse theorem" to the effect that if, instead of taking Hecke correspondences (which possess $\mathbf{Q}$-arithmetic uniformizations) we consider correspondences $\mathbf{X}$ possessing E-arithmetic uniformizations (E a real quadratic field) then certain spaces $H^0(\mathbf{X}, \mathbf{L})$ (which did not vanish in the $\mathbf{Q}$-arithmetic case) will vanish.

8.1. Review of Abelian schemes and their crystals

Before embarking onto the study of correspondences on modular and Shimura curves it will be useful to review the deformation theory and crystalline theory of Abelian schemes. The present Section is devoted to this review; it tacitly assumes familiarity with the concepts and its role is mainly to introduce the notation, conventions, and terminology that we shall be using.

8.1.1. Elliptic curves. Recall that an *elliptic curve* E over a Noetherian ring S is a smooth proper group scheme of relative dimension one with irreducible geometric fibers; for any such E the S-module $H^0(E, \Omega_{E/S})$ is locally free of rank one. If this module is free, a basis ω is called an *invertible* 1-*form*. If 6 is invertible in S then there is a bijection between the set

$$\{(E, \omega) | E \text{ an elliptic curve over } S,\ \omega \text{ an invertible } 1\text{-form on } E\} \tag{8.1}$$

and the set

$$\{(a, b) \in S \times S | 4a^3 + 27b^2 \in S^\times\} \tag{8.2}$$

given by $(a, b) \mapsto (E, \omega)$ where

$$E = Proj\ S[x_0, x_1, x_2]/(x_0 x_2^2 - (x_1^3 + a x_0^2 x_1 + b x_0^3)),$$

$$\omega = dx/y,\ x := x_1/x_0, y := x_2/x_0.$$

An *isogeny* of elliptic curves is a flat finite group scheme homomorphism. Sometimes elliptic curves will be referred to as *genuine elliptic curves* in order to distinguish them from *false elliptic curves* to be introduced presently.

8.1.2. Abelian schemes. Let S be a Noetherian ring. By an *Abelian scheme* we mean a smooth proper group scheme over S with irreducible geometric fibers. An *isogeny* between two Abelian schemes is a finite flat group scheme homomorphism. If A is an Abelian scheme over S we denote by $\check{A} = Pic^0(A/S)$ its dual Abelian scheme; that $\check{A}$ is an Abelian scheme is a delicate fact, cf. [**58**], p.3. (In our applications the Abelian schemes under consideration will be projective in which case the fact that $\check{A}$ is an Abelian scheme also follows from [**113**], p. 23.) The Lie algebras of A and $\check{A}$ identify with $H^0(A, T_{A/S})$ and $H^1(A, \mathcal{O})$ respectively. A *principal polarization* θ on A is an isomorphism $\theta : A \to \check{A}$ which, locally in the étale topology of S, is induced (in the well known sense) by an ample line bundle. Any elliptic curve has a canonical principal polarization. For an isogeny $u : A \to B$ of Abelian schemes we denote by $\check{u} : \check{B} \to \check{A}$ the dual isogeny. If k is an algebraically closed field of characteristic p and A_0 is an Abelian scheme (variety) over k we denote by

$$T_p(A_0) = \lim_{\leftarrow} A_0[p^n] \tag{8.3}$$

the *physical Tate module.*

8.1.3. Quaternion algebras. For the material below we refer to [**134**], [**30**], [**11**], [**48**], [**23**]. Let D be a quaternion algebra over $\mathbf{Q}$; recall that this means, by definition, that D is a ring whose center is $\mathbf{Q}$, such that D has dimension 4 as a $\mathbf{Q}-$vector space, and D has no non-trivial two-sided ideals. Assume D is non-split (i.e. $D \not\simeq Mat_2(\mathbf{Q})$; equivalently D is a division ring) and D is indefinite (i.e. $D \otimes_{\mathbf{Q}} \mathbf{R} \simeq Mat_2(\mathbf{R})$). Let $d = disc(D)$ be the discriminant of D (i.e. d is the product of all the primes l for which $D \otimes_{\mathbf{Q}} \mathbf{Q}_l \not\simeq Mat_2(\mathbf{Q}_l)$.) Fix a prime p not dividing d and fix an isomorphism

$$j : D \otimes_{\mathbf{Q}} \mathbf{Q}_p \to Mat_2(\mathbf{Q}_p). \tag{8.4}$$

It is known that $tr(j(\alpha)), det(j(\alpha)) \in \mathbf{Q}$ for $\alpha \in D$. Consider now the canonical involution

$$D \to D, \quad \alpha \mapsto \alpha' := det(j(\alpha)) \cdot \alpha^{-1} \quad (\alpha \neq 0) \tag{8.5}$$

So if $j(\alpha) = \left[\begin{array}{cc} a & b \\ c & d \end{array}\right]$ then $j(\alpha') = \left[\begin{array}{cc} d & -b \\ -c & a \end{array}\right]$. There exists a maximal order $\mathcal{O}_D$ of D (i.e. a maximal element in the set of all subrings $\mathcal{O} \subset D$ with $\mathcal{O} \simeq \mathbf{Z}^4$ as $\mathbf{Z}-$modules and $\mathcal{O} \otimes_{\mathbf{Z}} \mathbf{Q} = D$) which is stable under the canonical involution $\alpha \mapsto \alpha'$. It is known that there exists an element $\tau \in \mathcal{O}_D$ such that $\tau^2 = -d$. Note that $(\tau')^2 = -d$ hence, since τ, τ' commute and D is a division ring, $\tau' = -\tau$. Also $\tau^{-1} = -\frac{1}{d}\tau$. A key role in the theory is played by the involution

$$D \to D, \quad \alpha \mapsto \alpha^+ := \tau^{-1}\alpha'\tau = \tau\alpha'\tau^{-1}$$

which is known to induce a non-trivial involution of $\mathcal{O}_D$. Since $\mathcal{O}_D \otimes_{\mathbf{Z}} \mathbf{Z}_p$ is a maximal order in $D \otimes_{\mathbf{Q}} \mathbf{Q}_p$ it follows that $j(\mathcal{O}_D \otimes_{\mathbf{Z}} \mathbf{Z}_p)$ is a maximal order in $Mat_2(\mathbf{Q}_p)$. Since any maximal order of $Mat_2(\mathbf{Q}_p)$ is conjugated to $Mat_2(\mathbf{Z}_p)$ we may (and will) assume, by composing j with an inner automorphism of $Mat_2(\mathbf{Q}_p)$, that $j(\mathcal{O}_D \otimes_{\mathbf{Z}} \mathbf{Z}_p) = Mat_2(\mathbf{Z}_p)$. So we have an isomorphism

$$j : \mathcal{O}_D \otimes_{\mathbf{Z}} \mathbf{Z}_p \to Mat_2(\mathbf{Z}_p). \tag{8.6}$$

We can also assume (by further changing j) that

$$j(\tau) = \left[\begin{array}{cc} 0 & 1 \\ -d & 0 \end{array}\right]. \tag{8.7}$$

In all our discussion below D, τ, j are fixed and j satisfies Equations 8.6 and 8.7.

8.1.4. False elliptic curves. Assume S is a Noetherian ring. By a *false elliptic curve* over S we mean a pair (A, i) where A/S is an Abelian scheme of relative dimension two, and $i : \mathcal{O}_D \to End(A/S)$ is an injective ring homomorphism (preserving the unit element). Since D is a division algebra $i(\alpha)$ is automatically finite for all $\alpha \neq 0$. So any such $i(\alpha)$ is finite and flat fiber by fiber. So $i(\alpha)$ itself is flat, hence an isogeny, for all $\alpha \neq 0$. Cf. the argument in [**83**] p.77.

For any principal polarization $\theta : A \simeq \check{A}$ we have defined a *Rosati involution*

$$End(A/S) \to End(A/S), u \mapsto u^t := \theta^{-1} \circ \check{u} \circ \theta.$$

By a *polarized false elliptic curve* over S we will understand a triple (A, i, θ) where (A, i) is a false elliptic curve and $\theta : A \to \check{A}$ is a principal polarization such that the Rosati involution on $End(A/S)$ induces the involution $\alpha \mapsto \alpha^+$ on $\mathcal{O}_D$. By [**30**] if (A, i) is a false elliptic curve then there is a unique principal polarization θ on A such that for any geometric point $\bar{s}$ of S the Rosati involution on $End(A_{\bar{s}})$ is compatible with the involution $\alpha \mapsto \alpha^+$ on $\mathcal{O}_D$. So if (A, i, θ_1) and (A, i, θ_2) are polarized false elliptic curves then $\theta_1 = \theta_2$. Also, if S is reduced then for any false elliptic curve (A, i) there is a (unique) θ such that (A, i, θ) is a polarized false elliptic curve. Let (A_1, i_1, θ_1), (A_2, i_2, θ_2) be polarized false elliptic curves; by an *isogeny of false polarized elliptic curves* we mean an isogeny $u : A_1 \to A_2$ over S such that $i_2(\alpha) \circ u = u \circ i_1(\alpha)$ for all $\alpha \in \mathcal{O}_D$. (Note that $i_1(\alpha)$ for $\alpha \in \mathcal{O}_D$, $\alpha \neq 0$, is, by definition, an isogeny of Abelian schemes but it is *not* an isogeny of polarized false elliptic curves, unless α is in the center of D !) One knows that

$$u^t := \theta_1^{-1} \circ \check{u} \circ \theta_2 : A_2 \to A_1$$

is also an isogeny of false polarized elliptic curves and the composition $u^t \circ u$ is (the multiplication by) an integer $\partial(u)$ called the *false degree* of u. The actual degree $deg(u)$ of u is trivially seen to be $\partial(u)^2$.

8.1.5. False 1–forms. Assume in what follows that S is a $\mathbf{Z}_p$–algebra.

DEFINITION 8.1. A *false 1–form* ω on a false polarized elliptic curve (A, i, θ) over S is a column vector $\omega = (\omega^1, \omega^2)^t$ where $\omega^1, \omega^2 \in H^{10} := H^0(A, \Omega_{A/S})$ such that

$$i(\alpha)^* \omega = j(\alpha)\omega, \quad \alpha \in \mathcal{O}_D. \tag{8.8}$$

(Here the superscript t means, of course, matrix transpose.) We say an ω as above is *invertible* if ω^1, ω^2 form a basis of the S–module H^{10}. It is trivial to check that if ω is a false invertible 1–form then ω is a basis for the S–module of false 1–forms. Also, if $u : A_1 \to A_2$ is an isogeny of polarized false elliptic curves (A_1, i_1, θ_1) and (A_2, i_2, θ_2) whose degree is not divisible by p and if ω is a false invertible 1–form on (A_2, i_2, θ_2) then $u^*\omega$ is a false invertible 1–form on (A_1, i_1, θ_1).

Note that if (A, i, θ) is any polarized false elliptic curve over S then the S–module of false 1–forms has a useful alternative description as follows. We have a natural ring homomorphism

$$\mathcal{O}_D^{op} \to End_S(H^{10}), \quad \alpha \mapsto i(\alpha)^*$$

which induces a ring homomorphism

$$l : \mathcal{O}_D^{op} \otimes_{\mathbf{Z}} S \to End_S(H^{10}).$$

On the other hand the isomorphism j induces an isomorphism

$$j : \mathcal{O}_D^{op} \otimes_{\mathbf{Z}} S \to Mat_2(S)^{op}.$$

Consider the ring homomorphism

$$l \circ j^{-1} : Mat_2(S)^{op} \to \mathcal{O}_D^{op} \otimes_{\mathbf{Z}} S \to End_S(H^{10}).$$

Let $e_{11}, e_{12}, e_{21}, e_{22}$ be the images, via the latter homomorphism of the matrices

$$\begin{bmatrix} 1 & 0 \\ 0 & 0 \end{bmatrix}, \begin{bmatrix} 0 & 1 \\ 0 & 0 \end{bmatrix}, \begin{bmatrix} 0 & 0 \\ 1 & 0 \end{bmatrix}, \begin{bmatrix} 0 & 0 \\ 0 & 1 \end{bmatrix}.$$

Then we have a direct sum decomposition

$$H^{10} = e_{11}H^{10} \oplus e_{22}H^{10}$$

hence $e_{11}H^{10}$ is a locally free $S-$module of rank one. We claim that the map

$$\omega = (\omega^1, \omega^2)^t \mapsto \omega^1$$

is an isomorphism between the $S-$module of false $1-$forms and $e_{11}H^{10}$; we will sometimes view this isomorphism as an identification. Indeed, if $\omega^1 \in e_{11}H^{10}$ it is trivial to see that $\omega^2 := e_{12}\omega^1 \in e_{22}H^{10}$ and the column vector $\omega := (\omega^1, \omega^2)^t$ is a false $1-$form. Conversely, if $(\omega^1, \omega^2)^t$ is a false $1-$form then $\omega^1 \in e_{11}H^{10}$. The two maps, described above, between $e_{11}H^{10}$ and the module of false $1-$forms are inverse to each other. (By the way, if $e_{11}H^{10}$ happens to be free, which is the case locally in the Zariski topology, and ω^1 is a basis for this module, then it is trivial to check that $\omega^2 := e_{12}\omega^1$ is a basis for $e_{22}H^{10}$, hence $(\omega^1, \omega^2)^t$ is an invertible false $1-$form. Hence, if ω is a basis for the module of false $1-$forms then ω is invertible. We conclude that *invertible false $1-$forms always exist locally in the Zariski topology.*)

8.1.6. Serre-Tate parameters. For the material below we refer to [**80**]. Let A_0 be an ordinary Abelian scheme (hence variety) over $k := \mathbf{F}_p^a$; recall that *ordinary* means that the absolute Frobenius on $H^1(A_0, \mathcal{O})$ is bijective. The universal formal deformation space of A_0, [**80**], p.149, 159, (parameterizing liftings of A_0 to local Artin rings with residue field k) is canonically isomorphic as a functor to

$$Hom_{\mathbf{Z}_p}(T_p(A_0) \otimes T_p(\check{A}_0), \mathbf{G}_m^{for})$$

where $T_p(A_0)$ and $T_p(\check{A}_0)$ are the physical Tate modules of A_0 and its dual $\check{A}_0$. For any formal Abelian scheme A over a complete local ring S with residue field k which lifts A_0 we denote by

$$q(A/S; -, -) : T_p(A_0) \times T_p(\check{A}_0) \to \mathbf{G}_m^{for}(S) = 1 + m_S \tag{8.9}$$

the corresponding bilinear map, where m_S is the maximal ideal of S. In particular if we are given bases $b, \check{b}$ of $T_p(A_0)$ and $T_p(\check{A}_0)$ then

$$q_{ij} := q(A/S; b_i, \check{b}_j) \in 1 + m_S \tag{8.10}$$

are called the Serre-Tate parameters of A/S; $q = (q_{ij})$ is called the Serre-Tate matrix. If $q_{ij} = 1$ for all i and j one calls A/S the *canonical lift* of A_0/k (with respect to the bases b and $\check{b}$). If $R = \hat{\mathbf{Z}}_p^{ur}$ is the completion of the maximum unramified extension of $\mathbf{Z}_p$ then the universal formal deformation space can be identified, via $b, \check{b}$, with $Spf\ R[[t_{ij}]]$; any A/S as above corresponds to the map

$$R[[t_{ij}]] \to S, \quad t_{ij} \mapsto q_{ij} - 1. \tag{8.11}$$

Finally let us recall from [**80**], p. 161, that there is a functorial isomorphism (viewed sometimes later as an identification)

$$T_p(\check{A}_0) \simeq Hom_S(A^{for}, \mathbf{G}_m^{for}), \tag{8.12}$$

where $\mathbf{G}_m := Spec\ S[z, z^{-1}]$. Hence any basis $\check{b}$ of $T_p(\check{A}_0)$ induces, by pull-back of the standard invariant $1-$form $\frac{dz}{z}$ on $\mathbf{G}_m^{for}$, a basis $\omega = \omega(\check{b})$ of $1-$forms on A/S:

$$\omega = \check{b}^* \frac{dz}{z}, \tag{8.13}$$

where, as a rule we see ω and $\check{b}$ as column vectors.

The following result was proved in [**23**], Lemma 2.5.

LEMMA 8.2. *Let (A_0, i_0, θ_0) be a polarized false elliptic curve over $k = \mathbf{F}_p^a$ such that A_0 is ordinary. Then there exists a $\mathbf{Z}_p-$basis b of the physical Tate module $T_p(A_0)$ such that if $\breve{b}$ is the basis of the physical Tate module $T_p(\breve{A}_0)$ corresponding to b via θ_0 then for any false elliptic curve (A, i) over R equipped with an isomorphism $(A, i) \otimes k \simeq (A_0, i_0)$ the Serre-Tate matrix of (A, i) (with respect to $b, \breve{b}$) has the form*

$$\begin{bmatrix} q & 0 \\ 0 & q^d \end{bmatrix}$$

where $q \in 1 + pR$. Furthermore, viewing b as a column vector, we have:

$$i(\alpha)b = j(\alpha)b, \quad \alpha \in \mathcal{O}_D.$$

In particular, the pair of $1-$forms $\omega(\breve{b}) = \breve{b}^* \frac{dz}{z}$ is a false $1-$form.

DEFINITION 8.3. A b as in Lemma 8.2 will be called a *normalized* basis of $T_p(A_0)$.

The formal deformation space of (A_0, i_0, θ_0) (parameterizing liftings of the latter to polarized false elliptic curves over Artin local rings with residue field k) can be identified with $Spf\ R[[T]]$, embedded into $Spf\ R[[t_{ij}]]$ via

$$t_{11} \mapsto T,\ t_{22} \mapsto (T+1)^d - 1, \quad t_{12}, t_{21} \mapsto 0. \tag{8.14}$$

8.1.7. De Rham space. If X is a smooth scheme over a ring S one can consider the *De Rham $S-$modules* $H^i_{DR}(X/S)$ defined as the hypercohomology modules $\mathbf{H}^i(\Omega^*_{X/S})$ of the De Rham complex $(\Omega^*_{X/S}, d)$. If A/S is an Abelian scheme then one has a natural exact sequence (arising from the spectral sequence defining the hypercohomology):

$$0 \to H^{10} \to H^1 \to H^{01} \to 0, \tag{8.15}$$

$$H^{10} := H^0(A, \Omega^1_{A/S}),\ H^1 := H^1_{DR}(A/S),\ H^{01} := H^1(A, \mathcal{O}).$$

The modules above are locally free and their formation commutes with base change; cf. [**104**]. Let $\breve{H}^1 := H^1_{DR}(\breve{A}/S)$. By [**58**], p.81, there is a canonical bilinear pairing

$$\langle\ ,\ \rangle : \breve{H}^1 \times H^1 \to S$$

induced by the Poincaré bundle. Let $\theta : A \to \breve{A}$ be a principal polarization. Then θ defines an isomorphism $\theta_{DR} : \breve{H}^1 \to H^1$. Define

$$\langle\ ,\ \rangle_\theta : H^1 \times H^1 \to S \tag{8.16}$$

$$\langle \eta_1, \eta_2 \rangle_\theta := \langle \theta_{DR}^{-1}\eta_1, \eta_2 \rangle$$

The pairing $\langle\ ,\ \rangle_\theta$ is perfect, antisymmetric, and H^{10} is an isotropic subspace. Let

$$\theta_* : H^0(A, T_{A/S}) \to H^1(A, \mathcal{O}). \tag{8.17}$$

be the Lie algebra map induced by θ_*. If $\omega \in H^{10}$ and if $\eta \in H^1$ maps to $\xi \in H^{01}$ then

$$\langle \omega, \eta \rangle_\theta = \langle \omega, \theta_*^{-1}\xi \rangle \tag{8.18}$$

where the right hand side pairing is the canonical duality

$$\langle\ ,\ \rangle : H^0(A, \Omega) \times H^0(A, T) \to S.$$

On the other hand, if $A = Jac(C)$ is the Jacobian of a (smooth projective) curve C/S (possessing a section), where S is an integral domain and θ is induced by the theta divisor then, under the identification $H^1 \simeq H^1_{DR}(C/S)$ given by the Abel-Jacobi map, the pairing $\langle\ ,\ \rangle_\theta : H^1 \times H^1 \to S$ corresponds to the Poincaré duality pairing

$$\langle\ ,\ \rangle : H^1_{DR}(C/S) \times H^1_{DR}(C/S) \to S; \tag{8.19}$$

this can be checked, for instance, analytically.

Assume now that (A, i, θ) is a polarized false elliptic curve and $s : H^{01} \to H^1$ is an $\mathcal{O}_D$−equivariant right inverse of the projection $H^1 \to H^{01}$. Define

$$L := e_{11}H^{10},\ H := e_{11}H^1,\ Q := e_{11}H^{01}; \tag{8.20}$$

we get an induced map $s : Q \to H$ and set $L' := s(Q) \subset H$. The commutative diagrams

$$\begin{array}{ccc} A & \stackrel{\theta}{\to} & \check{A} \\ i(\alpha^+)\downarrow & & \downarrow i(\alpha)^{\check{}} \\ A & \stackrel{\theta}{\to} & \check{A} \end{array},$$

for $\alpha \in \mathcal{O}_D$, induce commutative diagrams

$$\begin{array}{ccc} H^0(A, T_{A/S}) & \stackrel{\theta_*}{\to} & H^1(A, \mathcal{O}) \\ i(\alpha^+)\downarrow & & \downarrow i(\alpha)^* \\ H^0(A, T_{A/S}) & \stackrel{\theta_*}{\to} & H^1(A, \mathcal{O}) \end{array}$$

at the level of Lie algebras. Let $\omega = (\omega^1, \omega^2)^t$ be an invertible false 1−form. Let $\xi = (\xi^1, \xi^2)^t$ be the unique local basis of $H^1(A, \mathcal{O})$ such that

$$\langle \theta_*^{-1}\xi, \omega^t \rangle = I, \tag{8.21}$$

where I is the identity matrix; we call ξ the $\theta-$*dual* of ω. We claim that the following formula holds:

$$i(\alpha)^*\xi = j(\alpha^+)^t\xi,\quad \alpha \in \mathcal{O}_D. \tag{8.22}$$

Indeed, if $i(\alpha)^*\xi = k(\alpha)\xi$ then

$$\begin{aligned} j(\alpha^+)^t &= \langle \theta_*^{-1}\xi, \omega^t j(\alpha^+)^t \rangle = \langle \theta_*^{-1}\xi, i(\alpha^+)^*\omega^t \rangle \\ &= \langle i(\alpha^+)\theta_*^{-1}\xi, \omega^t \rangle = \langle \theta_*^{-1} i(\alpha)^*\xi, \omega^t \rangle = k(\alpha), \end{aligned}$$

which proves our claim. Note that by our normalization 8.7 we have

$$j(\alpha^+)^t = \begin{pmatrix} d & 0 \\ 0 & 1 \end{pmatrix} j(\alpha) \begin{pmatrix} d^{-1} & 0 \\ 0 & 1 \end{pmatrix}, \tag{8.23}$$

where d is the discriminant of D. In particular

$$e_{11}\xi^1 = \xi^1,\ e_{22}\xi^2 = \xi^2,\ \ e_{12}\xi^1 = d\xi^2 \tag{8.24}$$

so ξ^1 is a local basis of Q. Define

$$\eta^1 := s(\xi^1),\ \ \eta^2 := s(\xi^2); \tag{8.25}$$

by $\mathcal{O}_D$−equivariance of s we have

$$e_{11}\eta^1 = \eta^1,\ e_{22}\eta^2 = \eta^2,\ \ e_{12}\eta^1 = d\eta^2, \tag{8.26}$$

in particular η^1 is a local basis of L'. Note that

$$\omega^1, \eta^1, \omega^2, \eta^2 \tag{8.27}$$

is a local basis of H^1.

DEFINITION 8.4. For a false elliptic curve over a Noetherian ring S, a local basis of H^1 of the form 8.27 will be called *normalized*; once the section s is fixed such a basis is uniquely determined by its first vector ω^1.

Note that

$$\langle \eta, \omega^t \rangle_\theta = I. \tag{8.28}$$

If A is an elliptic curve, and $s : H^{01} \to H^1$ is a section of the projection $H^1 \to H^{01}$ we set

$$L = H^{10},\ H = H^1,\ Q = H^{01},\ L' = s(Q). \tag{8.29}$$

If $\omega = \omega^1$ is a local basis of $L = H^{10}$, and $\xi = \xi^1$ satisfies $\langle \xi, \omega \rangle = 1$ (where $\langle\ ,\ \rangle$ is the Serre duality) then we set $\eta^1 := s(\xi)$.

DEFINITION 8.5. For an elliptic curve over a Noetherian ring S a local basis of H^1 of the form ω^1, η^1 above is called a *normalized basis*; once the section s is fixed such a basis is uniquely determined by its first component ω^1.

8.1.8. Gauss-Manin connection and Kodaira-Spencer map. Let S be a Noetherian ring and A/S an elliptic (or false elliptic) curve. Let Ω be either $\Omega^1_{S/R}$ or its $p-$adic completion in case S is $p-$adically complete or its m_S-adic completion if S is a local ring complete with respect to its maximal ideal m_S. Recall that we have at our disposal the Gauss-Manin connection

$$\nabla : H^1 \to H^1 \otimes \Omega; \tag{8.30}$$

here $\otimes$ stands for either the usual tensor product or for the appropriately completed tensor product. By the functoriality of the Gauss-Manin connection, in the false case, ∇ commutes with e_{ij} so ∇ induces a connection

$$\nabla : H \to H \otimes \Omega. \tag{8.31}$$

If we are in the false case then, with respect to a normalized basis of H^1, we may write

$$\begin{pmatrix} \nabla\omega^1 \\ \nabla\eta^1 \\ \nabla\omega^2 \\ \nabla\eta^2 \end{pmatrix} = \begin{pmatrix} \sigma_{11} & \sigma_{12} & 0 & 0 \\ \sigma_{21} & \sigma_{22} & 0 & 0 \\ 0 & 0 & \sigma_{11} & \sigma_{12}d \\ 0 & 0 & \sigma_{21}d^{-1} & \sigma_{22} \end{pmatrix} \begin{pmatrix} \omega^1 \\ \eta^1 \\ \omega^2 \\ \eta^2 \end{pmatrix}, \tag{8.32}$$

where $\sigma_{ij} \in \Omega$. Similarly, in the genuine elliptic curve case, we have

$$\begin{pmatrix} \nabla\omega^1 \\ \nabla\eta^1 \end{pmatrix} = \begin{pmatrix} \sigma_{11} & \sigma_{12} \\ \sigma_{21} & \sigma_{22} \end{pmatrix} \begin{pmatrix} \omega^1 \\ \eta^1 \end{pmatrix}. \tag{8.33}$$

In particular the *Kodaira-Spencer matrix*

$$\langle \nabla\omega, \omega^t \rangle_\theta \tag{8.34}$$

equals

$$\sigma_{12} \cdot \begin{pmatrix} 1 & 0 \\ 0 & d \end{pmatrix} \tag{8.35}$$

in the false elliptic case and equals σ_{12} in the genuine elliptic case. Consider the bilinear map

$$L \times L \to \Omega,\quad (x, y) \mapsto \langle \nabla x, y \rangle_\theta; \tag{8.36}$$

this map induces a map

$$\iota : L^{\otimes 2} \to \Omega. \tag{8.37}$$

The bilinear map above is, of course, given in a normalized basis by

$$(\omega^1, \omega^1) \mapsto \sigma_{12}. \tag{8.38}$$

DEFINITION 8.6. One says that A/S is *infinitesimally universal* if the Kodaira-Spencer matrix 8.34 is invertible; this is equivalent to asking that σ_{12} be a nowhere vanishing $1-$form. If X is a smooth scheme over R and A is a genuine respectively false elliptic curve over X which is infinitesimally universal then we call X a *modular curve* respectively a *Shimura curve.*

On the other hand note that the bilinear map

$$L' \times L \to S$$

$$(x, y) \mapsto \langle x, y\rangle_\theta$$

induces an isomorphism

$$L^{\otimes(-1)} \simeq L'$$

because, in a normalized basis, our bilinear maps sends

$$(\eta^1, \omega^1) \mapsto 1.$$

Note that if S is the $p-$adic completion of a smooth $R-$algebra ($R = \hat{\mathbf{Z}}_p^{ur}$), A/S is infinitesimally universal and P_0 is a $k-$point of $Spec\ S$ at which the fiber A_0 is ordinary then A/S is *formally universal* at P_0 in the sense that the completion of A along A_0 is isomorphic to the universal formal deformation of A_0 (as an elliptic respectively false elliptic curve).

8.1.9. Crystalline structure. Our main reference here is [**7**], page 184. Let S be a $p-$adically complete ring which is flat over $\mathbf{Z}_p$ and set $S_n := S/p^{n+1}S$, $n \geq 0$. Let X and Y be smooth schemes over S and set $X_n := X \otimes_S S_n$, $Y_n := Y \otimes_S S_n$. Then for any morphism $u : X_0 \to Y_0$ of schemes over S_0 there are naturally induced maps of S_n-modules

$$H^i_{crys}(u) : H^i_{DR}(Y_n/S_n) \to H^i_{DR}(X_n/S_n)$$

which are compatible as n varies. Assume in addition X, Y are Abelian schemes over S. Since the Hodge and De Rham cohomology modules of Abelian schemes are locally free and commute with base change ([**104**] p. 44) we get, by taking the inverse limit, a natural map of $S-$modules (still denoted by)

$$H^i_{crys}(u) : H^1_{DR}(Y/S) \to H^1_{DR}(X/S).$$

Assume A/S is an Abelian scheme and let $\varphi : S \to S'$ be a homomorphism of $p-$adically complete rings. Furthermore let $\phi : S \to S'$ be a ring homomorphism, let $\bar{\varphi}, \bar{\phi} : S/pS \to S'/pS'$ be the reductions modulo p of φ and ϕ respectively, and assume that $\bar{\phi} = \bar{\varphi} \circ F$, where F is the Frobenius endomorphism of S'/pS'. Let A^φ/S' be the pull-back of A/S via φ and similarly let A^ϕ/S' be the pull-back of A/S via ϕ. By the equality $\bar{\phi} = \bar{\varphi} \circ F$ it follows that $A^\phi \otimes (S'/pS')$ is *canonically* isomorphic to the pull-back $F^*(A^\varphi \otimes (S'/pS'))$ of $A^\varphi \otimes (S'/pS')$ via $F : S'/pS' \to S'/pS'$. Now the absolute $p-$power Frobenius endomorphism F_{abs} of $A^\varphi \otimes (S'/pS')$ induces a $S'/pS'-$morphism

$$F_{rel} : A^\varphi \otimes (S'/pS') \to F^*(A^\varphi \otimes (S'/pS'))$$

Composing the latter morphism with the canonical isomorphism

$$F^*(A^{\varphi} \otimes (S'/pS')) \simeq A^{\phi} \otimes (S'/pS')$$

we get a S'/pS'–morphism

$$F_{\varphi,\phi} : A^{\varphi} \otimes (S'/pS') \to A^{\phi} \otimes (S'/pS')$$

so we get induced morphisms of S'–modules

$$H^i_{crys}(F_{\varphi,\phi}) : H^1_{DR}(A^{\phi}/S') \to H^1_{DR}(A^{\varphi}/S').$$

Since

$$H^1_{DR}(A^{\phi}/S') = H^1_{DR}(A/S) \otimes_{S,\phi} S',\ H^1_{DR}(A^{\varphi}/S') = H^1_{DR}(A/S) \otimes_{S,\varphi} S',$$

we get an induced ϕ–linear map

$$\phi : H^i(A/S) \to H^i(A/S) \otimes_S S' \tag{8.39}$$

where, in the latter $\otimes_S S' = \otimes_{S,\varphi} S'$. (A ϕ–linear map from an S–module into an S'–module is an additive map f satisfying $f(ax) = \phi(a) f(x)$.) The most familiar case is when $\varphi = id$ in which case we have the ϕ–linear map

$$\phi : H^i(A/S) \to H^i(A/S). \tag{8.40}$$

However, in this book, an important role is played by situations in which φ is not the identity (not even an isomorphism).

8.1.10. Hasse invariant. Let $(\bar{A}, \bar{i}, \bar{\theta})$, $\bar{L}$, $\bar{\omega}$, $\bar{\xi}$ be obtained, from a polarized false elliptic curve over a Noetherian ring S, (A, i, θ), a false 1–form $\omega = (\omega^1, \omega^2)^t$, and its θ–dual ξ (cf. Equation 8.21), by base change to $\bar{S} := S/pS$. Note that the semilinear map

$$F^* : H^1(\bar{A}, \mathcal{O}) \to H^1(\bar{A}, \mathcal{O})$$

induced by the absolute p–power Frobenius on $\bar{A}$ commutes with all $\bar{S}$–linear endomorphisms

$$i(\alpha)^* : H^1(\bar{A}, \mathcal{O}) \to H^1(\bar{A}, \mathcal{O}).$$

Consider the *Hasse-Witt matrix* $\bar{h} \in Mat_2(\bar{S})$ of $\bar{A}$ with respect to the basis $\bar{\xi}$; hence, by definition,

$$F^* \bar{\xi} = \bar{h} \bar{\xi}. \tag{8.41}$$

Let

$$\mathbf{Z}_p \to \mathbf{F}_p, \quad x \mapsto c(x)$$

be the canonical projection. Using Equation 8.22 we have that, for all $\alpha \in \mathcal{O}_D$,

$$\begin{aligned} \bar{h} c(j(\alpha^+)^t) \bar{\xi} &= \bar{h} i(\alpha)^* \bar{\xi} = i(\alpha)^* \bar{h} \bar{\xi} = i(\alpha)^* F^* \bar{\xi} = F^* i(\alpha)^* \bar{\xi} \\ &= F^*(c(j(\alpha^+)^t) \bar{\xi}) = F(c(j(\alpha^+)^t)) F^* \bar{\xi} = c(j(\alpha^+)^t) \bar{h} \bar{\xi}. \end{aligned}$$

We conclude that

$$\bar{h} c(j(\alpha^+)^t) = c(j(\alpha^+)^t) \bar{h}$$

for all α. This forces $\bar{h}$ to be a scalar matrix, $\bar{h} \in \bar{S}$. Consider now the local section

$$\bar{H} := \bar{h} \cdot (\bar{\omega}^1)^{\otimes p-1} \in \bar{L}^{\otimes p-1}.$$

It is easy to see that the above section does not depend on the choice of ω; hence these local sections are well defined and glue together to give a "global section"

$$\bar{H} \in \bar{L}^{\otimes p-1} \tag{8.42}$$

which we call the *"Hasse invariant"*. (Cf., for instance, [**48**].) We choose now a section $H \in L^{\otimes(p-1)}$ whose reduction modulo p is $\bar{H}$ and we set

$$X_{ord} := Spec\{(\bigoplus_{m\geq 0} L^{\otimes m(p-1)})/(H-1)\}. \tag{8.43}$$

(The letter H was also used to denote a subspace of the de Rham space H^1; this should create no confusion.) If x is a local basis for L and $H = h \cdot x^{p-1}$ then, locally,

$$X_{ord} = Spec\ \mathcal{O}(X)[1/h] \tag{8.44}$$

so X_{ord} is an affine open subscheme of X. Note that X_{ord} depends on H so it is not entirely canonical; but the $p-$adic completion $\hat{X}_{ord}$ does not depend on H. If $X \otimes k$ contains a supersingular point (i.e. a point with non-invertible Hasse-Witt matrix) then, of course $X_{ord} \otimes k \neq X \otimes k$. As a rule the index *ord* will mean base change from X to X_{ord}.

If A is an elliptic curve over S then the *Hasse invariant* is, again, a section

$$\bar{H} \in \bar{L}^{\otimes p-1}$$

and we define X_{ord} by the same formula as above; this is of course the usual Hasse invariant as defined, say in [**81**].

8.1.11. Unit root space. Let $R := \hat{\mathbf{Z}}_p^{ur}$ and X be an affine smooth scheme over R. Let S be either the $p-$adic completion $\mathcal{O}(X)\hat{}$ or the completion $\mathcal{O}(X)^{for}$ of $\mathcal{O}(X)$ at a maximal ideal. Let A/S be an Abelian scheme. If $S = \mathcal{O}(X)\hat{}$ we assume $X_{ord} \otimes k \neq \emptyset$ and we set $S_{ord} := \mathcal{O}(X_{ord})\hat{}$; if $S = \mathcal{O}(X)^{for}$ we assume the reduction of A modulo the corresponding maximal ideal is ordinary and we set $S_{ord} = S$. It is known that

$$\phi H^{10} \subset pH^1 \tag{8.45}$$

for any lift of Frobenius $\phi : S \to S$; cf. [**82**], p.188. Then by [**82**] p.188 there exists a locally free $S-$submodule

$$U \subset H^1_{ord}$$

such that

$$H^1_{ord} = H^{10}_{ord} \oplus U$$

and

$$\phi U \subset U$$

for all liftings of Frobenius $\phi : S \to S$. Here, of course, the index *ord* stands for taking objects over S_{ord}. If $S = R[[t_{ij}]]$ is the formal deformation space of an ordinary Abelian variety A_0/k, $\check{b}$ is a basis of the Tate module of $\check{A}_0$, and $\omega = \omega(\check{b})$ is the basis of H^{10} attached to $\check{b}$ then one can find a basis u of U such that

$$\phi u = u \tag{8.46}$$

for any lifting of Frobenius $\phi : R[[t_{ij}]] \to R[[t_{ij}]]$ and such that

$$\nabla u = 0, \quad \nabla \omega = (d\ log\ q) \otimes u \tag{8.47}$$

where $q = (q_{ij})$, $q_{ij} := 1 + t_{ij}$. Cf. [**42**], 1.4.2., 2.1.3., [**80**] p. 175, [**54**], p. 116.

DEFINITION 8.7. A basis ω, u as above will be called a *canonical* $R[[t_{ij}]]-$basis for H^1 corresponding to $\check{b}$.

Assume in the above discussion that A_0 has dimension $g \in \{1,2\}$ and $(A_0, \theta_0) = (Jac(C_0), \theta_{C_0})$ for some curve C_0 of genus g. (The latter is automatic if $g = 1$.) Of course θ_{C_0} above denotes the theta polarization on $Jac(C_0)$. Under these assumptions formulae become more explicit as we shall see. Let $Spf\ R[[s_1, ..., s_\mu]]$ be the universal deformation space of C_0, where μ is 1 or 3 according as g is 1 or 2. By the symmetry property in [**80**], p. 149, the map $R[[t_{ij}]] \to R[[s_1, ..., s_\mu]]$ factors through $R[[t_{ij}]]/(t_{ij} - t_{ji})$. The reduction mod p of the map

$$S_{def} := R[[t_{ij}]]/(t_{ij} - t_{ji}) \to R[[s_1, ..., s_\mu]] \tag{8.48}$$

is surjective, hence an isomorphism. It follows that the map above itself is an isomorphism. (We may, and will, assume that $s_1, ..., s_\mu$ are the classes of t_{ij} for $i \leq j$.) Note that on H^1 we have available the cup product $\langle\ ,\ \rangle$ coming from the universal curve. Consider the matrix

$$\epsilon := -\langle \omega, u^t \rangle \in GL_g(R[[t_{ij}]]). \tag{8.49}$$

On the other hand one can consider the lifting of Frobenius ϕ on S_{def} induced by $\phi(q_{ij}) = q_{ij}^p$, where $q_{ij} = 1 + t_{ij}$; by [**42**], 2.1.3., we have, for this ϕ, that $\phi\omega = p\omega$. Using the equality

$$\langle \phi\omega, \phi u^t \rangle = p \cdot \phi\langle \omega, u^t \rangle$$

(cf. [**82**], pp. 190 and 184) we get that $\phi\epsilon = \epsilon$. Hence $\epsilon \in GL_2(\mathbf{Z}_p)$. Let C/R be any curve lifting C_0/k; such a curve corresponds to a matrix $t(C) = (t_{ij}(C))$ with entries in pR. We set $q(C) := 1 + t(C)$. For each such C we obtain a basis $\omega(C), u(C)$ of $H^1(C) := H^1_{DR}(C/R) = H^1_{DR}(Jac(C)/R)$. This basis is compatible with the Hodge filtration. We claim that $u(C)$ is fixed by the crystalline $\phi : H^1(C) \to H^1(C)$ induced by the unique lifting of Frobenius $\phi : R \to R$; i.e.

$$\phi u(C) = u(C).$$

Indeed one can easily construct a lifting of Frobenius ϕ on $R[[t_{ij}]]$ such that the homomorphism

$$R[[t_{ij}]] \to R, \quad t \mapsto t(C)$$

commutes with the action of ϕ on the two rings: just let

$$\phi t = t^p + \{\phi(t(C)) - t(C)^p\}.$$

In particular the induced homomorphism

$$H^1 \to H^1(C)$$

is ϕ-equivariant. Now one uses the fact that u is fixed by *any* lifting of Frobenius, in particular by $\phi : H^1 \to H^1$; it follows that $u(C)$ is fixed by $\phi : H^1(C) \to H^1(C)$ and our claim is proved. By [**82**], pp. 190-191, we have

$$\langle u(C), u(C)^t \rangle = 0. \tag{8.50}$$

As in loc. cit. we can consider, for each C, a column vector $\beta(C)$ of elements of $H^1(C)$ such that

$$\phi\beta(C) = p \cdot \beta(C)$$
$$\langle u(C), \beta(C)^t \rangle = I$$
$$\langle \beta(C), u(C)^t \rangle = -I$$
$$\langle \beta(C), \beta(C)^t \rangle = 0.$$

Express

$$\omega(C) = \Omega_1(C) \cdot \beta(C) + \Omega_2(C) \cdot u(C)$$

where $\Omega_i(C)$ are matrices with entries in R. We get that

$$\Omega_1(C) = \epsilon.$$

Also, by [**54**], p. 116, we have

$$\Omega_2(C) = log(q(C)).$$

Finally, from the equality $\langle \omega(C), \omega(C)^t \rangle = 0$ we get

$$0 = \epsilon \cdot log(q(C)) - log(q(C)) \cdot \epsilon^t.$$

Since this is true for any C it follows that ϵ must be a scalar matrix:

$$\epsilon \in \mathbf{Z}_p^\times \cdot I. \tag{8.51}$$

We conclude that

$$\omega(C) = \epsilon \cdot \beta(C) + log(q(C)) \cdot u(C). \tag{8.52}$$

If $g = 1$ one can choose $\epsilon = 1$ by [**42**], Introduction.

Assume now that S is the p–adic completion or the completion at a maximal ideal of a smooth R–algebra and (A, i, θ) is a polarized false elliptic curve over S. We claim that the unit root space U is an $\mathcal{O}_D$–submodule of H^1_{ord}. Indeed it is enough to check this in case S is the p–adic completion of a smooth R–algebra. It is then enough to check that U_P is an $\mathcal{O}_D$–submodule of H^1_P for all R–points P of $Spec\ S_{ord}$; here the index P means "base change from S to R via P". Now, by the functoriality of crystalline cohomology, the semilinear Frobenius ϕ acting on the de Rham module H^1_P commutes with the action of $\mathcal{O}_D$. By [**82**], p.172, Theorem 2, there is a basis of H^1_P of the form u^1, u^2, v^1, v^2, where u^1, u^2 form a basis of U_R, and

$$\phi u = u$$

$$\phi v = pBv + Cu$$

where $u := (u^1, u^2)^t$, $v := (v^1, v^2)^t$, and $B, C \in Mat_2(R)$. Write

$$i(\alpha)^* u = Mu + Nv, \quad \alpha \in \mathcal{O}_D$$

for some $M, N \in Mat_2(R)$. We get

$$Mu + Nv = i(\alpha)^*(\phi u) = \phi(i(\alpha)^* u) = M^\phi u + N^\phi (pBv + Cu)$$

hence $N = pN^\phi B$. By induction, N must be divisible by p^n for all n so $N = 0$. Consequently $i(\alpha)^* U \subset U$. This proves our claim. Next we define

$$\mathcal{U} := e_{11} U \quad \text{or} \quad \mathcal{U} = U \tag{8.53}$$

according as we are in the false elliptic or elliptic case. Then $\mathcal{U}$ is a locally free S_{ord}–module of rank one and, clearly, $H = L \oplus \mathcal{U}$.

REMARK 8.8. We claim that, if u^1 is a local basis of $\mathcal{U}$ and $\omega = (\omega^1, \omega^2)^t$ is an invertible false 1–form then

$$\langle u^1, \omega^1 \rangle_\theta \in S^\times_{ord}. \tag{8.54}$$

Indeed, if we consider a normalized basis of H^1 and write

$$\begin{pmatrix} \omega^1 \\ u^1 \end{pmatrix} = \begin{pmatrix} 1 & 0 \\ s_{21} & s_{22} \end{pmatrix} \begin{pmatrix} \omega^1 \\ \eta^1 \end{pmatrix} \tag{8.55}$$

with $s_{21}, s_{22} \in S_{ord}$ then s_{22} is invertible and

$$\langle u^1, \omega^1 \rangle_\theta = s_{22} \tag{8.56}$$

which proves our claim.

REMARK 8.9. We claim that if one is given a basis $u = (u^1, u^2)^t$ of U and a false 1–form $\omega = (\omega^1, \omega^2)^t$ such that

$$\langle u, \omega^t \rangle_\theta \in \mathbf{Z}_p^\times \cdot I$$

then $e_{11}u^1 = u^1$, $e_{22}u^2 = u^2$, and $e_{12}u^1 = du^2$; in particular u^1 must be a basis of $\mathcal{U}$. Indeed, we may assume the matrix $< u, \omega^t >_\theta$ is the identity. Let ξ^1, ξ^2 be the images of u^1, u^2 in H^{01} and set $\xi = (\xi^1, \xi^2)^t$. Then

$$\langle \theta_*^{-1}\xi, \omega^t \rangle = I.$$

So $u = s^U(\xi)$ where $s^U : H^{01}_{ord} \simeq U \subset H^1_{ord}$ is the $\mathcal{O}_D$–equivariant splitting defined by U and our claim follows from Equation 8.26.

Assume in what follows that (A_0, i_0, θ_0) is a polarized false elliptic curve over $k := \mathbf{F}_p^a$ such that (A_0, θ_0) is a Jacobian. Alternatively assume A_0 is an elliptic curve over k. Let b be a normalized basis of $T_p(A_0)$ (cf. Definition 8.3) and $\check{b}$ the basis of $T_0(\check{A}_0)$ corresponding to b via θ_0. Let ω, u be a canonical $R[[t_{ij}]]$–basis of H^1 corresponding to $\check{b}$ (cf. Definition 8.7). Let us still denote by ω, u the induced $R[[T]]$–basis of H^1 over $R[[T]]$, the formal deformation space of (A_0, i_0, θ_0) respectively A_0. Let (A, i, θ) respectively A be the universal polarized elliptic curve respectively elliptic curve over $R[[T]]$. So ω is a false 1–form in the false elliptic curve case. So ω^1 is an $R[[T]]$– basis of H. By Equations 8.49 and 8.51 and by Remark 8.9, u^1 is an $R[[T]]$– basis of $\mathcal{U}$ and $\langle u^1, \omega^1 \rangle_\theta = \epsilon$.

DEFINITION 8.10. A basis ω, u of H^1 as above will be called a *canonical* $R[[T]]$–basis.

8.2. Hecke correspondences over R: δ–line bundles and cycles

Assume X is an affine scheme, smooth over R, and A is a (genuine respectively false) elliptic curve over X (i.e. over $\mathcal{O}(X)$). We denote by $L = L_{A/X}$ the $\mathcal{O}(X)$–module of (genuine respectively false) 1–forms on A; cf. Definition 8.1. We also view L as line bundle on X. Recall from Definition 8.6 that A/X is called infinitesimally universal if the Kodaira-Spencer map $L^{\otimes 2} \to \Omega_X = K_X$ is an isomorphism; in that case we call X a modular or an (affine) Shimura curve depending on whether we are in the genuine or false elliptic case.

DEFINITION 8.11. A *Hecke correspondence* is a tuple

$$(\mathbf{X}, A, \tilde{u}, s), \tag{8.57}$$

where

i) $\mathbf{X} = (X, \tilde{X}, \sigma_1, \sigma_2)$ is an étale one dimensional correspondence of finite type and smooth over R, with irreducible base, and affine X,

ii) A/X is a (genuine respectively false) elliptic curve which is infinitesimally universal,

iii) $\tilde{u} : \sigma_1^* A \to \sigma_2^* A$ is an isogeny of (genuine respectively false) elliptic curves of (genuine respectively false) degree e not divisible by p, and

iv) s is an equivariant splitting of the exact sequence

$$0 \to H^0(A, \Omega^1) \to H^1_{DR}(A) \to H^1(A, \mathcal{O}) \to 0.$$

Here e, called the *degree* of $\mathbf{X}$, is a $\mathbf{Z}-$valued function on $\tilde{X}$ which is locally constant in the Zariski topology. If in addition $\tilde{X}$ is also affine we say $\mathbf{X}$ is *affine*. The splitting s is called *equivariant* if it is $\mathcal{O}_D-$equivariant in the "false" case; note that a natural splitting s always exists for genuine elliptic curves cf. [**81**], Appendix, and equivariant splittings exist (although they are not canonical) for false elliptic curves [**24**], Section 2. We will sometimes refer to the Hecke correspondence 8.57 simply as $\mathbf{X}$; in this case we will assume that $A, \tilde{u}, s$ are given or are clear from the context.

REMARK 8.12. Given a Hecke correspondence as in Equation 8.57 one can consider the natural projections $\sigma_i' : \sigma_i^* A \rightarrow A$, $i = 1, 2$, and the correspondence

$$\mathbf{A} := (A, \tilde{A}, \sigma_1', \sigma_2' \circ \tilde{u}), \tag{8.58}$$

where $\tilde{A} := \sigma_1^* A$. There is a natural morphism $\mathbf{A} \rightarrow \mathbf{X}$ which we would like to view, morally, as a "uniformization of $\mathbf{X}$ by $\mathbf{A}$". Alternatively, if

$$\mathbf{A}[p^\infty] := (A[p^\infty], \tilde{A}[p^\infty], \sigma_1', \sigma_2' \circ \tilde{u}) \tag{8.59}$$

is the induced correspondence (in the category of inductive systems of schemes) between the corresponding $p-$divisible groups then we have a morphism $\mathbf{A}[p^\infty] \rightarrow \mathbf{X}$ which can be viewed, even more appropriately, as a "uniformization of $\mathbf{X}$ by $\mathbf{A}[p^\infty]$".

Hecke correspondences typically arise from fine moduli spaces of elliptic or false elliptic curves. To pass from fine to coarse moduli spaces one can consider the following:

DEFINITION 8.13. Consider a Hecke correspondence as in Equation 8.57. An $R-$automorphism $\gamma : X \rightarrow X$ is *admissible* if we have an isomorphism $\gamma^* A \simeq A$ of elliptic (resp. false elliptic) curves over X. Let $\Gamma \subset Aut(X)$ be a finite subgroup consisting of admissible automorphisms, let $\pi : X \rightarrow X/\Gamma$ be the quotient map, and let $Y \subset X/\Gamma$ be an open subset above which π is étale. Then we say that the correspondence

$$\mathbf{Y} := (Y, \sigma_1^{-1}(\pi^{-1}(Y)) \cap \sigma_2^{-1}(\pi^{-1}(Y)), \pi \circ \sigma_1, \pi \circ \sigma_2) \tag{8.60}$$

is uniformized by the Hecke correspondence in Equation 8.57 and the group Γ. A correspondence such as $\mathbf{Y}$ can be called a $\mathbf{Q}-$*hyperbolic correspondence*.

REMARK 8.14. A $\mathbf{Q}-$hyperbolic correspondence is an analogue of a correspondence over $\mathbf{C}$ that admits a $\mathbf{Q}-$arithmetic hyperbolic uniformization. To define an analogue of correspondences over $\mathbf{C}$ that admit an $E-$arithmetic hyperbolic uniformization (E a totally real number field different from $\mathbf{Q}$), one must replace A in Definition 8.11 by a certain $p-$divisible group; we will not pursue this here.

The natural "vertical" conjecture, in our setting here, is the following:

CONJECTURE 8.15. *Let* $\mathbf{Y}$ *be a* $\mathbf{Q}-$*hyperbolic correspondence over* R*; cf. Equation 8.60. Then:*

1) $(\mathbf{Y}, \mathbf{K}^{-1})$ *is* $\delta-$*non-trivial and* $\delta-$*rational;*

2) the extension $k(Y) \cdot k\langle \mathbf{Y}, \mathbf{K}^{-1}\rangle \subset k\langle Y, K^{-1}\rangle$ *is algebraic.*

We will analyze, in this Chapter, the above Conjecture in case $\mathbf{X} = \mathbf{Y}$ and $\mathbf{X}$ possesses a special type of cycle which we will call *regular Hecke cycle*. While doing so we will investigate more problems including $\delta-$invariants and $\delta-$cohomology of certain natural $\delta-$line bundles polarization equivalent to $\mathbf{K}^{-1}$, Galois properties of the above extension, infinitesimal symmetries, $\delta-$tangent maps, converse results, etc.

In what follows we consider $\delta-$line bundles on Hecke correspondences. Consider a Hecke correspondence as in Equation 8.57. If $L = L_{A/X}$ then the isogeny $\tilde{u}$ induces an isomorphism $\tilde{u}^* : \sigma_2^* L \to \sigma_1^* L$. We then consider the line bundle $\mathbf{M} = (L, \tilde{u}^*)$ on $\mathbf{X}$. One can easily check that we have a commutative diagram

$$\begin{array}{ccc} \sigma_2^* K_X & \overset{e \cdot can}{\simeq} & \sigma_1^* K_X \\ \iota \uparrow & & \uparrow \iota \\ \sigma_2^* L^2 & \overset{\tilde{u}^*}{\to} & \sigma_1^* L^2 \end{array}$$

The diagram implies that

$$\mathbf{M}^2 \simeq \mathbf{K} \otimes \mathbf{F}(e).$$

For any $w \in W$ of even degree define

$$\mathbf{L}_w := \mathbf{M}^w \otimes \mathbf{F}(e)^{-deg(w)/2}; \tag{8.61}$$

then it follows that

$$\mathbf{L}_w^2 \simeq \mathbf{K}^w; \tag{8.62}$$

so if $-w \in W_+ \backslash \phi W_+$ then $\mathbf{L}_w$ is polarization equivalent to $\mathbf{K}^{-1}$. More generally, for $r \geq ord(w)$, we may consider the $\delta-$line bundle

$$\mathbf{L}_{r,w} := \mathbf{L}_w^{[r-ord(w)]}. \tag{8.63}$$

LEMMA 8.16. *The line bundle* $\mathbf{K}_{\mathbf{X}_{ord} \otimes k}^{(p-1)/2}$ *on* $\mathbf{X}_{ord} \otimes k$ *is trivial.*

This can be proved directly by showing that the pull-backs via σ_1 and σ_2 of the Hasse invariant $\bar{H}$ coincide, a fact due to Ihara; cf. [**73**], [**74**]. Since we will not use this Lemma in the sequel we postpone its proof; we will later derive the Lemma from a more general argument, cf. Remark 8.60.

Note that if $(\mathbf{X}, A, \tilde{u}, s)$ is a Hecke correspondence then for any $n \geq 1$ we have a naturally induced Hecke correspondence

$$(\mathbf{X}^{\circ n}, A, \tilde{u}^{\circ n}, s);$$

also the $\delta-$line bundle $\mathbf{L}_w^{\circ n}$ induced on $\mathbf{X}^{\circ n}$ from "$\mathbf{L}_w$ on $\mathbf{X}$" coincides with the $\delta-$line bundle "$\mathbf{L}_w$ on $\mathbf{X}^{\circ n}$".

DEFINITION 8.17. A *Hecke* $n-$*cycle* for a Hecke correspondence $(\mathbf{X}, A, \tilde{u}, s)$ is a tuple

$$(\tilde{P}_1, ..., \tilde{P}_n, P, b) \tag{8.64}$$

where $(\tilde{P}_1, ..., \tilde{P}_n, P)$ is an $n-$cycle on $\mathbf{X}$, the fiber A_0 of A at the reduction mod p, $P_0 \in X(k)$, of P is ordinary, the pull-back of A/X via P is the canonical lift A_1/R of A_0/k, and b is a basis of the physical Tate module $T_p(A_0)$; in the "false" situation we assume moreover that b is normalized (cf. Definition 8.3) and A_0 together with the unique principal polarization on A_0 compatible with the $\mathcal{O}_D$ action is isomorphic to a Jacobian. A Hecke $1-$ cycle is referred to as a *Hecke fixed point.*

Any Hecke $n-$cycle 8.64 for a Hecke correspondence $(\mathbf{X}, A, \tilde{u}, s)$ gives rise to a Hecke fixed point of $(\mathbf{X}^{\circ n}, A, \tilde{u}^{\circ n}, s)$, and actually to a Hecke fixed point on $(\mathbf{X}_1^{\circ n}, A, \tilde{u}^{\circ n}, s)$ provided that $\mathbf{X}_1^{\circ n}$ is an irreducible open sub-correspondence of $\mathbf{X}^{\circ n}$ "containing" the tuple $(\tilde{P}_1, ..., \tilde{P}_n)$.

Assume $(\tilde{P}, P, b)$ is a Hecke fixed point for a Hecke correspondence $(\mathbf{X}, A, \tilde{u}, s)$. The bases $b, \check{b}$ determine, via Serre-Tate theory (cf. Equation 8.14), an isomorphism $\mathcal{O}_{X,P_0}^{for} \simeq R[[T]]$. Let us compute the germ $\Sigma(T) \in R[[T]]$ of $\mathbf{X}$ using the Serre-Tate theory [**80**]; cf. also [**112**], [**24**]. Let $u_0 \in End(A_0)$ be the endomorphism induced by $\tilde{u}$, let $u_0^t \in End(A_0)$ be the image of u_0 under the Rosati involution (attached to θ_0), and let $\mu, \mu^+ \in \mathbf{Z}_p$ be the images of u_0 and u_0^t via the representation

$$End(A_0) \to End(T_p(A_0)).$$

So we have

$$u_0 b = \mu b, \quad u_0^t b = \mu^+ b, \quad \check{u}_0 \check{b} = \mu^+ \check{b}. \tag{8.65}$$

(The fact that, in the false case, μ and μ^+ are in the center of $End(T_p(A_0))$, i.e. in $\mathbf{Z}_p$, was checked in [**23**], Section 3.3.) Note also that

$$\mu\mu^+ = e,$$

so $\mu, \mu^+ \in \mathbf{Z}_p^\times$. Set

$$a := \frac{\mu^+}{\mu}, \quad m := -\frac{deg(w)}{2}. \tag{8.66}$$

LEMMA 8.18. *1) There is an equality $\Sigma(T) = (1+T)^a - 1$. So $\Sigma(T)$ is the multiplication by $a = \mu^+/\mu$ in the formal multiplicative group; in particular $a = \mu^+/\mu$ is the multiplier of $\mathbf{X}$ at $(\tilde{P}, P)$ and the canonical bundle $\mathbf{K}$ of $\mathbf{X}$ is formally flat at $(\tilde{P}, P)$.*

2) The line bundle $\mathbf{M}$ is formally flat at $(\tilde{P}, P)$ with multiplier μ^+; consequently the $\delta-$line bundle $\mathbf{L}_{r,w}$ is formally flat at $(\tilde{P}, P)$ with multiplier $(\mu^+/\mu)^{deg(w)/2} = a^{-m}$.

PROOF. To check assertion 1 let $\tilde{P}_0 \in \tilde{X}(k)$ be the reduction mod p of $\tilde{P}$. The isogeny $\tilde{u}$ pulled back to $\mathcal{O}_{\tilde{X},\tilde{P}_0}^{for}$ lifts u_0; so by [**80**], p.149 we must have

$$\mu \cdot log(1 + \Gamma(T)) = \mu^+ \cdot log(1+T);$$

and assertion 1 follows.

We check assertion 2 in the false elliptic case. (The genuine elliptic case is similar.) By Equation 8.12 the basis $\check{b}$ of $T_0(\check{A}_0)$ corresponding to b via θ_0 can be identified with a morphism $\check{b} : A^{for} \to (\mathbf{G}_m^{for})^2$ between the corresponding formal groups over $R[[T]]$; moreover the pair of $1-$forms

$$(\omega^1, \omega^2) := \check{b}^* \frac{dz}{z}$$

on A turns out to be a false $1-$form on $A \otimes_R R[[T]]$; cf. Lemma 8.2. Here dz/z is the invariant form on

$$\mathbf{G}_m = \mathbf{G}_{m/R[[T]]} = Spec\ R[[T]][z, z^{-1}].$$

Let $x_{for} = \omega^1$. We claim that

$$\tilde{u}^* \sigma_2^* x_{for} = \mu^+ \sigma_1^* x_{for}. \tag{8.67}$$

Indeed we need to show that

$$(\tilde{u}^{for})^*(\sigma_2^{for})^* = \mu^+(\sigma_1^{for})^* \tag{8.68}$$

as morphisms of formal groups $(\sigma_1^* A)^{for} \to (\mathbf{G}^2_{m/R[[T]]})^{for}$ over $R[[T]]$. Now $(\sigma_1^* A)^{for}$ itself is isomorphic to $(\mathbf{G}^2_{m/R[[T]]})^{for}$. On the other hand the morphism

$$End((\mathbf{G}^2_{m/R[[T]]})^{for}) \to End((\mathbf{G}^2_{m/k})^{for})$$

is an isomorphism (both rings being naturally isomorphic to $Mat_2(\mathbf{Z}_p)$). So in order to check the equality 8.68 holds it is enough to check it holds after tensorization with k; but this is clear by Equation 8.65. (Cf. [**23**], Section 3.1 for more details.) Equation 8.67 proves that $\mathbf{M}$ is formally flat with multiplier as in the statement of the Lemma and we are done. □

It is convenient to introduce the following terminology.

DEFINITION 8.19. A Hecke fixed point $(\tilde{P}, P, b)$ for a Hecke correspondence $(\mathbf{X}, A, \tilde{u}, s)$ will be called *regular* if μ^+/μ is not a root of unity, where μ^+, μ are defined by Equation 8.65. A Hecke $n-$cycle 8.64 for a Hecke correspondence $(\mathbf{X}, A, \tilde{u}, s)$ will be called *regular* if the induced Hecke fixed point of $(\mathbf{X}^{\circ n}, A, \tilde{u}^{\circ n}, s)$ is regular.

EXAMPLE 8.20. In what follows we would like to give examples of regular Hecke $n-$cycles, "at various p's", on the correspondences on modular curves in Equation 2.89. (The next example will address the case of Shimura curves.) Any such regular Hecke $n-$cycle will produce a regular Hecke fixed point on an irreducible correspondence as explained above.

Consider the affine modular curve $Y_1(N)/\mathbf{Z}[1/N]$, $N \geq 4$; in the discussion below we fix N. On the other hand consider a "variable" pair (p, l) of distinct primes not dividing N. For each such pair consider the correspondence

$$\mathbf{X}_{\mathcal{O}} := (X := Y_1(N), \tilde{X} := Y_1(N, l), \sigma_1, \sigma_2) \tag{8.69}$$

over $\mathcal{O} := \mathbf{Z}[1/Nl]$, defined in Equation 2.89, and the correspondence

$$\mathbf{X}_p := \mathbf{X}_p^l := (X_p, \tilde{X}_p, \sigma_1, \sigma_2) \tag{8.70}$$

over R_p obtained from $\mathbf{X}_{\mathcal{O}}$ by pull-back via $\mathcal{O} \to R_p := \hat{\mathbf{Z}}_p^{ur}$. Similarly we can consider the correspondence over $\mathbf{C}$,

$$\mathbf{X}_{\mathbf{C}} = (X_{\mathbf{C}}, \tilde{X}_{\mathbf{C}}, \sigma_1, \sigma_2); \tag{8.71}$$

this correspondence admits, of course, a $\mathbf{Q}-$arithmetic uniformization. If (A, α) is the universal pair over X and if $\tilde{u} : \tilde{A}_1 \to \tilde{A}_2$ is the universal isogeny over $\tilde{X}$ then $\tilde{A}_1 \simeq \sigma_1^* A$ and $\tilde{A}_2 \simeq \sigma_2^* A$ canonically so we have an induced isogeny

$$\tilde{u} : \sigma_1^* A \to \sigma_2^* A$$

of degree l; this, together with the natural splitting s of the de Rham space of elliptic curves [**81**], Appendix, gives a Hecke correspondence

$$(\mathbf{X}_p, A, \tilde{u}, s). \tag{8.72}$$

Let us show how to construct Hecke $n-$cycles for such Hecke correspondences.

As above, let (p, l) be a pair of distinct primes not dividing N. By a *periodic pair* for (p, l) we mean a pair (A_0, u_0) consisting of an ordinary elliptic curve A_0 over $k = R/pR$, where $R = R_p$, and a $k-$endomorphism u_0 of A_0 of degree l. To any such pair one can attach a Hecke n-cycle on the correspondence in Equation 8.72 as follows. Let A_1/R be the canonical lift of A_0; cf. [**80**]. By loc. cit. u_0 lifts to an endomorphism $u_1 : A_1 \to A_1$ of degree l. Let $\breve{u}_1 : A_1 \to A_1$ be the dual of u_1 and consider any R-point of $Y_1(N)$ of the form (A_1, α). Since the sequence $\breve{u}_1^j \circ \alpha$ is finite one can find $j < j'$ such that $\breve{u}_1^j \circ \alpha = \breve{u}_1^{j'} \circ \alpha$. Set $\alpha_1 := \breve{u}_1^j \circ \alpha$, $n := j' - j$; so $\breve{u}_1^n \circ \alpha_1 = \alpha_1$. Define

$$\alpha_n := \breve{u}_1 \circ \alpha_1, \alpha_{n-1} := \breve{u}_1 \circ \alpha_n, ..., \alpha_2 := \breve{u}_1 \circ \alpha_3.$$

Then $\breve{u}_1 \circ \alpha_2 = \breve{u}_1^n \circ \alpha_1 = \alpha_1$. Hence we have

$$u_1 \circ \alpha_1 = [l] \circ \alpha_2, u_1 \circ \alpha_2 = [l] \circ \alpha_3, ..., u_1 \circ \alpha_n = [l] \circ \alpha_1.$$

Let b be a basis of the physical Tate module of A_0. If $\tilde{P}_i$, $i = 1, ..., n$, are the R-points of $Y_1(N, l)$ represented by $u_1 : (A_1, \alpha_i) \to (A_1, [l] \circ \alpha_{i+1})$ (where $\alpha_{n+1} := \alpha_1$), and if P is the R-point of $Y_1(N)$ represented by $(A_1, [l] \circ \alpha_1)$ then

$$(\tilde{P}_1, ..., \tilde{P}_n, P, b) \tag{8.73}$$

is a Hecke n-cycle for the correspondence in Equation 8.72.

Note that if in the above construction one can arrange that $\breve{u}_0$ (and hence $\breve{u}_1$) fixes some α then we can take $n = 1$. This is always the case if, say, N is square free and divides $1 - Tr(u_0) + l$ but the latter is not divisible by p; indeed in this case $ker(u_0 - 1)$ is étale of cardinality divisible by N so it contains a point of exact order N. In proving regularity of Hecke cycles it is convenient to use the following elementary:

LEMMA 8.21. *If $l \geq 5$ is a prime integer and α is an integer in an imaginary quadratic extension of $\mathbf{Q}$ with trace $Tr(\alpha) \neq 0$ and norm $N(\alpha) = l$ then $\alpha/\bar{\alpha}$ is not a root of unity.*

PROOF. Set $a := Tr(\alpha)$ and assume $\alpha/\bar{\alpha} = \zeta_M$ where ζ_M is a primitive root of unity of order M. Since $\alpha\bar{\alpha} = l$ we get $\alpha^2 = \zeta_M l$. Since $a \neq 0$, α^2 has degree 2 over $\mathbf{Q}$. Hence $\phi(M) = 2$, so $M \in \{3, 4, 6\}$. Note that

$$a^2 = |a|^2 = |\alpha + \bar{\alpha}|^2 = l|1 + \zeta_M|^2.$$

If $M = 3$, $|1 + \zeta_M|^2 = 1$ so $a^2 = l$, a contradiction. If $M = 4$, $|1 + \zeta_M|^2 = 2$ so $a^2 = 2l$ which implies $l = 2$, a contradiction. If $M = 6$, $|1 + \zeta_M|^2 = 3$ so $a^2 = 3l$ which implies $l = 3$, a contradiction. □

PROPOSITION 8.22. *For any p not dividing N there exist infinitely many primes l such that the Hecke correspondence $\mathbf{X}_p^l$ in Equation 8.70 has a regular Hecke cycle.*

PROOF. We have the following general result, whose proof is a trivial adaptation of the argument in [**41**], p. 182. Let $\mathcal{O}$ be an order in an imaginary quadratic field $\mathcal{K}$, let $\mathcal{H}$ be the ring class field of $\mathcal{O}$, and let $l \in \mathbf{Z}$ be a prime that splits completely in $\mathcal{H}$ and does not divide the conductor of $\mathcal{O}$; then $l = \alpha\bar{\alpha}$ for some $\alpha \in \mathcal{O}$. Now fix an ordinary elliptic curve $A_0/\mathbf{F}_p^a$. By the general result above applied to $\mathcal{O} = End(A_0)$ we get that there are infinitely many primes l such that A_0 has an endomorphism $u_0 = \alpha \in \mathcal{O}$ of degree l. For each such l, (A_0, u_0) is periodic for (p, l) so we get a cycle as in Equation 8.73 on the correspondence in Equation 8.72. If $Tr(\alpha) = 0$ then $\alpha^2 = -l$ so $\mathbf{Q}(\sqrt{-l}) \subset \mathcal{K}$. There are only finitely many such l's. So for infinitely many l's we have, in addition, that $Tr(\alpha) \neq 0$. Since the Rosati involution on $End(A_0)$ is the complex conjugation it follows from

Lemma 8.21, that the quotient $u_0^t/u_0 = \mu^+/\mu$ is not a root of unity. So the cycle we constructed is regular. □

The above Proposition can be viewed as a "vertical" existence result for regular Hecke cycles on modular curves. Here is a "horizontal" existence result:

PROPOSITION 8.23. *For any l not dividing N there exist infinitely many p such that the Hecke correspondence $\mathbf{X}_p^l$ in Equation 8.70 possesses a regular Hecke cycle.*

PROOF. Assume first $l \geq 5$. Let A be a complex elliptic curve with complex multiplication by the ring of integers of $\mathbf{Q}(\sqrt{1-l})$. View A as defined over $\mathbf{Q}(j(A))$ and let A_0 be the reduction of A at a prime of $\mathbf{Q}(j(A))$ of good ordinary reduction that lies above a prime number p which is unramified in $\mathbf{Q}(j(A))$; then (A_0, u_0) is periodic for (p, l) where $u_0 = 1+\sqrt{1-l}$. There are infinitely many such p; cf. [**125**], pp. 122 and 184. Let p be such a prime. Then one can construct a Hecke $n-$cycle as in Equation 8.73 on the correspondence in Equation 8.72. Since $Tr(u_0) = 2 \neq 0$, it follows from Lemma 8.21, that the quotient $u_0^t/u_0 = \mu^+/\mu$ is not a root of unity. So the $n-$cycle we constructed is regular.

If $l = 2$ or $l = 3$ then, instead of the field $\mathbf{Q}(\sqrt{1-l})$, use $\mathbf{Q}(\sqrt{-7})$ and $u_0 = \frac{1+\sqrt{-7}}{2}$ or $\mathbf{Q}(\sqrt{-11})$ and $u_0 = \frac{1+\sqrt{-11}}{2}$ respectively. □

EXAMPLE 8.24. In what follows we would like to give examples of regular Hecke cycles for Hecke correspondences on Shimura curves. Any such cycle will produce a regular Hecke cycle on a suitable Hecke correspondence. We start by reviewing some of the constructions of models for Shimura curves due to Buzzard [**30**]; for details we refer to [**30**].

Fix a non-split indefinite quaternion algebra D over $\mathbf{Q}$, with discriminant d, and fix a maximal order $\mathcal{O}_D$. Let $\mathbf{Z}_f$ denote the projective limit of all $\mathbf{Z}/M\mathbf{Z}$, with M running through the set of all positive integers ordered by divisibilty and set $\mathcal{O}_{D,f} := \mathcal{O}_D \otimes \mathbf{Z}_f$. Also fix a compatible system of isomorphisms

$$\mathcal{O}_D \otimes \mathbf{Z}/M\mathbf{M} \to Mat_2(\mathbf{Z}/M\mathbf{Z})$$

where M runs through the positive integers coprime to d. Fix an open compact subgroup $U \subset \mathcal{O}_{D,f}^\times$ satisfying the following properties:

(i) $det(U) = \mathbf{Z}_f$,

(ii) U is "maximal at all primes dividing d"

(iii) there is a positive integer M coprime to d such that image of U in the group $GL_2(\mathbf{Z}/M\mathbf{Z})$ consists of matrices of the form $\begin{pmatrix} * & * \\ 0 & 1 \end{pmatrix}$; denote by M_U the smallest such M.

For any false elliptic curve (A, i) over a Noetherian ring S one defines in [**30**] a set $F_{A/S,U}$ whose elements are called *level U structures*. The formation of this set is functorial in S. For any isogeny of false elliptic curves $u : A \to B$, of false degree coprime to $M_U d$, there is an induced isomorphism

$$u_* : F_{A/S,U} \to F_{B/S,U}.$$

In particular $Aut(A, i)$ acts on $F_{A/S,U}$. There is a smooth projective scheme $X^D(U)$ over $\mathbf{Z}[1/M_U d]$ with geometrically irreducible fibers of dimension one whose $S-$points identify with the isomorphism classes of triples (A, i, α) over S, where $\alpha \in F_{A/S,U}$; cf. [**30**], Section 2. This is one of the prototypical Shimura curves introduced in [**30**]. Here is another one. It was proved in [**30**], Section 4, that, for any prime l not dividing $M_U d$, there is a scheme $X^D(U, \Gamma_0(l))$ over $\mathbf{Z}[1/M_U d]$ whose S-points identify with the isomorphism classes of tuples

$$(u : (A, i) \to (B, j), \alpha),$$

where $u : (A, i) \to (B, j)$ is an isogeny of false elliptic curves, of false degree l, and $\alpha \in F_{A/S,U}$. Moreover the morphisms

$$c_1 : X^D(U, \Gamma_0(l)) \to X^D(U),\ c_1(u : (A, i) \to (B, j), \alpha) = (A, i, \alpha),$$
$$c_2 : X^D(U, \Gamma_0(l)) \to X^D(U),\ c_2(u : (A, i) \to (B, j), \alpha) = (B, j, u_* \alpha),$$

are flat, finite, and become étale after base change to $\mathbf{Z}[1/M_U dl]$. All statements except the one about the étale property are proved in [**30**]. To check the étale property note that c_1 is generically étale and all geometric fibers in characteristic $\neq l$ have cardinality $l + 1$ (cf. the proof of Theorem 4.7 in [**30**]) which is, of course, enough to conclude our claim for c_1. A similar argument works for c_2. So we have at our disposal an étale correspondence

$$\mathbf{Y}_{\mathbf{Z}[1/M_U dl]} := (X^D(U), X^D(U, \Gamma_0(l)), c_1, c_2) \otimes \mathbf{Z}[1/M_U dl]. \tag{8.74}$$

For any prime p not dividing $M_U dl$ we can consider the induced correspondence

$$\mathbf{Y}_p := \mathbf{Y}_p^l := \mathbf{Y}_{\mathbf{Z}[1/M_U dl]} \otimes R_p, \tag{8.75}$$

$R_p := \hat{\mathbf{Z}}_p^{ur}$. Take any affine curve $X_p \subset X^D(U) \otimes R_p$ and consider the correspondence

$$\mathbf{X}_p = (X_p, \tilde{X}_p := c_1^{-1}(X_p) \cap c_2^{-1}(X_p), c_1, c_2) \tag{8.76}$$

which admits, of course, a structure of Hecke correspondence. Let us show how to construct regular Hecke cycles on such $\mathbf{X}_p$'s. There is an obvious notion of regular Hecke cycle on $\mathbf{Y}_p$. Any such cycle yields a regular Hecke cycle on an $\mathbf{X}_p$: one just has to let X_p be an affine open set containing our cycle.

We have the following "vertical" existence result for regular Hecke cycles in the Shimura curve context:

PROPOSITION 8.25. *There exists an integer $p_0 > M_U d$ such that for any prime $p \geq p_0$ there exist infinitely many primes l with the property that the correspondence $\mathbf{Y}_p^l$ in Equation 8.75 has a regular Hecke cycle.*

PROOF. According to Lemma 2.6 in [**23**] there exists $p_0 > M_U d$ such that if $p \geq p_0$ one can find a polarized false elliptic curve (A_0, i_0, θ_0) over $\mathbf{F}_p^a$ such that A_0 is ordinary and (A_0, θ_0) is isomorphic to a Jacobian plus its theta divisor. By Lemma 8.2 one can find a normalized basis b of the physical Tate module $T_p(A_0)$.

Claim. The center of $End^0(A_0) := End(A_0) \otimes \mathbf{Q}$ is an imaginary quadratic field $\mathbf{Q}(\pi)$.

To check the claim recall that, by results of Weil, Tate, and Honda, for any Abelian variety B over $\mathbf{F}_{p^\nu}$ which is simple over $\mathbf{F}_{p^\nu}$ (i.e. it does not contain proper non-zero Abelian subvarieties defined over $\mathbf{F}_{p^\nu}$) the ring $End(B)$ is a free $\mathbf{Z}$-module of finite rank and $End^0(B)$ is a division algebra with center $\mathbf{Q}(\pi)$, where π is an algebraic integer of absolute value $p^{\nu/2}$ under any embedding $\mathbf{Q}(\pi) \to \mathbf{C}$; moreover π corresponds to the p^ν power Frobenius endomorphism of B and

$$2 \cdot dim(B) = [\mathbf{Q}(\pi) : \mathbf{Q}][End^0(B) : \mathbf{Q}(\pi)]^{1/2}; \tag{8.77}$$

cf. Theorem 5.1 in [**108**].

In our case there are 2 possibilities for A_0:

1) A_0 is simple or

2) A_0 is isogenous to a product of two elliptic curves.

Assume first A_0 is simple. There is an integer ν such that (A_0, i_0, θ_0) and $End(A_0)$ are defined over $\mathbf{F}_{p^\nu}$. Since A_0 has multiplication by $\mathcal{O}_D$, which is non-commutative, $End^0(A_0) \neq \mathbf{Q}(\pi)$. It follows that $\mathbf{Q}(\pi)$ has degree ≤ 2 over $\mathbf{Q}$. Note that, since A_0 is ordinary, the images of π and its Rosati conjugate π^t via the map

$$End(A_0) \to End(T_p(A_0))$$

belong to $\mathbf{Z}_p^\times$ and $p^\nu \mathbf{Z}_p$ respectively; cf. Lemma 3.2 in [**23**] for details. In particular

$$\pi^t \neq \pi, -\pi.$$

Now, clearly, the Rosati involution on $End^0(A_0)$ induces an automorphism of the center $\mathbf{Q}(\pi)$. Since $\pi^t \neq \pi$ this automorphism is not the identity. In particular $\mathbf{Q}(\pi)$ is a quadratic extension of $\mathbf{Q}$. We claim it is an imaginary quadratic extension which will prove the Claim in this case. Indeed, if π were real, say $\pi = a + b\sqrt{c}$, $c > 0$, then, from the equality

$$|\pi| = p^{\nu/2} = |\pi^t|$$

we get either $a + b\sqrt{c} = a - b\sqrt{c}$, hence $b = 0$, hence $\pi \in \mathbf{Q}$, a contradiction, or $a + b\sqrt{c} = -a + b\sqrt{c}$, hence $a = 0$, hence $\pi^t = -\pi$, a contradiction.

Assume now A_0 is isogenous to a product of two elliptic curves $E_0 \times E_0'$. Since A_0 is ordinary, both E_0 and E_0' are ordinary. If E_0 and E_0' are not isogenous then $End^0(A) = End^0(E_0) \times End^0(E_0')$ is commutative, a contradiction. So E_0 and E_0' are isogenous hence $End^0(A_0) = Mat_2(\mathcal{K})$, where $\mathcal{K} := End^0(E_0) = \mathbf{Q}(\pi)$ is a quadratic imaginary field, and the Claim is proved in this case too.

Now consider the order $\mathcal{O} := \mathbf{Z}[\pi]$ in $\mathcal{K} = \mathbf{Q}(\pi)$. Exactly as in the proof of Proposition 8.22 we can find infinitely many primes l such that (A_0, i_0) has an endomorphism $u_0 \in \mathcal{O}$ of false degree l with $\bar{u_0}/u_0$ not a root of unity. From this point we can construct a regular Hecke cycle on $\mathbf{Y}_p$ exactly as in the elliptic case: take a canonical lifting A_1 of A_0, lift u_0 to $u_1 : A_1 \to A_1$, find a power u_1^n that fixes a level U structure, etc. We leave the (obvious) details to the reader. □

On the other hand we have the following "horizontal" existence result for regular Hecke cycles on Shimura curves.

PROPOSITION 8.26. *Fix $U \subset \mathcal{O}_{D,f}^\times$. There exists an infinite set of primes $\mathcal{L}$ such that for any $l \in \mathcal{L}$ there are infinitely many primes p with the property that the Hecke correspondence $\mathbf{Y}_p^l$ in Equation 8.75 has a regular Hecke cycle.*

PROOF. *Claim 1.* All except finitely many $\mathbf{C}$−points of $X^D(U)$ correspond to false elliptic curves which, as principally polarized Abelian varieties are isomorphic to Jacobians (of curves of genus 2).

Indeed start by considering an étale open set X' of $X^D(U)$ and a classifying map $\varphi : X'_{\mathbf{C}} \to \mathcal{A}_2$ into the moduli space of complex principally polarized Abelian varieties of dimension 2 (defined, for instance, by looking at the polarized false elliptic curve structure on the geometric generic fiber of the universal false elliptic curve over $X^D(U)$). Any $\mathbf{C}$−point of $\mathcal{A}_2$ corresponds to either a Jacobian (plus its theta polarization) or to a product of elliptic curves (with its product polarization). The locus of products of elliptic curves is a divisor $\mathcal{D}$ on $\mathcal{A}_2$. We cannot have $\varphi(X'_{\mathbf{C}}) \subset \mathcal{D}$ (because an inclusion $\mathcal{O}_D \subset End(E_1 \times E_2)$, E_i elliptic curves over $\mathbf{C}$, forces E_1 and E_2 to be isogenous and with CM so we would get that $\varphi(X'_{\mathbf{C}})$ is at most countable hence a point, a contradiction). We conclude that all except finitely many $\mathbf{C}$-points of $X'_{\mathbf{C}}$ correspond to (polarized) Jacobians, so the same holds for $X^D(U)$ which proves our Claim.

Claim 2. There are infinitely many $\mathbf{C}$-points of $X^D(U)$ corresponding to CM false elliptic curves; here an Abelian surface A over $\mathbf{C}$ is called CM if A is isogenous to a product $E \times E$ where E is a CM elliptic curve. Indeed start by considering a sequence K_n of quadratic imaginary extensions of $\mathbf{Q}$, with discriminants going to ∞, such that K_n embeds into D. (That one such K exists follows by taking $K = \mathbf{Q}(\alpha)$, $\alpha \in D \backslash \mathbf{Q}$, $Tr(\alpha)^2 - 4N(\alpha) < 0$; cf. [**95**], p.81. Now the condition that a quadratic field $\mathbf{Q}(\sqrt{-m})$ embeds into D is that $\mathbf{Q}(\sqrt{-m}) \otimes_{\mathbf{Q}} \mathbf{Q}_p$ is a field for all primes p dividing the discriminant d of D; cf. [**95**], p. 234. If $\mathbf{Q}(\alpha) = \mathbf{Q}(\sqrt{-m_1})$ the latter condition is expressible in terms of values of the Legendre symbols $\left(\frac{m_1}{p}\right)$ for $p|d$. These conditions will still be satisfied for infinitely many values $m_2, m_3, \ldots$) For each K_n we can find a CM false elliptic curve A_n with $End^0(A_n) = Mat_2(K_n)$. Indeed, since $K_n \subset D$, K_n splits D so we have an embedding

$$D \subset D \otimes_{\mathbf{Q}} K_n = Mat_2(K_n) = End^0(E_n \times E_n) \subset End(\mathbf{C}^2) \tag{8.78}$$

where $E_n = \mathbf{C}/\Lambda_n$ is an elliptic curve with $End^0(E_n) = K_n$, and the last embedding in Equation 8.78 identifies endomorphisms of $E_n \times E_n$ with linear maps that send $\Lambda_n \times \Lambda_n$ into itself. There is an induced embedding

$$\mathcal{O}' := \mathbf{Z} + N\mathcal{O}_D \to End(E_n \times E_n) \subset End(\mathbf{C}^2)$$

for some integer N. If $0 \neq \lambda_n \in \Lambda_n \times \Lambda_n$ then the map

$$D \to D \cdot \lambda_n \subset \mathbf{C}^2, \quad \alpha \mapsto \alpha\lambda_n$$

is injective. (Indeed the kernel of this $\mathbf{Q}$-linear map is a proper 2-sided ideal of D because if $\alpha\lambda_n = 0$, $\alpha, \beta \in D \backslash \{0\}$ then clearly $\beta\alpha\lambda_n = 0$ but also $\alpha\beta\lambda_n = (\alpha\beta\alpha^{-1})\alpha\lambda_n = 0$. On the other hand D has no such ideals.) So we have an isomorphism $\mathcal{O}' \to \mathcal{O}' \cdot \lambda_n$. In particular $\mathcal{O}' \cdot \lambda_n$ has rank 4 over $\mathbf{Z}$ and is contained in $\Lambda_n \times \Lambda_n$ so it has finite index in it. On the other hand $\mathcal{O}' \cdot \lambda_n$ has finite index in $\mathcal{O}_D \cdot \lambda_n$. Our Claim follows by taking $A_n = \mathbf{C}^2/\mathcal{O}_D \cdot \lambda_n$ which is clearly isogenous to $E_n \times E_n$, has multiplication by $\mathcal{O}_D$, and $End^0(A_n) = Mat_2(K_n)$.

From Claims 1 and 2 it follows that there exists a polarized false elliptic curve (A, i) which, as a polarized Abelian variety is a Jacobian, plus an isogeny $A \to E \times E$ where E is a CM elliptic curve. Let $\mathcal{K} = End^0(E)$. There is an order $\mathcal{O}$ in $\mathcal{K}$ which is contained in $End(A)$ and centralizes $\mathcal{O}_D$. By the argument in the proof of Proposition 8.22 there exist infinitely many primes l such that $l = \alpha_l\bar{\alpha}_l$, for some $\alpha_l \in \mathcal{O}$. So there exists an infinite set of primes $\mathcal{L}$ such that if $l \in \mathcal{L}$, (A, i) has an endomorphism α_l of false degree l. If $Tr(\alpha_l) = 0$ then $\sqrt{-l} \in \mathcal{K}$ which only can hold for finitely many l's. So we may and will assume that $Tr(\alpha_l) \neq 0$ for $l \in \mathcal{L}$. Now fix a model (over the ring of integers of some number field with some integer inverted) for the isogeny $A \to E \times E$ and note that for p sufficiently big the reduction modulo primes lying above p of A is ordinary if and only if the same holds for E. We get that for any $l \in \mathcal{L}$ there are infinitely many p's with the property that there exists a polarized false elliptic curve (A_0, i_0, θ_0) over $\mathbf{F}_p^a$ such that (A_0, θ_0) is a Jacobian and (A_0, i_0) has an endomorphism $\alpha_l \in \mathcal{O}$ of false degree l, with $Tr(\alpha_l) \neq 0$. From the latter we already know how to construct a Hecke cycle on $\mathbf{Y}_p$. $\square$

Let us address the problem of finding sequences of disjoint cycles, with $\mathbf{Z}_p-$ accumulating multipliers, on Hecke correspondences. We expect that, for any p, the correspondences $\mathbf{X}_p$ in Equation 8.70 and 8.76 have a sequence of pairwise disjoint cycles with $\mathbf{Z}_p-$ accumulating multipliers. However we are only able to prove a weaker statement which we now explain in the case of modular curves (more precisely for the correspondences $\mathbf{X}_{\mathbf{C}}$ in Equation 8.71). Let $\mathcal{O}_L$be the ring of integers in the union L of all number fields in $\mathbf{C}$ that are unramified above p and let $\mathcal{O}_{L,\wp}$ be the localization of $\mathcal{O}_L$ at a prime $\wp$ of $\mathcal{O}_L$ lying above p. Furthermore fix an embedding of $\mathcal{O}_{L,\wp}$ into $R_p = \hat{\mathbf{Z}}_p^{ur}$. We will say that a complex number is in $\mathbf{Z}_p$ if it is in $\mathcal{O}_{L,\wp}$ and its image in R_p lies in $\mathbf{Z}_p$. Let $\mathbf{X}_{\mathbf{C}}$ be an étale correspondence over $\mathbf{C}$, smooth over $\mathbf{C}$. A sequence of pairwise disjoint cycles (C_i) on $\mathbf{X}_{\mathbf{C}}$ is said to have $\mathbf{Z}_p-$ accumulating multipliers if all multipliers $a(C_i) \in \mathbf{C}$ belong to $\mathbf{Z}_p$ and satisfy $1 \neq a(C_i) \to 1$, $p-$adically, in $\mathbf{Z}_p$.

PROPOSITION 8.27. *For each p and l the correspondence $\mathbf{X}_{\mathbf{C}}$ in Equation 8.71 has a sequence of pairwise disjoint cycles with $\mathbf{Z}_p-$ accumulating multipliers.*

PROOF. We construct by induction a sequence of odd primes $\lambda_1, \lambda_2, \ldots$ and a sequence of positive integers $\nu_0, \nu_1, \nu_2, \ldots$ as follows. We let ν_0 be arbitrary and then, for each $i \geq 1$, we choose

$$\lambda_i > \max\{l^{\nu_{i-1}} - 1, p, l\}$$

and we choose ν_i satisfying the following conditions:
1) $v_{\lambda_i}(1 - l^{\nu_i})$ is odd,
2) $v_p(1 - l^{\nu_i})$ is even and $\geq i$,
3) $\left(\frac{K_i}{p}\right) = 1$ where $1 - l^{\nu_i} = p^{v_p(1-l^{\nu_i})} K_i$;

this is possible in view of Lemma 7.8. Consider the quadratic fields:

$$L_i := \mathbf{Q}(\sqrt{1 - l^{\nu_i}}) = \mathbf{Q}(\sqrt{K_i}).$$

Then the L_i's are pairwise distinct; for if $L_i = L_j$, $i < j$, then by condition 1 above λ_j divides the discriminant of L_j and hence λ_j divides the discriminant of L_i, so λ_j divides $l^{\nu_i} - 1$ which contradicts the inequality $\lambda_j > l^{\nu_i} - 1$. Also note that conditions 2 and 3 above imply that p splits completely in each L_i.

For each i we let A_i be a complex elliptic curve with complex multiplication by the ring of integers of L_i. Consider the endomorphism u_i of A_i defined by the complex number $z_i := 1 + \sqrt{1 - l^{\nu_i}}$. Then u_i has degree l^{ν_i}. Some power $u_i^{m_i}$ will fix a level structure; cf. the construction preceding Equation 8.73. Decomposing u_i into a product of ν_i isogenies of degree l we obtain an $m_i\nu_i-$ cycle C_i on $\mathbf{X}_{\mathbf{C}}$. The points of this cycle in $\tilde{X}_{\mathbf{C}}$ represent elliptic curves isogenous to A_i over $\mathbf{C}$ equipped with appropriate level structures. Since the L_i's are pairwise distinct it follows that the cycles C_i are pairwise disjoint. Finally we claim that all $a(C_i)$ belong to $\mathbf{Z}_p$ and $1 \neq a(C_i) \to 1$, $p-$adically. To check this note that

$$a(C_i) = (z_i/\bar{z}_i)^{\epsilon m_i}, \quad \epsilon \in \{1, -1\}. \tag{8.79}$$

This follows by applying Lemma 8.18 at some prime of good ordinary reduction for A_i that is unramified in a field of definition of both A_i and u_i. So it is enough to check that $z_i/\bar{z}_i$ belong to $\mathbf{Z}_p$ and converge to 1 in $\mathbf{Z}_p$. Now $z_i/\bar{z}_i = z_i^2/l^{\nu_i}$; by assertion 2 above $z_i \in \mathcal{O}_L$ where L is the union of all number fields unramified above p and $l \notin \wp$ where $\wp$ is any prime of $\mathcal{O}_L$ lying above p. So $z_i/\bar{z}_i \in \mathcal{O}_{L,\wp}$. By assertion 3 above we actually get $z_i/\bar{z}_i \in \mathbf{Z}_p$. Finally

$$v_p(z_i/\bar{z}_i - 1) = v_p(1 - l^{\nu_i})/2 \geq i/2,$$

and we are done. □

8.3. $\delta-$Serre-Tate expansion maps and $\delta-$Serre operators

The operators E, Δ, and ∂_r which we introduced in the general context of Chapter 3 have interesting incarnations in the specific case of Hecke correspondences.

Throughout this Section we fix the following data:

1) an affine irreducible Hecke correspondence $(\mathbf{X}, A, \tilde{u}, s)$, $\mathbf{X} = (X, \tilde{X}, \sigma_1, \sigma_2)$,

2) a regular Hecke fixed point $(\tilde{P}, P, b)$,

3) a $\delta-$line bundle $\mathbf{L}_{r,w}$, *cf. Equation 8.63,* $w = \sum_{i=0}^r a_i \phi^i$ *with* $deg(w)$ *even.*

As usual we denote by $P_0 \in X(k)$, $\tilde{P}_0 \in \tilde{X}(k)$ the reduction mod p of $P, \tilde{P}$ respectively. Let $X\hat{}$ and X^{for} denote, as usual, the $p-$adic completion of X and the completion of X at P_0 respectively. Similarly we let $\tilde{X}\hat{}$ and $\tilde{X}^{for}$ denote the $p-$adic completion of $\tilde{X}$ and the completion of $\tilde{X}$ at $\tilde{P}_0$ respectively. As a rule the subscript or superscript *for* for various objects over X will stand for *base change from X to X^{for}*. On the other hand the superscript *ord* for various objects over X will stand for *base change from X to X_{ord}*.

We still denote by A be the induced (genuine respectively false) elliptic curve over $S := \mathcal{O}(X\hat{})$ or $S_{for} = \mathcal{O}(X^{for})$. Similarly we denote by $\tilde{A}$ be the induced (genuine respectively false) elliptic curve over $\tilde{S} := \mathcal{O}(\tilde{X}\hat{})$ or $\tilde{S}_{for} = \mathcal{O}(\tilde{X}^{for})$ induced from A by pull back via σ_1. We also denote by A_0, $\tilde{A}_0$ the induced ordinary (genuine respectively false) elliptic curve over $k := R/pR$ corresponding to the points $P_0, \tilde{P}_0$; of course A_0 naturally identifies with $\tilde{A}_0$ via σ_1. Recall from Equation 8.14 that b gives a natural identification $S_{for} \simeq R[[T]]$; recall that this induces, via σ_1^*, an identification $\tilde{S}_{for} \simeq R[[T]]$.

We will later need to consider, for a Hecke correspondence $\mathbf{X} = (X, \tilde{X}, \sigma_1, \sigma_2)$ as above, the correspondence

$$\mathbf{X}_{ord} = (X_{ord}, \tilde{X}_{ord}, \sigma_1, \sigma_2)$$

where $\tilde{X}_{ord} := \sigma_1^{-1}(X_{ord}) \cap \sigma_1^{-1}(X_{ord})$.

8.3.1. $\delta-$Serre-Tate expansion maps E. In the study of modular curves a key role is played by Fourier expansions at cusps. Since Shimura curves don't have cusps one is lead to consider a different kind of expansions which can be called Serre-Tate expansion; cf. [**112**]. We will need here the "higher order analogue" of these expansions which turn out to be the incarnations, in our concrete situation, of the general expansion maps E at fixed points introduced earlier in a general setting. Consider, as in Equation 3.47, the "physical line bundle" over X with its zero section removed:

$$\mathbf{V}^*(L) = Spec \bigoplus_{n \in \mathbf{Z}} L^{\otimes n}.$$

For any $r \geq 0$ let us set $s = r - ord(w)$, and

$$S^r = S^r_X := \mathcal{O}^r(X), \quad M^r = M^r_X := \mathcal{O}^r(\mathbf{V}^*(L)), \tag{8.80}$$
$$M^r_X(w) := H^0(X, L^{w[s]}), \quad \bar{M}^r_X := M^r_X \otimes_R k, \quad \bar{M}^r_X(w) := M^r_X(w) \otimes_R k$$

$$S^r_{for} := R[[T]][T',...,T^{(r)}]\hat{}, \quad M^r_{X\,for} := S^r_{for}\hat{\otimes}_{\mathcal{O}^r(X)}M^r_X,$$

$$M^r_{X\,for}(w) := S^r_{for} \cdot x^w_{for}, \quad \bar{M}^r_{X\,for} := M^r_X \otimes_R k, \quad \bar{M}^r_{X\,for}(w) := M^r_X(w) \otimes_R k,$$

where $T', T'', ..., T^{(r)}$ are new variables, $\hat{}$ means, as usual, $p-$adic completion, and x_{for} is the (genuine respectively false) $1-$form introduced in the proof of Lemma 8.18. Note that if x is a local basis for L around P_0 then

$$M^r_{X\,for} = R[[T]][x, x^{-1}, x', ..., x^{(r)}, T', ..., T^{(r)}]\hat{}.$$

Since x gives rise to a basis of the module of (genuine respectively false) $1-$forms on the corresponding (genuine respectively false) elliptic curve over S_{for} we may write $x = x(T) \cdot x_{for}$ with $x(T) \in R[[T]]^{\times}$. Therefore:

$$M^r_{X\,for} = R[[T]][x_{for}, x^{-1}_{for}, x'_{for}, ..., x^{(r)}_{for}, T', ..., T^{(r)}]\hat{},$$

in particular $M^r_{X\,for}(w) \subset M^r_{X\,for}$.

Note that objects

$$M^r_X, \ M^r_X(w), \ \bar{M}^r_X, \ \bar{M}^r_X(w)$$

actually make sense (and will sometimes be used) for any (genuine or false) elliptic curve A/X over a smooth affine $R-$scheme X (not necessarily infinitesimally universal and not necessarily part of a Hecke correspondence). Moreover, by replacing X by X_{ord} we may introduce corresponding objects

$$M^r_{X_{ord}}, \ M^r_{X_{ord}}(w), \ \bar{M}^r_{X_{ord}}, \ \bar{M}^r_{X_{ord}}(w).$$

Since $\mathbf{V}^*(L_{ord}) := \mathbf{V}^*(L) \times_X X_{ord}$ is an open subscheme of $\mathbf{V}^*(L)$ we get, by Corollary 3.16 the following:

COROLLARY 8.28. *The conclusions of Corollary 3.16 are true for*

$$B^r = M^r_X, \quad C^r = M^r_{X_{ord}}.$$

In our Hecke correspondence case we define, in a similar way, objects

$$\tilde{S}^r, M^r_{\tilde{X}}, M^r_{\tilde{X}}(w), ...$$

analogous to the ones in list 8.80 by replacing X and L in the definitions above by $\tilde{X}$ and $\tilde{L} := \sigma_1^* L$; we also set $\tilde{x}_{for} := \sigma_1^* x_{for}$. There are natural ring homomorphisms

$$E : M^r_X \xrightarrow{\mathcal{E}} M^r_{X\,for} \to S^r_{for} \tag{8.81}$$

with $\mathcal{E}$ is given by $f \mapsto 1 \otimes f$ and $M^r_{X\,for} \to S^r_{for}$ given by

$$x_{for} \mapsto 1, \ x'_{for} \mapsto 0, \ x''_{for} \mapsto 0, ...$$

Both $\mathcal{E}$ and E will be referred to as *Serre-Tate expansion maps*; the map E is the incarnation, in our context here, of the expansion map defined, in the general context, in Equation 4.56. The maps in Equation 8.81 are compatible with weights so they induce maps

$$M^r_X(w) \xrightarrow{\mathcal{E}} M^r_{X\,for}(w) = S^r_{for} \cdot x^w_{for} \simeq S^r_{for}, \tag{8.82}$$

with $\mathcal{E}(f) = E(f) \cdot x^w$. Explicitly, for $f \in M^r_X(w)$ write locally $f = \beta \cdot x^w$ with $\beta \in \mathcal{O}^r(X)$. The natural map $\mathcal{O}^r(X) \to S^r_{for}$ sends β into an element $\beta(T, ..., T^{(r)}) \in S^r_{for}$. Then f gives rise to an element

$$\mathcal{E}(f) := \beta(T, ..., T^{(r)}) \cdot (x(T))^w x^w_{for} \in S^w_{for} \cdot x^w_{for}.$$

We have the following "$\delta-$Serre-Tate expansion principle" (analogue to the "Fourier expansion principle"):

PROPOSITION 8.29. *The maps*

$$\mathcal{E} : M_X^r(w) \to M_{X_{for}}^r(w)$$

are injective and so are their reductions mod p:

$$\bar{\mathcal{E}} : \bar{M}_X^r(w) \to \bar{M}_{X_{for}}^r(w).$$

PROOF. Since $(x(T))^w$ in the description of the map $\mathcal{E}$ is invertible one is reduced to Proposition 4.51. $\square$

Similarly, there are natural Serre-Tate expansion maps,

$$\tilde{\mathcal{E}} : M_{\tilde{X}}^r(w) \to M_{\tilde{X}_{for}}^r(w) = \tilde{S}_{for}^r \cdot \tilde{x}_{for}^w \tag{8.83}$$

and again the corresponding statement of Proposition 8.29 holds. (Here we use the fact that $\tilde{X} \otimes k$ is irreducible!) We also set $\tilde{\mathcal{E}}(f) = E(f) \cdot \tilde{x}_{for}^w$.

Next we consider the compatibility of the Serre-Tate expansion maps with the Δ operators that are used to define the spaces H^i. Recall that we have set $m = -deg(w)/2$; consider the map

$$\Delta_m = \Delta^{\mathbf{L}_{r,w}} : M_X^r(w) \to M_{\tilde{X}}^r(w), \quad \Delta_m(f) = \sigma_1^* f - e^m \cdot \tilde{u}^* \sigma_2^* f.$$

Then, of course,

$$Ker\ \Delta_m = H^0(\mathbf{X}, \mathbf{L}_{r,w}), \quad Coker\ \Delta_m = H^1(\mathbf{X}, \mathbf{L}_{r,w}).$$

Moreover if

$$\bar{\Delta}_m : \bar{M}_X^r(w) \to \bar{M}_{\tilde{X}}^r(w)$$

is the reduction mod p of the map Δ_m then

$$Ker\ \bar{\Delta}_m = H^0(\mathbf{X}, \mathbf{L}_{r,w}; k).$$

On the other hand, by Lemma 8.18, the germ of $\mathbf{X}$ identifies with

$$\Sigma(T) = \Sigma_a(T) := (1+T)^a - 1,$$

where $a = \mu^+/\mu$. If $\Lambda := \Lambda_a := \Lambda(r, a^{-1})$ then the germ of $\mathbf{L}_{r,w}$ identifies with the δ-line bundle

$$\Lambda^m = \Lambda(r, a^{-m}).$$

Define

$$\Delta_m : M_{X_{for}}^r(w) = R[[T]][T', ..., T^{(r)}]\hat{}\cdot x_{for}^w \to M_{\tilde{X}_{for}}^r(w) = R[[T]][T', ..., T^{(r)}]\hat{}\cdot \tilde{x}_{for}^w$$

by the formula

$$\begin{aligned} &\Delta_m(F(T, ..., T^{(r)}) \cdot x_{for}^w) \\ &= F(T, ..., T^{(r)}) \cdot \tilde{x}_{for}^w - e^m \cdot F(\Sigma(T), ..., \delta^r(\Sigma(T))) \cdot (\mu^+ \tilde{x}_{for})^w \\ &= [F(T, ..., T^{(r)}) - a^{-m} \cdot F(\Sigma(T), ..., \delta^r(\Sigma(T)))] \cdot \tilde{x}_{for}^w. \end{aligned}$$

So, under the identification of $M_{X_{for}}^r$ with S_{for}^r the operator Δ_m corresponds to the operator Δ^{Λ^m} introduced in Equation 4.10 and

$$Ker(\Delta_m : M_{X_{for}}^r(w) \to M_{\tilde{X}_{for}}^r(w)) \simeq H^0(\Sigma, \Lambda^m).$$

So, by Proposition 4.36 we have:

PROPOSITION 8.30. *The following diagram is commutative:*

$$\begin{array}{ccccc} E: M^r_X(w) & \stackrel{\mathcal{E}}{\to} & M^r_{X_{for}}(w) & \simeq & S^r_{for} \\ \downarrow \Delta_m & & \downarrow \Delta_m & & \downarrow \Delta^{\Lambda^m} \\ E: M^r_{\tilde{X}}(w) & \stackrel{\tilde{\mathcal{E}}}{\to} & M^r_{\tilde{X}_{for}}(w) & \simeq & \tilde{S}^r_{for} \end{array}$$

In particular, for $f \in M^r_X(w)$ *we have* $f \in H^0(\mathbf{X}, \mathbf{L}_{r,w})$ *if and only if* $E(f) \in H^0(\Sigma, \Lambda^m)$.

So, by Proposition 4.36 for any element $f \in H^0(\mathbf{X}, \mathbf{L}_{r,w})$ there exists a homogeneous polynomial $\Phi \in K[x_1, ..., x_r]$ of degree $m = -deg(w)/2$, such that

$$E(f) = \Phi(\Psi, ..., \phi^{r-1}\Psi),$$

where $\Psi = \Psi_{\mathbf{G}_m}$, cf. Equation 4.40. In particular we have natural maps

$$H^0(\mathbf{X}, \mathbf{L}_{r,w}) \to H^0(\Sigma, \Lambda^m). \tag{8.84}$$

The next Proposition computes the image of the Hasse invariant $\bar{H} \in \bar{M}^0_X(p-1)$ via the Serre-Tate expansion map $\bar{E} : \bar{M}^0_X(p-1) \to k[[T]]$; the result is similar to the well known result for the Fourier expansion of the usual Hasse invariant on modular curves.

PROPOSITION 8.31. $\bar{E}(\bar{H}) = 1$.

PROOF. This follows directly from Equation 8.46 because the projection $U \to H^{01}$ there is an isomorphism and the crystalline Frobenius $\phi : U \to U$ transported to H^{01} reduces modulo p to the $p-$power Frobenius map $F^* : \bar{H}^{01} \to \bar{H}^{01}$. □

8.3.2. $\delta-$Serre operators ∂_j. The classical Serre operator ∂ acting on modular forms [**88**] was introduced by Serre in "coordinates" given by the Eisenstein series a_4 and a_6 and then was interpreted in a coordinate free manner by Katz. Our next aim is to construct $\delta-$Serre operators ∂_j acting on $\delta-$sections of order r. The construction was pioneered by Barcau [**2**] who introduced these operators, in case $j = r$, for modular curves of level one using the "coordinates" a_4 and a_6. Our approach here is coordinate free and applies to the Shimura curves as well. The special case $j = r$ of this approach was used in [**24**].

Recall from Equations 8.20 and 8.29 that we set $H = H^1 = H^1_{DR}(A/S)$, $Q = H^{01}$, $L' = s(Q)$ in the genuine elliptic case and $L = e_{11}H^{10}$, $H := e_{11}H^1$, $Q := e_{11}H^{01}$, $L' := s(Q)$ in the false elliptic case. By Equations 8.30 and 8.31 the Gauss-Manin connection $\nabla : H^1 \to H^1 \otimes \Omega$ induces a connection $\nabla : H \to H \otimes \Omega$ hence a connection

$$\nabla : Symm(H) \to Symm(H) \otimes \Omega \tag{8.85}$$

on the symmetric algebra of H. Using the natural inclusion $L \subset H$, its right inverse given by the first projection $\pi : H = L \oplus L' \to L$, and the isomorphism $L^{\otimes 2} \simeq K_X = \Omega$ we get an $R-$derivation ∂, depending on the splitting map s:

$$\partial : \bigoplus_{n \geq 0} L^{\otimes n} = Symm(L) \to Symm(H) \stackrel{\nabla}{\to} Symm(H) \otimes \Omega \simeq Symm(H) \otimes L^{\otimes 2} \to$$

$$\to Symm(H) \otimes Symm(H) \stackrel{mult}{\to} Symm(H) \stackrel{\pi}{\to} Symm(L) = \bigoplus_{n \geq 0} L^{\otimes n}.$$

If we choose a normalized local basis $\omega^1, \eta^1, \omega^2, \eta^2$ (resp. ω^1, η^1) of H^1 (in the sense of Definitions 8.4 and 8.5) according as X is a Shimura curve or a modular curve then $\bigoplus_{n\geq 0}\hat{L}^{\otimes n}$ is locally isomorphic, over S, to a ring of polynomials $S[x]$, where the variable x corresponds to ω^1. If we write

$$\begin{pmatrix} \nabla\omega^1 \\ \nabla\eta^1 \end{pmatrix} = \begin{pmatrix} \sigma_{11} & \sigma_{12} \\ \sigma_{21} & \sigma_{22} \end{pmatrix} \begin{pmatrix} \omega^1 \\ \eta^1 \end{pmatrix},$$

with $\sigma_{ij}\in\Omega$, then the isomorphism $L^{\otimes 2}\simeq\Omega$ is given by $(\omega^1)^{\otimes 2}\mapsto\sigma_{12}$ and the derivation ∂ satisfies

$$\partial(x) = \frac{\sigma_{11}}{\sigma_{12}}x^3 \tag{8.86}$$

$$\partial(s) = \frac{ds}{\sigma_{12}}x^2, \quad s\in S. \tag{8.87}$$

(The above formulae make sense because σ_{12} is an invertible 1–form.) In particular

$$\partial(L^{\otimes n})\subset L^{\otimes(n+2)}. \tag{8.88}$$

If X is a modular curve then the R–derivation ∂ is, of course the Serre operator for modular curves; cf. the Appendix to [**81**]. Note that we have an induced derivation

$$\partial:\bigoplus_{n\in\mathbf{Z}}L^{\otimes n}\to\bigoplus_{n\in\mathbf{Z}}L^{\otimes n}$$

and the formulae 8.88 still hold. By Proposition 3.56 we have, for $r,j\geq 0$, naturally induced derivations $\partial_j: M^r_X\to M^r_X$ satisfying the properties listed in that Proposition; in particular

$$\partial_j(M^r_X(w))\subset M^r_X(w+2\phi^j).$$

All that was said so far in this section can be done for the Gauss-Manin connection

$$\nabla^{for}: H_{for}\to H_{for}\otimes\Omega^{for}$$

in place of ∇, where "*for*" stands for the situation over S_{for}. A similar picture holds, of course, for X, L replaced by $\tilde{X},\tilde{L}$.

The following follows directly from the functoriality of our constructions:

PROPOSITION 8.32. *We have commutative diagrams:*

$$\begin{array}{ccc} M^r_X(w) & \stackrel{\partial_j}{\to} & M^r_X(w+2\phi^j) \\ \mathcal{E}\downarrow & & \downarrow\mathcal{E} \\ M^r_{X^{for}}(w) & \stackrel{\partial_j}{\to} & M^r_{X^{for}}(w+2\phi^j) \end{array}$$

8.3.3. Ramanujan form P and Hecke form $P^{(e)}$. In what follows we need to recall the basic definition of Ramanujan's P function as a "p–adic modular form"; this is due to Serre and Katz [**81**] in the modular case and can be immediately adapted to the Shimura curve case as follows (cf. [**24**]).

DEFINITION 8.33. Consider a normalized local basis ω^1,η^1 (resp. $\omega^1,\eta^1,\omega^2,\eta^2$) of $H^1_{ord}=H^1_{DR}(A\otimes X_{ord}/X_{ord})$ according as A is a genuine or false elliptic curve respectively; cf. Definitions 8.4 and 8.5. Also choose a basis u^1 of $\mathcal{U}$ (the unit root space or its e_{11} projection; cf. Equation 8.53). Consider the local section

$$P:=\frac{\langle u^1,\eta^1\rangle_\theta}{\langle u^1,\omega^1\rangle_\theta}\cdot(\omega^1)^{\otimes 2}\in\hat{L}^{\otimes 2}_{ord}=M^0_{X_{ord}}(2).$$

These various local sections glue together to give a global section

$$P \in \hat{L}_{ord}^{\otimes 2} = M^0_{X_{ord}}(2),$$

called the *Ramanujan form.*

The above construction applies for *any* genuine or false elliptic curve over X, not necessarily infinitesimally universal.

Let ω, u be a canonical $R[[T]]-$ basis of H^1_{for} in the sense of Definition 8.10. Also complete ω to a normalized basis $\omega^1, \eta^1, \omega^2, \eta^2$ of H^1_{for} in the sense of Definition 8.4.

The following is well known in the genuine elliptic curve case [**81**]; cf. [**24**] in the false elliptic case.

PROPOSITION 8.34. *The following holds:*

$$\nabla \omega^1 = \frac{dT}{1+T} u^1 = -\epsilon E(P) \frac{dT}{1+T} \omega^1 + \epsilon \frac{dT}{1+T} \eta^1. \tag{8.89}$$

In particular, with $x = \omega^1$,

$$\begin{aligned} \partial(x) &= -E(P) x^3, \\ \partial(T) &= \epsilon^{-1}(1+T) x^2. \end{aligned} \tag{8.90}$$

PROOF. Write $u^1 = s_{21}\omega^1 + s_{22}\eta^1$, with $s_{21} \in R[[T]]$, $s_{22} = R[[T]]^{\times}$; cf. Equation 8.55. Note that

$$\epsilon = \langle u^1, \omega^1 \rangle = s_{22}.$$

Consequently

$$\langle \eta^1, u^1 \rangle = s_{21},$$

hence

$$s_{21} x^2 = -\epsilon \frac{\langle u^1, \eta^1 \rangle_\theta}{\langle u^1, \omega^1 \rangle_\theta} x^2 = -\epsilon E(P) x^2,$$

hence

$$s_{21} = -\epsilon E(P).$$

The assertions about ∂ follow from Equation 8.89 plus the *for* analogues of the formulae 8.86 and 8.87. □

REMARK 8.35. Equation 8.90 shows that the reduction mod p, $\bar{\partial} : \bar{M}^0_X \to \bar{M}^0_X$ of ∂ and the reduction mod p, $\bar{\mathcal{D}} : \bar{M}^0_X \to \bar{M}^0_X$ of the Euler operator $\mathcal{D} : M^0_X \to M^0_X$ (cf. Remark 3.58) are linearly independent. Consequently, by Remark 3.58, there is a Zariski open set of X above which

$$\partial_0, \mathcal{D}_0, ..., \partial_r, \mathcal{D}_r$$

form a basis of the tangent sheaf of $Spf\ M^r_X$.

The following construction goes back to Hecke [**47**], p. 47, in the modular case; cf. [**2**] for the modular description in the modular curve case and [**24**] for the Shimura curve case.

DEFINITION 8.36. Choose local *normalized* bases ω_i, η_i, $i = 1, 2$, for the H^1 spaces of $\sigma_i^* A$, $i = 1, 2$, respectively (cf. Definitions 8.4 and 8.5) and let $P^{(e)}$ be defined by

$$\langle \eta_1^1, \tilde{u}^* \eta_2^1 \rangle_\theta \cdot \omega_1^1 \cdot \tilde{u}^* \omega_2^1$$

where the 1 superscript means "the first component" of a pair. These local sections glue to give a global section

$$P^{(e)} \in M^0_{\tilde{X}}(2),$$

called the *Hecke form.*

PROPOSITION 8.37. *The Serre-Tate expansion of $P^{(e)}$ is given by:*

$$E(P^{(e)}) = e[a \cdot E(P) \circ \Sigma - E(P)]\tilde{x}^2_{for} = -e \cdot \Delta_{-1}(E(P)) \in M^0_{\tilde{X}_{for}}(2).$$

In particular we have an equality

$$P^{(e)} = -e \cdot \Delta_{-1}(P) \in M^0_{\tilde{X}_{ord}}(2).$$

PROOF. This follows as in [**81**], p. 177, (cf. also a Remark in [**2**]); for convenience we provide some details. Start with the normalized basis ω, η of the de Rham space of A over $R[[T]]$ as in Proposition 8.34 and recall from that Proposition that

$$\nabla\omega^1 = -\epsilon E(P)\frac{dT}{1+T}\omega^1 + \epsilon\frac{dT}{1+T}\eta^1.$$

Consider the forms $\omega_i^1 := \sigma_i^* x$, $x := \omega^1$, and the de Rham classes $\eta_i^1 := \sigma_i^*\eta^1$. Using 8.67 and the fact that ∇ commutes with σ_1^* and $u^*\sigma_2^*$, it follows that

$$\langle \eta_1^1, \tilde{u}^*\eta_2^1\rangle \cdot \omega_1^1 \cdot \tilde{u}^*\omega_2^1$$

equals the right hand side of the equality in the statement of our Proposition. □

8.3.4. Commutation relations between Δ and ∂_j.

THEOREM 8.38. *For any $w = \sum_{i=0}^r a_i\phi^i$ of even degree $deg(w) = -2m$ and any $j \geq 0$ we have the following equality*

$$\Delta_{m-1} \circ \partial_j - \partial_j \circ \Delta_m = a_j p^j e^{m-1}(P^{(e)})^{\phi^j} \circ \tilde{u}^*\sigma_2^* \tag{8.91}$$

of maps $M^r_X(w) \to M^r_{\tilde{X}}(w + 2\phi^j)$. Equivalently we have an equality

$$\Delta_{m-1} \circ \partial_j - (\partial_j - a_j p^j e^{-1}(P^{(e)})^{\phi^j}) \circ \Delta_m = a_j p^j e^{-1}(P^{(e)})^{\phi^j} \circ \sigma_1^* \tag{8.92}$$

of maps $M^r_X(w) \to M^r_{\tilde{X}}(w + 2\phi^j)$.

Here $\tilde{u}^*\sigma_2^*$ and σ_1^* are viewed as maps from $M^r_X(w)$ to $M^r_{\tilde{X}}(w)$. Also the element

$$a_j p^j e^{-1}(P^{(e)})^{\phi^j} \in M^r_{\tilde{X}}(2\phi^j)$$

is identified with the multiplication map

$$M^r_{\tilde{X}}(w) \to M^r_{\tilde{X}}(w + 2\phi^j),$$

$$f \mapsto a_j p^j e^{-1}(P^{(e)})^{\phi^j} \cdot f.$$

PROOF. By the injectivity of the Serre-Tate map

$$\tilde{\mathcal{E}} : M^r_{\tilde{X}}(w + 2\phi^j) \to M^r_{\tilde{X}_{for}}(w + 2\phi^j),$$

(cf. Proposition 8.29) together with the compatibility of $\mathcal{E}$, $\tilde{\mathcal{E}}$ with the maps ∂_j and Δ (cf. Propositions 8.32 and 8.30), it is enough to check the following equality

$$\Delta_{m-1} \circ \partial_j - \partial_j \circ \Delta_m = a_j p^j e^{m-1} \cdot (E(P^{(e)}))^{\phi^j} \cdot \tilde{u}^*\sigma_2^*,$$

of maps

$$M^r_{X_{for}}(w) \to M^r_{\tilde{X}_{for}}(w + 2\phi^j)$$

where $\tilde{u}^*\sigma_2^*$ is viewed as a map $M^r_{X_{for}}(w) \to M^r_{\tilde{X}_{for}}(w)$. Let $F \cdot x^w_{for} \in S^r_{for} \cdot x^w_{for}$. A straightforward but tedious computation (using the formulae 8.90, the proof of Lemma 8.18, formula 3.31, and Proposition 3.45) yields:

$$\Delta_{m-1}(\partial_j(F \cdot x^w_{for})) - \partial_j(\Delta_m(F \cdot x^w_{for}))$$

$$= a_j p^j e^m (\mu^+)^{deg(w)} \cdot (F \circ \Sigma) \cdot [a \cdot (E(P) \circ \Sigma) - E(P)]^{\phi^j} \cdot \tilde{x}^{w+2\phi^j}_{for}.$$

We may conclude by formula 8.67 and Proposition 8.37. □

Consider, in what follows, the class $[P^{(e)}] \in H^1(\mathbf{X}, \mathbf{L}_2)$ of $P^{(e)} \in M^0_{\tilde{X}}(2)$; we get classes $[P^{(e)}]^{\phi^j} \in H^1(\mathbf{X}, \mathbf{L}_{2\phi^j})$.

COROLLARY 8.39. *If $w = \sum_{i=0}^r a_i\phi^i$, $j \geq 0$, $a_j \in p^\nu R \backslash p^{\nu+1}R$, and $f \in H^0(\mathbf{X}, \mathbf{L}_w)$ then the class*

$$\xi := f \cup [P^{(e)}]^{\phi^j} \in H^1(\mathbf{X}, \mathbf{L}_{r,w+2\phi^j})$$

is $p^{j+\nu}$–torsion, i.e. $p^{j+\nu}\xi = 0$. Assume in addition that $j + \nu \geq 1$ and

$$H^0(\mathbf{X}, \mathbf{L}_{r,w+2\phi^j}) = 0, \quad \partial_j f \notin pM^r_X(w + 2\phi^j); \tag{8.93}$$

then $p^{j+\nu-1}\xi \neq 0$.

PROOF. By Theorem 8.38 we have

$$\Delta_{m-1}(\partial_j f) = a_j p^j e^{-1} (P^{(e)})^{\phi^j} \cdot \sigma_1^* f;$$

so $p^{j+\nu}\xi = 0$. Assume now that conditions 8.93 are satisfied. Assume

$$p^{j+\nu-1}\xi = 0$$

and seek a contradiction. So there exists $g \in M^r_X(w + 2\phi^j)$ such that

$$\Delta_{m-1}(g) = a_j p^{j-1} e^{-1} (P^{(e)})^{\phi^j} \cdot \sigma_1^* f.$$

So $\Delta_{m-1}(\partial_j f - pg) = 0$. So, by the first condition in 8.93 we get $\partial_j f - pg = 0$; but this contradicts the second condition in 8.93. □

COROLLARY 8.40. *If $a_j = 0$ (e.g. if $ord(w) < j$) then we have an equality*

$$\Delta_{m-1} \circ \partial_j = \partial_j \circ \Delta_m$$

of maps $M^r_X(w) \to M^r_{\tilde{X}}(w + 2\phi^j)$. In particular if $a_j = 0$ and $f \in H^0(\mathbf{X}, \mathbf{L}_{r,w})$ then $\partial_j f \in H^0(\mathbf{X}, \mathbf{L}_{r,w+2\phi^j})$.

It is convenient to consider, in what follows, the following derivation

$$\partial^* := \partial + P\mathcal{D} : \bigoplus_{m \in \mathbf{Z}} L^{\otimes m}_{X_{ord}} \to \bigoplus_{m \in \mathbf{Z}} L^{\otimes m}_{X_{ord}}, \tag{8.94}$$

where $\mathcal{D}$ is the Euler derivation; cf. Equation 3.53. If $w = \sum a_i\phi^i$, the conjugates $\partial_j^* : M^r_{X_{ord}} \to M^r_{X_{ord}}$ define the maps

$$\partial^*_{j,w} := \partial_j + a_j p^j P^{\phi^j} : M^r_{X_{ord}}(w) \to M^r_{X_{ord}}(w + 2\phi^j),$$

$$\tilde{\partial}^*_{j,w} := \partial_j + a_j p^j \sigma_1^* P^{\phi^j} : M^r_{\tilde{X}_{ord}}(w) \to M^r_{\tilde{X}_{ord}}(w + 2\phi^j),$$

and R–derivations

$$\partial_j^* : \bigoplus_{w \in W(r)} M^r_{X_{ord}}(w) \to \bigoplus_{w \in W(r)} M^r_{X_{ord}}(w), \tag{8.95}$$

$$\tilde{\partial}_j^* : \bigoplus_{w\in W(r)} M^r_{\tilde{X}_{ord}}(w) \to \bigoplus_{w\in W(r)} M^r_{\tilde{X}_{ord}}(w).$$

Putting together Theorem 8.38 and Proposition 8.37 we get, by a straightforward computation:

COROLLARY 8.41. *We have the following equalities*

$$\Delta_{m-1} \circ \partial_{j,w}^* = \tilde{\partial}_{j,w}^* \circ \Delta_m$$

of maps $M^r_{X_{ord}}(w) \to M^r_{\tilde{X}_{ord}}(w+2\phi^j)$. *In particular*

$$\Delta_{m-1} \circ \partial_j^* = \tilde{\partial}_j^* \circ \Delta_m$$

as endomorphisms of $\bigoplus_{w\in W(r)} M^r_{X_{ord}}(w)$ *and*

$$\partial_j^*(H^0(\mathbf{X}_{ord}, \mathbf{L}_{r,w})) \subset H^0(\mathbf{X}_{ord}, \mathbf{L}_{r,w+2\phi^j}).$$

Moreover the class

$$[P^{(e)}] \in H^1(\mathbf{X}_{ord}, \mathbf{L}_2)$$

vanishes.

8.3.5. Commutation relations between ∂_j^* and E. Consider the canonical derivation

$$\partial^{can} := (1+T)\frac{d}{dT} : R[[T]] \to R[[T]]$$

attached to (the endomorphisms of) the formal multiplicative group and let

$$\partial_j^{can} : R[[T]][T', ..., T^{(r)}]\hat{} \to R[[T]][T', ..., T^{(r)}]\hat{}$$

be the conjugate operators.

PROPOSITION 8.42. *We have an equality*

$$E \circ \partial_j^* = \epsilon^{-1} \cdot \partial_j^{can} \circ E$$

of maps $\bigoplus_{w\in W(r)} M^r_{X_{ord}}(w) \to R[[T]][T', ..., T^{(r)}]\hat{}$. *Explicitly, for* $w = \sum_{i=0}^r a_i\phi^i$ *and* $f \in M^r_{X_{ord}}(w)$, *the following formula holds*

$$E(\partial_j(f)) = \epsilon^{-1}(1+T^{\phi^j})\sum_{s=0}^{\infty} \frac{\partial(E(f))}{\partial T^{(s)}} \cdot \Lambda_{sj}(T, ..., T^{(s-1)}) - a_j p^j E(f)(E(P))^{\phi^j}.$$

PROOF. We may write

$$\mathcal{E}(f) = F\cdot x^w,\ F := E(f),\ x := x_{for}.$$

Then, using Proposition 8.32, Proposition 8.34, Proposition 3.45, and formula 3.31, we may compute:

$$\begin{aligned}
\mathcal{E}(\partial_j f) = \partial_j(\mathcal{E}(f)) &= \partial_j(Fx^w) = (\partial_j F)x^w + F\partial_j(x^w) \\
&= \sum_{s=0}^{\infty} \frac{\partial F}{\partial T^{(s)}}\Lambda_{sj}\cdot(\partial T)^{\phi^j} x^w + F a_j x^{w-\phi^j}\partial_j(x^{\phi^j}) \\
&= \sum_{s=0}^{\infty} \frac{\partial F}{\partial T^{(s)}}\Lambda_{sj}\cdot\epsilon^{-1}[(1+T)x^2]^{\phi^j} x^w + F a_j x^{w-\phi^j} p^j[-E(P)x^3]^{\phi^j} \\
&= \{\sum_{s=0}^{\infty} \frac{\partial F}{\partial T^{(s)}}\Lambda_{sj}\epsilon^{-1}(1+T^{\phi^j}) - F a_j p^j E(P)^{\phi^j}\} x^{w+2\phi^j}.
\end{aligned}$$

This concludes the proof. $\square$

PROPOSITION 8.43. *$\bar{P} \cdot \bar{H} = \bar{\partial}\bar{H}$ in $\bar{M}^0_{X_{ord}}(p+1)$.*

PROOF. (cf. [**2**] for the case of modular curves). Shrinking X we may assume $\bar{H}$ lifts to an element $H \in M^0_{X_{ord}}(p+1)$. By Proposition 8.42 the form $PH - \partial H \in M^0_{X_{ord}}(p+1)$ has Serre-Tate expansion

$$E(PH - \partial H) = E(P)E(H) - \left[\epsilon^{-1}(1+T)\frac{\partial(E(H))}{\partial T} - (p-1)E(H)E(P)\right].$$

Hence by Proposition 8.31

$$\bar{E}(\bar{P}\bar{H} - \bar{\partial}\bar{H}) = \bar{E}(\bar{P}) - \bar{E}(\bar{P}) = 0.$$

By the Serre-Tate expansion principle $\bar{P}\bar{H} - \bar{\partial}\bar{H} = 0$. □

8.4. Constructions of δ–invariants

8.4.1. Smooth δ–modular forms. In constructing elements of the spaces $H^0(\mathbf{X}, \mathbf{L})$ for Hecke correspondences it will be useful to study a "stacky version" of these spaces; these stacky spaces will consist of "rules" that we call *smooth δ–modular forms*. These rules are δ–analogues of the modular forms defined in [**81**]; they are a "smooth" version of the δ–modular forms in [**21**], [**23**], [**24**], and [**26**]. (Restricting ourselves to this "smooth version" is sufficient for all our purposes here although it doesn't seem to be enough for the theory in the above mentioned papers.) We start by introducing some notation.

For any affine smooth scheme X over R denote by $\mathbf{M}_1(X)$ the set of all pairs (A, ω) consisting of an elliptic curve A over X and an invertible 1–form ω on A. In the elliptic case we make the following:

DEFINITION 8.44. A *smooth δ–modular function* (of genus 1) and order $\leq r$ is a rule f that associates to any pair $(A, \omega) \in \mathbf{M}_1(X)$ an element

$$f(A, \omega) \in \mathcal{O}^r(X)$$

satisfying the following properties:

1) $f(A, \omega)$ depends on the isomorphism class of (A, ω) only.

2) The formation of $f(A, \omega)$ is functorial in X.

We denote by M^r the R–algebra of smooth δ–modular functions of order $\leq r$. Let us fix $w \in W$ of even degree $deg(w)$ and order $ord(w) \leq r$. A *smooth δ–modular form* of weight w and order $\leq r$ is an element $f \in M^r$ satisfying the following property:

3) If $\lambda \in \mathcal{O}(X)^{\times}$ then

$$f(A, \lambda\omega) = \lambda^{-w} \cdot f(A, \omega).$$

We denote by $M^r(w)$ the R–module of smooth δ–modular forms of weight w and order $\leq r$. A *smooth isogeny covariant δ–modular form* of weight w and order $\leq r$ is an element $f \in M^r(w)$ satisfying the following property:

4) If $(A_1, \omega_1), (A_2, \omega_2) \in \mathbf{M}_1(X)$, and $u : A_1 \to A_2$ is any isogeny of degree $deg(u)$ prime to p, such that $u^*\omega_2 = \omega_1$, then:

$$f(A_1, \omega_1) = deg(u)^{-deg(w)/2} \cdot f(A_2, \omega_2).$$

We denote by $I^r(w)$ the R–module of smooth isogeny covariant δ–modular forms of weight w and order $\leq r$.

Similar definitions may be given in the false elliptic case. Let D be a non-split indefinite quaternion algebra over $\mathbf{Q}$. We denote by $\mathbf{M}_D(X)$ the set of all tuples (A, i, θ, ω) where (A, i, θ) is a polarized false elliptic curve over X and ω is a false invertible 1–form on it.

DEFINITION 8.45. A *smooth δ–modular function on D* of order $\leq r$ is a rule f that associates to any smooth affine scheme X over R and any $(A, i, \theta, \omega) \in \mathbf{M}_D(X)$ an element $f(A, i, \theta, \omega) \in \mathcal{O}^r(X)$ satisfying the following properties:

D1) $f(A, i, \theta, \omega)$ depends on the isomorphism class of (A, i, θ, ω) only,

D2) The formation of $f(A, i, \theta, \omega)$ is functorial in X.

We denote by M^r the R–algebra of all smooth δ–modular function on D of order $\leq r$. Fix $w \in W$ with $deg(w)$ even and $ord(w) \leq r$. A *smooth δ– modular form* of weight w and order $\leq r$ on D is an element $f \in M^r$ satisfying the following property:

D3) If $\lambda \in \mathcal{O}(X)^\times$ then

$$f(A, i, \theta, \lambda\omega) = \lambda^{-w} \cdot f(A, i, \theta, \omega).$$

Denote by $M^r(w)$ the R–module of smooth δ– modular forms of weight w and order $\leq r$ on D. A *smooth isogeny covariant δ– modular form* of weight w and order $\leq r$ on D is an element $f \in M^r(w)$ satisfying the following property:

D4) If $(A_1, i_1, \theta_1, \omega_1), (A_2, \theta_2, i_2, \omega_2) \in \mathbf{M}_D(X)$, and $u : A_1 \to A_2$ is any isogeny of polarized false elliptic curves, of false degree $\partial(u)$ prime to p, such that $u^*\omega_2 = \omega_1$ then

$$f(A_1, i_1, \theta_1, \omega_1) = \partial(u)^{-deg(w)/2} \cdot f(A_2, i_2, \theta_2, \omega_2). \tag{8.96}$$

Denote by $I^r(w)$ the R–module of smooth isogeny covariant δ– modular forms of weight w and order $\leq r$ on D.

If $r = ord(w)$ we simply write $M(w)$ and $I(w)$ in place of $M^r(w)$ and $I^r(w)$ respectively. Note that the δ–modular functions and forms in [**21**], [**2**], [**23**], [**24**] give rise, in particular, to smooth δ–modular functions respectively forms. So the word *smooth* here is not an adjective ! Also our notation here is at variance with that in [**21**], [**2**], [**23**], [**24**], where the symbols M^r, $M^r(w)$, $I^r(w)$ stand for "δ–modular functions and forms" rather than for "smooth δ–modular functions and forms".

If in the above definitions of smooth δ–modular functions, smooth δ–modular forms, and isogeny covariant smooth δ–modular forms we ask that our rules f are only defined for X's such that $X_{ord} \otimes k = X \otimes k$ then f will be called an *ordinary smooth δ–modular function, ordinary smooth δ–modular form, or ordinary isogeny covariant smooth δ–modular form* respectively. The corresponding spaces will be denoted by M^r_{ord}, $M^r_{ord}(w)$, and $I^r_{ord}(w)$, with the superscript r for the last two spaces dropped if $r = ord(w)$. Any smooth δ–modular function, smooth δ–modular form, respectively isogeny covariant smooth δ–modular form gives rise, by restriction, to an ordinary smooth δ–modular function, an ordinary smooth δ–modular form, respectively an ordinary isogeny covariant smooth δ–modular form; so the word *ordinary* here is not an adjective !

REMARK 8.46. The formation of the Ramanujan forms P for various X's shows that P defines an ordinary smooth δ–modular form of order 0 and weight 2: $P \in M^0_{ord}(2)$. Cf. also [**81**].

REMARK 8.47. Let a_4 and a_6 be variables over R and let

$$\Delta := -2^6 a_4^3 - 2^4 3^3 a_6^2. \tag{8.97}$$

Using the bijection between the sets in Equations 8.1 and 8.2 we get a natural identification, in the elliptic case,

$$M^r \simeq R[a_4^{(\leq r)}, a_6^{(\leq r)}, \Delta^{-1}]\hat{}$$

where $a_4^{(\leq r)}$ is the collection of variables $a_4, a_4', ..., a_4^{(r)}$ and similarly for $a_6^{(\leq r)}$. The identification is given as follows. One considers the smooth affine scheme

$$X = Spec\ M_{Weier} := Spec\ R[a_4, a_6, \Delta^{-1}], \tag{8.98}$$

the Weierstrass elliptic curve

$$E_{Weier} = Proj\ R[a_4, a_6, \Delta^{-1}][x_0, x_1, x_2]/(x_0 x_2^2 - (x_1^3 + a_4 x_0^2 x_1 + a_6 x_0^3)), \tag{8.99}$$

and the invertible 1–form

$$\omega_{Weier} = dx/y,\ x := x_1/x_0, y := x_2/x_0, \tag{8.100}$$

and one attaches to each $f \in M^r$ the value

$$f(E_{Weier}, \omega_{Weier}) \in \mathcal{O}^r(X) = R[a_4^{(\leq r)}, a_6^{(\leq r)}, \Delta^{-1}]\hat{}.$$

In what follows we relate the stacky spaces, $M^r(w), I^r(w)$, introduced above with our spaces of sections of δ–line bundles. Note that if A (respectively (A, i, θ)) is an elliptic curve (respectively a polarized false elliptic curve) over a smooth affine R–scheme X and if $w \in W$, $ord(w) \leq r$, $s := r - ord(w)$, then there are natural maps

$$M^r(w) \to M^r_X(w) := H^0(X, L^{w[s]}),\ f \mapsto f(A), \tag{8.101}$$

in the elliptic case and

$$M^r(w) \to M^r_X(w) := H^0(X, L^{w[s]}),\ f \mapsto f(A, i, \theta), \tag{8.102}$$

in the false elliptic curve case, defined as follows. If $f \in M^r(w)$ and if ω is a local basis of L on an open set $U \subset X$ then one can can consider the section

$$f(A, \omega) \cdot \omega^{\otimes w} \in \mathcal{O}^r(X) \cdot \omega^{\otimes w} = H^0(U, L^{w[s]})$$

in the elliptic case or

$$f(A, i, \theta, \omega) \cdot \omega^{\otimes w} \in \mathcal{O}^r(X) \cdot \omega^{\otimes w} = H^0(U, L^{w[s]})$$

in the false elliptic case respectively. Due to the conditions 3) respectively D3) in the definition of δ–modular forms, these local sections glue together to give a global section $f(A)$ respectively $f(A, i, \theta)$.

As a rule, if X is a smooth affine R–scheme and f is a smooth δ–modular form in $M^r(w)$ then we will often continue to denote by f the image of f in either $M^r_X(w)$ or M^r_X. We also often denote by $\bar{f}$ the image of f in either $\bar{M}^r_X(w)$ or $\bar{M}^r_X$. Each time this will be done we will specify which space containing f or $\bar{f}$ we are using.

It is trivial to check that if $(\mathbf{X}, A, \tilde{u}, s)$ is a Hecke correspondence then the maps 8.101 and 8.102 induce maps

$$I^r(w) \to H^0(\mathbf{X}, \mathbf{L}_{r,w}),\ f \mapsto f(A), \tag{8.103}$$

in the elliptic case and

$$I^r(w) \to H^0(\mathbf{X}, \mathbf{L}_{r,w}),\ f \mapsto f(A, i, \theta), \tag{8.104}$$

in the false elliptic curve case.

8.4.2. Jet construction. There is a relative version of the construction of $f^r_{jet}(E,\omega)$ in Equation 7.13 which we now explain; this construction will yield isogeny covariant smooth δ-modular forms

$$f^r_{jet} \in I^r(-1-\phi^r) \tag{8.105}$$

in the elliptic curve case which will play a key role later. A similar construction can be made in the false elliptic curve case but, for simplicity, we will not consider this case. We start with an elliptic curve A/X over a smooth affine R-scheme X plus an invertible 1-form ω on A and we let $S^r := \mathcal{O}^r(X) = \mathcal{O}(J^r(X))$. Our aim is to define a δ-function $f^r_{jet}(A,\omega) \in S^r$. We proceed as follows. As in Equation 7.8 the projection $\pi_r : J^r(A) \to J^0(A) \times_{S^0} S^r$ has sections locally in the Zariski topology; let

$$A = \bigcup U_i, \quad s_i : \hat{U}_i \times_{S^0} S^r \to \pi_r^{-1}(\hat{U}_i) \tag{8.106}$$

be a Zariski open covering and a corresponding family of sections of π_r. Let $N^r := Ker\ \pi_r$. The sections in Equation 8.106 induce maps

$$s_i - s_j : \hat{U}_{ij} \times_{S^0} S^r = (\hat{U}_i \cap \hat{U}_j) \times_{S^0} S^r \to J^r(A)$$

that obviously factor through maps

$$s_{ij} : \hat{U}_{ij} \times_{S^0} S^r \to N^r. \tag{8.107}$$

Fix an isomorphism between the completion of $A \otimes S^0$ along the zero section and $Spec\ S^0[[T]]$ and let $\mathcal{F} \in S^0[[T_1,T_2]]$ be the corresponding formal group law. Exactly as in Proposition 4.45 one has a natural isomorphism

$$N^r \times_{S^0} S^r \simeq (\hat{\mathbf{A}}^r_{S^r}, [+])$$

which we view, in what follows as an identification; recall that $[+]$ is group law induced by $\mathcal{F}$. So we have an identification

$$s_{ij} = (\alpha^1_{ij}, \ldots, \alpha^r_{ij}) \in \mathcal{O}(\hat{U}_{ij} \times_{S^0} S^r)^r,$$

Recall from Proposition 4.42 that we have, at our disposal, series $L^r \in S^r[T', \ldots, T^{(r)}]\hat{}$ that define homomorphisms

$$L^r : N^r \times_{S^0} S^r \simeq (\hat{\mathbf{A}}^r_{S^r}, [+]) \to \hat{\mathbf{G}}_{a,S^r} = (\hat{\mathbf{A}}^1_{S^r}, +)$$

Define

$$\varphi^r_{ij} := L^r(\alpha^1_{ij}, \ldots, \alpha^r_{ij}) \in \mathcal{O}(\hat{U}_{ij} \times_{S^0} S^r).$$

The collection of these sections, for varying i,j, defines a cohomology class

$$\varphi^r \in H^1(\hat{A} \times_{S^0} S^r, \mathcal{O}) = H^1(A, \mathcal{O}) \otimes_{S^0} S^r.$$

Denote by

$$<\ ,\ >: H^1(A,\mathcal{O}) \otimes_{S^0} S^r \times H^0(A, \Omega^1) \otimes_{S^0} S^r \to S^r$$

the Serre duality pairing and define

$$f^r_{jet}(A,\omega) := \langle \varphi^r, \omega \rangle \in S^r. \tag{8.108}$$

PROPOSITION 8.48. *The rule f^r_{jet} is a smooth isogeny covariant δ-modular form of order $\leq r$ and weight $-1-\phi^r$; in symbols, $f^r_{jet} \in I^r(-1-\phi^r)$.*

PROOF. *Step 1.* First we claim that $f^r_{jet}(A,\omega)$ depends on the isomorphism class of (A,ω) only (i.e. it does not depend on the choice of the isomorphism between the completion of $A\otimes S^0$ along the zero section with $Spec\ S^0[[T]]$). We will show that φ^r_{ij} depend on (A,ω) only. Start by noting that the image of ω in $S^0[[T]]dT$ is given by

$$\omega(T)=d(l(T))=\frac{\partial l}{\partial T}(T)dT$$

where $l(T)\in S^0\otimes K[[T]]$ is the logarithm of $\mathcal{F}$. On the other hand, if $\tilde{T}=u(T)\in S^0[[T]]$, $u(0)=0$,

$$\frac{\partial u}{\partial T}(0)\in(S^0)^\times$$

and $\tilde{l}$ is the logarithm with respect to $\tilde{T}$ then we have

$$\tilde{l}(u(T))=\frac{\partial u}{\partial T}(0)\cdot l(T).$$

If we assume, in addition, that $\omega\equiv d\tilde{T}$ mod (T), then it follows that $\frac{\partial u}{\partial T}(0)=1$ hence $\tilde{l}(u(T))=l(T)$, in other words $[\tilde{l}]\circ[u(T)]=[l]$. Let $\tilde{L}^r$, $\tilde{\alpha}^r_{ij}$, and $\tilde{\varphi}^r_{ij}$ be the corresponding quantities constructed starting with the parameters $\tilde{T}$. We have

$$\begin{aligned}
\tilde{\varphi}^r_{ij} &:= \tilde{L}^r(\tilde{\alpha}^1_{ij},\dots,\tilde{\alpha}^r_{ij})\\
&= \frac{1}{p}[\tilde{l}^{(\phi^r)}]\circ[\phi^r(T)]\circ[0,\tilde{\alpha}^1_{ij},\dots,\tilde{\alpha}^r_{ij}]\\
&= \frac{1}{p}[\tilde{l}^{(\phi^r)}]\circ[\phi^r(T)]\circ[u,\delta u,\dots,\delta^r u]\circ[0,T',\dots,T^{(r)}]\circ[0,\alpha^1_{ij},\dots,\alpha^r_{ij}]\\
&= \frac{1}{p}[\tilde{l}^{(\phi^r)}]\circ[\phi^r(u(T))]\circ[0,T',\dots,T^{(r)}]\circ[0,\alpha^1_{ij},\dots,\alpha^r_{ij}]\\
&= \frac{1}{p}[\tilde{l}^{(\phi^r)}]\circ[u^{(\phi^r)}(T)]\circ[\phi^r(T)]\circ[0,T',\dots,T^{(r)}]\circ[0,\alpha^1_{ij},\dots,\alpha^r_{ij}]\\
&= \frac{1}{p}[l^{(\phi^r)}]\circ[\phi^r(T)]\circ[0,T',\dots,T^{(r)}]\circ[0,\alpha^1_{ij},\dots,\alpha^r_{ij}]\\
&= L^r(\alpha^1_{ij},\dots,\alpha^r_{ij})=\varphi^r_{ij}
\end{aligned}$$

and our Claim follows.

Step 2. A computation similar to the one above shows that if $\tilde{\omega}=\lambda\omega$ where $\lambda\in(S^0)^\times$ and if $\tilde{\varphi}^r_{ij}$ are the quantities corresponding to $(A,\tilde{\omega})$ then we have

$$\tilde{\varphi}^r_{ij}=\lambda^{\phi^r}\cdot\varphi^r_{ij};$$

so we get:

$$f^r_{jet}(A,\lambda\omega)=\langle\lambda^{\phi^r}\varphi^r,\lambda\omega\rangle=\lambda^{1+\phi^r}\langle\varphi^r,\omega\rangle=\lambda^{1+\phi^r}\cdot f^r_{jet}(A,\omega).$$

This shows that $f^r_{jet}\in M^r(-1-\phi^r)$.

Step 3. The following computation shows that f^r_{jet} is isogeny covariant. Let (A_1,ω_1), $(A_2,\omega_2)\in\mathbf{M}_1(X)$, and $u:A_1\to A_2$ be an isogeny of degree d prime to p such that $u^*\omega_2=\omega_1$. If φ^r_1,φ^r_2 are the corresponding cohomology classes then our construction implies that $\varphi^r_1=u^*\varphi^r_2$ Hence we have

$$f^r_{jet}(A_1,\omega_1)=\langle\varphi^r_1,\omega_1\rangle=\langle u^*\varphi^r_2,u^*\omega_2\rangle=d\cdot\langle\varphi^r_2,\omega_2\rangle=d\cdot f^r_{jet}(A_2,\omega_2).$$

This concludes the proof of isogeny covariance. □

The construction of the quantities $f^{1,2}_{jet}(E,\omega)$ in Remark 7.21 can also be made relative; here is the result (whose proof, similar to that in the case of $f^r_{jet}(E,\omega)$, is left to the reader):

PROPOSITION 8.49. *The map*

$$\mathbf{M}_1(R) \to R, \quad (E,\omega) \mapsto f^{1,2}_{jet}(E,\omega)$$

defined in Remark 7.21 is induced by a smooth isogeny covariant δ-modular form

$$f^{1,2}_{jet} \in I^2(-\phi - \phi^2). \tag{8.109}$$

In what follows we explain how one can compute the reduction mod p,

$$\bar{f}^1_{jet} \in k[a_4, a_6, a'_4, a'_6, \Delta^{-1}] \tag{8.110}$$

of f^1_{jet}. The idea is to relate this reduction mod p to an "internal Kodaira-Spencer class" that relativizes the one in Definition 3.10, cf. [**21**], and then to compute this class explicitly, cf. [**71**]. Again let A/X be an elliptic curve over a smooth affine scheme over R. Let the index 0 denote, as usual, tensorization with k; in particular $S^r_0 := S^r \otimes k$. Exactly as in Proposition 3.9 the morphism

$$J^1(A)_0 \to A_0 \otimes_{S^0_0} S^1_0 \tag{8.111}$$

is a (Zariski locally trivial) principal homogeneous space for the Frobenius tangent bundle

$$FT(A_0/X_0) \otimes_{S^0_0} S^1_0 = Spec(Symm(F^*\Omega_{A_0/X_0})) \otimes_{S^0_0} S^1_0 \to A_0 \otimes_{S^0_0} S^1_0.$$

DEFINITION 8.50. The cohomology class in $H^1(A_0, F^*T_{A_0/X_0}) \otimes_{S^0_0} S^1_0$ of the principal homogeneous space in Equation 8.111 is called the *(internal) Kodaira-Spencer class* of A/X and is denoted by $KS(A/X)$.

This class is of course compatible with specialization in the sense that if $P \in X(R)$ is an R-point then the class $KS(A_P)$ of A_P (in the sense of Definition 3.10) is the image of $KS(A)$ under the natural morphism induced by the map $S^1 \to R$ (where the latter is the map obtained from P via the universality property of $J^1(X)$). Explicitly $KS(A/X)$ is computed as follows. One covers A by affine open subsets U_i, one lifts the p-derivation $\delta : S^0 \to S^1$ to p-derivations

$$\delta_i : \mathcal{O}(U_i) \to \mathcal{O}(U_i) \otimes_{S^0_0} S^1_0,$$

one considers the differences

$$\delta_i - \delta_j : \mathcal{O}(U_{ij}) \to \mathcal{O}(U_{ij}) \otimes_{S^0_0} S^1_0$$

which turn out to induce usual derivations from $\mathcal{O}(U_{ij})_0$ to $\mathcal{O}(U_{ij}) \otimes_{S^0_0} S^1_0$ relative to the ring homomorphism obtained by composing the absolute Frobenius of $\mathcal{O}(U_{ij})_0$ with the canonical map of the latter into $\mathcal{O}(U_{ij})_0 \otimes_{S^0_0} S^1_0$. One takes $KS(A/X)$ to be the cohomology class defined by this cocycle. For any invertible 1-form ω on A_0 we let

$$\bar{f}_{def}(A/X,\omega) \in S^1_0$$

to be the image of $KS(A/X)$ under the identification

$$H^1(A_0, F^*T_{A_0/X_0}) \otimes_{S^0_0} S^1_0 \overset{\omega}{\simeq} H^1(A_0, \mathcal{O}) \otimes_{S^0_0} S^1_0 \overset{\langle\ ,\omega\rangle}{\longrightarrow} S^1_0$$

where $\langle\ ,\ \rangle$ is the Serre duality.

The computation of $\bar{f}_{def}(A/X,\omega)$ has been carried out explicitly by C. Hurlburt for $X = Spec\ M_{Weier}$, $M_{Weier} := R[a_4, a_6, \Delta^{-1}]$, $A = E_{Weier}$ the Weierstrass curve over X, and $\omega = \omega_{Weier} = dx/y$; cf. [**71**]. Set

$$\bar{f}_{def} := \bar{f}_{def}(E_{Weier}/M_{Weier}, \omega_{Weier}) \in k[a_4, a_6, a_4', a_6', \Delta^{-1}]. \tag{8.112}$$

The result is the following:

PROPOSITION 8.51. [**71**]. *Up to multiplication by a scalar in* $\mathbf{F}_p^{\times}$ *we have*

$$\bar{f}_{def} = \bar{H} \cdot \bar{\Delta}^{-p} \cdot (2a_4^p a_6' - 3a_6^p a_4') + \bar{F}_0$$

where $\bar{F}_0 \in \mathbf{F}_p[a_4, a_6, \Delta^{-1}]$.

Here $\bar{H}, \bar{\Delta} \in \mathbf{F}_p[a_4, a_6]$ denote the Hasse invariant and the reduction mod p of Δ respectively.

PROOF. Cf. [**71**]. Note however that the difficult part of the computation in [**71**] is the explicit determination of $\bar{F}_0$ which we do not need here. □

On the other hand we have the following:

PROPOSITION 8.52. [**21**]. *There is a constant* $c \in k^{\times}$ *and an integer* $m \in \mathbf{Z}$ *such that the reduction mod* p *of* f_{jet}^1 *is given by*

$$\bar{f}_{jet}^1 = c \cdot \bar{\Delta}^m \cdot \bar{f}_{def}$$

in the ring $k[a_4, a_6, a_4', a_6', \Delta^{-1}]$.

We will review the proof of the above Proposition presently. Before that let us note, however, that Propositions 8.51 and 8.52 imply the following determination of $\bar{f}_{jet}^1$:

COROLLARY 8.53. *There is a constant* $c \in k^{\times}$ *and an integer* $N \in \mathbf{Z}$ *such that the reduction mod* p *of* f_{jet}^1 *is given by*

$$\bar{f}_{jet}^1 = c \cdot \bar{H} \cdot \bar{\Delta}^N \cdot (2a_4^p a_6' - 3a_6^p a_4') + \bar{f}_0$$

where $\bar{f}_0 \in k[a_4, a_6, \Delta^{-1}]$.

We will prove later that, in the above formula, $N = -p$. In what follows we review the proof of Proposition 8.52 following [**21**]. We need the following:

LEMMA 8.54. *Let* Z *be an integral scheme and let* Y *be a scheme over* Z. *Assume* Y *admits two structures of principal homogeneous space under* $\mathbf{G}_{a,Z}$, *that are locally trivial in the Zariski topology. Then the two structures differ by a unit in* $\mathcal{O}(Z)$.

PROOF. We may assume, of course, that Z is affine, say $Z = Spec\ B$, and that $Y \to X$ has a section. So we may assume $Y = \mathbf{G}_{a,B}$ and the first structure of principal homogeneous space

$$\rho_1 : \mathbf{G}_{a,B} \times Y \to Y$$

is given by the multiplication map of $\mathbf{G}_a$. Let

$$\rho_2 : \mathbf{G}_{a,B} \times Y \to Y$$

be the action defining the second structure of principal homogeneous space on Y. Since for any integral domain B' which is a $B-$algebra we have

$$Aut_{B'}(Y \otimes B') = \mathbf{G}_{m,B}(B') \times_{\epsilon} \mathbf{G}_{a,B}(B')$$

(where "$\times_\epsilon$" means here semidirect product via the natural action ϵ of $\mathbf{G}_m$ on $\mathbf{G}_a$) and since the schemes $\mathbf{G}_{a,B}$ and $\mathbf{G}_{m,B} \times_\epsilon \mathbf{G}_{a,B}$ are integral, we get that ρ_2 induces a homomorphism

$$\mathbf{G}_{a,B} \to \mathbf{G}_{m,B} \times_\epsilon \mathbf{G}_{a,B}$$

over B. The composition of the latter with the projection to $\mathbf{G}_{m,B}$ must be trivial so the action of $\mathbf{G}_{a,B}$ on Y via ρ_2 is via translations $\rho(t,x) = x + b(t)$, $t \in B'$, $x \in B'$, $b(t) \in B'$. Since $b(t_1+t_2) = b(t_1) + b(t_2)$ it follows that $b(t) = \sum_{j \geq 0} \beta_j t^{p^j}$ for some $\beta_j \in B$, where $p = char\ B$ and $\beta_j = 0$ for $j \geq 1$ if $p = 0$. From the condition that ρ_2 defines a principal homogeneous space, i.e. from the condition that $(t,x) \mapsto (x + \sum \beta_j t^{p^j}, x)$ is an isomorphism of schemes, we get that $\beta_j = 0$ for $j \geq 1$ and $\beta_0 \in B^\times$ which concludes the proof of the Lemma. □

PROOF. (of Proposition 8.52). The morphism 8.111 has two structures of principal homogeneous space for

$$\mathbf{G}_{a, A_0 \otimes_{S_0^0} S_0^r}.$$

One structure is obtained by looking at the morphism 8.111 as a principal homogeneous space for the Frobenius tangent bundle as in Equation 8.111 and noting that the Frobenius tangent bundle is trivial (because A_0 is an elliptic curve). Another structure is obtained by viewing the morphism 8.111 as a morphism of group schemes which is a principal homogeneous space for its kernel times the base. By their very construction $\bar{f}^1_{jet}$ and $\bar{f}_{def}$ are the cohomology classes corresponding to these two principal homogeneous space structures. By Lemma 8.54 $\bar{f}^1_{jet}$ and $\bar{f}_{def}$ differ by a unit in $k[a_4, a_6, a_4', a_6', \Delta^{-1}]$ hence by an element in $k \cdot \Delta^{\mathbf{Z}}$. □

Next we compute the reduction mod p, $\bar{f}^r_{jet}$ of f^r_{jet}. Recall that $\bar{H} \in \mathbf{F}_p[a_4, a_6]$ is the Hasse invariant.

PROPOSITION 8.55. *We have*

$$\bar{f}^r_{jet} = \bar{H}^{1+p+\ldots+p^{r-2}} \cdot (\bar{f}^1_{jet})^{p^{r-1}} \in k[a_4^{(\leq r)}, a_6^{(\leq r)}, \Delta^{-1}].$$

PROOF. Let ω be an invertible 1–form and ξ the dual class. Let φ^r be as in Equation 8.108. By Proposition 4.41 $\bar{\varphi}^r \equiv (\bar{\varphi}^1)^{p^{r-1}}$, where the upper bar denotes, as usual, the reduction mod p. Write $\varphi^1 = \mu \cdot \xi$ for some μ. We have:

$$\bar{f}^1_{jet} = \langle \bar{\varphi}^1, \bar{\omega} \rangle = \langle \bar{\mu}\bar{\xi}, \bar{\omega} \rangle = \bar{\mu} \langle \bar{\xi}, \bar{\omega} \rangle = \bar{\mu}.$$

Since $F^*(\bar{\xi}) = \bar{H} \cdot \bar{\xi}$ it follows that

$$(F^*)^k(\bar{\mu}\bar{\xi}) = \bar{\mu}^{p^k} \cdot \bar{H}^{1+p+\ldots+p^{k-1}} \cdot \bar{\xi}.$$

We then have

$$\bar{f}^r_{jet} = \langle \bar{\varphi}^r, \bar{\omega} \rangle = \langle (F^*)^{r-1} \bar{\varphi}^1, \bar{\omega} \rangle = \langle (F^*)^{r-1}(\bar{\mu}\bar{\xi}), \bar{\omega} \rangle$$

$$= \bar{\mu}^{p^{r-1}} \cdot \bar{H}^{1+p+\ldots+p^{r-2}} \cdot \langle \bar{\xi}, \bar{\omega} \rangle = \bar{H}^{1+p+\ldots+p^{r-2}} \cdot (\bar{f}^1_{jet})^{p^{r-1}}.$$

□

We end by noting that if $\mathbf{X}$ is a Hecke correspondence then we can consider the image

$$f^r_{jet} \in H^0(\mathbf{X}, \mathbf{L}_{-1-\phi^r}) \tag{8.113}$$

of f^r_{jet} in the corresponding H^0 space; this section plays an important role in the theory.

8.4.3. Crystalline construction. We will construct isogeny covariant smooth δ–modular forms

$$f^r_{crys} \in I^r(-1-\phi^r) \tag{8.114}$$

in both the elliptic and false elliptic curve case which will play a key role later. Let X be a smooth affine scheme over R and $(A,\omega) \in \mathbf{M}_1(X)$ or $(A,i,\theta) \in \mathbf{M}_D(X)$ respectively. Set $S^r := \mathcal{O}^r(X) = \mathcal{O}(J^r(X))$. Let H^1 denote, as usual, the de Rham module $H^1_{DR}(A/X)$. By crystalline theory there are ϕ–linear maps

$$H^1 \otimes S^0 \stackrel{\phi}{\to} H^1 \otimes S^1 \stackrel{\phi}{\to} H^1 \otimes S^2 \stackrel{\phi}{\to} ...;$$

cf. Equation 8.39. Recall also that we have at our disposal a pairing

$$\langle\ ,\ \rangle_\theta : H^1 \otimes S^r \times H^1 \otimes S^r \to S^r;$$

cf. Equation 8.16. We then define

$$f^r_{crys}(A,\omega) := \frac{1}{p} \cdot \langle \phi^r \omega, \omega \rangle \in S^r, \tag{8.115}$$

$$f^r_{crys}(A,i,\theta,\omega) := \frac{1}{p} \cdot \langle \phi^r \omega^1, \omega^1 \rangle_\theta \in S^r. \tag{8.116}$$

Note that the definitions above are correct. Indeed the rules $p \cdot f^r_{crys}$ clearly define smooth δ–modular functions, i.e. elements of M^r. On the other hand, by Equation 8.45, the value of $p \cdot f^r_{crys}(A,\omega)$ (respectively $p \cdot f^r_{crys}(A,i,\theta,\omega)$) at any point $P \in X(R)$ lies in pR. Hence, by Proposition 3.19, $p \cdot f^r_{crys}(A,\omega)$ (respectively $p \cdot f^r_{crys}(A,i,\theta,\omega)$) belong to $p \cdot S^r$ so $f^r_{crys}(A,\omega)$ (respectively $f^r_{crys}(A,i,\theta,\omega)$) belong to S^r which proves correctness of our definitions.

PROPOSITION 8.56. *The rule f^r_{crys} is a smooth isogeny covariant δ–modular form of order $\leq r$ and weight $-1-\phi^r$; in symbols, $f^r_{crys} \in I^r(-1-\phi^r)$.*

PROOF. Let's give the proof in the false elliptic curve case following [**3**]; the case of genuine elliptic curves is similar and easier. Let $\varphi^r \in H^{01}$ be the projection of $\phi^r\omega^1 \in H^1$. Then

$$p \cdot f^r_{crys}(A,i,\theta,\omega) = \langle \theta_*^{-1}\varphi^r, \omega^1 \rangle,$$

cf. Equation 8.18. To check that f^r_{crys} has weight $-1-\phi^r$ let $\lambda \in (S^0)^\times$ and note that

$$p \cdot f^r_{crys}(A,i,\theta,\lambda\omega) = \langle \theta_*^{-1}\lambda^{\phi^r}\varphi^r, \lambda\omega^1 \rangle = \lambda^{1+\phi^r} \cdot \langle \theta_*^{-1}\varphi^r, \omega^1 \rangle.$$

The next computation shows that f^r_{crys} is isogeny covariant. Let $(A_1,i_1,\theta_1,\omega_1)$, $(A_2,i_2,\theta_2,\omega_2) \in \mathbf{M}_D(X)$, and $u : A_1 \to A_2$ be an isogeny of false elliptic curves of false degree e prime to p such that $u^*\omega_2 = \omega_1$. Note that the isomorphism

$$\gamma = u \circ u^t : A_2 \stackrel{\theta_2}{\to} \check{A}_2 \stackrel{\check{u}}{\to} \check{A}_1 \stackrel{\theta_1^{-1}}{\to} A_1 \stackrel{u}{\to} A_2$$

induces an isomorphism at the level of Lie algebras

$$\gamma_* : H^0(A_2, T_{A_2/X}) \stackrel{\theta_{2*}}{\to} H^1(A_2,\mathcal{O}) \stackrel{\check{u}_*=u^*}{\to} H^1(A_1,\mathcal{O})$$

$$\stackrel{\theta_{1*}^{-1}}{\to} H^0(A_1, T_{A_1/X}) \stackrel{u_*}{\to} H^0(A_2, T_{A_2/X}).$$

Then γ_* above is dual to $\gamma^* : H^0(A_2,\Omega) \to H^0(A_2,\Omega)$. On the other hand γ is simply the multiplication by the false degree e of u. So $\gamma^*\omega_2 = e\omega_2$. Also let $u^* : H^0(A_2,\Omega^1) \to H^0(A_1,\Omega^1)$ be the pull-back induced by u; it is dual to the Lie algebra map u_* above. If φ_i^r is the class that corresponds to $(A_i, i_i, \theta_i, \omega_i)$ then, our construction implies that $\varphi_1^r = u^*\varphi_2^r = \check{u}_*\varphi_2^r$. Hence we have

$$\langle(\theta_{1*})^{-1}\varphi_1^r, \omega_1^1\rangle = \langle(\theta_{1*})^{-1}\check{u}_*\varphi_2^r, u^*\omega_2^1\rangle = \langle u_*(\theta_{1*})^{-1}\check{u}_*\varphi_2^r, \omega_2^1\rangle$$
$$= \langle\gamma_*(\theta_{2*})^{-1}\varphi_2^r, \omega_2^1\rangle = \langle(\theta_{2*})^{-1}\varphi_2^r, \gamma^*\omega_2^1\rangle = e \cdot \langle(\theta_{2*})^{-1}\varphi_2^r, \omega_2^1\rangle.$$

This concludes the proof of isogeny covariance. □

As in the jet construction we make the following notation. Assume $\mathbf{X}$ is a Hecke correspondence; then we can consider the image

$$f_{crys}^r \in H^0(\mathbf{X}, \mathbf{L}_{-1-\phi^r}) \tag{8.117}$$

of f_{crys}^r in the corresponding H^0 space; this section plays an important role in the theory and will be related to f_{jet}^r.

In what follows we define, following [**2**] in the elliptic case and [**24**] in the false elliptic case, a basic smooth ordinary δ-modular form of weight $-1+\phi$,

$$f^\partial \in M_{ord}^1(-1+\phi). \tag{8.118}$$

Let X be any smooth affine scheme over R and $(A,\omega) \in \mathbf{M}_1(X)$ or $(A,i,\theta,\omega) \in \mathbf{M}_D(X)$. Consider a normalized local basis of H^1 in the sense of Definitions 8.4 and 8.5 respectively. Also let u^1 be a basis of $\mathcal{U}$ over $\mathcal{O}(X_{ord})\hat{}$ (see Equation 8.53) and consider the local sections

$$f^\partial := \frac{\langle\phi u^1, \omega^1\rangle_\theta}{\phi(\langle u^1, \omega^1\rangle_\theta)}(\omega^1)^{\otimes(-1+\phi)} \in M_{X_{ord}}^1(-1+\phi).$$

Clearly the above local sections are independent of the choices of ω^1 and u^1 so they glue together to give a global section $f^\partial(A)$ or $f^\partial(A,i,\theta)$ in $M_{X_{ord}}^1(-1+\phi)$.

Fix in what follows a smooth affine scheme X together with a genuine or false elliptic curve A/X. Let us view f^∂ and H as elements of $M_{X_{ord}}^1$ and let us denote by $\bar{f}^\partial, \bar{H} \in \bar{M}_{X_{ord}}^1$ their images.

PROPOSITION 8.57. [**2**], [**24**]. *$\bar{f}^\partial = \bar{H}$ in $\bar{M}_{X_{ord}}^1$.*

PROOF. We give the proof in the false elliptic curve case following [**24**]; the proof in [**2**] of the elliptic case (which inspired our proof in [**24**]) is similar. Let ω be an invertible false 1-form, ξ the θ-dual basis of H_{ord}^{01} and u a basis of the unit root space U such that $\pi(u) = \xi$ where $\pi : H_{ord}^1 \to H_{ord}^{01}$ is the canonical projection. In particular $\langle u, \omega^t\rangle_\theta = I$, so u^1 is a basis of $\mathcal{U}$; cf. Remark 8.9. Write $F^*\bar{\xi} = \bar{h}\cdot\bar{\xi}$ as in 8.41. For any $g \in M_{X_{ord}}^1$ let $\bar{g} \in \bar{M}_{X_{ord}}^1$ denote the image of g. Also, let $\langle\ ,\ \rangle_{\bar{\theta}}$ be the pairing on $\bar{H}^1 := H^1 \otimes k$ induced by θ, denote by $v \mapsto \bar{v}$ the natural map $H^1 \to \bar{H}^1$, and denote by F^* the semilinear map on $\bar{H}^1$ induced by the p-power Frobenius. We get

$$\bar{f}^\partial = \langle F^*\bar{u}^1, \bar{\omega}^1\rangle_{\bar{\theta}}(\bar{\omega}^1)^{\otimes(\phi-1)} = \langle\bar{\theta}_*^{-1}\pi F^*\bar{u}^1, \bar{\omega}^1\rangle(\bar{\omega}^1)^{\otimes(p-1)}$$
$$= \langle\theta_*^{-1}F^*\pi\bar{u}^1, \bar{\omega}^1\rangle(\bar{\omega}^1)^{\otimes(p-1)} = \langle\bar{\theta}_*^{-1}F^*\bar{\xi}^1, \bar{\omega}^1\rangle(\bar{\omega}^1)^{\otimes(p-1)}$$
$$= \bar{h}\cdot\langle\bar{\theta}_*^{-1}\bar{\xi}^1, \bar{\omega}^1\rangle(\bar{\omega}^1)^{\otimes(p-1)} = \bar{h}\cdot(\bar{\omega}^1)^{\otimes(p-1)} = \bar{H}.$$

□

COROLLARY 8.58. *There is a form $f_\partial \in M_{X_{ord}}^1(1-\phi)$ such that $f_\partial \cdot f^\partial = 1$.*

PROOF. By Proposition 8.57 it follows that $\bar{f}^{\partial}$ is invertible in $\bar{M}^1_{X_{ord}}$. Since $M^1_{X_{ord}}$ is p-adically complete, f^{∂} itself must be invertible in $M^1_{X_{ord}}$. We may conclude by assertion iii) in Proposition 3.56. $\square$

In view of this Corollary it is reasonable to define, for any

$$v = v_+ + v_- \in W, \quad v_+, -v_- \in W_+,$$

the form

$$(f^{\partial})^v := (f^{\partial})^{v_+} \cdot (f_{\partial})^{-v_-} \in M^1_{X_{ord}}(v(\phi - 1)).$$

For the rest of this Section $\mathbf{X}$ is an affine Hecke correspondence with a regular Hecke fixed point and view f^{∂} as an element of $M^1_{X_{ord}}(-1+\phi)$; then its image under the Serre-Tate expansion map

$$E : M^1_{X_{ord}}(-1+\phi) \to R[[T]][T']\hat{}$$

is given by:

PROPOSITION 8.59. *We have*

$$E(f^{\partial}) = E(f_{\partial}) = 1. \tag{8.119}$$

In particular, by Proposition 8.30

$$f^{\partial} \in H^0(\mathbf{X}_{ord}, \mathbf{L}_{\phi-1}), \quad f_{\partial} \in H^0(\mathbf{X}_{ord}, \mathbf{L}_{\phi-1}).$$

PROOF. It is enough to prove $E(f^{\partial}) = 1$. Choose a pair of bases $\omega^1, \eta^1, \omega^2, \eta^2$ and ω, u of H^1_{for} where the first is normalized and the second is canonical. Also set $x = x_{for} = \omega^1$. In particular recall that

$$\phi u = u, \quad \langle u, \omega^t \rangle = \epsilon \cdot I, \quad \epsilon \in \mathbf{Z}_p^{\times}. \tag{8.120}$$

Using Equation 8.120, we get that the image of f^{∂} in $M^1_{X for}$ equals

$$\frac{\langle \phi u^1, \omega^1 \rangle}{\phi(\langle u^1, \omega^1 \rangle)} x^{-1} x^{\phi} = \frac{\epsilon}{\phi(\epsilon)} x^{-1} x^{\phi} = x^{-1} x^{\phi}$$

which ends our proof. $\square$

REMARK 8.60. At this point we are ready to give a proof of Lemma 8.16. Indeed by Propositions 8.59 and 8.57 it follows that $\bar{H}$, viewed as an element of

$$H^0(X_{ord} \otimes k, K^{(p-1)/2})$$

has the property that

$$\sigma_1^* \bar{H} = \sigma_2^* \bar{H}.$$

The latter implies, of course, that $\mathbf{K}^{(p-1)/2}_{\mathbf{X}_{ord} \otimes k}$ is trivial.

In the next Proposition we view f^r_{crys} as an element of $H^0(\mathbf{X}, \mathbf{L}_{-1-\phi^r})$, where $\mathbf{X}$ is, as before, an affine Hecke correspondence with a regular Hecke fixed point; cf. Equation 8.117.

PROPOSITION 8.61.

$$E(f^r_{crys}) = \epsilon(\phi^{r-1}\Psi + p\phi^{r-2}\Psi + \dots + p^{r-1}\Psi) = \frac{\epsilon}{p}(l_{\mathbf{G}_m}(T)^{\phi^r} - p^r l_{\mathbf{G}_m}(T)),$$

where $\Psi = \Psi_{\mathbf{G}_m}$, cf. Equation 4.40, and $l_{\mathbf{G}_m}(T) = log(1+T)$.

PROOF. Consider the injective map

$$R[[T]][T',...,T^{(r)}]\hat{} \to \mathcal{F}(1+pR,R) = \{\text{maps } 1+pR \to R\}$$

mapping a series $F \in R[[T]][T',...,T^{(r)}]\hat{}$ into the function

$$q \mapsto F\langle q\rangle := F(q-1,\delta(q-1),...,\delta^r(q-1)), \quad q \in 1+pR.$$

Also assume that the Hecke fixed point is given by $(A_0,\theta_0) = (Jac(C_0),\theta_{C_0})$ for some curve C_0/k, where θ_{C_0} is the theta divisor. Then, for any $q \in 1+pR$, we have

$$E(f^r_{crys})\langle q\rangle = f(Jac(C),\theta_C,\omega(C)), \tag{8.121}$$

where C is a lifting of C_0 such that $Jac(C)$ has Serre-Tate parameter q, θ_C is the theta divisor on $Jac(C)$ and $\omega(C) = (\omega^1(C),\omega^2(C))^t$ is as in Equation 8.52. Now, using Equations 8.121 and 8.52 and the notation there we have the following computation:

$$E(f^r_{crys})\langle q\rangle = f^r_{crys}(Jac(C),\theta_C,\omega(C)) = \frac{1}{p}\langle\phi^r\omega^1(C),\omega^1(C)\rangle$$

$$= \frac{1}{p}\langle\epsilon\cdot p^r\cdot\beta^1(C) + (log\ \phi^r(q))\cdot u^1(C), \epsilon\cdot\beta^1(C) + u^1(C)\cdot(log\ q)\rangle$$

$$= \frac{\epsilon}{p}\{log\ \phi^r(q) - p^r\cdot log\ q\}$$

$$= \frac{\epsilon}{p}\{[log\ \phi^r(q) - p\cdot log\ \phi^{r-1}(q)] + ... + [p^{r-1}\cdot log\ \phi(q) - p^r\cdot log\ q\}$$

$$= \epsilon\{\phi^{r-1}\Psi + ... + p^{r-1}\Psi\}\langle q\rangle$$

which ends the proof. □

COROLLARY 8.62. *We have*

$$(f^1_{crys})^{\phi}f^k_{crys} - f^2_{crys}(f^{k-1}_{crys})^{\phi} + pf^1_{crys}(f^{k-2}_{crys})^{\phi^2} = 0$$

in $H^0(\mathbf{X},\mathbf{L}_{-1-\phi-\phi^2-\phi^k})$.

PROOF. Using Proposition 8.61 one checks immediately that the Serre-Tate expansion of the left hand side is 0. □

The relations in Corollary 8.62 are strikingly similar (but not identical) to the relations 6.17, holding in a rather different context.

REMARK 8.63. The germ $E(f^r) \in R[[T]][T',...,T^{(r)}]\hat{}$ of f^r at our Hecke fixed point satisfies

$$E(f^r)(0,...,0,T) = \frac{\epsilon}{p}log(1+p^rT) \in p^{r-1}\cdot T\cdot R[[T]]^{\times};$$

in particular f^r is a $1-$ regular section and f^1 is a $1-$ coregular section at our Hecke fixed point.

In the next Proposition we view $f^1_{crys}, f^{\partial}, P^{\phi}$ as elements in $M^1_{X_{ord}}$.

PROPOSITION 8.64. *We have*

$$\partial_1 f^1_{crys} = f^{\partial} + pP^{\phi}f^1_{crys} \quad in \quad M^1_{X_{ord}}. \tag{8.122}$$

Equivalently

$$\partial_1^* f^1_{crys} = f^{\partial} \quad in \quad H^0(\mathbf{X}_{ord},\mathbf{L}_{\phi-1}). \tag{8.123}$$

PROOF. Both members of the equality in the Proposition belong to

$$M^1_{X_{ord}}(-1+\phi);$$

cf. Proposition 3.56. By the Serre-Tate expansion principle in Proposition 8.29 it is enough to show that the images of the two members of the above equality under the Serre-Tate map

$$E : M^1_{X_{ord}}(-1+\phi) \to R[[T]][T']\hat{}$$

coincide. Now, by Proposition 8.61 $E(f^1_{crys}) = \epsilon\Psi$ while, by Proposition 8.59, we have $E(f^{\partial}) = 1$. A trivial computation, using the latter two formulae plus Proposition 8.42 shows that $E(\partial_1 f^1_{crys})$ and $E(f^{\partial} + pP^{\phi} f^1_{crys})$ coincide. □

COROLLARY 8.65. *$\bar{\partial}_1 \bar{f}^1_{crys} = \bar{H}$ in $\bar{M}^1_X$.*

PROOF. By Proposition 8.64 and Proposition 8.57 we have

$$\bar{\partial}_1 \bar{f}^1_{crys} = \bar{f}^{\partial} = \bar{H} \in \bar{M}^1_{X_{ord}}.$$

We conclude by the injectivity of $\bar{M}^1_X \to \bar{M}^1_{X_{ord}}$; cf. Corollary 8.28. □

PROPOSITION 8.66. *We have*

$$\partial_0 f^1_{crys} = -f_{\partial} + P \cdot f^1_{crys} \quad in \quad M^1_{X_{ord}}. \tag{8.124}$$

Equivalently

$$\partial_0^* f^1_{crys} = -f_{\partial} \quad in \quad H^0(\mathbf{X}_{ord}, \mathbf{L}_{1-\phi}). \tag{8.125}$$

PROOF. Both members belong to

$$M^1_{X_{ord}}(1-\phi).$$

On the other hand

$$E(\partial_0 f^1_{crys}) = \epsilon^{-1}(1+T)\left[\frac{\partial(\epsilon\Psi)}{\partial T} - T^{p-1}\frac{\partial(\epsilon\Psi)}{\partial T'}\right] + E(P)\epsilon\Psi$$
$$= -1 + E(P)\epsilon\Psi = E(-f_{\partial} + P \cdot f^1_{crys}),$$

and we conclude as in the proof of Proposition 8.64. □

COROLLARY 8.67. *We have*

$$\bar{\partial}_0 \bar{f}^1_{crys} = -\bar{H}^{-1} + \bar{P} \cdot \bar{f}^1_{crys} \quad in \quad \bar{M}^1_{X_{ord}}. \tag{8.126}$$

Equivalently, by Proposition 8.43,

$$\bar{f}^1_{crys} \cdot \bar{\partial}\bar{H} - \bar{H} \cdot \bar{\partial}_0 \bar{f}^1_{crys} = 1 \quad in \quad \bar{M}^1_{X_{ord}} \quad hence \ in \quad \bar{M}^1_X. \tag{8.127}$$

So we get, in particular,

COROLLARY 8.68. *The schemes Spec $\bar{M}^1_X/(\bar{H})$ and Spec $\bar{M}^1_X/(\bar{f}^1_{crys})$ are smooth over k. Moreover*

$$Spec\ \bar{M}^1_X/(\bar{H}) \cap Spec\ \bar{M}^1_X/(\bar{f}^1_{crys}) = \emptyset,$$
$$Spec\ \bar{M}^1_X/(\bar{H}) \cap Spec\ \bar{M}^1_X/(\bar{\partial}\bar{H}) = \emptyset,$$
$$Spec\ \bar{M}^1_X/(\bar{\partial}_0 \bar{f}^1_{crys}) \cap Spec\ \bar{M}^1_X/(\bar{f}^1_{crys}) = \emptyset,$$
$$Spec\ \bar{M}^1_X/(\bar{\partial}\bar{H}) \cap Spec\ \bar{M}^1_X/(\bar{\partial}_0 \bar{f}^1_{crys}) = \emptyset,$$

in Spec$\bar{M}^1_X$. In particular, if E/R is an elliptic curve with supersingular reduction then $KS(E) \neq 0$.

The statement that $Spec\ \bar{M}^1_X/(\bar{H})$ is smooth and disjoint from $Spec\ \bar{M}^1_X/(\bar{\partial}\bar{H})$ can also be deduced, in the case X is a modular curve, from known results about the Hasse invariant, cf. [**88**], p. 167. The fact that $Spec\ \bar{M}^1_X/(\bar{H})$ does not contain $Spec\ \bar{M}^1_X/(\bar{f}^1_{crys})$ was proved by Hurlburt [**72**], in the elliptic case, for infinitely many primes p; Hurlburt's proof relies on her explicit computation of $\bar{f}^1_{jet}$ in [**71**].

PROOF. Equation 8.127 shows that none of the elements $\bar{H}$ and $\bar{f}^1_{crys}$ can belong to the square of a maximal ideal of $\bar{M}^1_X$; this proves smoothness of the schemes $Spec\ \bar{M}^1_X/(\bar{H})$ and $Spec\ \bar{M}^1_X/(\bar{f}^1_{crys})$. The claims about the intersections follow directly from Equation 8.127. □

COROLLARY 8.69. *The following holds*

$$\bar{\partial}\bar{P} + \bar{P}^2 = \bar{\partial}\bar{H}\cdot\bar{\partial}_0^2\bar{f}^1_{crys} - \bar{\partial}^2\bar{H}\cdot\bar{\partial}_0\bar{f}^1_{crys} \quad in \quad \bar{M}^1_{X_{ord}}.$$

In particular

$$\bar{\partial}\bar{P} + \bar{P}^2 \in \bar{M}^1_X \cap \bar{M}^0_{X_{ord}} = \bar{M}^0_X. \tag{8.128}$$

The statement in Equation 8.128 says that $\bar{\partial}\bar{P}+\bar{P}^2$ has no singularity along the zero locus of $\bar{H}$; this is, of course, well known in the case of modular curves for in that situation $\bar{\partial}\bar{P}+\bar{P}^2$ is actually a $\mathbf{F}_p^\times$-multiple of the modular form a_4, cf. [**88**], p. 161.

PROOF. Taking $\bar{\partial}_0$ in Equation 8.127 we get

$$\bar{\partial}^2\bar{H}\cdot\bar{f}^1_{crys} - \bar{H}\cdot\bar{\partial}_0^2\bar{f}^1_{crys} = 0. \tag{8.129}$$

Using Proposition 8.43 we get

$$\bar{\partial}^2\bar{H} = (\bar{\partial}\bar{P}+\bar{P}^2)\cdot\bar{H}. \tag{8.130}$$

Using Equations 8.129 and 8.130 and dividing by $\bar{H}$ we get

$$(\bar{\partial}\bar{P}+\bar{P}^2)\cdot\bar{f}^1_{crys} = \bar{\partial}_0^2\bar{f}^1_{crys}. \tag{8.131}$$

To conclude multiply Equation 8.131 by $\bar{\partial}\bar{H}$, multiply Equation 8.130 by $-\bar{\partial}_0\bar{f}^1_{crys}$, add the results, and use Equation 8.127. □

REMARK 8.70. By Equation 3.54 we have another set of differential equations, this time with respect to the Euler operator, satisfied by f^1_{crys}; they simply express the fact that f^1_{crys} has weight $-1-\phi$:

$$\mathcal{D}_0 f^1_{crys} = -f^1_{crys}, \quad \mathcal{D}_1 f^1_{crys} = -pf^1_{crys}. \tag{8.132}$$

PROPOSITION 8.71. *The following formulae hold:*

$$\partial_1 f^\partial = -p\cdot P^\phi\cdot f^\partial, \quad \partial_0 f^\partial = P\cdot f^\partial. \tag{8.133}$$

Equivalently,

$$\partial_1^* f^\partial = 0, \quad \partial_0^* f^\partial = 0. \tag{8.134}$$

In particular

$$\partial^*_{**} f^\partial = 0,$$

so ∂^* is a variational infinitesimal symmetry of f^∂. Also (by Propositions 8.64 and 8.66) ∂^* is an infinitesimal symmetry of $K\cdot f^\partial + K\cdot f_\partial + K\cdot f^1_{crys}$.

PROOF. Let's check the first of the equalities 8.133. (The other equality is proved similarly.) Both members of this equality belong to

$$M^1_{X_{ord}}(1+\phi);$$

cf. Proposition 3.56. By the Serre-Tate expansion principle in Proposition 8.29 it is enough to show that the images of the two members of the above equality under the Serre-Tate map

$$E : M^1_{X_{ord}}(1+\phi) \to R[[T]][T']\hat{}$$

coincide. Now, by Proposition 8.59, we have $E(f^\partial) = 1$. A trivial computation, using the latter formula plus Proposition 8.42 shows that $E(\partial_1 f^\partial)$ and $E(-pP^\phi f^\partial)$ coincide. □

COROLLARY 8.72. *The following formulae hold:*

$$f^1_{crys} \cdot \partial_0 f^\partial - f^\partial \cdot \partial_0 f^1_{crys} = 1, \tag{8.135}$$

$$\partial_1(f^\partial f^1_{crys}) = f^\partial \cdot \partial_1 f^1_{crys} + f^1_{crys} \cdot \partial_1 f^\partial = (f^\partial)^2, \tag{8.136}$$

$$\partial_0^2 f^1_{crys} = (\partial P + P^2) f^1_{crys}, \quad \partial_1^2 f^1_{crys} = p^2(\partial P + P^2)^\phi f^1_{crys}, \tag{8.137}$$

$$\frac{\partial_1^2 f^1_{crys}}{f^1_{crys}} = p^2 \left(\frac{\partial_0^2 f^1_{crys}}{f^1_{crys}} \right)^\phi . \tag{8.138}$$

Formula 8.135 is, of course, a lifting to characteristic zero of the formula 8.127 and is equivalent to the formula

$$\partial_0(f_\partial f^1_{crys}) = -(f_\partial)^2. \tag{8.139}$$

On the other hand formula 8.136 is a lifting to characteristic zero of the formula in Corollary 8.65.

PROOF. Plugging in the value

$$P = \frac{\partial_0 f^\partial}{f^\partial}$$

obtained from Equation 8.133 in the Equation of Proposition 8.66 we get Equation 8.135. Plugging in the value

$$p \cdot P^\phi = -\frac{\partial_1 f^\partial}{f^\partial}$$

obtained from Equation 8.133 in the Equation of Proposition 8.64 we get Equation 8.136. Equation 8.137 follows directly from Equation 8.133 plus Propositions 8.64 and 8.66. Equation 8.138 follows from Equation 8.137. □

REMARK 8.73. By Equation 3.54 we have another set of differential equations, this time with respect to the Euler operator, satisfied by f^∂; they simply express the fact that f^∂ has weight $\phi - 1$:

$$\mathcal{D}_0 f^\partial = -f^\partial, \quad \mathcal{D}_1 f^\partial = p f^\partial. \tag{8.140}$$

Equations 8.140 and 8.132 together with Equations 8.133, 8.122, and 8.124 provide a computation of the derivatives of f^∂ and f^1_{crys} with respect to

$$\partial_0, \mathcal{D}_0, \partial_1, \mathcal{D}_1 \tag{8.141}$$

which form a basis of the tangent sheaf of $Spf\ M^1_{X_1}$ where $X_1 \subset X$ is an affine open subset; this list of equations can therefore be viewed as a Pfaffian system of equations satisfied by f^∂, f_∂, and f^1_{crys} as follows.

COROLLARY 8.74. *Let $\omega_0, \eta_0, \omega_1, \eta_1$ be the basis of the 1–forms on $Spf\ M^1_{X_1}$ dual to the basis in Equation 8.141 and let $\nu := P\omega_0 - \eta_0$. Then $(U, V, W) := (f^\partial, f_\partial, f^1_{crys})$ satisfies the following system of equations.*

$$dU = U \cdot (\nu - \phi^*\nu),$$

$$dV = V \cdot (-\nu + \phi^*\nu),$$

$$dW = W \cdot (\nu + \phi^*\nu) - V \cdot \omega_0 + U \cdot \omega_1.$$

In particular the 1–form $\nu - \phi^*\nu$ is closed.

8.5. δ–invariants in the ordinary case

In this Section we consider a Hecke correspondence $\mathbf{X}$ equipped with a regular Hecke fixed point. We will not need to assume $\mathbf{X}$ is affine since, as we shall see, the non-affine case will reduce immediately to the affine case.

Let us fix $w \in W$, $deg(w) = -2m$, $m \in \mathbf{Z}$, $ord(w) = r \geq 1$. Note that for each $v \in W_+$ with $deg(v) = m$ and $ord(v) \leq r - 1$ there is a unique $v' \in W$ with $ord(v') \leq r - 1$ such that

$$(-1 - \phi)v + (\phi - 1)v' = w.$$

We call v' the w–complement of v.

PROPOSITION 8.75. *Let $\mathbf{X}$ be an irreducible Hecke correspondence that admits a regular Hecke fixed point and let $w \in W$ have even degree $deg(w) = -2m$. Then the K–linear space*

$$H^0(\mathbf{X}_{ord}, \mathbf{L}_{r,w}) \otimes K$$

has a basis consisting of all monomials

$$(f^1_{crys})^v (f^\partial)^{v'}$$

where v runs through the set $W_+(r-1, m)$ of all elements of W_+ of degree m and order $\leq r - 1$ and v' is the w–complement of v. In particular

$$rank\ H^0(\mathbf{X}_{ord}, \mathbf{L}_{r,w}) = \frac{(r+m-1)!}{(r-1)!m!} \quad for \quad deg(w) \leq 0,$$

$$H^0(\mathbf{X}, \mathbf{L}_{r,w}) = H^0(\mathbf{X}_{ord}, \mathbf{L}_{r,w}) = 0, \quad for \quad deg(w) > 0.$$

PROOF. By replacing $\tilde{X}$ with an affine neighborhood of the point $\tilde{P}$ defining the Hecke fixed point we may assume $\mathbf{X}$ is affine. Recall that we have an injective map

$$E : H^0(\mathbf{X}, \mathbf{L}_{r,w}) \to H^0(\Sigma, \Lambda^m).$$

By Proposition 4.36 the dimension of $H^0(\Sigma, \Lambda^m) \otimes K$ is the dimension of the space of polynomials of degree m in r variables. The latter equals, however, the cardinality of $W_+(r-1, m)$. Since the monomials $(f^1_{crys})^v (f^\partial)^{v'}$ are linearly independent the map $E \otimes K$ is an isomorphism. □

PROPOSITION 8.76. *The following formula holds*

$$f^r_{crys} = \sum_{i=0}^{r-1} p^{r-1-i} (f^1_{crys})^{\phi^i} (f^\partial)^{\frac{-1-\phi^r+\phi^i+\phi^{i+1}}{\phi-1}}.$$

PROOF. By Proposition 8.75 f^r_{crys} is a linear combination with $K-$coefficients of the monomials

$$(f^1_{crys})^{\phi^i} (f^\partial)^{\frac{-1-\phi^r+\phi^i+\phi^{i+1}}{\phi-1}};$$

by Proposition 8.61 we get that the coefficients of these monomials equal p^{r-1-i}. □

The following Proposition can be viewed as an analogue of Propositions 6.13 and 7.25.

PROPOSITION 8.77. *We have*

$$\partial_i^* \partial_j^* f^r_{crys} = 0 \tag{8.142}$$

for all i, j, r. So if

$$V_r^{ij} := \partial_i^* f^r_{crys} \cdot \partial_j^* - \partial_j^* f^r_{crys} \cdot \partial_i^* \in Der_R(M^r_{X_{ord}}, M^r_{X_{ord}})$$

is the $(i,j)-$Hamiltonian flow defined by f^r_{crys} and ∂^*, then any two Hamiltonian flows V_r^{ij} and V_s^{mn} commute. Moreover, for any $\alpha \in pR$, we have

$$E((\alpha\partial^*)_{**} f^r_{crys}) = \epsilon^{-1} E(f^r_{crys})(e_{\mathbf{G}_m}(\alpha)) \tag{8.143}$$

and the equality $(\alpha\partial^*)_{**} f^r_{crys} = 0$ implies $\alpha = 0$. Consequently we have the following equality in $K[[T, T', ..., T^{(r)}]]$:

$$\left(l_{\mathbf{G}_m}(T) \cdot (1+T) \cdot \frac{d}{dT}\right)_{**} E(f^r_{crys}) = E(f^r_{crys}); \tag{8.144}$$

in particular $l_{\mathbf{G}_m}(T) \cdot (1+T) \cdot \frac{d}{dT}$ is an infinitesimal symmetry for each of the spaces $K \cdot E(f^r_{crys})$.

Finally, the Euler-Lagrange equation, $\langle \partial^*, EL_1((f^1_{crys})^2)\rangle_1$, attached to $(f^1_{crys})^2$ and ∂^* equals

$$2[f^1_{crys} f^\partial - p(f^1_{crys})^\phi f^\phi_\partial].$$

In the statement above $e_{\mathbf{G}_m}$ is the exponential series of the multiplicative formal group $\mathbf{G}_m$ (viewed as a bijection $pR \to pR$).

PROOF. Let's check Equation 8.142. By Proposition 8.76 f^r_{crys} is a linear combination of products of the form $(f^1_{crys})^{\phi^a}(f^\partial)^{v'}$, $v' \in W$. By Equations 8.134

$$\partial_j^*((f^1_{crys})^{\phi^a}(f^\partial)^{v'}) = (f^\partial)^{v'} \partial_j^*((f^1_{crys})^{\phi^a}) = (f^\partial)^{v'} \cdot p^a (\partial^*_{j-a} f^1_{crys})^{\phi^a}.$$

The latter is 0 if $j-a \in \mathbf{Z}\backslash\{0,1\}$ and equals $(f^\partial)^{v''}$ for some $v'' \in W$ if $j-a \in \{0,1\}$; cf. Equations 8.123 and 8.125. We conclude by Equations 8.134. To check the assertion about $(\alpha\partial^*)_{**} f^r_{crys}$ we compute using Proposition 8.61:

$$E((\alpha\partial^*)_{**} f^r_{crys}) = E(\sum_{j=0}^{r} p^{-j} \alpha^{\phi^j} \partial_j^* f^r_{crys})$$

$$= \frac{1}{p}(\alpha^{\phi^r} - p^r \alpha) = \epsilon^{-1} E(f^r_{crys})(e_{\mathbf{G}_m}(\alpha)).$$

Clearly the above expression can only vanish if $\alpha = 0$. The assertion about the Euler-Lagrange equation is left to the reader. □

EXAMPLE 8.78. As in the spherical and flat case one can explicitly compute δ–tangent maps; here is a typical example. Let $X(R)'$ be the set of all points $Q \in X(R)$ such that $f^1_{crys}(Q) \not\equiv 0 \bmod p$, let $f : X(R)' \to \mathbf{P}^1(R)$ be the map defined by the sections $(s_0, s_1) = ((f^1_{crys})^{1+\phi^2}, (f^2_{crys})^{1+\phi})$, let $Q \in X(R)'$, and let $T_Q f : T_Q X \to T_{f(Q)}\mathbf{P}^1$ be the δ–tangent map at Q; cf. Equation 3.44. Note that $P^{-1}\partial$ is a basis of the tangent sheaf $T_{X/R}$ on some Zariski open set $X_1 \subset X$ so if $\alpha \in R$ then the δ–tangent map $T_Q f$ is given by

$$(T_Q f)(\alpha P^{-1}\partial_Q) = \left[(\alpha P^{-1}\partial)_* \left(\frac{(f^2_{crys})^{1+\phi}}{(f^1_{crys})^{1+\phi^2}}\right)\right](Q).$$

It is a trivial exercise to explicitly compute the latter by using our computations so far. As a result, for instance, the equality $(T_Q f)(\alpha P^{-1}\partial_Q) = 0$ implies that Q is a zero of the reduction mod p of the form $P^{p-1} - \alpha^{p-1}H^2$.

THEOREM 8.79. *Let $\mathbf{X}$ be an irreducible Hecke correspondence that admits a regular Hecke fixed point. For each $w \in W$ of even degree and order ≥ 1 there is a natural isomorphism of R–modules*

$$H^0(\mathbf{X}_{ord}, \mathbf{L}_w) \simeq R[x^{(\leq ord(w)-1)}]_{-deg(w)/2}.$$

Consequently

$$R(\mathbf{X}_{ord}, \mathbf{L}_w)^* \simeq (R\{x_*, y\}^*)^{((w_0))}, \quad w_0 := (-deg(w)/2)\phi^{ord(w)-1}.$$

As the proof will show, if $f \in H^0(\mathbf{X}_{ord}, \mathbf{L}_w)$ corresponds to a polynomial $\Phi \in R[x^{(\leq ord(w)-1)}]_{-deg(w)/2}$ then the Serre-Tate expansion of f is given by

$$E(f) = \Phi(\Psi, \delta\Psi, ..., \delta^{ord(w)-1}\Psi).$$

PROOF. Again, we may assume $\mathbf{X}$ is affine. Set $r = ord(w)$. By Proposition 8.75 we may consider the map

$$B : H^0(\mathbf{X}_{ord}, \mathbf{L}_w) \otimes K \to R[x^{(\leq r-1)}]_{-deg(w)/2} \otimes K = K[x, x^\phi, ..., x^{\phi^{r-1}}]_{-deg(w)/2}$$

$$\sum a_v (\epsilon^{-1} f^1_{crys})^v (f^\partial)^{v'} \mapsto \sum a_v x^v,$$

defined for all $a_v \in K$, where all summations here are over $v \in W_+(r-1, m)$. We will prove that

$$B(H^0(\mathbf{X}_{ord}, \mathbf{L}_w)) = R[x^{(\leq r-1)}]_{-deg(w)/2}. \tag{8.145}$$

Since the source and the target of B have the same rank the conclusion of the Theorem will follow. To check the equality 8.145 start with

$$f = \sum a_v (\epsilon^{-1} f^1_{crys})^v (f^\partial)^{v'} \in H^0(\mathbf{X}_{ord}, \mathbf{L}_w).$$

Since $E(f) \in S^r = R[[T]][T', ..., T^{(r)}]\hat{}$ it follows from Propositions 8.61 and 8.59 that

$$\sum a_v \Psi^v \in R[[T]][T', ..., T^{(r)}]\hat{}. \tag{8.146}$$

Consider the unique R–algebra homomorphism

$$R[[T]][T', ..., T^{(r)}]\hat{} \to R[[T]][T', ..., T^{(r)}]\hat{} \tag{8.147}$$

that commutes with δ and sends

$$T \mapsto exp(pT) - 1.$$

Taking the image of the element in Equation 8.146 via the map in Equation 8.147 we get

$$\sum a_v(T^\phi - pT)^v \in R[[T]][T',...,T^{(r)}]\hat{}. \tag{8.148}$$

Since the element in Equation 8.148 also belongs to $K[T,T',...,T^{(r)}]$ we get

$$\sum a_v(T^\phi - pT)^v \in R[T,T',...,T^{(r)}]. \tag{8.149}$$

Set

$$G := \sum a_v x^v \in K[x^{(\leq r-1)}] \tag{8.150}$$

and let $\mu \in \mathbf{Z}_+$ be minimal such that

$$F := p^\mu G \in R[x^{(\leq r-1)}].$$

Assume $\mu \geq 1$. Then the image $\bar{F}$ of F in $k[x^{(\leq r-1)}]$ is non-zero and, on the other hand, by Equation 8.149,

$$F(T^\phi - pT, \delta(T^\phi - pT), ..., \delta^{r-1}(T^\phi - pT)) \in p^\mu R[T,T',...,T^{(r)}];$$

this contradicts Proposition 5.24. So $\mu = 0$, in other words

$$B(f) \in R[x^{(\leq r-1)}]_{-deg(w)/2}.$$

Let's assume now that

$$\sum a_v x^v \in R[x^{(\leq r-1)}]_{-deg(w)/2}$$

and set

$$G := \sum a_v x_1^v x_2^{\frac{w+(\phi+1)v}{\phi-1}} \in K[x_1^{(\leq r-1)}, x_2^{(\leq r-1)}, x_2^{-1-...-\phi^{r-1}}].$$

Let $\mu \in \mathbf{Z}_+$ be minimum such that

$$F := p^\mu G \in R[x_1^{(\leq r-1)}, x_2^{(\leq r-1)}, x_2^{-1-...-\phi^{r-1}}].$$

If $\mu \geq 1$ then note that the image $\bar{F}$ of F in $k[x_1^{(\leq r-1)}, x_2^{(\leq r-1)}, x_2^{-1-...-\phi^{r-1}}]$ is non-zero and

$$F(x_1, x_2^\phi/x_2, ..., x_1^{(r-1)}, \delta^{r-1}(x_2^\phi/x_2)) = p^\mu \sum a_v x_1^v x_2^{w+(\phi+1)v}$$

$$= p^\mu x_2^w \sum a_v (x_1 x_2^{\phi+1})^v \in p^\mu R[x_1^{(\leq r-1)}, x_2^{(\leq r-1)}, x_2^{-1-...-\phi^{r-1}}].$$

This contradicts Proposition 5.14. So $\mu = 0$ and hence

$$G(\epsilon^{-1} f^1_{crys}, f^\partial, ..., \delta^{r-1}(\epsilon^{-1} f^1_{crys}), \delta^{r-1}(f^\partial))$$

defines an element of $H^0(\mathbf{Y}_{ord}, \mathbf{L}_w)$; consequently $\sum a_v x^v$ is in $B(H^0(\mathbf{X}_{ord}, \mathbf{L}_w))$ which concludes the proof. □

Putting together the above Theorem and Proposition 5.14 we have:

COROLLARY 8.80. *Let* $\mathbf{X}$ *be an irreducible Hecke correspondence that admits a regular Hecke fixed point. For any* $w \in W$ *of even, negative degree there is a natural isomorphism of* $R-$*algebras*

$$R\langle \mathbf{X}_{ord}, \mathbf{L}_w \rangle \simeq (R\{x\}_*)_{((p))},$$

where x^ϕ/x *corresponds to* $(f^1_{crys})^{\phi-1}(f^\partial)^{\phi+1}$*; so in particular* $(\mathbf{X}_{ord}, \mathbf{L}_w)$ *is* $\delta-$*rational.*

8.6. δ-invariants in the non-ordinary case

THEOREM 8.81. *Let* $\mathbf{X}$ *be an irreducible Hecke correspondence that admits a regular Hecke fixed point and let*

$$w = -1 - \phi - \phi^2 - \phi^3.$$

Then there is a natural isomorphism of R-*algebras*

$$R\langle \mathbf{X}, \mathbf{L}_w \rangle\hat{} \simeq R\langle \mathbf{X}_{ord}, \mathbf{L}_w \rangle\hat{}.$$

Moreover

$$\frac{(f^2_{crys})^{1+\phi}}{(f^1_{crys})^{1+\phi^2}}$$

is a δ-*basis of* $R\langle \mathbf{X}, \mathbf{L}_w \rangle$; *in particular,* $(\mathbf{X}, \mathbf{L}_w)$ *is* δ-*non-trivial and* δ-*rational.*

PROOF. As usual, we may assume $\mathbf{X}$ is affine. It is easy to check that the natural homomorphism

$$R(\mathbf{X}, \mathbf{L}_w)^* \to R(\mathbf{X}_{ord}, \mathbf{L}_w)^* \tag{8.151}$$

is telescopic. (By the way, identifying $R(\mathbf{X}_{ord}, \mathbf{L}_w)^*$ with $(R\{x_*, y\}^*)^{((w_0))}$ as in Theorem 8.79 and composing the map in Equation 8.151 with the telescopic map

$$(R\{x_*, y\}^*)^{((w_0))} \to R\{x\}_*$$

we get a telescopic map

$$R(\mathbf{X}, \mathbf{L}_w)^* \to R\{x\}_*.) \tag{8.152}$$

Now the map 8.151 induces a field homomorphism

$$k\langle \mathbf{X}, \mathbf{L}_w \rangle \to k\langle \mathbf{X}_{ord}, \mathbf{L}_w \rangle; \tag{8.153}$$

it is enough to show that the latter is an isomorphism. Note that, by Corollary 8.80 and Proposition 5.14,

$$k\langle \mathbf{X}_{ord}, \mathbf{L}_w \rangle = k(\eta_0, \eta_1, \eta_2, \ldots),$$

where η_i is the image of $\delta^i(x^\phi/x) \in (R\{x\}_*)_{((p))}$ in $(R\{x\}_*)_{((p))} \otimes k$. On the other hand, by Proposition 8.76, we have

$$f^2_{crys} = p f^1_{crys} (f^\partial)^{-\phi} + (f^1_{crys})^\phi f^\partial.$$

Consequently, if

$$t := \frac{(f^2_{crys})^{1+\phi}}{(f^1_{crys})^{1+\phi^2}} - p \in R\langle \mathbf{X}, \mathbf{L}_w \rangle,$$

then

$$t = (f^1_{crys})^{\phi-1} (f^\partial)^{\phi+1} + p (f^1_{crys})^{2\phi-1-\phi^2} (f^\partial)^{1-\phi^2} + p^2 (f^1_{crys})^{\phi-\phi^2} (f^\partial)^{-\phi-\phi^2};$$

so the image of t in $(R\{x\}_*)_{((p))}$, which we still denote by t, is

$$t = z + pz z^{-\phi} + p^2 z^{-\phi},$$

where $z = x^\phi/x$. Note that one can write

$$t = z + pG_0(z, z^{-1}) + p^2 H_0(z, z^{-1}, \delta z)$$

in the ring $((R\{x\}_*)_{((p))})\hat{}$, where $G_0 \in \mathbf{Z}_p[Z, Z_1]\hat{}$, $H_0 \in \mathbf{Z}_p[Z, Z_1, Z']\hat{}$, are restricted Laurent power series in variables Z, Z_1, and Z, Z_1, Z' respectively. It is then easy to check by induction that, for any $i \geq 0$, we have an equality in the ring $((R\{x\}_*)_{((p))})\hat{}$ of the form

$$\delta^i t = \delta^i z + F_i(z, z^{-1}, \delta z, ..., \delta^{i-1} z) + pG_i(z, z^{-1}, \delta z, ..., \delta^i z)$$
$$+p^2 H_i(z, z^{-1}, \delta z, ..., \delta^{i+1} z),$$

where F_i, G_i, H_i are restricted Laurent power series. So if β_i is the image of $\delta^i t \in (R\{x\}_*)_{((p))}$ in $(R\{x\}_*)_{((p))} \otimes k$ then clearly

$$k(\beta_0, \beta_1, \beta_2, ...) = k(\eta_0, \eta_1, \eta_2, ...)$$

which implies that the extension in Equation 8.153 is an equality. □

REMARK 8.82. Theorem 8.81 clearly continues to hold if instead of assuming $\mathbf{X}$ irreducible we assume $\mathbf{X}$ is a disjoint union of irreducible Hecke correspondences.

Unlike in the case of $\mathbf{X}_{ord}$ where the structure of $H^0(\mathbf{X}_{ord}, \mathbf{L})$ was entirely determined, the case of $\mathbf{X}$ is still mysterious in general. Here is a sample of known facts about it.

THEOREM 8.83. *Let $\mathbf{X}$ be an irreducible Hecke correspondence that admits a regular Hecke fixed point and assume X possesses a supersingular $k-$point. Let $w \in W$, $deg(w) = -2m$, $m \leq p$, $ord(w) = r \geq 1$.*

1) Define

$$\delta_w := \sharp\{v \in W_+(r-1, m) | v(p) < \frac{w(p)}{p+1}\}. \tag{8.154}$$

Then

$$rank\ H^0(\mathbf{X}, \mathbf{L}_w) \leq rank\ H^0(\mathbf{X}_{ord}, \mathbf{L}_w) - \delta_w = \frac{(r+m-1)!}{(r-1)!m!} - \delta_w. \tag{8.155}$$

2) If $i, j \geq 0$, the spaces

$$H^0(\mathbf{X}, \mathbf{L}_{-\phi^i - \phi^j}) \tag{8.156}$$

have rank one for $i \neq j$ and vanish for $i = j$.

3) We have

$$H^0(\mathbf{X}, \mathbf{L}_{r,w}) = 0 \tag{8.157}$$

if $deg(w) = 0$ and either $w(p) < 0$ or $0 < w(p) \leq p^{ord(\bar{w})} - p^{ord(\bar{w})-1}$. In particular

$$H^0(\mathbf{X}, \mathbf{L}_{\phi - 1}) = H^0(\mathbf{X}, \mathbf{L}_{1-\phi}) = 0.$$

4) If $\mathbf{X}$ is affine, the class

$$\xi := f^1_{crys} \cup [P^{(e)}]^\phi \in H^1(\mathbf{X}, \mathbf{L}_{\phi-1})$$

is $p-$torsion (i.e. $p\xi = 0$) but $\xi \neq 0$.

5) If $\mathbf{X}$ is affine, the class $[P^{(e)}]^\phi \in H^1(\mathbf{X}, \mathbf{L}_{2\phi})$ is non-torsion.

PROOF. For assertions 1, 2, 3 we may assume $\mathbf{X}$ is affine. Let us prove assertion 1. Let $V \subset H^0(\mathbf{X}_{ord}, \mathbf{L}_w)$ be the $R-$submodule with basis

$$B := \{(f^1_{crys})^v (f^\partial)^{v'} | v \in W_+(r-1, m), (-1-\phi)v + (\phi - 1)v' = w, v(p) < -\frac{w(p)}{p+1}\}.$$

Since $rank\ V = \delta_w$ it is enough to show that

$$V \cap H^0(\mathbf{X}, \mathbf{L}_w) = 0.$$

Assume $0 \neq \tilde{f} \in V \cap H^0(\mathbf{X}, \mathbf{L}_w)$,

$$\tilde{f} = \sum_{v \in B} \tilde{\lambda}_v (f^1_{crys})^v (f^\partial)^{v'}, \quad \tilde{\lambda}_v \in R.$$

Write $\tilde{f} = p^n f$, $f \in H^0(\mathbf{X}, \mathbf{L}_w)$, $f \notin p \cdot H^0(\mathbf{X}, \mathbf{L}_w)$, and set $\lambda_v = p^{-n} \tilde{\lambda}_v \in K$; so

$$f = \sum_{v \in B} \lambda_v (f^1_{crys})^v (f^\partial)^{v'}.$$

By the proof of Theorem 8.79 the polynomial

$$\sum_{v \in B} \lambda_v \epsilon^v x^v$$

belongs to $R\{x\}_m$. By Lemma 5.2 we must have $\lambda_v \in R$ for all $v \in B$. By Proposition 8.57 we have

$$\bar{f} = \sum_{v \in B} (\bar{f}^1_{crys})^{v(p)} \bar{H}^{v'(p)} \quad \text{in} \quad \bar{M}^r_{X_{ord}}.$$

Note that

$$v'(p) = \frac{w(p) + (p+1)v(p)}{p-1} < 0 \quad \text{for} \quad v \in B.$$

For each $k \geq 0$ set

$$\bar{\mu}_k := \sum_{v(p)=k} \bar{\lambda}_v;$$

clearly

$$\bar{\mu}_k = 0 \quad \text{if} \quad \frac{w(p) + (p+1)v(p)}{p-1} \notin \mathbf{Z}.$$

We may write

$$\bar{f} = \sum_{k \geq 0} \bar{\mu}_k (\bar{f}^1_{crys})^k \bar{H}^{\frac{w(p)+(p+1)v(p)}{p-1}}. \tag{8.158}$$

Note that $\bar{f} \neq 0$ in $\bar{M}^r_{X_{ord}}$; for if $\bar{f} = 0$ it follows that $f \in pM^r_{X_{ord}} \cap M^r_X = pM^r_X$ hence $f \in p \cdot H^0(\mathbf{X}, \mathbf{L}_w)$, a contradiction. Let

$$k_1 := \min\{k | \bar{\mu}_k \neq 0\}, \quad k'_1 := \frac{w(p) + (p+1)k_1}{p-1}.$$

Multiplying Equation 8.158 by $\bar{H}^{-k'_1}$ we get an equation of the form

$$(\bar{f}^1_{crys})^{k_1} = \bar{H} \cdot \bar{G}, \tag{8.159}$$

with $\bar{G} \in \bar{M}^r_X$. On the other hand, raising Equation 8.127 to the power k_1 we get an equation of the form

$$(\bar{f}^1_{crys})^{k_1} (\bar{\partial} \bar{H})^{k_1} + \bar{H} \cdot \bar{G}_1 = 1, \tag{8.160}$$

for some $\bar{G}_1 \in \bar{M}^1_X$. Combining Equations 8.159 and 8.160 we get

$$\bar{H} \cdot [\bar{G} \cdot (\bar{\partial} \bar{H})^{k_1} + \bar{G}_1] = 1.$$

Hence $\bar{H}$ is invertible in $\bar{M}^r_X$, hence $\bar{H}$ is invertible in $\bar{M}^0_X$, cf. Corollary 8.28 and Corollary 3.16, assertion 1. We deduce that $X_{ord} \otimes k = X \otimes k$, a contradiction. This ends the proof of assertion 1.

To check assertion 2 note that the spaces in Equation 8.156 are non-zero for $i \neq j$ because they contain the forms

$$(f_{crys}^{j-i})^{\phi^i}.$$

On the other hand

$$\delta_{-\phi^i-\phi^j} = j \quad \text{or} \quad j-1$$

according as $i = j$ or $i \neq j$ and we may conclude by assertion 1 of the Theorem.

Let's check assertion 3. Assume first $w(p) < 0$ and let $g \in H^0(\mathbf{X}, \mathbf{L}_w)$. By Theorem 8.79 we may write

$$g = (f^{\partial})^{v'}$$

where $v'(p) < 0$. So

$$\bar{g} = (\bar{f}^{\partial})^{v'(p)} = \bar{H}^{v'(p)}$$

which implies that $\bar{H}$ is invertible, a contradiction. Let's assume now that

$$deg(w) = 0, \quad 0 < w(p) \leq p^{ord(\bar{w})} - p^{ord(\bar{w})-1}. \tag{8.161}$$

In particular $w \notin pW$. Let's examine the case $w = \phi - 1$ first. Take $g \in H^0(\mathbf{X}, \mathbf{L}_{\phi-1})$. So $E(g) = \lambda \in R$. Consequently the form

$$h := g \cdot (f_{crys}^1)^{\phi}$$

belongs to $H^0(\mathbf{X}, \mathbf{L}_{-1-\phi^2})$ and has Serre-Tate expansion $E(h) = \lambda\epsilon\Psi^{\phi}$. On the other hand the form $f_{crys}^2 \in H^0(\mathbf{X}, \mathbf{L}_{-1-\phi^2})$ has Serre-Tate expansion

$$E(f_{crys}^2) = \epsilon(\Psi^{\phi} + p\Psi).$$

Since, by assertion 2, $H^0(\mathbf{X}, \mathbf{L}_{-1-\phi^2})$ has rank one it follows that h and f_{crys}^2 are linearly dependent; hence $E(h)$ and $E(f_{crys}^2)$ are linearly dependent. This forces $\lambda = 0$ hence $g = 0$. This proves assertion 3 for $w = \phi - 1$, i.e.

$$H^0(\mathbf{X}, \mathbf{L}_{\phi-1}) = 0. \tag{8.162}$$

Let us note next that $f^{\partial} \notin M_X^1$; for if $f^{\partial} \in M_X^1$, since $f^{\partial} \in M_{X_{ord}}^1(\phi - 1)$, it would follow that $f^{\partial} \in M_X^1(\phi - 1)$. Cf. Corollary 8.28 and Corollary 3.16. Since $E(f^{\partial}) = 1$ we would get $f^{\partial} \in H^0(\mathbf{X}, \mathbf{L}_{\phi-1})$ which contradicts Equation 8.162. Let us prove now assertion 3 for arbitrary w satisfying condition 8.161. Let $g \in H^0(\mathbf{X}, \mathbf{L}_w)$. So again $E(g) = \lambda \in R$. Since $deg(w) = 0$ we may write $w = (\phi - 1)v'$. Hence

$$E(g) = E(\lambda \cdot (f^{\partial})^{v'}),$$

hence $g = \lambda \cdot (f^{\partial})^{v'}$. Note that $m' := ord(\bar{v}') = ord(\bar{w}) - 1 =: m - 1$ and $v'(p) = w(p)/(p-1)$ so $v' \notin pW$ and

$$1 \leq v'(p) \leq (p^m - p^{m-1})/(p-1) = p^{m-1} = p^{m'}.$$

If $\lambda \neq 0$ we get

$$g \in M_X^r \cap \lambda M_{X_{for}}^r = \lambda M_X^r;$$

cf. Corollary 8.28 and Corollary 3.16, assertion 5. So $(f^{\partial})^{v'} \in M_X^r$. Since $f^{\partial} \in (M_{X_{ord}}^1)^{\times}$, by Corollary 8.28 and Corollary 3.16, assertion 6, we get $f^{\partial} \in M_X^1$, a contradiction. So $\lambda = 0$ and, hence, $g = 0$.

To check assertion 4 note that $\partial_1 f^1$ does not belong to $pM_X^1(\phi - 1)$ by Corollary 8.65. In view of Equation 8.162 we may conclude by Corollary 8.39.

Let us check assertion 5. Assume $p^N(P^{(e)})^\phi = \Delta_{-1}Q$, $N \geq 1$, $Q \in M^1_X(2\phi)$. Since, by Proposition 8.37, $(P^{(e)})^\phi = -e\Delta_{-1}(P^\phi)$ in $M^1_{\tilde{X}_{ord}}(2\phi)$ it follows that

$$\Delta_{-1}(Q + ep^N P^\phi) = 0.$$

So $Q + ep^N P^\phi \in H^0(\mathbf{X}, \mathbf{L}_{2\phi})$; by Proposition 8.75 the latter space vanishes, so $Q = -ep^N P^\phi$ in $M^1_{X_{ord}}(2\phi)$. Since $M^1_X(2\phi) \otimes k \to M^1_{X_{ord}}(2\phi) \otimes k$ is injective and since $M^1_X(2\phi)$ is torsion free it follows that $[P^{(e)}]^\phi = 0$ in $H^1(\mathbf{X}, \mathbf{L}_{2\phi})$. This would imply $\xi = 0$ in statement 4, a contradiction. □

COROLLARY 8.84. *Assume* $\mathbf{X}$ *is a Hecke correspondence with a regular Hecke fixed point and assume we are in the elliptic case. Then, for* $r \geq 1$, *there are equalities*

$$f^r_{crys} = c_r \cdot f^r_{jet} \quad in \quad H^0(\mathbf{X}, \mathbf{L}_{-1-\phi^r}) \quad with \quad c_r \in R^\times,$$
$$f^{1,2}_{jet} = c_{1,2} \cdot (f^1_{jet})^\phi \quad in \quad H^0(\mathbf{X}, \mathbf{L}_{-\phi-\phi^2}) \quad with \quad c_{1,2} \in R.$$

PROOF. We may assume $\mathbf{X}$ is affine. Since $H^0(\mathbf{X}, \mathbf{L}_{-1-\phi^r})$ has rank one the forms f^r_{crys} and f^r_{jet} differ by a constant in K. On the other hand we claim that none of these forms is in $p \cdot H^0(\mathbf{X}, \mathbf{L}_{-1-\phi^r})$. For f^r_{crys} the claim follows from the fact that $E(f^r_{crys}) \notin pR[[T]][T', ..., T^{(r)}]\hat{}$. For f^1_{jet} the claim follows from Corollary 8.53. For f^r_{jet}, $r \geq 2$, the claim follows from Proposition 8.55. Finally, since $H^0(\mathbf{X}, \mathbf{L}_{-\phi-\phi^2})$ has rank one we have

$$f^{1,2}_{jet} = c_{1,2} \cdot (f^1_{jet})^\phi, \quad c_{1,2} \in K.$$

We conclude by the fact that $(f^1_{jet})^\phi \notin p \cdot H^0(\mathbf{X}, \mathbf{L}_{-\phi-\phi^2})$; the latter is the case because

$$E((f^1_{jet})^\phi) \in R^\times \cdot \Psi^\phi.$$

□

REMARK 8.85. It actually follows from the main result in [**19**] that, in Corollary 8.84, we have $c_{1,2} = 1$. We will not use this fact here.

In particular we reobtain the following result due to M. Barcau:

COROLLARY 8.86. [**2**] *Assume we are in the elliptic case. Then, for* $r \geq 1$, *there are equalities*

$$f^r_{crys} = c_r \cdot f^r_{jet} \quad in \quad M^r = R[a_4^{(\leq r)}, a_6^{(\leq r)}, \Delta^{-1}]\hat{} \quad with \quad c_r \in R^\times.$$

PROOF. Let $M := M_{Weier} = R[a_4, a_6, \Delta^{-1}]$. By Proposition 8.22 one can find a Hecke correspondence $(\mathbf{X}, A, \tilde{u}, s)$ with a regular Hecke fixed point. (One takes $N = 4$ in loc. cit.) There are induced maps

$$Spec\ M \xrightarrow{\pi} Spec\ R[j] \xleftarrow{\rho} X$$

where $j = -2^{12}3^3a_4^3/\Delta$. One can choose an open set $U \subset Spec\ R[j]$ such that

$$\rho : X' := \rho^{-1}(U) \to U$$

is étale and

$$\mathbf{V}(L) \times_X X' \simeq \pi^{-1}(U) \times_U X'. \tag{8.163}$$

Corollary 8.84 implies that we have equalities $f^r_{crys} = c_r \cdot f^r_{jet}$ ($c_r \in R^\times$) in $H^0(\mathbf{X}, \mathbf{L}_{-1-\phi^r})$ hence in $\mathcal{O}^r(\mathbf{V}(L))$ hence in $\mathcal{O}^r(\mathbf{V}(L) \times_X X')$ hence in

$$\mathcal{O}^r(\pi^{-1}(U) \times_U X')$$

hence in $\mathcal{O}^r(\pi^{-1}(U))$ (because $\pi^{-1}(U)\times_U X' \to \pi^{-1}(U)$ is étale) hence in

$$\mathcal{O}^r(Spec\ M) = M^r$$

and we are done. $\square$

Here are more consequences of our theory related to Hurlburt's formula in Corollary 8.53. We place ourselves in the genuine elliptic curves. Recall from [**88**], p. 161, that we have at our disposal the *classical Serre operator*

$$\partial : R[a_4, a_6] \to R[a_4, a_6],$$

which is defined to be the unique R-derivation satisfying

$$\partial a_4 = -2^3 3^2 a_6, \quad \partial a_6 = 2^4 a_4^2, \quad \text{(hence also } \partial\Delta = 0). \tag{8.164}$$

As shown by Katz [**81**] this classical Serre operator is compatible (via base changes as in Equation 8.163) with the operator ∂ that we have been using so far. We denote by

$$\partial_j : R[a_4^{(\leq r)}, a_6^{(\leq r)}, \Delta^{-1}]\hat{} \to R[a_4^{(\leq r)}, a_6^{(\leq r)}, \Delta^{-1}]\hat{}$$

the conjugate operators attached to the classical Serre operator ∂. Furthermore we denote by

$$\bar{\partial} : k[a_4, a_6] \to k[a_4, a_6],$$
$$\bar{\partial}_j : k[a_4^{(\leq r)}, a_6^{(\leq r)}, \Delta^{-1}] \to k[a_4^{(\leq r)}, a_6^{(\leq r)}, \Delta^{-1}]$$

the corresponding reductions mod p of the above operators. Set

$$\Phi = 2a_4^p a_6' - 3a_6^p a_4'.$$

Hence Hurlburt's formula reads

$$\bar{f}^1_{crys} = cc_1 \bar{H} \cdot \bar{\Delta}^N \cdot \bar{\Phi} + c_1 \bar{f}_0, \tag{8.165}$$

$c, c_1 \in k^\times$, $N \in \mathbf{Z}$. Using the formulae

$$\bar{\partial}_1 a_4' = -2^3 3^2 a_6^p,$$
$$\bar{\partial}_1 a_6' = 2^4 a_4^{2p},$$

one gets

$$\bar{\partial}_1 \bar{f}^1_{crys} = 2^7 cc_1 \cdot \bar{H} \cdot \bar{\Delta}^{N+p}. \tag{8.166}$$

As in the proof of Corollary 8.86, combining Equation 8.166 with the Equation in Corollary 8.65 we get $cc_1 = 2^{-7}$ and, more importantly,

COROLLARY 8.87. *In Hurlburt's formula of Corollary 8.53 we have $N = -p$.*

On the other hand, using formula 3.31 we have

$$\bar{\partial}_0 a_4' = -a_4^{p-1}\bar{\partial} a_4 = 2^3 3^2 a_4^{p-1} a_6,$$
$$\bar{\partial}_0 a_6' = -a_6^{p-1}\bar{\partial} a_6 = -2^4 a_6^{p-1} a_4^2,$$

and an easy computation yields:

$$\bar{\partial}_0 \bar{\Phi} = 2^{-1}(a_4 a_6)^{p-1}\bar{\Delta}. \tag{8.167}$$

Applying $\bar{\partial}_0$ to Equation 8.165 and using Equations 8.164 and 8.167 we get

$$\bar{\partial}_0 \bar{f}^1_{crys} = cc_1 \bar{\Delta}^{-p} \cdot \bar{\Phi} \cdot \bar{\partial}\bar{H} + 2^{-1} cc_1 \bar{H} \cdot \bar{\Delta}^{1-p} \cdot (a_4 a_6)^{p-1} + c_1 \bar{\partial}_0 \bar{f}_0. \tag{8.168}$$

By Corollary 8.67 we get

$$\bar{\partial}_0 \bar{f}^1_{crys} = -\bar{H}^{-1} + cc_1 \bar{P} \cdot \bar{H} \cdot \bar{\Delta}^{-p} \cdot \bar{\Phi} + c_1 \bar{P} \bar{f}_0. \tag{8.169}$$

Identifying the coefficients of a'_4 and a'_6 in Equations 8.168 and 8.169 we get the following formulae in the ring $k[a_4, a_6, \Delta^{-1}, H^{-1}]$:

$$\bar{\partial}\bar{H} = \bar{P} \cdot \bar{H}, \tag{8.170}$$

which we already knew, and

$$2^{-1}cc_1\bar{H} \cdot \bar{\Delta}^{1-p} \cdot (a_4a_6)^{p-1} + c_1\bar{\partial}_0 f_0 = -\bar{H}^{-1} + c_1\bar{P}\bar{f}_0. \tag{8.171}$$

COROLLARY 8.88. *The following formula holds in* $k[a_4, a_6, \Delta^{-1}]$*:*

$$\bar{H} \cdot \bar{\partial}\bar{f}_0 - \bar{f}_0 \cdot \bar{\partial}\bar{H} = \gamma_1 + \gamma_2 \cdot \bar{H}^2 \cdot \bar{\Delta}^{1-p} \cdot (a_4a_6)^{p-1},$$

with $\gamma_1, \gamma_2 \in k^\times$.

PROOF. As in the proof of Corollary 8.86, combining Equations 8.170 and 8.171 we get that our formula holds in $k[a_4, a_6, \Delta^{-1}, H^{-1}]$; since both terms of the equality are in $k[a_4, a_6, \Delta^{-1}]$, the formula holds in $k[a_4, a_6, \Delta^{-1}]$. □

COROLLARY 8.89. *Let E be an elliptic curve over R, having supersingular reduction. Then, for some (equivalently any) invertible 1–form ω we have*

$$f^1_{crys}(E, \omega) \not\equiv 0 \quad mod \quad p,$$

$$f^2_{crys}(E, \omega) \equiv 0 \quad mod \quad p.$$

PROOF. Let (E, ω) be represented by $(\alpha_4, \alpha_6) \in R^2$, $\Delta(\alpha_4, \alpha_6) \in R^\times$. Since E has supersingular reduction $\bar{H}$ vanishes at (α_4, α_6). By Corollary 8.88 $\bar{f}_0$ does not vanish at (α_4, α_6). By Equation 8.165 $\bar{f}^1_{crys}$ does not vanish at (α_4, α_6). On the other hand $\bar{f}^2_{crys}$ vanishes at (α_4, α_6) by Corollary 8.86 and Proposition 8.55. □

8.7. δ–cohomology

THEOREM 8.90. *Let* $\mathbf{X}$ *be an affine irreducible Hecke correspondence that admits a regular Hecke fixed point. For each $w \in W$ of even degree and order 1 the canonical map*

$$\cup[1] : H^0(\mathbf{X}, \mathbf{L}_{1,w}) \otimes K \to H^1(\mathbf{X}, \mathbf{L}_{1,w}) \otimes K$$

is injective.

PROOF. We may assume $\mathbf{X} = \mathbf{X}_{ord}$. Let $m := -deg(w)/2$. By Proposition 8.75 the Theorem is true if $m < 0$. Now we proceed by induction on $m \geq 0$. If $m = 0$ then by Proposition 8.75 $H^0(\mathbf{X}, \mathbf{L}_{1,w}) \otimes K$ has basis $(f^\partial)^{m'}$ where $m' = w/(\phi - 1)$. Assume

$$\lambda\sigma_1^*(f^\partial)^{m'} = \Delta_0 g$$

for some $\lambda \in R$, $\lambda \neq 0$, and some $g \in M^1_X(w)$, and let us derive a contradiction. Taking Serre-Tate expansions we get (by Propositions 8.30 and 8.59) that

$$\lambda = G(\tilde{T}, \tilde{T}') - G(\Gamma(\tilde{T}), \delta(\Gamma(\tilde{T})),$$

for some series G. Picking out the constant term in the above equality we get $\lambda = 0$, a contradiction. For the induction step, assume $m > 0$. By Proposition 8.75 $H^0(\mathbf{X}, \mathbf{L}_{1,w}) \otimes K$ has basis $(f^1_{crys})^m(f^\partial)^{m'}$ where $(-1 - \phi)m + (\phi - 1)m' = w$. Assume

$$\lambda\sigma_1^*[(f^1_{crys})^m(f^\partial)^{m'}] = \Delta_m g$$

for some $\lambda \in R$, $\lambda \neq 0$, and some $g \in M^1_X(w)$, and let us derive a contradiction. By Corollary 8.41 and Proposition 8.75 we get

$$\Delta_{m-1}\partial_1^* g = \tilde{\partial}_1^* \Delta_m g = \lambda\tilde{\partial}_1^*\sigma_1^*[(f^1_{crys})^m (f^{\partial})^{m'}]$$

$$= \lambda\sigma_1^*\{\partial_1^*[(f^1_{crys})^m (f^{\partial})^{m'}]\} = \lambda m\sigma_1^*[(f^1_{crys})^{m-1}(f^{\partial})^{m'+1}].$$

By the induction hypothesis $\lambda m = 0$, a contradiction. □

We now look at the p–torsion in H^1. Let us denote by $\Psi_n \in k[[T]][T', ..., T^{(n+1)}]$ the class of $\delta^n\Psi \in R[[T]][T', ..., T^{(n+1)}]\hat{}$.

PROPOSITION 8.91. $k[[T]][T', ..., T^{(n+1)}] = k[[T]][\Psi_0, ..., \Psi_n]$.

PROOF. It is enough to prove that, for any $n \geq 0$, we have

$$\delta^n\Psi \in \frac{T^{(n+1)}}{1 + T^{p^{n+1}}} + R[[T]][T', ..., T^{(n)}]\hat{} + pR[[T]][T', ..., T^{(n+1)}]\hat{};$$

the latter can be easily checked by induction on n. □

For $a \in \mathbf{Z}_p$ let $\Sigma(T) = (1+T)^a - 1 \in R[[T]]$. Let

$$\Sigma_n(T, ..., T^{(n)}) \in k[[T]][T', ..., T^{(n)}]$$

denote the class of $\delta^n(\Sigma(T)) \in R[[T]][T', ..., T^{(n+1)}]\hat{}$. The map

$$\mathbf{Z}_p \to k[[T]], \quad \gamma \mapsto \Sigma_0(T)$$

defines an injective ring homomorphism

$$\mathbf{Z}_p \to End(\mathbf{G}^{for}_{m,k})$$

into the endomorphism ring of the multiplicative formal group over k.

LEMMA 8.92. *Assume $a \in \mathbf{Z}_p^\times$ is not a root of unity. If $f \in k[[T]]$ satisfies a functional equation of the form*

$$f(\Sigma_0(T)) = c \cdot f(T),$$

for some $c \in \mathbf{F}_p$ then $f \in k$.

PROOF. If $c = 0$ we are done. So we may assume $c \neq 0$. Note that f^{p-1} satisfies a functional equation as above with $c = 1$; if we show that $f^{p-1} \in k$ then it follows that $f \in k$. So we may assume $c = 1$. Assume $f = f(T) \notin k$. Replacing f by $f - f(0)$ we may assume $f(0) = 0$, $f \neq 0$. Let e be the order of the series $f(T)$. Let $k[[f]]$ be the subring of $k[[T]]$ consisting of all series of the form $\sum_{i=0}^{\infty} \lambda_i f^i$, ($\lambda_i \in k$); it is trivial to see that $k[[T]]$ is generated as a $k[[f]]$–module by $1, T, ..., T^{e-1}$. Denote by $k((f))$, $k((T))$ the quotient fields of $k[[f]]$, $k[[T]]$. Now, since a is invertible in $\mathbf{Z}_p$ but is not a root of unity, it follows that the substitution homomorphism $T \mapsto \Sigma_0(T)$ defines an automorphism, Σ_0, of $k[[T]]$ of infinite order. On the other hand since the induced automorphism Σ_0 of $k((T))$ fixes the subfield $k((f))$ and since the field extension $k((T))/k((f))$ is finite, it follows that the automorphism Σ_0 of $k((T))$ has finite order, a contradiction. □

For any $m \in \mathbf{Z}$ set

$$Z(m, a) := \{j \in \mathbf{Z}_+ | a^{j-m} \equiv 1 \ \ mod \ \ p\}.$$

PROPOSITION 8.93. *Assume $a \in \mathbf{Z}_p^\times$ is not a root of unity. An element $f \in k[[T]][T']$ satisfies a functional equation of the form*

$$f(\Sigma_0(T), \Sigma_1(T, T')) = a^m \cdot f(T, T'),$$

for some $m \in \mathbf{Z}$ if and only if it has the form

$$f(T, T') = \sum_{j \in Z(m,a)} c_j \Psi_0^j, \quad c_j \in k.$$

PROOF. The *if* part of the Proposition is clear. Assume now f satisfies the functional equation in the Proposition. By Proposition 8.91 we may write

$$f = \sum_{j=0}^{N} a_j(T) \Psi_0^j, \quad a_j(T) \in k[[T]].$$

Since

$$\Psi_0(\Sigma_0(T), \Sigma_1(T, T')) = a \cdot \Psi_0,$$

the functional equation in our Proposition implies

$$a_j(\Sigma_0(T)) = a^{m-j} \cdot a_j(T)$$

for all j. By Lemma 8.92 we must have $c_j := a_j(T) \in k$ for all j. Hence we also must have $c_j = 0$ if $a^{m-j} \neq 1$. □

Consider the reduction mod p,

$$\bar{E} : \bar{M}_X^r \to \bar{S}_{for}^r := k[[T]][T', ..., T^{(r)}]$$

of the Serre-Tate expansion map $E : M_X^r \to S_{for}^r$.

LEMMA 8.94. *Let $w \in W$ have even degree $deg(w) = -2m$, $ord(w) \leq r$, and let*

$$C = max\{0, -\frac{w(p)}{p+1}\}.$$

Let j be an integer such that

$$j \equiv m \quad mod \quad p-1, \quad j \geq C.$$

Then there exists a unique form $\bar{g} = \bar{g}_{w,j} \in \bar{M}_X^r(w)$ such that $\bar{E}(\bar{g}) = \bar{\epsilon}^j \Psi_0^j$.

Note that if $j \in Z(m, a)$ and the image $\bar{a} \in \mathbf{F}_p$ of a is a primitive root mod p then $j \equiv m$ mod $p - 1$.

PROOF. Uniqueness follows from Proposition 8.29. To prove existence we are going to construct $\bar{g}$ locally on X; then we will show that our definition allows gluing the various local elements. (Uniqueness cannot be directly used to glue the local forms because uniqueness only applies a priori to pieces of X that contain P_0.) So assume L is trivial with basis x. Write $f^1 = f^1_{crys}$,

$$f^1 = \alpha \cdot x^{-1-\phi}, \quad \alpha \in \mathcal{O}^1(X),$$

$$\partial_1 f^1 - \beta \cdot x^{\phi-1}, \quad \beta \in \mathcal{O}^1(X).$$

Let $\bar{\alpha}, \bar{\beta} \in \mathcal{O}^1(X) \otimes k$, $(x^v)^- \in \bar{M}_X^r(v)$ be the classes mod p of α, β, x^v. Note that

$$(x^w)^- = \bar{x}^{w(p)} \quad in \quad \bar{M}_X^r.$$

Note that

$$(p+1)j + w(p) \equiv (p+1)m + w(p) \equiv 2m + w(1) = 0 \quad mod \quad p-1.$$

so there exists an integer j' such that

$$(p+1)j + w(p) = (p-1)j'.$$

By the assumptions of the Proposition $j' \geq 0$. Now define

$$\bar{g} := \bar{\alpha}^j \bar{\beta}^{j'} (x^w)^- \in \bar{M}_X^r(w).$$

The above definition does not depend on x; indeed if $x = uy$, $u \in \mathcal{O}(X)^\times$, then $f^1 = \alpha u^{-1-\phi} y^{-1-\phi}$, $\partial_1 f^1 = \beta u^{\phi-1} y^{\phi-1}$, so

$$((\alpha u^{-1-\phi})^-)^j ((\beta u^{\phi-1})^-)^{j'} (y^w)^- = \bar{\alpha}^j \bar{\beta}^{j'} \bar{u}^{(-1-p)j+(p-1)j'-w(p)} (x^w)^- = \bar{g}.$$

This shows that our local definition of $\bar{g}$ globalizes. On the other hand, by equations 8.119 and 8.61 we have

$$\bar{E}(\bar{g}) = \bar{E}(\bar{\alpha}^j \bar{\beta}^{j'} \bar{x}^{w(p)}) = \bar{E}((\bar{\alpha}(x^{-1-\phi})^-)^j (\bar{\beta}(x^{\phi-1})^-)^{j'}) = \bar{E}(\bar{f}^1)^j \bar{E}(\bar{f}^\partial)^{j'} = \bar{\epsilon}^j \Psi_0^j.$$

□

THEOREM 8.95. *Let* $\mathbf{X}$ *be an irreducible Hecke correspondence that admits a regular Hecke fixed point. Assume* w *has even degree* $deg(w) = -2m$, $ord(w) \leq r$, *and let*

$$C := max\{0, -\frac{w(p)}{p+1}\}.$$

Then the image of the space $H^0(\mathbf{X}, \mathbf{L}_{r,w}; k)$ *under the (injective) Serre-Tate map*

$$\bar{E} : \bar{M}_X^r(w) \to k[[T]][T', ..., T^{(r)}]$$

contains the space

$$V := \sum_{\substack{j \equiv m \ mod \ p-1 \\ j \geq C}} k \cdot \Psi^j.$$

In particular $H^0(\mathbf{X}, \mathbf{L}_{r,w}; k)$ *is infinite dimensional. Assume moreover that* $r = 1$, $w(p) \geq 0$ *(so* $C = 0$*), and the class of* μ^+/μ *in* $\mathbf{F}_p$ *is a primitive root mod* p. *Then* $\bar{E} : H^0(\mathbf{X}, \mathbf{L}_{r,w}; k) \to V$ *is an isomorphism.*

PROOF. The first assertion follows from Lemma 8.94 plus Proposition 8.29. The second assertion follows from Lemma 8.93. □

COROLLARY 8.96. *Let* $\mathbf{X}$ *be an irreducible Hecke correspondence that admits a regular Hecke fixed point. Assume* w *has even degree. Then the* $k-$*linear space* $H^1(\mathbf{X}, \mathbf{L}_w)[p]$ *is infinite dimensional.*

PROOF. This follows from the exact sequence 2.52; indeed, in that sequence the middle term is infinite dimensional by Theorem 8.95 while the left term is finite dimensional by Theorem 8.79. □

8.8. The relative generic $\delta-$fiber

In this section we assume that our Hecke correspondences are attached to genuine elliptic (rather than false elliptic) curves. Our aim is to prove the following:

THEOREM 8.97. *Let* $\mathbf{X}$ *be an irreducible (genuine elliptic) Hecke correspondence that possesses a regular Hecke fixed point and let* $w = -1 - \phi - \phi^2 - \phi^3$. *Then the field extension*

$$k(X) \cdot k\langle \mathbf{X}, \mathbf{L}_w \rangle \subset k\langle X, L^w \rangle$$

is prosolvable.

The following two lemmas will be used in the proof.

LEMMA 8.98. *Let* S *be a* δ-*ring and* $S^0 \subset S^1 \subset S^2 \subset ... \subset S$ *subrings such that* $\delta S^n \subset S^{n+1}$. *Let* $a, b, d \in S^0$, $A \in S^0$, $B \in S^1$, $C \in S^2$, $D \in S^1$. *Define*

$$F := a(\delta b)^{\phi} + A + pB + p^2 C \in S^2, \quad G := \delta b + d + pD \in S^1.$$

Set $\omega := \phi + p$ *and consider the induced* p-*derivations*

$$\delta : S^{i+1}[G^{-\omega^{i-1}}] \rightarrow S^{i+2}[G^{-\omega^i}], \quad i \geq 1.$$

Then, for any $n \geq 1$, *we have* $\delta^n(F/G) = F_n/G^{\omega^n}$, *where*

$$F_n = a^{p^n} G^{(2^n-1)p^n}(\delta^{n+1}b)^{\phi} - (a(\delta b)^p + A)^{p^n} G^{2(2^{n-1}-1)p^n}(\delta^{n+1}b) + A_n + pB_n + p^2 C_n,$$

$$A_n \in S^n, \quad B_n \in S^{n+1}, \quad C_n \in S^{n+2}.$$

PROOF. Induction on n. The details are somewhat tedious; cf. [**28**] for a similar situation. □

LEMMA 8.99. *Let* B *be a ring and* $t^{(\leq n)}$, $\Lambda^{(\leq n)}$ *tuples of variables* $t, t', ..., t^{(n)}$ *and* $\Lambda, \Lambda', ..., \Lambda^{(n)}$. *Consider the* B-*algebra homomorphism*

$$\mu : B[t^{(\leq n)}, t^{-1}] \rightarrow B[t^{(\leq n)}, \Lambda^{(\leq n)}, t^{-1}, \Lambda^{-1}],$$

$$\mu(t^{(i)}) := \delta^i(\Lambda t).$$

Then $Ker(\mu - id) = B$.

We shall apply the above statement when B has characteristic p; in this case, of course, the expressions $\delta^i(\Lambda t)$ will be viewed modulo p; for instance $\delta(\Lambda t) = \Lambda^p t' + t^p \lambda'$.

PROOF. Assume $f \in B[t^{(\leq n)}, t^{-1}]$ is in $Ker(\mu - id)$ i.e. it satisfies

$$f(\Lambda t, \delta(\Lambda t), ..., \delta^n(\Lambda t)) = f(t, t', ..., t^{(n)}).$$

Setting $t = 1, t' = 0, ..., t^{(n)} = 0$ we get

$$f(\Lambda, \Lambda', ..., \Lambda^{(n)}) = f(1, 0, ..., 0) \in B.$$

The conclusion follows. □

For the proof of the Theorem above recall that the space of δ-modular functions has the description

$$M^r = R[a_4^{(\leq r)}, a_6^{(\leq r)}, \Delta^{-1}]\hat{}.$$

If $g \in M^r$, $g \notin pM^r$, we define

$$M^r_{\{g\}} := M^r[g^{-1}]\hat{} = R[a_4^{(\leq r)}, a_6^{(\leq r)}, \Delta^{-1}, g^{-1}]\hat{}.$$

Let $j : -2^{12}3^3 a_4^3/\Delta$, $i := 2^6 3^3 - j$, $t := a_6/a_4$. A simple computation (cf., say, [**21**], Proposition 3.10), shows that

$$M^n_{\{a_4 a_6\}} = R[j^{(\leq n)}, j^{-1}, i^{-1}, t^{(\leq n)}, t^{-1}]\hat{}.$$

If $w \in W$ the space $M^r(w)$ identifies with the set of all $f \in M^r$ such that

$$f(\lambda^4 a_4, \lambda^6 a_6, \delta(\lambda^4 a_4), \delta(\lambda^6 a_6), ...) = \lambda^w f(a_4, a_6, a_4', a_6', ...), \tag{8.172}$$

for all $\lambda \in R$. If $g \in M^r(w)$ then we let $M^r_{\{g\}}(w)$ be the set of all $f \in M^r_{\{g\}}$ that satisfy Equation 8.172. One can easily show (cf. [**21**], Corollary 3.11), that we have

$$M^r_{\{a_4 a_6\}}(0) = R[j^{(\leq r)}, j^{-1}, i^{-1}]\hat{}.$$

More generally, if $g \in M^0(m)$, $m \in 2\mathbf{Z}$, then we have

$$M^r_{\{a_4 a_6 g\}}(0) = R[j^{(\leq r)}, j^{-1}, i^{-1}, \tilde{g}^{-1}]\hat{},$$

where $\tilde{g} := gt^{-m/2}$. It will be useful to also consider the form

$$b := a_6^2 / a_4^3. \tag{8.173}$$

Note that

$$j = \frac{2^8 3^3}{4 + 27b}, \; j^{-1} = 2^{-6}3^{-3} + 2^{-8}b, \; i = \frac{2^6 3^6 b}{4+27b}, \; i^{-1} = 2^{-4}3^{-6}b^{-1} + 2^{-6}3^{-3};$$

in particular we have

$$M^r_{\{a_4 a_6\}}(0) = R[b^{(\leq r)}, b^{-1}, (4 + 27b)^{-1}]\hat{}.$$

We are ready to prove Theorem 8.97.

PROOF. We may assume $\mathbf{X}$ is affine. Consider the ring homomorphisms

$$M^r \to M^r_X, \quad M^r(w) \to M^r_X(w).$$

If $g \in M^0(m)$, $g \not\equiv 0$ mod p, then there are induced homomorphisms $M^r_{\{g\}} \to M^r_{X_g}$ inducing homomorphisms $M^r_{\{g\}}(w) \to M^r_{X_g}(w)$, where X_g is the locus in X where g is invertible. In particular, if $H \in M^0(p-1)$ is a lifting of the Hasse invariant then we have natural homomorphisms

$$M^r_{\{H\}} \to M^r_{X_{ord}}, \quad M^r_{\{H\}}(w) \to M^r_{X_{ord}}(w).$$

Let's view $f^1 := f^1_{crys}$ as an element of $M^1_X(-1-\phi)$. Also view f^∂ as an element of $M^1_{\{H\}}(\phi - 1)$. On the other hand by Corollary 8.86, Corollary 8.53 and Proposition 8.57,

$$f^1 = cH\Delta^N(2a_4^p a_6' - 3a_6^p a_4') + f_0 + pf_1, \tag{8.174}$$

$$c \in R^\times, f_0 \in M^0(-1-p), f_1 \in M^1,$$

$$f^\partial = H + pg, \quad g \in M^1_{\{H\}}. \tag{8.175}$$

It is trivial to check that one can write

$$b' = a_4^{-4p} a_6^p (2a_4^p a_6' - 3a_6^p a_4') + ph, \quad h \in M^1_{\{a_4\}}. \tag{8.176}$$

Set $a_0 := cH\Delta^N a_4^{4p} a_6^{-p}$ and combining equations 8.174 and 8.176 we get

$$f^1 = a_0 b' + f_0 + ph_1, \quad h_1 \in M^1_{\{a_4\}}. \tag{8.177}$$

Consider the following elements

$$a := a_0^{p-1} H^{p+1}, \quad A := a_0^{-1} f_0^p H^{p+1},$$
$$d := a_0^{-1} f_0, \quad D := a_0^{-1} h_1,$$

and set

$$S^r(0) := R[j^{(\leq r)}, j^{-1}, i^{-1}, \tilde{H}^{-1}]\hat{} = R[b^{(\leq r)}, b^{-1}, (4+27b)^{-1}, \tilde{H}^{-1}]\hat{},$$

$$S^r := M^r_{\{Ha_4a_6\}} = S^r(0)[t^{(\leq r)}, t^{-1}]\hat{}.$$

Then it is trivial to check that $D \in S^1$ and

$$a, A, d \in M^0_{\{Ha_4a_6\}}(0) = S^0(0).$$

Using equations 8.177 and 8.175 an easy computation yields:

$$u := (f^1)^{\phi-1}(f^{\partial})^{\phi+1} = F/G,$$

where

$$F = a(b')^{\phi} + A + pB + p^2C \in S^2, B \in S^1, \quad C \in S^2,$$
$$G = a_0^{-1}f^1 = b' + d + pD \in S^1.$$

By Lemma 8.98 we have $\delta^r u = F_r/G^{\omega^r}$ where

$$F_r = a^{p^r}G^{(2^r-1)p^r}(b^{(r+1)})^{\phi} - (a(b')^p + A)^{p^r}G^{(2^r-2)p^r}b^{(r+1)} + A_r + pB_r + p^2C_r,$$

with $A_r \in S^r$, $B_r \in S^{r+1}$, $C_r \in S^{r+2}$. Set

$$S^r_0(0) = S^r(0) \otimes k = k[j^{(\leq r)}, j^{-1}, i^{-1}, \tilde{H}^{-1}] = k[b^{(\leq r)}, b^{-1}, (4+27b)^{-1}, \tilde{H}^{-1}],$$
$$S^r_0 := S^r \otimes k = S^r_0(0)[t^{(\leq r)}, t^{-1}].$$

Let

$$u_i \in S^{i+2}[G^{-\omega^i}] \otimes k = S^{i+2}_0[(b' + \bar{d})^{-1}]$$

be the image of $\delta^i u$ where we denote by upper bars reduction mod p. Then we have

$$u_0 = \frac{\bar{a}(b')^p + \bar{A}}{b' + \bar{d}}, \tag{8.178}$$

$$u_r = \bar{a}^{p^r}(b' + \bar{d})^{-p^r}(b^{(r+1)})^p - (\bar{a}(b')^p + \bar{A})^{p^r}(b' + \bar{d})^{-2p^r}b^{(r+1)} + (b' + \bar{d})^{-2^r p^r}\bar{A}_r. \tag{8.179}$$

for $r \geq 1$. Here $\bar{a}, \bar{A}, \bar{d} \in S^0_0(0)$ and $\bar{A}_r \in S^r_0$. Note that $u_r \in S^{r+1}_0$.

Claim. $\bar{A}_r \in S^r_0(0)$.

Indeed consider the homomorphism

$$\mu : S^{r+1}_0(0)[t^{(\leq r+1)}, t^{-1}] \to S^{r+1}_0(0)[t^{(\leq r+1)}, \Lambda^{(\leq r+1)}, t^{-1}, \Lambda^{-1}],$$

that sends $t^{(i)}$ into $\delta^i(\Lambda t)$ mod p. Note that $\mu(u_r) = u_r$. By Lemma 8.99 we have $u_r \subset S^{r+1}_0(0)$. By Equation 8.179 we get $\bar{A}_r \subset S^{r+1}_0(0)$. So $\bar{A}_r \subset S^{r+1}_0(0) \cap S^r_0 - S^r_0(0)$ and our claim is proved. We conclude that the extension

$$k(b, u_0) \subset k(b, b')$$

is generated by a root of the equation

$$Z^p - \bar{a}^{-1}u_0 Z + \bar{a}^{-1}(\bar{A} - u_0\bar{d}) = 0. \tag{8.180}$$

On the other hand, for each $r \geq 1$, the extension

$$k(b, b', ..., b^{(r)}, u_r) \subset k(b, b', ..., b^{(r)}, b^{(r+1)})$$

is generated by a root of

$$Z^p - \left(\frac{(b')^p + \bar{a}^{-1}A}{b' + \bar{d}}\right)^{p^r} Z + \frac{A_r}{\bar{a}^{p^r}(b' + \bar{d})^{(2^r-1)p^r}} = 0. \tag{8.181}$$

Let ξ be a $(p-1)$- root of $\bar{a}^{-1}u_0$, let η be a $(p-1)$- root of

$$\frac{(b')^p + \bar{a}^{-1}\bar{A}}{b' + \bar{d}},$$

and set

$$L_r := k(b, u_0, u_1, ..., u_r), \quad E_r := k(b, b', ..., b^{(r+1)}, \xi, \eta).$$

Then the extension $L_r \subset E_r$ is a tower of Abelian extensions; indeed $\xi^{p-1}, \eta^{p-1} \in k(b, u_0)$, and the equations 8.180 and 8.181 are (after a change of variable involving ξ, η) Artin-Schreier equations. Set

$$L := k(X)(u_0, u_1, u_2, ...) = k(X) \cdot k\langle \mathbf{X}_{ord}, \mathbf{L}_w \rangle = k(X) \cdot k\langle \mathbf{X}, \mathbf{L}_w \rangle;$$

cf. Corollary 8.80 and Theorem 8.81. Note that, for each r, $L \subset LE_r$ is a tower of Abelian extensions. This implies that $E := \cup LE_r$ can be embedded into a field E' that is algebraic Galois over L with prosolvable Galois group. On the other hand E contains the field

$$k(X)(b', b'', ...) = k(X_1)(b', b''...) \tag{8.182}$$

where $X_1 \subset X$ is any affine open set. Choose X_1 such that the line bundle L is trivial on X_1 and X_1 is étale over $Spec\ R[b]$. Then the field 8.182 identifies with $k\langle X_1, L^w \rangle$ which, in its turn, contains $k\langle X, L^w \rangle$. So $k\langle X, L^w \rangle$ can be embedded into E' and we are done. □

8.9. Hecke correspondences with a regular Hecke $n-$cycle

Some of our results about Hecke correspondences with a Hecke fixed point generalize to Hecke correspondences with a Hecke cycle and to correspondences uniformized by these. Here is an example.

THEOREM 8.100. *Let* $\mathbf{X}$ *be an irreducible genuine elliptic Hecke correspondence that possesses a regular Hecke* $n-$*cycle and let* $w = -1 - \phi - \phi^2 - \phi^3$. *Then:*

1) $(\mathbf{X}, \mathbf{L}_w)$ *is* $\delta-$*non-trivial and* $\delta-$*rational,*

2) The field extension

$$k(X) \cdot k\langle \mathbf{X}, \mathbf{L}_w \rangle \subset k\langle X, L^w \rangle$$

is (algebraic) prosolvable.

Moreover, if $\mathbf{Y}$ *is a correspondence uniformized by* $\mathbf{X}$ *and by some subgroup of* $Aut(X)$ *of admissible automorphisms then*

1′) $(\mathbf{Y}, \mathbf{K}^{-1})$ *is* $\delta-$*non-trivial and* $\delta-$*rational,*

2′) The field extension

$$k(Y) \cdot k\langle \mathbf{Y}, \mathbf{K}^{-1} \rangle \subset k\langle Y, K^{-1} \rangle$$

is algebraic.

If, instead of the genuine elliptic case we are in the false elliptic case then assertions 1 and 1′ still hold.

Of course, we expect that, in the false elliptic case, assertions 2 and 2′ hold as well.

PROOF. The regular Hecke cycle on $\mathbf{X}$ gives rise to a regular Hecke fixed point on an irreducible Hecke correspondence $\mathbf{X}_1^{\circ n}$, where $\mathbf{X}_1^{\circ n}$ is an irreducible open subcorrespondence of $\mathbf{X}^{\circ n}$ with base X. Note that we have telescopic inclusion

$$R(\mathbf{X}, \mathbf{L}_w)^* \subset R(\mathbf{X}_1^{\circ n}, \mathbf{L}_w^{\circ n})^*, \tag{8.183}$$

hence an inclusion of local $p-$rings

$$R\langle \mathbf{X}, \mathbf{L}_w \rangle \subset R\langle \mathbf{X}_1^{\circ n}, \mathbf{L}_w^{\circ n} \rangle.$$

By Theorem 8.81 we have that $R\langle \mathbf{X}_1^{\circ n}, \mathbf{L}_w^{\circ n}\rangle$ has a $\delta-$basis consisting of

$$(f^2_{crys})^{1+\phi}(f^1_{crys})^{-1-\phi^2}. \tag{8.184}$$

But the forms f^i_{crys} belong to the space $H^0(\mathbf{X}, \mathbf{L}_{-1-\phi^i})$ because they "come from" isogeny covariant forms. So we must have

$$k\langle \mathbf{X}, \mathbf{L}_w\rangle = k\langle \mathbf{X}_1^{\circ n}, \mathbf{L}_w^{\circ n}\rangle \tag{8.185}$$

and $(\mathbf{X}, \mathbf{L}_w)$ is $\delta-$rational. Finally, in the genuine elliptic case, by Theorem 8.97 applied to $\mathbf{X}_1^{\circ n}$ the field $k\langle X, L^w\rangle$ can be embedded into a Galois extension E of

$$k(X) \cdot k\langle \mathbf{X}_1^{\circ n}, \mathbf{L}_w^{\circ n}\rangle$$

with Galois group a projective limit of solvable groups which settles the assertions about $\mathbf{X}$ in the Theorem. To check the assertions about $\mathbf{Y}$ one notes that the $\delta-$basis in Equation 8.184 can be written as

$$\frac{(f^2_{crys})^{1+\phi}(f^1_{crys})^{1+\phi^2}}{(f^1_{crys})^{2+2\phi^2}}. \tag{8.186}$$

Now, using Proposition 3.27, both the numerator and the denominator of the fraction in Equation 8.186 are easily seen to come from sections in $H^0(Y, K^{-2w})$ and we are done. □

In what follows we will prove our (horizontal) Main Theorem 2.62 about modular curves. We need the following well known result about the set of $CM-$points,

$$CM := \{j(E) \in \mathbf{C} \mid E/\mathbf{C} \text{ an elliptic curve with CM}\}.$$

LEMMA 8.101. *For any integer N the set*

$$\{j \in CM \mid [\mathbf{Q}(j) : \mathbf{Q}] \leq N\}$$

is finite.

PROOF. For convenience we recall the argument. The basic reference for the facts below is Shimura's book [**122**]. Let $E/\mathbf{C}$ be a CM elliptic curve with

$$[\mathbf{Q}(j(E)) : \mathbf{Q}] \leq N.$$

Then $End(E) = \mathbf{Z} + c\mathcal{O}_K$ for some positive integer $c \in \mathbf{Z}$ and some quadratic field K with ring of integers $\mathcal{O}_K$. Let h_K be the class number of K. By [**122**], pp. 106 and 123, we have

$$[\mathbf{Q}(j(E)) : \mathbf{Q}] = h_K \cdot c \cdot [\mathcal{O}_K^\times : \mathcal{O}^\times]^{-1} \cdot \prod_{p|c} \left[1 - \left(\frac{K}{p}\right) p^{-1}\right],$$

where $\left(\frac{K}{p}\right)$ is 1, -1, or 0 according as p decomposes in K, stays prime in K or is ramified in K. Consequently, since $\sharp \mathcal{O}_K^\times \leq 6$, we get

$$N \geq [\mathbf{Q}(j(E)) : \mathbf{Q}] \geq \frac{1}{12} \cdot h_K \cdot c^{1/2}.$$

So both h_K and c are bounded. By Siegel [**123**] there are only finitely many K's with a given h_K. Since $E \simeq \mathbf{C}/\Lambda$ for some proper $\mathcal{O}-$ideal Λ [**122**] there are only finitely many possibilities for $j(E)$. □

For the next proof we need to recall Elkies' theorem on supersingular reduction.

THEOREM 8.102. [**56**] [**57**] *Let L be a number field possessing at least one real embedding and let E be an elliptic curve over L. Then there exist infinitely many places of L where E has supersingular reduction.*

We are ready to prove Theorem 2.62.

PROOF. Consider the δ−line bundles $\mathbf{L}_w$ on $\mathbf{X}_p$ for all primes p. Recall that $\mathbf{L}_w$ is polarization equivalent to $\mathbf{K}_p^{-1}$ if $-w \in W_+\backslash\phi W_+$. The forms $f^i = f^i_{crys}$ belong to the space $H^0(\mathbf{X}_p, \mathbf{L}_{-1-\phi^i,p})$. This immediately implies the δ−non-triviality of $(\mathbf{X}_p, \mathbf{K}_p^{-1})$. By Proposition 8.23 there are infinitely many p such that $\mathbf{X}_p$ has a regular Hecke cycle. So assertions 1 and 2 follow from Theorem 8.100.

Let us prove assertion 3 (in the stronger form appearing in Remark 2.63).

We first check that the δ−base locus $\overline{BS}(\mathbf{X}_{\mathcal{O}'}, \mathbf{K}_{\mathcal{O}'}^{-1})$ is finite. Let P be in this locus hence there exists a place $\wp$ of F' not lying above places in S or above 2 or 3 such that P is an $\mathcal{O}'_{\wp}$−point of $Y_1(N)$ belonging to the δ−base locus $BS(\mathbf{X}_{\wp}, \mathbf{K}_{\wp}^{-1})$. So P is represented by a pair (E, α) where E is an elliptic curve over $\mathcal{O}_{\wp}$ (in particular over F') and α is a corresponding level structure. Since $P \otimes R_{\wp}$ is a strict δ−base point for $(\mathbf{X}_{\wp}, \mathbf{K}_{\wp}^{-1})$ (where $p\mathbf{Z} = \mathbf{Z} \cap \wp$) it follows that $f^1_{crys}(E_{\wp}, \omega_{\wp}) = 0$ for some (equivalently any) invertible 1−form $\omega_{\wp}$ on $E_{\wp} := E \otimes R_{\wp}$. By Proposition 7.15 and Corollary 8.86 it follows that $E_{\wp}$ has a lift of Frobenius. We claim that $E/\mathbf{C}$ has CM. Indeed let us assume $E/\mathbf{C}$ doesn't have CM and let us derive a contradiction. By Corollary 8.89, E has ordinary reduction at $\wp$. Let $\sharp(\mathcal{O}/\wp) = p^{\nu}$. Then the p^{ν}-power Frobenius on the reduction $\bar{E}_{\wp}$ of E mod $\wp$ lifts to an endomorphism of $E_{\wp}$. The latter endomorphism is in $\mathbf{Z}$ so $\mathbf{Q}(\pi) = \mathbf{Q}$. By the Weil-Honda-Tate equality in Equation 8.77 $End^0(\bar{E}_{\wp})$ is non-commutative. But then $\bar{E}_{\wp}$ cannot be ordinary (cf. [**125**], p. 137), a contradiction. Our Claim is proved. Now by Lemma 8.101 there are only finitely many possible values for the j−invariant $j(E)$. So there are only finitely many possibilities for the point P and we are done.

Let us check that $\underline{B}(\mathbf{X}_{F'}, \mathbf{K}_{F'}^{-1}) = \emptyset$ provided F' has a real embedding. Assume $P \in \underline{B}(\mathbf{X}_{F'}, \mathbf{K}_{F'}^{-1})$. Then P is represented by an elliptic curve E/F' with a level structure. By Elkies' Theorem 8.102 there exist infinitely many places $\wp$ of F' where E has supersingular reduction. By Corollary 8.89, for any such unramified $\wp$, we have $f^1_{crys}(E_{\wp}, \omega_{\wp}) \not\equiv 0$ mod p ($\omega_{\wp}$ any invertible 1−form on $E_{\wp}$). So $P \notin BS(\mathbf{X}_{\wp}, \mathbf{K}_{\wp})$ for infinitely many $\wp$, a contradiction.

Finally let us check that $\underline{BS}(\mathbf{X}_{F'}, \mathbf{K}_{F'})$ is empty for any finite field extension F' of F. Indeed if the latter locus has a point this point is represented by an elliptic curve E/F' with some level structure and by our hypothesis, $f^1_{crys}(E_{\wp}, \omega_{\wp}) = 0$ for almost all $\wp$. As above, the latter vanishing (for even a single $\wp$!) implies that E has CM. But a CM elliptic curve has infinitely many places of supersingular reduction; cf. [**125**], pp. 122, 178, 184. So as before $f^1_{crys}(E_{\wp}, \omega_{\wp}) \not\equiv 0$ mod p for infinitely many $\wp$, a contradiction. □

Theorem 2.62 has a Shimura analogue which we would like to explain now. Fix a subgroup $U \subset \mathcal{O}_{D,f}^{\times}$ as in Example 8.24 and for each prime l not dividing $M_U d$ consider the correspondences $\mathbf{Y}_{\mathbf{Z}[1/M_U dl]}$ and $\mathbf{Y}_p$ in Equations 8.74 and 8.75 respectively.

THEOREM 8.103. *There is an infinite set of primes $\mathcal{L}$ such that for any $l \in \mathcal{L}$ there exist infinitely many primes p with the property that $(\mathbf{Y}_p, \mathbf{K}_p^{-1})$ is δ−non-trivial and δ−rational.*

PROOF. Same argument as in the above proof of Theorem 2.62. Briefly, by Proposition 8.26, there exists an infinite set of primes $\mathcal{L}$ such that for any $l \in \mathcal{L}$ there is an infinite set $\mathcal{P}_l$ of primes with the property that if $p \in \mathcal{P}_l$ then $\mathbf{Y}_p$ has a regular Hecke $n-$cycle. Let $\mathbf{X}_p$ be a Hecke correspondence obtained by removing from the basis of $\mathbf{Y}_p$ finitely many points away from our cycle. Then one shows, using Theorem 8.100 (and its proof), that the ring extension

$$R\langle \mathbf{Y}_p, \mathbf{K}^{-1}\rangle \rightarrow R\langle \mathbf{X}_p, \mathbf{K}^{-1}\rangle$$

induces an isomorphism between the corresponding residue fields. □

8.10. A converse theorem: non-rational hyperbolic uniformization

In this section we will prove the following:

THEOREM 8.104. *For any real quadratic field E there exists an étale geometrically irreducible correspondence $\mathbf{X}$ of finite type and smooth over a ring $\mathcal{O}$ of $S-$integers in a number field F, and there exists an infinite set Σ of places of $\mathcal{O}$ such that the following hold:*

1) $\mathbf{X}$ admits an $E-$arithmetic uniformization;

2) For any $\wp \in \Sigma$ and any $\delta-$line bundle $\mathbf{L}_\wp$ on $\mathbf{X}_\wp$ with $\mathbf{L}_\wp^2 = \mathbf{K}^{-1-\phi}$ we have $H^0(\mathbf{X}_\wp, \mathbf{L}_\wp) = 0$.

Note that for Hecke correspondences (which, by the way, have $\mathbf{Q}-$ arithmetic uniformizations) the spaces $H^0(\mathbf{X}_\wp, \mathbf{L}_\wp)$ are always non-zero since they contain the form f^1_{crys}; so if Conjecture 2.54 is to hold in the non-rational arithmetic case then powers other than $\mathbf{K}^{-1-\phi}$ should be involved.

We start by considering the picture over the complex numbers. Consider a tuple $(\mathbf{H}, \Gamma, \tilde{\Gamma}, id, \tau)$ where $\tilde{\Gamma}$ and Γ are Fuchsian lattices, $\tau\tilde{\Gamma}\tau^{-1}$ a subgroup of finite index in Γ. Let $\pi : \mathbf{H} \rightarrow \mathbf{H}/\Gamma$, $\tilde{\pi} : \mathbf{H} \rightarrow \mathbf{H}/\tilde{\Gamma}$ be the projections which, for simplicity, **we assume here to be étale**. So we may consider the étale induced correspondence

$$\mathbf{X} := (\mathbf{H}/\Gamma, \mathbf{H}/\tilde{\Gamma}, \sigma_1, \sigma_2).$$

Also fix a subgroup $G \subset GL_2^+(\mathbf{R})$ which projects onto Γ and fix an element $b \in GL_2^+(\mathbf{R})$ which projects onto τ. Set

$$T(b) := tr(b), \quad D(b) := det(b), \quad \Delta(b) := T(b)^2 - 4D(b). \tag{8.187}$$

Note that if $\Delta(b) < 0$ then there exists a unique $z_0 \in \mathbf{H}$ such that $\tau(z_0) = z_0$; so we get a fixed point on $\mathbf{X}$ defined by

$$(\tilde{P}, P) := (\tilde{\pi}(z_0), \pi(z_0)).$$

One can define, in the same way as in the $p-$adic context, the *multiplier* $a = a(\tilde{P}, P)$ of $\mathbf{X}$ at $(\tilde{P}, P)$. Throughout the discussion below we make the convention that, if d is a positive real number then $\sqrt{-d}$ is the unique complex number in $\mathbf{H}$ whose square is $-d$.

LEMMA 8.105. *Assume $\Delta := \Delta(b) < 0$, and let $a = a(\tilde{P}, P)$ the multiplier of the associated fixed point. Then there exists $\epsilon \in \{1, -1\}$ such that*

$$a^\epsilon = \frac{T - \sqrt{\Delta}}{T + \sqrt{\Delta}}.$$

PROOF. We have $a := \tau'(z_0)$ where

$$\tau(z) = \frac{b_{11}z + b_{12}}{b_{21}z + b_{22}},$$

$b = (b_{ij})$, $\tau(z_0) = z_0$, and one concludes by a trivial computation; the value of ϵ will depend on the sign of b_{21}. □

In the next Lemma, for a quaternion algebra B over a field E we denote by $T, D : B \to E$ the reduced trace and reduced norm (which coincide with the usual trace and determinant once we view B as a subset of $B \otimes_E E^a = Mat_2(E^a)$). We let $\Delta := T^2 - 4D : B \to E$. We denote by $x \mapsto x'$ the canonical involution $B \to B$; so, for all $b \in B$,

$$T(b) = b + b', \quad D(b) = bb' = b'b, \quad \Delta(b) = (b - b')^2.$$

LEMMA 8.106. *Assume $E := \mathbf{Q}(\sqrt{d}) \subset \mathbf{R}$ is a real quadratic extension of $\mathbf{Q}$ (d a positive square free integer). Then there exists a quaternion $E-$algebra B, an element $b \in B$, and an automorphism σ of $\mathbf{Q}^a$ lifting the involution ρ of E, such that following properties are satisfied.*

1) $B \otimes_E \mathbf{R} \simeq Mat_2(\mathbf{R})$ and $B \otimes_{E,\rho} \mathbf{R} \simeq \mathcal{H}$;

2) $D(b) > 0$ (hence $b \in B^+$) and $\Delta(b) < 0$;

3) If one sets

$$a := \frac{T(b) - \sqrt{\Delta(b)}}{T(b) + \sqrt{\Delta(b)}}$$

then $\sigma^2 a = a$, $a \neq \sigma a$, and $a\sigma a$ is not a root of unity; hence a is quadratically $\sigma-$independent to order 1.

PROOF. Choose $\alpha, \beta \in F$ with the following properties:

1) $\alpha > 0$, $\rho\alpha > 0$, $\alpha\rho\alpha$ is not a square in E;

2) $\beta < 0$, $\rho\beta > 0$.

(For instance one can take $\beta := 1 - \sqrt{d}$ and $\alpha := k - \sqrt{d}$, where k is any integer $> d$ and coprime to d.) Let B be the (associative unitary) $E-$ algebra generated by elements I and J with relations

$$I^2 = -\alpha, \quad J^2 = -\beta, \quad IJ + JI = 0.$$

Clearly B is a quaternion $E-$algebra with involution satisfying $I' = -I$, $J' = -J$, and condition 2) in the statement of the Theorem holds; indeed a quaternion algebra over $\mathbf{R}$ is isomorphic to $Mat_2(\mathbf{R})$ or $\mathcal{H}$ according as the real quadratic form $x \mapsto D(x)$ on it is indefinite or definite respectively. For any positive integer n set

$$b_n := 1 + nI \in B.$$

We have $T(b_n) = 2$, $D(b_n) = 1 + n^2\alpha > 0$, $\Delta(b_n) = -4n^2\alpha < 0$. So $\sqrt{\Delta(b_n)} = 2ni\sqrt{\alpha}$, $\sqrt{\rho\Delta(b_n)} = 2ni\sqrt{\rho\alpha}$, where $i = \sqrt{-1} \in \mathbf{C}$. The condition that $\alpha\rho\alpha$ is not a square in E implies that each inclusion in the tower below has degree 2:

$$E \subset E(i\sqrt{\alpha}) \subset E(i\sqrt{\alpha}, i\sqrt{\rho\alpha}).$$

So one can prolong $\rho : E \to E$ to an automorphism σ of $\mathbf{Q}^a$ such that

$$\sigma(i\sqrt{\alpha}) = i\sqrt{\rho\alpha}, \quad \sigma(i\sqrt{\rho\alpha}) = i\sqrt{\alpha}.$$

Note that if

$$a_n := \frac{T(b_n) - \sqrt{\Delta(b_n)}}{T(b_n) + \sqrt{\Delta(b_n)}}$$

then

$$a_n = \frac{1 - in\sqrt{\alpha}}{1 + in\sqrt{\alpha}}, \ \sigma a_n = \frac{1 - in\sqrt{\rho\alpha}}{1 + in\sqrt{\rho\alpha}}, \ \sigma^2 a_n = a_n.$$

Note that, trivially, $\sigma a_n \neq a_n$. On the other hand

$$a_n \sigma a_n = \frac{1 - in(\sqrt{\alpha} + \sqrt{\rho\alpha}) + n^2\sqrt{\alpha\rho\alpha}}{1 + in(\sqrt{\alpha} + \sqrt{\rho\alpha}) + n^2\sqrt{\alpha\rho\alpha}}$$

so the sequence $a_n \sigma a_n$ tends to 1 as $n \to \infty$ and no term of this sequence equals 1. It follows that there are infinitely many distinct numbers among the terms of the above sequence. But there are at most finitely many roots of unity among the numbers $a_n \sigma a_n$ because any such root of unity, ζ, satisfies the condition $\mathbf{Q}(\zeta) \subset E(i\sqrt{\alpha}, i\sqrt{\rho\alpha})$ so ζ has degree $2, 4$, or 8. It follows that, for n big enough, $a_n \sigma a_n$ is not a root of unity; we conclude by letting $b = b_n$ for n large enough. $\square$

We are ready to prove Theorem 8.104.

PROOF. Let B,b, and σ be as in Lemma 8.106. With the help of B and b one can construct, using Proposition 1.37, an étale correspondence

$$\mathbf{X} = \mathbf{X}_{b,N} := (\mathbf{H}/\Gamma(N), \mathbf{H}/\tilde{\Gamma}, \sigma_1, \sigma_2)$$

where N is a suitable positive integer, $\Gamma(N)$ is the image in $PSL_2(\mathbf{R})$ of $G(N) := \{g \in \mathcal{O}_B^1 | g - 1 \in N\mathcal{O}_B\}$, $\tilde{\Gamma}$ is the image in $PSL_2(\mathbf{R})$ of $G(N) \cap b^{-1} \cdot G(N) \cdot b$, and σ_1 and σ_2 are induced by id and b respectively. Since $\Delta(b) < 0$ there is an associated fixed point of $\mathbf{X}$, $(\tilde{P}, P)$. By Lemma 8.105 the multiplier, a, of this fixed point satisfies

$$a^{\epsilon} = \frac{T(b) - \sqrt{\Delta(b)}}{T(b) + \sqrt{\Delta(b)}}$$

for some $\epsilon \in \{1, -1\}$. By Lemma 8.106 a is quadratically $\sigma-$ independent to order 1. By the work of Shimura (last Chapter of [**122**]) $\mathbf{X}$ is defined over a number field F. Select a model of $\mathbf{X}$ over a ring of $S-$integers of F, $\mathcal{O}$, such that $\tilde{P}$ and P are induced by $\mathcal{O}-$points. Let M be any (finite) Galois extension of $\mathbf{Q}$ containing F, a, and all coordinates of the points $\tilde{P}$ and P; then σ induces an automorphism of M which we still denote by σ. By Chebotarev's density theorem (applied to the extension $M/\mathbf{Q}$) there exist infinitely many places $\mathcal{P}$ of M, unramified over $\mathbf{Q}$, such that

$$\sigma = (\mathcal{P}, M/\mathbf{Q}) \in G(M/\mathbf{Q}) \tag{8.188}$$

where (,) is the Artin symbol. Let $\wp$ denote the place of F below $\mathcal{P}$. The natural map $R_{\wp} = (\hat{\mathcal{O}}_{\wp}^{ur})\hat{} \to (\hat{\mathcal{O}}_{\mathcal{P}}^{ur})\hat{}$ is an isomorphism which is equivariant with respect to the action of ϕ on $R_{\wp}$ and the action of σ on $(\hat{\mathcal{O}}_{\mathcal{P}}^{ur})\hat{}$. Consequently the image of a in $R_{\wp}$ is quadratically $\phi-$independent to order 1. We conclude by Corollary 4.67. $\square$

List of Results

Below is a list of the "main implications" proved in this book. The implications are loosely stated. For the precise statements we refer to the results in the text. References for the latter are indicated in parentheses.

Uniformization implies $\delta-$non-triviality / $\delta-$rationality. (Cf. Theorems 2.59, 2.60, 2.62, 6.20, 6.24, 7.34, 8.81, 8.100, 8.103.)

Uniformization implies algebraicity of the relative generic $\delta-$fiber. (Cf. Theorems 2.59, 2.60, 2.62, 6.24, 7.60, 8.97, 8.100.)

Uniformization implies finiteness of $\delta-$base loci. (Cf. Theorem 2.60, 2.62, 7.45.)

Finiteness of the $\delta-$base loci implies critical finiteness and hence (in the case of dynamical systems with non-negative Euler characteristic) flat analytic uniformization. (Cf. Theorem 2.64.)

Uniformization implies postcritical finiteness. (Cf. Proposition 1.48.)

Spherical or hyperbolic analytic uniformization imply non-flat $\mathbf{K}^{\nu}$ for $\nu \neq 0$. (Cf. Proposition 1.24.)

Flat uniformization implies flat $\mathbf{K}^2$ and trivial $\mathbf{K}^{p-1}_{\mathbf{X}\otimes k}$. (Cf. Lemma 7.4.)

Spherical uniformization implies trivial $\mathbf{K}^{\nu}_{\mathbf{X}\otimes k}$. (Cf. Lemma 6.4.)

Hyperbolic uniformization implies trivial $\mathbf{K}^{\nu}_{\mathbf{X}_{ord}\otimes k}$. (Cf. Lemma 8.16.)

Uniformization implies $\mathbf{Z}_p-$accumulating multipliers. (Cf. Propositions 6.5, 7.7, 8.27.)

$\mathbf{Z}_p-$accumulating multipliers plus $d-$regular $\delta-$ invariants implies accumulating Schwarzians. (Cf. Corollary 4.60.)

Existence of $\delta-$invariants for W_+-powers of $\mathbf{K}$ implies symmetry of the multiplier set. (Cf. Corollary 4.63.)

Existence of $\delta-$invariants for W_+-powers of $\mathbf{K}^{-1}$ implies quadratic $\phi-$dependence of multipliers. (Cf. Corollary 4.67.)

Existence of $\delta-$invariants for W_+-powers of $\mathbf{K}_{\mathbf{X}}^{-1}$ implies $\mathbf{K}^{\nu}_{\mathbf{U}\otimes k}$ trivial for some $\mathbf{U} \subset \mathbf{X}$. (Cf. Theorems 2.65, 4.57.)

Existence of $\delta-$invariants for W_+-powers of $\mathbf{K}_{\mathbf{X}}^{-1}$ for correspondences attached to dynamical systems on $\mathbf{P}^1$ implies flat analytic uniformization. (Cf. Theorem 2.65.)

Multiplier not a generalized root of unity implies finite generation of the module of $\delta-$invariants. (Cf. Theorem 4.61.)

Existence of local $d-$regular $\delta-$invariants implies local $(d, a, r)-$flatness of $\mathbf{K}$. (Cf. Proposition 4.16.)

Existence of local $d-$regular $\delta-$invariants plus local flatness of $\mathbf{K}$ implies local uniformization by a formal group. (Cf. Proposition 4.24.)

Local uniformization by a formal group implies the existence of local $\delta-$invariants. (Cf. Proposition 4.26.)

Bibliography

[1] M. Artin, *Versal deformations and algebraic stacks*, Invent. Math. 27 (1974), 165-189.

[2] M.Barcau, *Isogeny covariant differential modular forms and the space of elliptic curves up to isogeny*, Compositio Math., 137 (2003), 237-273.

[3] M. Barcau, A. Buium, *Siegel differential modular forms*, International Math. Res. Notices, 28 (2002), 1459-1503.

[4] A. F. Beardon, *Iteration of Rational Functions*, GTM, Springer, 2000.

[5] R.L.Benedetto, *p−adic dynamics and Sullivan's no wandering domains theorem*, Compositio Math., 122 (2000), 281-298.

[6] R.L.Benedetto, *Components and periodic points in non-Archimedian dynamics*, Proc. London Math. Soc. 84 (2002), 231-256.

[7] P. Berthelot, A. Ogus, *F-isocrystals and de Rham cohomology*, Invent. Math. 72 (1983), 159-199.

[8] D. Bertrand, W. Zudilin, *On the transcendence degree of the differential field generated by Siegel modular forms*, Crelle J. 554 (2003), 47-68.

[9] B. Bielefeld, Y. Fischer, J. H. Hubbard, *The classification of critically periodic polynomials as dynamical systems*, J. Amer. Math. Soc., 5 (1992), 721-762.

[10] S. Bosch, W. Lütkebohmert, M. Raynaud, *Néron Models*, Ergebnisse der Mathematik und ihrer Grenzgebiete, 3, 21, Springer, 1990.

[11] J.F.Boutot, H.Carayol, *Uniformisation p−adique des courbes de Shimura; les théorèmes de Cerednik et de Drinfeld*, Astérisque 196-197 (1991), 45-158.

[12] A.Buium, *Intersections in jet spaces and a conjecture of S.Lang*, Annals of Math. 136 (1992) 557-567.

[13] A. Buium, *Differential Algebra and Diophantine Geometry*, Hermann, Paris, 1994.

[14] A.Buium, *Geometry of differential polynomial functions I: algebraic groups*, Amer. J. Math. 115, 6 (1993), 1385-1444.

[15] A.Buium, *Geometry of differential polynomial functions II: algebraic curves*, Amer. J. Math. 116, 4 (1994), 785-818.

[16] A. Buium, *Geometry of differential polynomial functions III: moduli spaces*, Amer J. Math. 117 (1995), 1-73.

[17] A.Buium, *Differential characters of Abelian varieties over p−adic fields*, Invent. Math. 122 (1995), 309-340.

[18] A. Buium, *Geometry of p−jets*, Duke J. Math. 82, 2 (1996), 349-367.

[19] A.Buium, *Differential characters and characteristic polynomial of Frobenius*, J. reine angew. Math., 485 (1997), 209-219.

[20] A. Buium, *Arithmetic analogues of derivations*, J. Algebra, 198, (1997), 290-299.

[21] A.Buium, *Differential modular forms*, J. reine angew. Math., 520 (2000), 95-167.

[22] A. Buium, *Quotients of algebraic varieties by Zariski dense equivalence relations*, in: Parshin Festschrift, Contemp. Math. 300 (2002), AMS, pp. 59-97.

[23] A. Buium, *Differential modular forms on Shimura curves,I*, Compositio Math. 139 (2003), 197-237.

[24] A. Buium, *Differential modular forms on Shimura curves,II: Serre operators*, Compositio Math. 140 (2004), 1113-1134.

[25] A. Buium, *Pfaffian equations satisfied by differential modular forms*, Math. Research Letters 11 (2004), 453-466.

[26] A. Buium, *Geometry of Fermat adeles*, Trans. AMS 357 (2004), 901-964.

[27] A. Buium, *Complex dynamics and invariant forms mod p*, International Math. Res. Notices, 2005, to appear.

[28] A. Buium, K. Zimmerman, *Homogeneous p−differential polynomials*, J. of Algebra 269 (2003), 492-507.

[29] A. Buium, K. Zimmerman, *Differential orbit spaces of discrete dynamical systems*, J. reine angew. Math. 580 (2005), 201-230.

[30] K. Buzzard, *Integral models of certain Shimura curves*, Duke Math J. 87,3 (1997), 591-612.

[31] H. Carayol, *Sur la mauvaise reduction des courbes de Shimura*, Compositio Math. 59 (1986), 151-230.

[32] P.Cassidy, *Differential algebraic groups*, Amer.J.Math. 94 (1972), 891-954.

[33] Z. Chatzidakis, E. Hrushovski, *Model theory of difference fields*, Trans. AMS, 351 (1999), 2997-3071.

[34] L. Clozel, E. Ullmo, *Correspondences modulaires et mesures invariantes*, Crelle J. 558 (2003), 47-83.

[35] R. Cohn, *Difference Algebra*, Interscience, New York, 1965.

[36] A. Connes, *Non-commutative Geometry*, Academic Press, 1994.

[37] A. Connes, M. Marcolli, **Q**−*lattices: quantum statistical mechanics and Galois theory*, to appear in Journal of Geometry and Physics.

[38] A. Connes, M. Marcolli, N. Ramachandran, *KMS states and complex multiplication*, preprint 2004.

[39] A. Connes, H. Moscovici, *Modular Hecke algebras and their Hopf symmetries*, Moscow Math. J. Vol. 4, No. 1 (2004), 67-109.

[40] A. Connes, M. Rieffel, *Yang-Mills for non-commutative two-tori*, Contemp. Math. 62 (1987), 237-266.

[41] D. A. Cox, *Primes of the form $x^2 + ny^2$*, John Wiley and Sons, New York, 1989.

[42] P. Deligne, L. Illusie, *Cristaux ordinaires et coordonnées canoniques*, in: Surfaces Algébriques, LNM 686, Springer 1981.

[43] P. Deligne, L. Illusie, *Revêtements modulo p^2 et décomposition de complexe de de Rham*, Invent. Math. 89 (1987), 247-270.

[44] P. Deligne, D. Mumford, *The irreducibility of the space of curves of given genus*, Publ. Math IHES, 36 (1969), 75-109.

[45] J. Denef, F. Loeser, *Geometry of arc spaces of algebraic varieties*, Proceedings of the Third European Congress of Mathematics, Barcelona 2000, Progr. Math., 201, Birkhäuser, Basel, 2001, pp. 327-348.

[46] C. Deninger, *Local L-factors of motives and regularized determinants*, Invent Math. 107,1 (1992), 135-150.

[47] F. Diamond and J. Im, *Modular forms and modular curves*, in:Seminar on Fermat's Last Theorem, Conference Proceedings, Volume 17, Canadian Mathematical Society, 1995, pp. 39-134.

[48] F. Diamond, R. Taylor, *Non-optimal levels for mod l modular representations of* $Gal(\bar{\mathbf{Q}}/\mathbf{Q})$, Invent. Math. 115 (1994), 435-462.

[49] J. Dieudonné, *Groupes de Lie et hyperalgèbres de Lie sur un corps de caractéristique $p > 0$, (VIII)*, Math.Ann. 134 (1957), 114-133.

[50] L. E. Dickson, *On invariants and the theory of numbers*, Dover, N.Y. 2004.

[51] A. Douady, J. H. Hubbard, *A proof of Thurston's topological characterization of rational functions*, Acta Math., 171 (1993), 263-297.

[52] V. G. Drinfeld, *Elliptic modules*, Math. Sbornik 94 (1974), 594-627.

[53] B. Dwork, *p−adic cycles*, Publ. Math. IHES 37 (1969), 27-115.

[54] B. Dwork, A. Ogus, *Canonical liftings of Jacobians*, Compositio Math. 58 (1986), 111-131.

[55] B. Dwork, G. Gerotto, F. J. Sullivan, *An Introduction to G−functions*, Annals of Math. Studies 133, Princeton, 1994.

[56] N. Elkies, *The existence of infinitely many supersingular primes for every elliptic curve over* **Q**, Invent. Math 89 (1987), 561-567.

[57] N. Elkies, *Supersingular primes for elliptic curves over real fields*, Compositio Math. 72 (1989), 165-172.

[58] G.Faltings, C-L. Chai, *Degeneration of Abelian varieties*, Springer, Heidelberg, New York, 1990.

[59] H.M.Farkas, I.Kra, *Riemann Surfaces*, GTM, Springer, 1992.

[60] M. Fried, *On a conjecture of Schur*, Michigan Math. J. 17 (1970), 41-55.

[61] R. Goodman, N. R. Wallach, *Representations and Invariants of Classical Groups*, Cambridge Univ. Press, 1999.

[62] D. Goss, *Basic Structures in Function Field Arithmetic*, Springer, 1998.

[63] M.Greenberg, *Schemata over local rings*, Annals of Math. 73 (1961), 624-648.

[64] A. Grothendieck, *Eléments de Géométrie Algébrique*, III, Publ. Math. IHES 11 (1961).

[65] R. M. Guralnick, P. Muller, J. Saxl, *The rational function analogue of a question of Schur and exceptionality of permutation representations*, Memoirs A.M.S. 773 (2003).

[66] R. Hartshorne, *Algebraic Geometry*, Graduate Texts in Math., Springer, 1977.

[67] M. Hazewinkel, *On formal groups. The functional equation lemma and some of its applications*, Astérisque 63 (1979), 73-82.

[68] W.L.Hill, *Formal groups and zeta functions of elliptic curves*, Invent. Math. 12 (1971), 321-336.

[69] E. Hrushovski, *Geometric model theory*, Doc. Math., Extra Vol ICM 1998, 281-302.

[70] E. Hrushovski, *The elementary theory of the Frobenius automorphisms*, manuscript.

[71] C. Hurlburt, *Isogeny covariant differential modular forms modulo p*, Compositio Math., 128, (2001), no. 1, 17-34.

[72] C. Hurlburt, *Zero loci of differential modular forms*, J. Number Theory, 98, 1 (2003), 47-54.

[73] Y. Ihara, *An invariant multiple differential attached to the field of elliptic modular functions of characteristic p*, Amer. J. Math., XCIII, 1, (1971), 139-147.

[74] Y. Ihara, *On the differentials associated to congruence relations and the Schwarzian equations defining uniformizations*, J. Fac. Sci. Univ. Tokyo, Sec. IA, Vol. 21, No. 3, (1974), 309-332.

[75] Y. Ihara, *On Fermat quotient and differentiation of numbers*, RIMS Kokyuroku 810 (1992), 324-341, (In Japanese). English translation by S. Hahn, Univ. of Georgia preprint.

[76] K. Ireland, M. Rosen, *A Classical Introduction to Modern Number Theory*, GTM 84, Springer 1990.

[77] H. Iwaniec, *Topics in classical automorphic forms*, GTM, Vol. 17, Amer. Math. Soc., Providence, 1997.

[78] J.Johnson, *Prolongations of integral domains*, J.Algebra 94,1 (1985), 173-211.

[79] A. Joyal, *$\delta-$anneaux et vecteurs de Witt*, C.R. Acad. Sci. Canada, Vol. VII, No. 3, (1985), 177-182.

[80] N. Katz, *Serre-Tate local moduli*, Springer LNM 868 (1981), 138-202.

[81] N. Katz, *$p-$adic properties of modular schemes and modular forms*, LNM 350, Springer 1973, 69-190.

[82] N. Katz, *Travaux de Dwork*, Expose 409, Sem. Bourbaki 1971/72, Springer LNM 317 (1973), 167-200.

[83] N.Katz, B.Mazur, *Arithmetic moduli of elliptic curves*, Annals of Math. Studies, Princeton Univ. Press, 1985.

[84] E. R. Kolchin, *Differential Algebra and Algebraic Groups*, Academic Press , New York, 1973.

[85] D. Krammer, *An example of an arithmetic Fuchsian group*, Crelle J. 473(1996), 69-85.

[86] N. Kurokawa, H. Ochiai, M. Wakayama, *Absolute derivations and zeta functions*, Documenta Math., Extra Volume Kato (2003), 565-584.

[87] S. Lang, *Algebraic Number Theory*, Springer, 1986.

[88] S. Lang, *Introduction to Modular forms*, Springer, 1976.

[89] S. Lattès, *Sur l'iteration des substitutions rationelles et les fonctions de Poincaré*, C.R.Acad. Sci. Paris 16 (1918), 26-28.

[90] M.Lazard, *Sur les groupes de Lie formels à un paramètre*, Bull. Soc. Math. France, 83 (1955), 251-274.

[91] Hua-Cieh Li, *$p-$adic dynamical systems and formal groups*, Compositio Math. 104 (1996), 41-54.

[92] J. Lubin, *One parameter formal Lie groups over $p-$adic integer rings*, Ann of Math. 80 (1964), 464-484.

[93] J. Lubin, J. Tate, *Formal complex multiplication in local fields*, Ann. of Math. 81 (1965), 380-387.

[94] J. Lubin, *Non Archimedian dynamical systems*, Compositio Math., 94 (1994), 321-346.

[95] C. Maclachlan, A. W. Reid, *The Arithmetic of Hyperbolic $3-$Manifolds*, GTM, Springer, 2003.

[96] R. Mañé, P. Sad, D. Sullivan, *On the dynamics of rational maps*, Ann. Sci. Ecole Norm. Sup. 16 (1983), 193-217.

[97] Yu.I.Manin, *Algebraic curves over fields with differentiation*, Izv. Akad. Nauk SSSR, Ser. Mat. 22 (1958), 737-756 = AMS translations Series 2, 37 (1964), 59-78.

[98] Yu.I.Manin, *Lectures on zeta functions and motives (according to Denniger and Kurokawa)*, Astérisque 228 (1995), 121-163.

[99] Yu. I. Manin, *Real multiplication and noncommutative geometry*, preprint.

[100] Yu. I. Manin, M. Marcolli, *Continued fractions, modular symbols, and non-commutative geometry*, Selecta Math. (New Series) 8, 3 (2002), 475-520.

[101] M.Marcolli, *Lectures on Arithmetic Non-commutative Geometry*, to appear.

[102] G.A.Margulis, *Discrete Subgroups of Semisimple Lie groups*, Ergebnisse der Mathematik and ihrer Grenzgebiete, 3. Folge, Bd 17, Springer, 1991.

[103] H. Matsumura, *Commutative ring theory*, Cambridge studies in advanced mathematics, Cambridge University Press, 1986.

[104] B. Mazur, W. Messing, *Universal extensions and one dimensional crystalline cohomology*, LNM 370, Springer, 1974.

[105] C. McMullen, *Families of rational maps and iterative root-finding algorithms*, Ann. of Math. 125 (1987), 467-493.

[106] L. Merel, *Bornes pour la torsion des courbes elliptiques sur les corps des nombres*, Invent. Math. 124 (1996), 437-449.

[107] W. Messing, *The crystals associated to Barsotti-Tate groups: with applications to Abelian schemes*, LNM 264, Springer, 1972.

[108] J. S. Milne, *Points on Shimura varieties modulo p*, Proc. Symp. in Pure Math., 33 (1979), Part 2, 165-184.

[109] J. Milnor, *Dynamics in One Complex Variable*, Vieweg, 1999.

[110] J. Milnor, *On Lattès maps*, preprint: www.math.sunysb.edu/jack/PREPRINTS/index.html.

[111] S. Mochizuchi, *Foundations of $p-$adic Teichmüller theory*, Studies in Advanced Mathematics, AMS/IP, 1999.

[112] A. Mori, *Power series expansions of modular forms at CM points*, Rend. Sem. Mat. Univ. Pol. Torino, Vol. 53, 4 (1995), 361-374.

[113] D. Mumford, J. Fogarty, F. Kirwan, *Geometric invariant theory*, Springer, 1994.

[114] P.J.Olver, *Applications of Lie Groups to Differential Equations*, GTM 107, Springer, 2000.

[115] M. Raynaud, *Courbes sur une variété abélienne et points de torsion*, Invent.Math. 71 (1983), 207-235.

[116] J. F. Ritt, *Permutable rational functions*, Trans AMS 25 (1923), 399-448.

[117] J. F. Ritt, *Differential Algebra*, AMS Coll. Pub 33, AMS, 1950.

[118] J-P. Serre, *Lectures on the Mordell-Weil theorem*, Vieweg, 1997.

[119] J-P. Serre, *Algebraic groups and class fields*, GTM 117, Springer, 1988.

[120] J. P. Serre, *Local fields*, Springer, 1995.

[121] G. Shimura, *Automorphic functions and number theory*, LNM 54, Springer, 1968.

[122] G. Shimura, *Introduction to the arithmetic theory of automorphic functions*, Princeton University Press, 1971.

[123] C.L.Siegel, *Uber die Classenzahl quadratischer Zahlkörper*, Acta Arithmetica 1 (1935), 83-86.

[124] J.H.Silverman, *The Arithmetic of Elliptic Curves*, Springer, Berlin, New York, 1986.

[125] J. H. Silverman, *Advanced Topics in the Arithmetic of Elliptic Curves*, GTM 151, Springer, 1994.

[126] J. H. Silverman, *Wieferich's criterion and the abc-conjecture*, J. Number Theory, 30 (1988), 226-237.

[127] Y. Soibelman, V. Vologodski, *Non-commutative compactifications and elliptic curves*, preprint, arXiv.math.AG/0205117 v3.

[128] C. Soulé, *Les variétés sur le corps a un élément*, Moscow Math. J., 4,1, (2004), 217-244.

[129] J. Tate, Letter to Dwork.

[130] D. Thakur, *Function Field Arithmetic*, World Scientific, 2004.

[131] P. Vojta, *Jets via Hasse-Schmidt derivations*, preprint, arXiv.math.AG / 0407113v1, 7 July 2004.

[132] J.F.Voloch, *On a question of Buium*, Canad. Math. Bull. 43,2 (2000), 236-238.

[133] J.F.Voloch, *Elliptic Wieferich primes*, J. Number Theory 81(2000),2, 205-209.

[134] A.Weil, *Basic Number Theory*, Springer, 1995.

[135] E.J.Wilczynski, *Projective Differential Geometry of Curves and Ruled Surfaces*, Teubner, Leipzig 1906.

[136] A. Zdunick, *Parabolic orbifolds and the dimension of the maximal measure for rational maps*, Invent. Math. 99 (1990), 627-649.

Index

INDEX OF TERMINOLOGY

INDEX OF NOTATION

Titles in This Series

121 **Anton Zettl,** Sturm-Liouville Theory, 2005
120 **Barry Simon,** Trace Ideals and Their Applications, 2005
119 **Tian Ma and Shouhong Wang,** Geometric theory of incompressible flows with applications to fluid dynamics, 2005
118 **Alexandru Buium,** Arithmetic differential equations, 2005
117 **Volodymyr Nekrashevych,** Self-similar groups, 2005
116 **Alexander Koldobsky,** Fourier analysis in convex geometry, 2005
115 **Carlos Julio Moreno,** Advanced analytic number theory: L-functions, 2005
114 **Gregory F. Lawler,** Conformally invariant processes in the plane, 2005
113 **William G. Dwyer, Philip S. Hirschhorn, Daniel M. Kan, and Jeffrey H. Smith,** Homotopy limit functors on model categories and homotopical categories, 2004
112 **Michael Aschbacher and Stephen D. Smith,** The classification of quasithin groups II. Main theorems: The classification of simple QTKE-groups, 2004
111 **Michael Aschbacher and Stephen D. Smith,** The classification of quasithin groups I. Structure of strongly quasithin K-groups, 2004
110 **Bennett Chow and Dan Knopf,** The Ricci flow: An introduction, 2004
109 **Goro Shimura,** Arithmetic and analytic theories of quadratic forms and Clifford groups, 2004
108 **Michael Farber,** Topology of closed one-forms, 2004
107 **Jens Carsten Jantzen,** Representations of algebraic groups, 2003
106 **Hiroyuki Yoshida,** Absolute CM-periods, 2003
105 **Charalambos D. Aliprantis and Owen Burkinshaw,** Locally solid Riesz spaces with applications to economics, second edition, 2003
104 **Graham Everest, Alf van der Poorten, Igor Shparlinski, and Thomas Ward,** Recurrence sequences, 2003
103 **Octav Cornea, Gregory Lupton, John Oprea, and Daniel Tanré,** Lusternik-Schnirelmann category, 2003
102 **Linda Rass and John Radcliffe,** Spatial deterministic epidemics, 2003
101 **Eli Glasner,** Ergodic theory via joinings, 2003
100 **Peter Duren and Alexander Schuster,** Bergman spaces, 2004
99 **Philip S. Hirschhorn,** Model categories and their localizations, 2003
98 **Victor Guillemin, Viktor Ginzburg, and Yael Karshon,** Moment maps, cobordisms, and Hamiltonian group actions, 2002
97 **V. A. Vassiliev,** Applied Picard-Lefschetz theory, 2002
96 **Martin Markl, Steve Shnider, and Jim Stasheff,** Operads in algebra, topology and physics, 2002
95 **Seiichi Kamada,** Braid and knot theory in dimension four, 2002
94 **Mara D. Neusel and Larry Smith,** Invariant theory of finite groups, 2002
93 **Nikolai K. Nikolski,** Operators, functions, and systems: An easy reading. Volume 2: Model operators and systems, 2002
92 **Nikolai K. Nikolski,** Operators, functions, and systems: An easy reading. Volume 1: Hardy, Hankel, and Toeplitz, 2002
91 **Richard Montgomery,** A tour of subriemannian geometries, their geodesics and applications, 2002
90 **Christian Gérard and Izabella Łaba,** Multiparticle quantum scattering in constant magnetic fields, 2002
89 **Michel Ledoux,** The concentration of measure phenomenon, 2001

TITLES IN THIS SERIES

88 **Edward Frenkel and David Ben-Zvi,** Vertex algebras and algebraic curves, second edition, 2004

87 **Bruno Poizat,** Stable groups, 2001

86 **Stanley N. Burris,** Number theoretic density and logical limit laws, 2001

85 **V. A. Kozlov, V. G. Maz′ya, and J. Rossmann,** Spectral problems associated with corner singularities of solutions to elliptic equations, 2001

84 **László Fuchs and Luigi Salce,** Modules over non-Noetherian domains, 2001

83 **Sigurdur Helgason,** Groups and geometric analysis: Integral geometry, invariant differential operators, and spherical functions, 2000

82 **Goro Shimura,** Arithmeticity in the theory of automorphic forms, 2000

81 **Michael E. Taylor,** Tools for PDE: Pseudodifferential operators, paradifferential operators, and layer potentials, 2000

80 **Lindsay N. Childs,** Taming wild extensions: Hopf algebras and local Galois module theory, 2000

79 **Joseph A. Cima and William T. Ross,** The backward shift on the Hardy space, 2000

78 **Boris A. Kupershmidt,** KP or mKP: Noncommutative mathematics of Lagrangian, Hamiltonian, and integrable systems, 2000

77 **Fumio Hiai and Dénes Petz,** The semicircle law, free random variables and entropy, 2000

76 **Frederick P. Gardiner and Nikola Lakic,** Quasiconformal Teichmüller theory, 2000

75 **Greg Hjorth,** Classification and orbit equivalence relations, 2000

74 **Daniel W. Stroock,** An introduction to the analysis of paths on a Riemannian manifold, 2000

73 **John Locker,** Spectral theory of non-self-adjoint two-point differential operators, 2000

72 **Gerald Teschl,** Jacobi operators and completely integrable nonlinear lattices, 1999

71 **Lajos Pukánszky,** Characters of connected Lie groups, 1999

70 **Carmen Chicone and Yuri Latushkin,** Evolution semigroups in dynamical systems and differential equations, 1999

69 **C. T. C. Wall (A. A. Ranicki, Editor),** Surgery on compact manifolds, second edition, 1999

68 **David A. Cox and Sheldon Katz,** Mirror symmetry and algebraic geometry, 1999

67 **A. Borel and N. Wallach,** Continuous cohomology, discrete subgroups, and representations of reductive groups, second edition, 2000

66 **Yu. Ilyashenko and Weigu Li,** Nonlocal bifurcations, 1999

65 **Carl Faith,** Rings and things and a fine array of twentieth century associative algebra, 1999

64 **Rene A. Carmona and Boris Rozovskii, Editors,** Stochastic partial differential equations: Six perspectives, 1999

63 **Mark Hovey,** Model categories, 1999

62 **Vladimir I. Bogachev,** Gaussian measures, 1998

61 **W. Norrie Everitt and Lawrence Markus,** Boundary value problems and symplectic algebra for ordinary differential and quasi-differential operators, 1999

60 **Iain Raeburn and Dana P. Williams,** Morita equivalence and continuous-trace C^*-algebras, 1998

59 **Paul Howard and Jean E. Rubin,** Consequences of the axiom of choice, 1998

For a complete list of titles in this series, visit the AMS Bookstore at **www.ams.org/bookstore/**.